Transamerica Delaval
Engineering Handbook

OTHER McGRAW-HILL HANDBOOKS OF INTEREST

American Society of Mechanical Engineers: ASME Handbooks:
    Engineering Tables
    Metals Engineering—Design
    Metals Engineering—Processes
    Metals Properties
Beeman • Industrial Power Systems Handbook
Bovay • Handbook of Mechanical and Electrical Systems for Buildings
Brady and Clauser • Materials Handbook
Brater and King • Handbook of Hydraulics
Burington • Handbook of Mathematical Tables and Formulas
Burington and May • Handbook of Probability and Statistics with Tables
Callender • Time-Saver Standards for Architectural Design Data
Carrier Air Conditioning Company • Handbook of Air Conditioning System Design
Carroll • Industrial Instrument Servicing Handbook
Considine • Energy Technology Handbook
Considine • Process Instruments and Controls Handbook
Considine and Ross • Handbook of Applied Instrumentation
Crocker and King • Piping Handbook
Davis • Handbook of Applied Hydraulics
Dudley • Gear Handbook
Fink and Beaty • Standard Handbook for Electrical Engineers
Fink and Christiansen • Electronics Engineer's Handbook
Gaylord and Gaylord • Structural Engineering Handbook
Harris • Handbook of Noise Control
Harris and Crede • Shock and Vibration Handbook
Harper • Handbook of Electronic System Design
Hetsroni • Handbook of Multiphase Systems
Hicks • Standard Handbook of Engineering Calculations
Hopf • Handbook of Building Security Planning and Design
Juran • Quality Control Handbook
Lyman, Reehl, and Rosenblatt • Handbook of Chemical Property Estimation Methods
Machol • System Engineering Handbook
McPartland • McGraw-Hill's National Electrical Code® Handbook
Mantell • Engineering Materials Handbook
Maynard • Industrial Engineering Handbook
Morrow • Maintenance Engineering Handbook
Parmley • Standard Handbook of Fastening and Joining
Peckner • Handbook of Stainless Steels
Perry • Chemical Engineers' Handbook
Perry • Engineering Manual
Rohsenow and Hartnett • Handbook of Heat Transfer
Rosaler and Rice • Standard Handbook of Plant Engineering
Rothbart • Mechanical Design and Systems Handbook
Shugar • Chemical Technician's Ready Reference Handbook
Smeaton • Motor Application and Maintenance Handbook
Smeaton • Switchgear and Control Handbook
Society of Manufacturing Engineers: •
    Die Design Handbook
    Manufacturing Planning and
      Estimating Handbook
    Handbook of Fixture Design
    Tool and Manufacturing Engineers
      Handbook
Streeter • Handbook of Fluid Dynamics
Truxal • Control Engineers' Handbook
Tuma • Engineering Mathematics Handbook
Tuma • Handbook of Physical Calculations
Turner and Malloy • Handbook of Thermal Insulation Design Economics for Pipes and
    Equipment
Turner and Malloy • Thermal Insulation Handbook

Transamerica Delaval Engineering Handbook

Compiled by
The Engineering Staff of
Transamerica Delaval Inc.

Edited by

Harry J. Welch

Fourth Edition

McGRAW-HILL BOOK COMPANY

New York   St. Louis   San Francisco   Auckland   Bogotá   Hamburg
Johannesburg   London   Madrid   Mexico   Montreal   New Delhi
Panama   Paris   São Paulo   Singapore   Sydney   Tokyo   Toronto

Library of Congress Cataloging in Publication Data

Engineering handbook.
    Transamerica Delaval engineering handbook.

    Fourth ed. of: De Laval engineering handbook.
1970 (McGraw-Hill handbooks)
    Includes index.
    1. Turbomachines—Handbooks, manuals, etc.   2. Mechanical engineering—
Handbooks, manuals, etc.   I. Welch, Harry J.   II. Transamerica Delaval
Inc.   III. Title.
TJ266.E53   1983   621.406′0212   82-17975

1 2 3 4 5 6 7 8 9 0   KGPKGP   8 9 8 7 6 5 4 3

ISBN 0-07-016250-6

The editors for this book were Harold B. Crawford and Beatrice E. Eckes, the
designer was Mark E. Safran, and the production supervisor was Paul A. Malchow.
It was set in Times Roman by University Graphics, Inc.
Printed and bound by The Kingsport Press.

Contents

Preface to the Fourth Edition

The fourth edition of the *Transamerica Delaval Engineering Handbook* retains the objective of previous editions, which is to provide ready-reference material to engineers and technicians who are working with pumps, turbines, gears, compressors, engines, condensers, and filters in utility, industrial, and marine fields. The handbook has been updated to convey progress that has been made in those product lines and industries.

Because of the increasing use of the metric (SI) system of units, a more extensive table for converting to the metric system was added. The use of U.S. customary units was retained in the text to facilitate employment of the handbook by many users for some future period.

As with previous editions, the revisions and additions that were made in writing this fourth edition were contributed by members of the engineering staff of Transamerica Delaval Inc. The editor is sincerely appreciative of their efforts. A thank-you to members of the Delaval secretarial staff is also given for their special and needed contribution.

The address of the corporate offices of Transamerica Delaval Inc. is 3450 Princeton Pike, Lawrenceville, New Jersey 08648.

Acknowledgment

Many members of the Transamerica Delaval engineering staff contributed to this edition and previous editions of this handbook. We particularly want to thank those who provided additions and revisions to this fourth edition:

George C. Angelovich
Pentti I. Arent
Stephen B. Bennett
Henry W. Braun
Edward A. Bulanowski
Marie E. Fares
Alfred R. Fleischer
Harry B. Gayley
Alan S. Gertz
Marmaduke V. Grove
John S. Haverstick
William G. Hoppock
George C. Hud
Jerry K. Koo

Darrell McAfee
Alan H. Miller
Vincent J. Polignano
Josef M. Pollich
Vasil Rusak
Richard M. Salzmann
Joseph A. Silvaggio
Rene Simonin, Jr.
Frank A. Stevens
Frederic A. Thoma
Albert E. Truran
Kenneth G. Van Bramer
Harry J. Welch
Joseph J. Worek

Transamerica Delaval
Engineering Handbook

Mathematical Data and Conversion Tables

TABLE 1-1 SQUARES AND CUBES OF NUMBERS

No.	Square	Cube	No.	Square	Cube
1	1	1	51	2,601	132,651
2	4	8	52	2,704	140,608
3	9	27	53	2,809	148,877
4	16	64	54	2,916	157,464
5	25	125	55	3,025	166,375
6	36	216	56	3,136	175,616
7	49	343	57	3,249	185,193
8	64	512	58	3,364	195,112
9	81	729	59	3,481	205,379
10	100	1,000	60	3,600	216,000
11	121	1,331	61	3,721	226,981
12	144	1,728	62	3,844	238,328
13	169	2,197	63	3,969	250,047
14	196	2,744	64	4,096	262,144
15	225	3,375	65	4,225	274,625
16	256	4,096	66	4,356	287,496
17	289	4,913	67	4,489	300,763
18	324	5,832	68	4,624	314,432
19	361	6,859	69	4,761	328,509
20	400	8,000	70	4,900	343,000
21	441	9,261	71	5,041	357,911
22	484	10,648	72	5,184	373,248
23	529	12,167	73	5,329	389,017
24	576	13,824	74	5,476	405,224
25	625	15,625	75	5,625	421,875
26	676	17,576	76	5,776	438,976
27	729	19,683	77	5,929	456,533
28	784	21,952	78	6,084	474,552
29	841	24,389	79	6,241	493,039
30	900	27,000	80	6,400	512,000
31	961	29,791	81	6,561	531,441
32	1,024	32,768	82	6,724	551,368
33	1,089	35,937	83	6,889	571,787
34	1,156	39,304	84	7,056	592,704
35	1,225	42,875	85	7,225	614,125
36	1,296	46,656	86	7,396	636,056
37	1,369	50,653	87	7,569	658,503
38	1,444	54,872	88	7,744	681,472
39	1,521	59,319	89	7,921	704,969
40	1,600	64,000	90	8,100	729,000
41	1,681	68,921	91	8,281	753,571
42	1,764	74,088	92	8,464	778,688
43	1,849	79,507	93	8,649	804,357
44	1,936	85,184	94	8,836	830,584
45	2,025	91,125	95	9,025	857,375
46	2,116	97,336	96	9,216	884,736
47	2,209	103,823	97	9,409	912,673
48	2,304	110,592	98	9,604	941,192
49	2,401	117,649	99	9,801	970,299
50	2,500	125,000	100	10,000	1,000,000

TABLE 1-1 SQUARES AND CUBES OF NUMBERS *(Continued)*

No.	Square	Cube	No.	Square	Cube
101	10,201	1,030,301	151	22,801	3,442,951
102	10,404	1,061,208	152	23,104	3,511,008
103	10,609	1,092,727	153	23,409	3,581,577
104	10,816	1,124,864	154	23,716	3,652,264
105	11,025	1,157,625	155	24,025	3,723,875
106	11,236	1,191,016	156	24,336	3,796,416
107	11,449	1,225,043	157	24,649	3,869,893
108	11,664	1,259,712	158	24,964	3,944,312
109	11,881	1,295,029	159	25,281	4,019,679
110	12,100	1,331,000	160	25,600	4,096,000
111	12,321	1,367,631	161	25,921	4,173,281
112	12,544	1,404,928	162	26,244	4,251,528
113	12,769	1,442,897	163	26,569	4,330,747
114	12,996	1,481,544	164	26,896	4,410,944
115	13,225	1,520,875	165	27,225	4,492,125
116	13,456	1,560,896	166	27,556	4,574,296
117	13,689	1,601,613	167	27,889	4,657,463
118	13,924	1,643,032	168	28,224	4,741,632
119	14,161	1,685,159	169	28,561	4,826,809
120	14,400	1,728,000	170	28,900	4,913,000
121	14,641	1,771,561	171	29,241	5,000,211
122	14,884	1,815,848	172	29,584	5,088,448
123	15,129	1,860,867	173	29,929	5,177,717
124	15,376	1,906,624	174	30,276	5,268,024
125	15,625	1,953,125	175	30,625	5,359,375
126	15,876	2,000,376	176	30,976	5,451,776
127	16,129	2,048,383	177	31,329	5,545,233
128	16,384	2,097,152	178	31,684	5,639,752
129	16,641	2,146,689	179	32,041	5,735,339
130	16,900	2,197,000	180	32,400	5,832,000
131	17,161	2,248,091	181	32,761	5,929,741
132	17,424	2,299,968	182	33,124	6,028,568
133	17,689	2,352,637	183	33,489	6,128,487
134	17,956	2,406,104	184	33,856	6,229,504
135	18,225	2,460,375	185	34,225	6,331,625
136	18,496	2,515,456	186	34,596	6,434,856
137	18,769	2,571,353	187	34,969	6,539,203
138	19,044	2,628,072	188	35,344	6,644,672
139	19,321	2,685,619	189	35,721	6,751,269
140	19,600	2,744,000	190	36,100	6,859,000
141	19,881	2,803,221	191	36,481	6,967,871
142	20,164	2,863,288	192	36,864	7,077,888
143	20,449	2,924,207	193	37,249	7,189,057
144	20,736	2,985,984	194	37,636	7,301,384
145	21,025	3,048,625	195	38,025	7,414,875
146	21,316	3,112,136	196	38,416	7,529,536
147	21,609	3,176,523	197	38,809	7,645,373
148	21,904	3,241,792	198	39,204	7,762,392
149	22,201	3,307,949	199	39,601	7,880,599
150	22,500	3,375,000	200	40,000	8,000,000

TABLE 1-1   SQUARES AND CUBES OF NUMBERS *(Continued)*

No.	Square	Cube	No.	Square	Cube
201	40,401	8,120,601	251	63,001	15,813,251
202	40,804	8,242,408	252	63,504	16,003,008
203	41,209	8,365,427	253	64,009	16,194,277
204	41,616	8,489,664	254	64,516	16,387,064
205	42,025	8,615,125	255	65,025	16,581,375
206	42,436	8,741,816	256	65,536	16,777,216
207	42,849	8,869,743	257	66,049	16,974,593
208	43,264	8,998,912	258	66,564	17,173,512
209	43,681	9,129,329	259	67,081	17,373,979
210	44,100	9,261,000	260	67,600	17,576,000
211	44,521	9,393,931	261	68,121	17,779,581
212	44,944	9,528,128	262	68,644	17,984,728
213	45,369	9,663,597	263	69,169	18,191,447
214	45,796	9,800,344	264	69,696	18,399,744
215	46,225	9,938,375	265	70,225	18,609,625
216	46,656	10,077,696	266	70,756	18,821,096
217	47,089	10,218,313	267	71,289	19,034,163
218	47,524	10,360,232	268	71,824	19,248,832
219	47,961	10,503,459	269	72,361	19,465,109
220	48,400	10,648,000	270	72,900	19,683,000
221	48,841	10,793,861	271	73,441	19,902,511
222	49,284	10,941,048	272	73,984	20,123,648
223	49,729	11,089,567	273	74,529	20,346,417
224	50,176	11,239,424	274	75,076	20,570,824
225	50,625	11,390,625	275	75,625	20,796,875
226	51,076	11,543,176	276	76,176	21,024,576
227	51,529	11,697,083	277	76,729	21,253,933
228	51,984	11,852,352	278	77,284	21,484,952
229	52,441	12,008,989	279	77,841	21,717,639
230	52,900	12,167,000	280	78,400	21,952,000
231	53,361	12,326,391	281	78,961	22,188,041
232	53,824	12,487,168	282	79,524	22,425,768
233	54,289	12,649,337	283	80,089	22,665,187
234	54,756	12,812,904	284	80,656	22,906,304
235	55,225	12,977,875	285	81,225	23,149,125
236	55,696	13,144,256	286	81,796	23,393,656
237	56,169	13,312,053	287	82,369	23,639,903
238	56,644	13,481,272	288	82,944	23,887,872
239	57,121	13,651,919	289	83,521	24,137,569
240	57,600	13,824,000	290	84,100	24,389,000
241	58,081	13,997,521	291	84,681	24,642,171
242	58,564	14,172,488	292	85,264	24,897,088
243	59,049	14,348,907	293	85,849	25,153,757
244	59,536	14,526,784	294	86,436	25,412,184
245	60,025	14,706,125	295	87,025	25,672,375
246	60,516	14,886,936	296	87,616	25,934,336
247	61,009	15,069,223	297	88,209	26,198,073
248	61,504	15,252,992	298	88,804	26,463,592
249	62,001	15,438,249	299	89,401	26,730,899
250	62,500	15,625,000	300	90,000	27,000,000

TABLE 1-1   SQUARES AND CUBES OF NUMBERS *(Continued)*

No.	Square	Cube	No.	Square	Cube
301	90,601	27,270,901	351	123,201	43,243,551
302	91,204	27,543,608	352	123,904	43,614,208
303	91,809	27,818,127	353	124,609	43,986,977
304	92,416	28,094,464	354	125,316	44,361,864
305	93,025	28,372,625	355	126,025	44,738,875
306	93,636	28,652,616	356	126,736	45,118,016
307	94,249	28,934,443	357	127,449	45,499,293
308	94,864	29,218,112	358	128,164	45,882,712
309	95,481	29,503,629	359	128,881	46,268,279
310	96,100	29,791,000	360	129,600	46,656,000
311	96,721	30,080,231	361	130,321	47,045,881
312	97,344	30,371,328	362	131,044	47,437,928
313	97,969	30,664,297	363	131,769	47,832,147
314	98,596	30,959,144	364	132,496	48,228,544
315	99,225	31,255,875	365	133,225	48,627,125
316	99,856	31,554,496	366	133,956	49,027,896
317	100,489	31,855,013	367	134,689	49,430,863
318	101,124	32,157,432	368	135,424	49,836,032
319	101,761	32,461,759	369	136,161	50,243,409
320	102,400	32,768,000	370	136,900	50,653,000
321	103,041	33,076,161	371	137,641	51,064,811
322	103,684	33,386,248	372	138,384	51,478,848
323	104,329	33,698,267	373	139,129	51,895,117
324	104,976	34,012,224	374	139,876	52,313,624
325	105,625	34,328,125	375	140,625	52,734,375
326	106,276	34,645,976	376	141,376	53,157,376
327	106,929	34,965,783	377	142,129	53,582,633
328	107,584	35,287,552	378	142,884	54,010,152
329	108,241	35,611,289	379	143,641	54,439,939
330	108,900	35,937,000	380	144,400	54,872,000
331	109,561	36,264,691	381	145,161	55,306,341
332	110,224	36,594,368	382	145,924	55,742,968
333	110,889	36,926,037	383	146,689	56,181,887
334	111,556	37,259,704	384	147,456	56,623,104
335	112,225	37,595,375	385	148,225	57,066,625
336	112,896	37,933,056	386	148,996	57,512,456
337	113,569	38,272,753	387	149,769	57,960,603
338	114,244	38,614,472	388	150,544	58,411,072
339	114,921	38,958,219	389	151,321	58,863,869
340	115,600	39,304,000	390	152,100	59,319,000
341	116,281	39,651,821	391	152,881	59,776,471
342	116,964	40,001,688	392	153,664	60,236,288
343	117,649	40,353,607	393	154,449	60,698,457
344	118,336	40,707,584	394	155,236	61,162,984
345	119,025	41,063,625	395	156,025	61,629,875
346	119,716	41,421,736	396	156,816	62,099,136
347	120,409	41,781,923	397	157,609	62,570,773
348	121,104	42,144,192	398	158,404	63,044,792
349	121,801	42,508,549	399	159,201	63,521,199
350	122,500	42,875,000	400	160,000	64,000,000

TABLE 1-1 SQUARES AND CUBES OF NUMBERS (Continued)

No.	Square	Cube	No.	Square	Cube
401	160,801	64,481,201	451	203,401	91,733,851
402	161,604	64,964,808	452	204,304	92,345,408
403	162,409	65,450,827	453	205,209	92,959,677
404	163,216	65,939,264	454	206,116	93,576,664
405	164,025	66,430,125	455	207,025	94,196,375
406	164,836	66,923,416	456	207,936	94,818,816
407	165,649	67,419,143	457	208,849	95,443,993
408	166,464	67,917,312	458	209,764	96,071,912
409	167,281	68,417,929	459	210,681	96,702,579
410	168,100	68,921,000	460	211,600	97,336,000
411	168,921	69,426,531	461	212,521	97,972,181
412	169,744	69,934,528	462	213,444	98,611,128
413	170,569	70,444,997	463	214,369	99,252,847
414	171,396	70,957,944	464	215,296	99,897,344
415	172,225	71,473,375	465	216,225	100,544,625
416	173,056	71,991,296	466	217,156	101,194,696
417	173,889	72,511,713	467	218,089	101,847,563
418	174,724	73,034,632	468	219,024	102,503,232
419	175,561	73,560,059	469	219,961	103,161,709
420	176,400	74,088,000	470	220,900	103,823,000
421	177,241	74,618,461	471	221,841	104,487,111
422	178,084	75,151,448	472	222,784	105,154,048
423	178,929	75,686,967	473	223,729	105,823,817
424	179,776	76,225,024	474	224,676	106,496,424
425	180,625	76,765,625	475	225,625	107,171,875
426	181,476	77,308,776	476	226,576	107,850,176
427	182,329	77,854,483	477	227,529	108,531,333
428	183,184	78,402,752	478	228,484	109,215,352
429	184,041	78,953,589	479	229,441	109,902,239
430	184,900	79,507,000	480	230,400	110,592,000
431	185,761	80,062,991	481	231,361	111,284,641
432	186,624	80,621,568	482	232,324	111,980,168
433	187,489	81,182,737	483	233,289	112,678,587
434	188,356	81,746,504	484	234,256	113,379,904
435	189,225	82,312,875	485	235,225	114,084,125
436	190,096	82,881,856	486	236,196	114,791,256
437	190,969	83,453,453	487	237,169	115,501,303
438	191,844	84,027,672	488	238,144	116,214,272
439	192,721	84,604,519	489	239,121	116,930,169
440	193,600	85,184,000	490	240,100	117,649,000
441	194,481	85,766,121	491	241,081	118,370,771
442	195,364	86,350,888	492	242,064	119,095,488
443	196,249	86,938,307	493	243,049	119,823,157
444	197,136	87,528,384	494	244,036	120,553,784
445	198,025	88,121,125	495	245,025	121,287,375
446	198,916	88,716,536	496	246,016	122,023,936
447	199,809	89,314,623	497	247,009	122,763,473
448	200,704	89,915,392	498	248,004	123,505,992
449	201,601	90,518,849	499	249,001	124,251,499
450	202,500	91,125,000	500	250,000	125,000,000

TABLE 1-1   SQUARES AND CUBES OF NUMBERS *(Continued)*

No.	Square	Cube	No.	Square	Cube
501	251,001	125,751,501	551	303,601	167,284,151
502	252,004	126,506,008	552	304,704	168,196,608
503	253,009	127,263,527	553	305,809	169,112,377
504	254,016	128,024,064	554	306,916	170,031,464
505	255,025	128,787,625	555	308,025	170,953,875
506	256,036	129,554,216	556	309,136	171,879,616
507	257,049	130,323,843	557	310,249	172,808,693
508	258,064	131,096,512	558	311,364	173,741,112
509	259,081	131,872,229	559	312,481	174,676,879
510	260,100	132,651,000	560	313,600	175,616,000
511	261,121	133,432,831	561	314,721	176,558,481
512	262,144	134,217,728	562	315,844	177,504,328
513	263,169	135,005,697	563	316,969	178,453,547
514	264,196	135,796,744	564	318,096	179,406,144
515	265,225	136,590,875	565	319,225	180,362,125
516	266,256	137,388,096	566	320,356	181,321,496
517	267,289	138,188,413	567	321,489	182,284,263
518	268,324	138,991,832	568	322,624	183,250,432
519	269,361	139,798,359	569	323,761	184,220,009
520	270,400	140,608,000	570	324,900	185,193,000
521	271,441	141,420,761	571	326,041	186,169,411
522	272,484	142,236,648	572	327,184	187,149,248
523	273,529	143,055,667	573	328,329	188,132,517
524	274,576	143 877,824	574	329,476	189,119,224
525	275,625	144,703,125	575	330,625	190,109,375
526	276,676	145,531,576	576	331,776	191,102,976
527	277,729	146,363,183	577	332,929	192,100,033
528	278,784	147,197,952	578	334,084	193,100,552
529	279,841	148,035,889	579	335,241	194,104,539
530	280,900	148,877,000	580	336,400	195,112,000
531	281,961	149,721,291	581	337,561	196,122,941
532	283,024	150,568,768	582	338,724	197,137,368
533	284,089	151,419,437	583	339,889	198,155,287
534	285,156	152,273,304	584	341,056	199,176,704
535	286,225	153,130,375	585	342,225	200,201,625
536	287,296	153,990,656	586	343,396	201,230,056
537	288,369	154,854,153	587	344,569	202,262,003
538	289,444	155,720,872	588	345,744	203,297,472
539	290,521	156,590,819	589	346,921	204,336,469
540	291,600	157,464,000	590	348,100	205,379,000
541	292,681	158,340,421	591	349,281	206,425,071
542	293,764	159,220,088	592	350,464	207,474,688
543	294,849	160,103,007	593	351,649	208,527,857
544	295,936	160,989,184	594	352,836	209,584,584
545	297,025	161,878,625	595	354,025	210,644,875
546	298,116	162,771,336	596	355,216	211,708,736
547	299,209	163,667,323	597	356,409	212,776,173
548	300,304	164,566,592	598	357,604	213,847,192
549	301,401	165,469,149	599	358,801	214,921,799
550	302,500	166,375,000	600	360,000	216,000,000

TABLE 1-1   SQUARES AND CUBES OF NUMBERS *(Continued)*

No.	Square	Cube	No.	Square	Cube
601	361,201	217,081,801	651	423,801	275,894,451
602	362,404	218,167,208	652	425,104	277,167,808
603	363,609	219,256,227	653	426,409	278,445,077
604	364,816	220,348,864	654	427,716	279,726,264
605	366,025	221,445,125	655	429,025	281,011,375
606	367,236	222,545,016	656	430,336	282,300,416
607	368,449	223,648,543	657	431,649	283,593,393
608	369,664	224,755,712	658	432,964	284,890,312
609	370,881	225,866,529	659	434,281	286,191,179
610	372,100	226,981,000	660	435,600	287,496,000
611	373,321	228,099,131	661	436,921	288,804,781
612	374,544	229,220,928	662	438,244	290,117,528
613	375,769	230,346,397	663	439,569	291,434,247
614	376,996	231,475,544	664	440,896	292,754,944
615	378,225	232,608,375	665	442,225	294,079,625
616	379,456	233,744,896	666	443,556	295,408,296
617	380,689	234,885,113	667	444,889	296,740,963
618	381,924	236,029,032	668	446,224	298,077,632
619	383,161	237,176,659	669	447,561	299,418,309
620	384,400	238,328,000	670	448,900	300,763,000
621	385,641	239,483,061	671	450,241	302,111,711
622	386,884	240,641,848	672	451,584	303,464,448
623	388,129	241,804,367	673	452,929	304,821,217
624	389,376	242,970,624	674	454,276	306,182,024
625	390,625	244,140,625	675	455,625	307,546,875
626	391,876	245,314,376	676	456,976	308,915,776
627	393,129	246,491,883	677	458,329	310,288,733
628	394,384	247,673,152	678	459,684	311,665,752
629	395,641	248,858,189	679	461,041	313,046,839
630	396,900	250,047,000	680	462,400	314,432,000
631	398,161	251,239,591	681	463,761	315,821,241
632	399,424	252,435,968	682	465,124	317,214,568
633	400,689	253,636,137	683	466,489	318,611,987
634	401,956	254,840,104	684	467,856	320,013,504
635	403,225	256,047,875	685	469,225	321,419,125
636	404,496	257,259,456	686	470,596	322,828,856
637	405,769	258,474,853	687	471,969	324,242,703
638	407,044	259,694,072	688	473,344	325,660,672
639	408,321	260,917,119	689	474,721	327,082,769
640	409,600	262,144,000	690	476,100	328,509,000
641	410,881	263,374,721	691	477,481	329,939,371
642	412,164	264,609,288	692	478,864	331,373,888
643	413,449	265,847,707	693	480,249	332,812,557
644	414,736	267,089,984	694	481,636	334,255,384
645	416,025	268,336,125	695	483,025	335,702,375
646	417,316	269,586,136	696	484,416	337,153,536
647	418,609	270,840,023	697	485,809	338,608,873
648	419,904	272,097,792	698	487,204	340,068,392
649	421,201	273,359,449	699	488,601	341,532,099
650	422,500	274,625,000	700	490,000	343,000,000

TABLE 1-1 SQUARES AND CUBES OF NUMBERS (Continued)

No.	Square	Cube	No.	Square	Cube
701	491,401	344,472,101	751	564,001	423,564,751
702	492,804	345,948,408	752	565,504	425,259,008
703	494,209	347,428,927	753	567,009	426,957,777
704	495,616	348,913,664	754	568,516	428,661,064
705	497,025	350,402,625	755	570,025	430,368,875
706	498,436	351,895,816	756	571,536	432,081,216
707	499,849	353,393,243	757	573,049	433,798,093
708	501,264	354,894,912	758	574,564	435,519,512
709	502,681	356,400,829	759	576,081	437,245,479
710	504,100	357,911,000	760	577,600	438,976,000
711	505,521	359,425,431	761	579,121	440,711,081
712	506,944	360,944,128	762	580,644	442,450,728
713	508,369	362,467,097	763	582,169	444,194,947
714	509,796	363,994,344	764	583,696	445,943,744
715	511,225	365,525,875	765	585,225	447,697,125
716	512,656	367,061,696	766	586,756	449,455,096
717	514,089	368,601,813	767	588,289	451,217,663
718	515,524	370,146,232	768	589,824	452,984,832
719	516,961	371,694,959	769	591,361	454,756,609
720	518,400	373,248,000	770	592,900	456,533,000
721	519,841	374,805,361	771	594,441	458,314,011
722	521,284	376,367,048	772	595,984	460,099,648
723	522,729	377,933,067	773	597,529	461,889,917
724	524,176	379,503,424	774	599,076	463,684,824
725	525,625	381,078,125	775	600,625	465,484,375
726	527,076	382,657,176	776	602,176	467,288,576
727	528,529	384,240,583	777	603,729	469,097,433
728	529,984	385,828,352	778	605,284	470,910,952
729	531,441	387,420,489	779	606,841	472,729,139
730	532,900	389,017,000	780	608,400	474,552,000
731	534,361	390,617,891	781	609,961	476,379,541
732	535,824	392,223,168	782	611,524	478,211,768
733	537,289	393,832,837	783	613,089	480,048,687
734	538,756	395,446,904	784	614,656	481,890,304
735	540,225	397,065,375	785	616,225	483,736,625
736	541,696	398,688,256	786	617,796	485,587,656
737	543,169	400,315,553	787	619,369	487,443,403
738	544,644	401,947,272	788	620,944	489,303,872
739	546,121	403,583,419	789	622,521	491,169,069
740	547,600	405,224,000	790	624,100	493,039,000
741	549,081	406,869,021	791	625,681	494,913,671
742	550,564	408,518,488	792	627,264	496,793,088
743	552,049	410,172,407	793	628,849	498,677,257
744	553,536	411,830,784	794	630,436	500,566,184
745	555,025	413,493,625	795	632,025	502,459,875
746	556,516	415,160,936	796	633,616	504,358,336
747	558,009	416,832,723	797	635,209	506,261,573
748	559,504	418,508,992	798	636,804	508,169,592
749	561,001	420,189,749	799	638,401	510,082,399
750	562,500	421,875,000	800	640,000	512,000,000

TABLE 1-1   SQUARES AND CUBES OF NUMBERS *(Continued)*

No.	Square	Cube	No.	Square	Cube
801	641.601	513,922,401	851	724.201	616.295,051
802	643.204	515 849,608	852	725.904	618,470,208
803	644,809	517.781 627	853	727,609	620,650,477
804	646,416	519,718,464	854	729,316	622,835,864
805	648,025	521,660,125	855	731,025	625,026,375
806	649,636	523,606,616	856	732,736	627,222,016
807	651,249	525,557,943	857	734,449	629,422,793
808	652,864	527.514.112	858	736,164	631,628,712
809	654,481	529,475,129	859	737,881	633,839,779
810	656,100	531,441,000	860	739,600	636,056,000
811	657,721	533,411,731	861	741,321	638,277,381
812	659,344	535,387,328	862	743,044	640.503,928
813	660,969	537,367,797	863	744,769	642,735,647
814	662,596	539,353,144	864	746,496	644,972,544
815	664,225	541,343,375	865	748,225	647,214,625
816	665,856	543.338.496	866	749,956	649,461,896
817	667,489	545,338,513	867	751.689	651,714 363
818	669,124	547,343,432	868	753,424	653,972,032
819	670,761	549,353,259	869	755,161	656,234,909
820	672,400	551,368,000	870	756,900	658,503,000
821	674,041	553,387,661	871	758,641	660,776.311
822	675,684	555,412,248	872	760,384	663,054,848
823	677,329	557,441,767	873	762.129	665,338 617
824	678,976	559,476,224	874	763,876	667,627,624
825	680,625	561,515,625	875	765,625	669,921,875
826	682,276	563.559 976	876	767,376	672,221,376
827	683.929	565,609,283	877	769.129	674,526,133
828	685,584	567,663.552	878	770,884	676,836,152
829	687,241	569,722,789	879	772,641	679,151,439
830	688,900	571,787,000	880	774,400	681,472,000
831	690,561	573,856,191	881	776,161	683,797,841
832	692.224	575,930,368	882	777,924	686.128.968
833	693,889	578,009,537	883	779,689	688,465,387
834	695,556	580,093,704	884	781,456	690,807,104
835	697,225	582,182,875	885	783,225	693,154,125
836	698,896	584,277.056	886	784,996	695,506,456
837	700,569	586,376,253	887	786,769	697,864,103
838	702,244	588,480,472	888	788,544	700,227,072
839	703,921	590,589,719	889	790,321	702,595,369
840	705,600	592,704,000	890	792,100	704,969,000
841	707,281	594,823,321	891	793,881	707,347,971
842	708.964	596,947,688	892	795,664	709,732,288
843	710,649	599,077,107	893	797,449	712,121,957
844	712,336	601,211,584	894	799,236	714,516,984
845	714,025	603,351,125	895	801,025	716,917,375
846	715,716	605,495,736	896	802,816	719,323,136
847	717,409	607,645,423	897	804,609	721,734,273
848	719,104	609,800 192	898	806,404	724,150,792
849	720,801	611,960,049	899	808,201	726,572,699
850	722,500	614,125,000	900	810,000	729,000,000

TABLE 1-1   SQUARES AND CUBES OF NUMBERS (Continued)

No.	Square	Cube	No.	Square	Cube
901	811,801	731,432,701	951	904,401	860,085,351
902	813,604	733,870,808	952	906,304	862,801,408
903	815,409	736,314,327	953	908,209	865,523,177
904	817,216	738,763,264	954	910,116	868,250,664
905	819,025	741,217,625	955	912,025	870,983,875
906	820,836	743,677,416	956	913,936	873,722,816
907	822,649	746,142,643	957	915,849	876,467,493
908	824,464	748,613,312	958	917,764	879,217,912
909	826,281	751,089,429	959	919,681	881,974,079
910	828,100	753,571,000	960	921,600	884,736,000
911	829,921	756,058,031	961	923,521	887,503,681
912	831,744	758,550,528	962	925,444	890,277,128
913	833,569	761,048,497	963	927,369	893,056,347
914	835,396	763,551,944	964	929,296	895,841,344
915	837,225	766,060,875	965	931,225	898,632,125
916	839,056	768,575,296	966	933,156	901,428,696
917	840,889	771,095,213	967	935,089	904,231,063
918	842,724	773,620,632	968	937,024	907,039,232
919	844,561	776,151,559	969	938,961	909,853,209
920	846,400	778,688,000	970	940,900	912,673,000
921	848,241	781,229,961	971	942,841	915,498,611
922	850,084	783,777,448	972	944,784	918,330,048
923	851,929	786,330,467	973	946,729	921,167,317
924	853,776	788,889,024	974	948,676	924,010,424
925	855,625	791,453,125	975	950,625	926,859,375
926	857,476	794,022,776	976	952,576	929,714,176
927	859,329	796,597,983	977	954,529	932,574,833
928	861,184	799,178,752	978	956,484	935,441,352
929	863,041	801,765,089	979	958,441	938,313,739
930	864,900	804,357,000	980	960,400	941,192,000
931	866,761	806,954,491	981	962,361	944,076,141
932	868,624	809,557,568	982	964,324	946,966,168
933	870,489	812,166,237	983	966,289	949,862,087
934	872,356	814,780,504	984	968,256	952,763,904
935	874,225	817,400,375	985	970,225	955,671,625
936	876,096	820,025,856	986	972,196	958,585,256
937	877,969	822,656,953	987	974,169	961,504,803
938	879,844	825,293,672	988	976,144	964,430,272
939	881,721	827,936,019	989	978,121	967,361,669
940	883,600	830,584,000	990	980,100	970,299,000
941	885,481	833,237,621	991	982,081	973,242,271
942	887,364	835,896,888	992	984,064	976,191,488
943	889,249	838,561,807	993	986,049	979,146,657
944	891,136	841,232,384	994	988,036	982,107,784
945	893,025	843,908,625	995	990,025	985,074,875
946	894,916	846,590,536	996	992,016	988,047,936
947	896,809	849,278,123	997	994,009	991,026,973
948	898,704	851,971,392	998	996,004	994,011,992
949	900,601	854,670,349	999	998,001	997,002,999
950	902,500	857,375,000	1000	1,000,000	1,000,000,000

TABLE 1-2   AREAS OF CIRCLES

Diameter	Area	Diameter	Area	Diameter	Area
1/32	0.00077	2	3.1416	5	19.635
1/16	0.00307				
3/32	0.00690	2-1/16	3.3410	5-1/16	20.129
1/8	0.01227	2-1/8	3.5466	5-1/8	20.629
5/32	0.01917	2-3/16	3.7583	5-3/16	21.135
3/16	0.02761	2-1/4	3.9761	5-1/4	21.648
7/32	0.03758	2-5/16	4.2000	5-5/16	22.166
1/4	0.04909	2-3/8	4.4301	5-3/8	22.691
9/32	0.06213	2-7/16	4.6664	5-7/16	23.221
5/16	0.07670	2-1/2	4.9087	5-1/2	23.758
11/32	0.09281	2-9/16	5.1572	5-9/16	24.301
3/8	0.11045	2-5/8	5.4119	5-5/8	24.850
13/32	0.12962	2-11/16	5.6727	5-11/16	25.406
7/16	0.15033	2-3/4	5.9396	5-3/4	25.967
15/32	0.17257	2-13/16	6.2126	5-13/16	26.535
		2-7/8	6.4918	5-7/8	27.109
		2-15/16	6.7771	5-15/16	27.688
½	0.19635				
17/32	0.22166	3	7.0686	6	28.274
9/16	0.24850	3-1/16	7.3662	6-1/16	28.866
19/32	0.27688	3-1/8	7.6699	6-1/8	29.465
5/8	0.30680	3-3/16	7.9798	6-3/16	30.069
21/32	0.33824	3-1/4	8.2958	6-1/4	30.680
11/16	0.37122	3-5/16	8.6179	6-5/16	31.296
23/32	0.40574	3-3/8	8.9462	6-3/8	31.919
3/4	0.44179	3-7/16	9.2806	6-7/16	32.548
25/32	0.47937	3-1/2	9.6211	6-1/2	33.183
13/16	0.51849	3-9/16	9.9678	6-9/16	33.824
27/32	0.55914	3-5/8	10.321	6-5/8	34.472
7/8	0.60132	3-11/16	10.680	6-11/16	35.125
29/32	0.64504	3-3/4	11.045	6-3/4	35.785
15/16	0.69029	3-13/16	11.416	6-13/16	36.450
31/32	0.73708	3-7/8	11.793	6-7/8	37.122
		3-15/16	12.177	6-15/16	37.800
1	0.7854	4	12.566	7	38.485
1-1/16	0.8866	4-1/16	12.962	7-1/16	39.175
1-1/8	0.9940	4-1/8	13.364	7-1/8	39.871
1-3/16	1.1075	4-3/16	13.772	7-3/16	40.574
1-1/4	1.2272	4-1/4	14.186	7-1/4	41.282
1-5/16	1.3530	4-5/16	14.607	7-5/16	41.997
1-3/8	1.4849	4-3/8	15.033	7-3/8	42.718
1-7/16	1.6230	4-7/16	15.466	7-7/16	43.445
1-1/2	1.7671	4-1/2	15.904	7-1/2	44.179
1-9/16	1.9175	4-9/16	16.349	7-9/16	44.918
1-5/8	2.0739	4-5/8	16.800	7-5/8	45.664
1-11/16	2.2365	4-11/16	17.257	7-11/16	46.415
1-3/4	2.4053	4-3/4	17.721	7-3/4	47.173
1-13/16	2.5802	4-13/16	18.190	7-13/16	47.937
1-7/8	2.7612	4-7/8	18.665	7-7/8	48.707
1-15/16	2.9483	4-15/16	19.147	7-15/16	49.483

TABLE 1-2   AREAS OF CIRCLES *(Continued)*

Diameter	Area	Diameter	Area	Diameter	Area
8	50.265	**13**	132.73	**18**	254.47
8-1/8	51.849	13-1/8	135.30	18-1/8	258.02
8-1/4	53.456	13-1/4	137.89	18-1/4	261.59
8-3/8	55.088	13-3/8	140.50	18-3/8	265.18
8-1/2	56.745	13-1/2	143.14	18-1/2	268.80
8-5/8	58.426	13-5/8	145.80	18-5/8	272.45
8-3/4	60.132	13-3/4	148.49	18-3/4	276.12
8-7/8	61.862	13-7/8	151.20	18-7/8	279.81
9	63.617	**14**	153.94	**19**	283.53
9-1/8	65.397	14-1/8	156.70	19-1/8	287.27
9-1/4	67.201	14-1/4	159.48	19-1/4	291.04
9-3/8	69.029	14-3/8	162.30	19-3/8	294.83
9-1/2	70.882	14-1/2	165.13	19-1/2	298.65
9-5/8	72.760	14-5/8	167.99	19-5/8	302.49
9-3/4	74.662	14-3/4	170.87	19-3/4	306.35
9-7/8	76.589	14-7/8	173.78	19-7/8	310.24
10	78.540	**15**	176.71	**20**	314.16
10-1/8	80.516	15-1/8	179.67	20-1/8	318.10
10-1/4	82.516	15-1/4	182.65	20-1/4	322.06
10-3/8	84.541	15-3/8	185.66	20-3/8	326.05
10-1/2	86.590	15-1/2	188.69	20-1/2	330.06
10-5/8	88.664	15-5/8	191.75	20-5/8	334.10
10-3/4	90.763	15-3/4	194.83	20-3/4	338.16
10-7/8	92.886	15-7/8	197.93	20-7/8	342.25
11	95.033	**16**	201.06	**21**	346.36
11-1/8	97.205	16-1/8	204.22	21-1/8	350.50
11-1/4	99.402	16-1/4	207.39	21-1/4	354.66
11-3/8	101.62	16-3/8	210.60	21-3/8	358.84
11-1/2	103.87	16-1/2	213.82	21-1/2	363.05
11-5/8	106.14	16-5/8	217.08	21-5/8	367.28
11-3/4	108.43	16-3/4	220.35	21-3/4	371.54
11-7/8	110.75	16-7/8	223.65	21-7/8	375.83
12	113.10	**17**	226.98	**22**	380.13
12-1/8	115.47	17-1/8	230.33	22-1/8	384.46
12-1/4	117.86	17-1/4	233.71	22-1/4	388.82
12-3/8	120.28	17-3/8	237.10	22-3/8	393.20
12-1/2	122.72	17-1/2	240.53	22-1/2	397.61
12-5/8	12519	17-5/8	243.98	22-5/8	402.04
12-3/4	127.68	17-3/4	247.45	22-3/4	406.49
12-7/8	130.19	17-7/8	250.95	22-7/8	410.97

TABLE 1-2 AREAS OF CIRCLES *(Continued)*

Diameter	Area	Diameter	Area	Diameter	Area
23	415.48	**28**	615.75	**33**	855.30
23-1/8	420.00	28-1/8	621.26	33-1/8	861.79
23-1/4	424.56	28-1/4	626.80	33-1/4	868.31
23-3/8	429.13	28-3/8	632.36	33-3/8	874.85
23-1/2	433.74	28-1/2	637.94	33-1/2	881.41
23-5/8	438.36	28-5/8	643.55	33-5/8	888.00
23-3/4	443.01	28-3/4	649.18	33-3/4	894.62
23-7/8	447.69	28-7/8	654.84	33-7/8	901.26
24	452.39	**29**	660.52	**34**	907.92
24-1/8	457.11	29-1/8	666.23	34-1/8	914.61
24-1/4	461.86	29-1/4	671.96	34-1/4	921.32
24-3/8	466.64	29-3/8	677.71	34-3/8	928.06
24-1/2	471.44	29-1/2	683.49	34-1/2	934.82
24-5/8	476.26	29-5/8	689.30	34-5/8	941.61
24-3/4	481.11	29-3/4	695.13	34-3/4	948.42
24-7/8	485.98	29-7/8	700.98	34-7/8	955.25
25	490.87	**30**	706.86	**35**	962.11
25-1/8	495.79	30-1/8	712.76	35-1/8	969.00
25-1/4	500.74	30-1/4	718.69	35-1/4	975.91
25-3/8	505.71	30-3/8	724.64	35-3/8	982.84
25-1/2	510.71	30-1/2	730.62	35-1/2	989.80
25-5/8	515.72	30-5/8	736.62	35-5/8	996.78
25-3/4	520.77	30-3/4	742.64	35-3/4	1003.8
25-7/8	525.84	30-7/8	748.69	35-7/8	1010.8
26	530.93	**31**	754.77	**36**	1017.9
26-1/8	536.05	31-1/8	760.87	36-1/8	1025.0
26-1/4	541.19	31-1/4	766.99	36-1/4	1032.1
26-3/8	546.35	31-3/8	773.14	36-3/8	1039.2
26-1/2	551.55	31-1/2	779.31	36-1/2	1046.3
26-5/8	556.76	31-5/8	785.51	36-5/8	1053.5
26-3/4	562.00	31-3/4	791.73	36-3/4	1060.7
26-7/8	567.27	31-7/8	797.98	36-7/8	1068.0
27	572.56	**32**	804.25	**37**	1075.2
27-1/8	577.87	32-1/8	810.54	37-1/8	1082.5
27-1/4	583.21	32-1/4	816.86	37-1/4	1089.8
27-3/8 ·	588.57	32-3/8	823.21	37-3/8	1097.1
27-1/2	593.96	32-1/2	829.58	37-1/2	1104.5
27-5/8	599.37	32-5/8	835.97	37-5/8	1111.8
27-3/4	604.81	32-3/4	842.39	37-3/4	1119.2
27-7/8	610.27	32-7/8	848.83	37-7/8	1126.7

TABLE 1-2 AREAS OF CIRCLES *(Continued)*

Diameter	Area	Diameter	Area	Diameter	Area
38	1134.1	**43**	1452.2	**48**	1809.6
38-1/8	1141.6	43-1/8	1460.7	48-1/8	1819.0
38-1/4	1149.1	43-1/4	1469.1	48-1/4	1828.5
38-3/8	1156.6	43-3/8	1477.6	48-3/8	1837.9
38-1/2	1164.2	43-1/2	1486.2	48-1/2	1847.5
38-5/8	1171.7	43-5/8	1494.7	48-5/8	1857.0
38-3/4	1179.3	43-3/4	1503.3	48-3/4	1866.5
38-7/8	1186.9	43-7/8	1511.9	48-7/8	1876.1
39	1194.6	**44**	1520.5	**49**	1885.7
39-1/8	1202.3	44-1/8	1529.2	49-1/8	1895.4
39-1/4	1210.0	44-1/4	1537.9	49-1/4	1905.0
39-3/8	1217.7	44-3/8	1546.6	49-3/8	1914.7
39-1/2	1225.4	44-1/2	1555.3	49-1/2	1924.4
39-5/8	1233.2	44-5/8	1564.0	49-5/8	1934.2
39-3/4	1241.0	44-3/4	1572.8	49-3/4	1943.9
39-7/8	1248.8	44-7/8	1581.6	49-7/8	1953.7
40	1256.6	**45**	1590.4	**50**	1963.5
40-1/8	1264.5	45-1/8	1599.3	50-1/8	1973.3
40-1/4	1272.4	45-1/4	1608.2	50-1/4	1983.2
40-3/8	1280.3	45-3/8	1617.0	50-3/8	1993.1
40-1/2	1288.2	45-1/2	1626.0	50-1/2	2003.0
40-5/8	1296.2	45-5/8	1634.9	50-5/8	2012.9
40-3/4	1304.2	45-3/4	1643.9	50-3/4	2022.8
40-7/8	1312.2	45-7/8	1652.9	50-7/8	2032.8
41	1320.3	**46**	1661.9	**51**	2042.8
41-1/8	1328.3	46-1/8	1670.9	51-1/8	2052.8
41-1/4	1336.4	46-1/4	1680.0	51-1/4	2062.9
41-3/8	1344.5	46-3/8	1689.1	51-3/8	2073.0
41-1/2	1352.7	46-1/2	1698.2	51-1/2	2083.1
41-5/8	1360.8	46-5/8	1707.4	51-5/8	2093.2
41-3/4	1369.0	46-3/4	1716.5	51-3/4	2103.3
41-7/8	1377.2	46-7/8	1725.7	51-7/8	2113.5
42	1385.4	**47**	1734.9	**52**	2123.7
42-1/8	1393.7	47-1/8	1744.2	52-1/8	2133.9
42-1/4	1402.0	47-1/4	1753.5	52-1/4	2144.2
42-3/8	1410.3	47-3/8	1762.7	52-3/8	2154.5
42-1/2	1418.6	47-1/2	1772.1	52-1/2	2164.8
42-5/8	1427.0	47-5/8	1781.4	52-5/8	2175.1
42-3/4	1435.4	47-3/4	1790.8	52-3/4	2185.4
42-7/8	1443.8	47-7/8	1800.1	52-7/8	2195.8

TABLE 1-2   AREAS OF CIRCLES *(Continued)*

Diameter	Area	Diameter	Area	Diameter	Area
53	2206.2	**61**	2922.5	**71**	3959.2
53-1/8	2216.6	61-1/4	2946.5	71-1/4	3987.1
53-1/4	2227.0	61-1/2	2970.6	71-1/2	4015.2
53-3/8	2237.5	61-3/4	2994.8	71-3/4	4043.3
53-1/2	2248.0				
53-5/8	2258.5	**62**	3019.1	**72**	4071.5
53-3/4	2269.1	62-1/4	3043.5	72-1/4	4099.8
53-7/8	2279.6	62-1/2	3068.0	72-1/2	4128.2
		62-3/4	3092.6	72-3/4	4156.8
54	2290.2				
		63	3117.2	**73**	4185.4
54-1/8	2300.8	63-1/4	3142.0	73-1/4	4214.1
54-1/4	2311.5	63-1/2	3166.9	73-1/2	4242.9
54-3/8	2322.1	63-3/4	3191.9	73-3/4	4271.8
54-1/2	2332.8				
54-5/8	2343.5	**64**	3217.0	**74**	4300.8
54-3/4	2354.3	64-1/4	3242.2	74-1/4	4329.9
54-7/8	2365.0	64-1/2	3267.5	74-1/2	4359.2
		64-3/4	3292.8	74-3/4	4388.5
55	2375.8				
55-1/4	2397.5	**65**	3318.3	**75**	4417.9
55-1/2	2419.2	65-1/4	3343.9	75-1/4	4447.4
55-3/4	2441.1	65-1/2	3369.6	75-1/2	4477.0
		65-3/4	3395.3	75-3/4	4506.7
56	2463.0				
56-1/4	2485.0	**66**	3421.2	**76**	4536.5
56-1/2	2507.2	66-1/4	3447.2	76-1/4	4566.4
56-3/4	2529.4	66-1/2	3473.2	76-1/2	4596.3
		66-3/4	3499.4	76-3/4	4626.4
57	2551.8				
57-1/4	2574.2	**67**	3525.7	**77**	4656.6
57-1/2	2596.7	67-1/4	3552.0	77-1/4	4686.9
57-3/4	2619.4	67-1/2	3578.5	77-1/2	4717.3
		67-3/4	3605.0	77-3/4	4747.8
58	2642.1				
58-1/4	2664.9	**68**	3631.7	**78**	4778.4
58-1/2	2687.8	68-1/4	3658.4	78-1/4	4809.0
58-3/4	2710.9	68-1/2	3685.3	78-1/2	4839.8
		68-3/4	3712.2	78-3/4	4870.7
59	2734.0				
		69	3739.3	**79**	4901.7
59-1/4	2757.2	69-1/4	3766.4	79-1/4	4932.7
59-1/2	2780.5	69-1/2	3793.7	79-1/2	4963.9
59-3/4	2803.9	69-3/4	3821.0	79-3/4	4995.2
60	2827.4	**70**	3848.5	**80**	5026.5
60-1/4	2851.0	70-1/4	3876.0	80-1/4	5058.0
60-1/2	2874.0	70-1/2	3903.6	80-1/2	5089.6
60-3/4	2898.6	70-3/4	3931.4	80-3/4	5121.2

TABLE 1-2   AREAS OF CIRCLES *(Continued)*

Diameter	Area	Diameter	Area
81	5153.0	**91**	6503.9
81-1/4	5184.9	91-1/4	6539.7
81-1/2	5216.8	91-1/2	6575.5
81-3/4	5248.9	91-3/4	6611.5
82	5281.0	**92**	6647.6
82-1/4	5313.3	92-1/4	6683.8
82-1/2	5345.6	92-1/2	6720.1
82-3/4	5378.1	92-3/4	6756.4
83	5410.6	**93**	6792.9
83-1/4	5443.3		
83-1/2	5476.0	93-1/4	6829.5
83-3/4	5508.8	93-1/2	6866.1
84	5541.8	93-3/4	6902.9
84-1/4	5574.8	**94**	6939.8
84-1/2	5607.9		
84-3/4	5641.2	94-1/4	6976.7
85	5674.5	94-1/2	7013.8
		94-3/4	7051.0
85-1/4	5707.9		
85-1/2	5741.5	**95**	7088.2
85-3/4	5775.1	95-1/4	7125.6
86	5808.8	95-1/2	7163.0
		95-3/4	7200.6
86-1/4	5842.6		
86-1/2	5876.5	**96**	7238.2
86-3/4	5910.6	96-1/4	7276.0
87	5944.7	96-1/2	7313.8
		96-3/4	7351.8
87-1/4	5978.9		
87-1/2	6013.2	**97**	7389.8
87-3/4	6047.6	97-1/4	7428.0
88	6082.1	97-1/2	7466.2
		97-3/4	7504.5
88-1/4	6116.7		
88-1/2	6151.4	**98**	7543.0
88-3/4	6186.2	98-1/4	7581.5
89	6221.1	98-1/2	7620.1
		98-3/4	7658.9
89-1/4	6256.1		
89-1/2	6291.2	**99**	7697.7
89-3/4	6326.4	99-1/4	7736.6
90	6361.7	99-1/2	7775.6
		99-3/4	7814.8
90-1/4	6397.1		
90-1/2	6432.6		
90-3/4	6468.2	**100**	7854.0

LOGARITHMS

If $x = \log_e n$, then $e^x = n$ $\log n^c = c \log n$ $\log 1 = 0$

$\log ab = \log a + \log b$ $\log \sqrt[c]{n} = (1/c) \log n$ $\pi = 3.141593$

$\log a/b = \log a - \log b$ $\log_e x = 2.3026 \log_{10} x$ $\log_{10} \pi = .497150$

       $= -\log b/a$ $\log_{10} x = 0.4343 \log_e x$

$\log 1/n = -\log n$ $e = 2.718282$

TABLE 1-3   LOGARITHMS TO BASE 10

Number	0	1	2	3	4	5	6	7	8	9	Proportional Parts 1 2 3	4 5 6	7 8 9
1.0	0000	0043	0086	0128	0170	0212	0253	0294	0334	0374	4 8 12	17 21 25	29 33 37
1.1	0414	0453	0492	0531	0569	0607	0645	0682	0719	0755	4 8 11	15 19 23	26 30 34
1.2	0792	0828	0864	0899	0934	0969	1004	1038	1072	1106	3 7 10	14 17 21	24 28 31
1.3	1139	1173	1206	1239	1271	1303	1335	1367	1399	1430	3 6 10	13 16 19	23 26 29
1.4	1461	1492	1523	1553	1584	1614	1644	1673	1703	1732	3 6 9	12 15 18	21 24 27
1.5	1761	1790	1818	1847	1875	1903	1931	1959	1987	2014	3 6 8	11 14 17	20 22 25
1.6	2041	2068	2095	2122	2148	2175	2201	2227	2253	2279	3 5 8	11 13 16	18 21 24
1.7	2304	2330	2355	2380	2405	2430	2455	2480	2504	2529	2 5 7	10 12 15	17 20 22
1.8	2553	2577	2601	2625	2648	2672	2695	2718	2742	2765	2 5 7	9 12 14	16 19 21
1.9	2788	2810	2833	2856	2878	2900	2923	2945	2967	2989	2 4 7	9 11 13	16 18 20
2.0	3010	3032	3054	3075	3096	3118	3139	3160	3181	3201	2 4 6	8 11 13	15 17 19
2.1	3222	3243	3263	3284	3304	3324	3345	3365	3385	3404	2 4 6	8 10 12	14 16 18
2.2	3424	3444	3464	3483	3502	3522	3541	3560	3579	3598	2 4 6	8 10 12	14 15 17
2.3	3617	3636	3655	3674	3692	3711	3729	3747	3766	3784	2 4 6	7 9 11	13 15 17
2.4	3802	3820	3838	3856	3874	3892	3909	3927	3945	3962	2 4 5	7 9 11	12 14 16
2.5	3979	3997	4014	4031	4048	4065	4082	4099	4116	4133	2 3 5	7 9 10	12 14 15
2.6	4150	4166	4183	4200	4216	4232	4249	4265	4281	4298	2 3 5	7 8 10	11 13 15
2.7	4314	4330	4346	4362	4378	4393	4409	4425	4440	4456	2 3 5	6 8 9	11 13 14
2.8	4472	4487	4502	4518	4533	4548	4564	4579	4594	4609	2 3 5	6 8 9	11 12 14
2.9	4624	4639	4654	4669	4683	4698	4713	4728	4742	4757	1 3 4	6 7 9	10 12 13
3.0	4771	4786	4800	4814	4829	4843	4857	4871	4886	4900	1 3 4	6 7 9	10 11 13
3.1	4914	4928	4942	4955	4969	4983	4997	5011	5024	5038	1 3 4	6 7 8	10 11 12
3.2	5051	5065	5079	5092	5105	5119	5132	5145	5159	5172	1 3 4	5 7 8	9 11 12
3.3	5185	5198	5211	5224	5237	5250	5263	5276	5289	5302	1 3 4	5 6 8	9 10 12
3.4	5315	5328	5340	5353	5366	5378	5391	5403	5416	5428	1 3 4	5 6 8	9 10 11
3.5	5441	5453	5465	5478	5490	5502	5514	5527	5539	5551	1 2 4	5 6 7	9 10 11
3.6	5563	5575	5587	5599	5611	5623	5635	5647	5658	5670	1 2 4	5 6 7	8 10 11
3.7	5682	5694	5705	5717	5729	5740	5752	5763	5775	5786	1 2 3	5 6 7	8 9 10
3.8	5798	5809	5821	5832	5843	5855	5866	5877	5888	5899	1 2 3	5 6 7	8 9 10
3.9	5911	5922	5933	5944	5955	5966	5977	5988	5999	6010	1 2 3	4 5 7	8 9 10
4.0	6021	6031	6042	6053	6064	6075	6085	6096	6107	6117	1 2 3	4 5 6	8 9 10
4.1	6128	6138	6149	6160	6170	6180	6191	6201	6212	6222	1 2 3	4 5 6	7 8 9
4.2	6232	6243	6253	6263	6274	6284	6294	6304	6314	6325	1 2 3	4 5 6	7 8 9
4.3	6335	6345	6355	6365	6375	6385	6395	6405	6415	6425	1 2 3	4 5 6	7 8 9
4.4	6435	6444	6454	6464	6474	6484	6493	6503	6513	6522	1 2 3	4 5 6	7 8 9
4.5	6532	6542	6551	6561	6571	6580	6590	6599	6609	6618	1 2 3	4 5 6	7 8 9
4.6	6628	6637	6646	6656	6665	6675	6684	6693	6702	6712	1 2 3	4 5 6	7 7 8
4.7	6721	6730	6739	6749	6758	6767	6776	6785	6794	6803	1 2 3	4 5 6	6 7 8
4.8	6812	6821	6830	6839	6848	6857	6866	6875	6884	6893	1 2 3	4 4 5	6 7 8
4.9	6902	6911	6920	6928	6937	6946	6955	6964	6972	6981	1 2 3	4 4 5	6 7 8
5.0	6990	6998	7007	7016	7024	7033	7042	7050	7059	7067	1 2 3	3 4 5	6 7 8
5.1	7076	7084	7093	7101	7110	7118	7126	7135	7143	7152	1 2 3	3 4 5	6 7 8
5.2	7160	7168	7177	7185	7193	7202	7210	7218	7226	7235	1 2 2	3 4 5	6 7 7
5.3	7243	7251	7259	7267	7275	7284	7292	7300	7308	7316	1 2 2	3 4 5	6 6 7
5.4	7324	7332	7340	7348	7356	7364	7372	7380	7388	7396	1 2 2	3 4 5	6 6 7

TABLE 1-3   LOGARITHMS TO BASE 10 *(Continued)*

Num-ber	0	1	2	3	4	5	6	7	8	9	Proportional Parts								
											1	2	3	4	5	6	7	8	9
5.5	7404	7412	7419	7427	7435	7443	7451	7459	7466	7474	1	2	2	3	4	5	5	6	7
5.6	7482	7490	7497	7505	7513	7520	7528	7536	7543	7551	1	2	2	3	4	5	5	6	7
5.7	7559	7566	7574	7582	7589	7597	7604	7612	7619	7627	1	2	2	3	4	5	5	6	7
5.8	7634	7642	7649	7657	7664	7672	7679	7686	7694	7701	1	1	2	3	4	4	5	6	7
5.9	7709	7716	7723	7731	7738	7745	7752	7760	7767	7774	1	1	2	3	4	4	5	6	7
6.0	7782	7789	7796	7803	7810	7818	7825	7832	7839	7846	1	1	2	3	4	4	5	6	6
6.1	7853	7860	7868	7875	7882	7889	7896	7903	7910	7917	1	1	2	3	4	4	5	6	6
6.2	7924	7931	7938	7945	7952	7959	7966	7973	7980	7987	1	1	2	3	3	4	5	6	6
6.3	7993	8000	8007	8014	8021	8028	8035	8041	8048	8055	1	1	2	3	3	4	5	5	6
6.4	8062	8069	8075	8082	8089	8096	8102	8109	8116	8122	1	1	2	3	3	4	5	5	6
6.5	8129	8136	8142	8149	8156	8162	8169	8176	8182	8189	1	1	2	3	3	4	5	5	6
6.6	8195	8202	8209	8215	8222	8228	8235	8241	8248	8254	1	1	2	3	3	4	5	5	6
6.7	8261	8267	8274	8280	8287	8293	8299	8306	8312	8319	1	1	2	3	3	4	5	5	6
6.8	8325	8331	8338	8344	8351	8357	8363	8370	8376	8382	1	1	2	3	3	4	4	5	6
6.9	8388	8395	8401	8407	8414	8420	8426	8432	8439	8445	1	1	2	2	3	4	4	5	6
7.0	8451	8457	8463	8470	8476	8482	8488	8494	8500	8506	1	1	2	2	3	4	4	5	6
7.1	8513	8519	8525	8531	8537	8543	8549	8555	8561	8567	1	1	2	2	3	4	4	5	5
7.2	8573	8579	8585	8591	8597	8603	8609	8615	8621	8627	1	1	2	2	3	4	4	5	5
7.3	8633	8639	8645	8651	8657	8663	8669	8675	8681	8686	1	1	2	2	3	4	4	5	5
7.4	8692	8698	8704	8710	8716	8722	8727	8733	8739	8745	1	1	2	2	3	4	4	5	5
7.5	8751	8756	8762	8768	8774	8779	8785	8791	8797	8802	1	1	2	2	3	3	4	5	5
7.6	8808	8814	8820	8825	8831	8837	8842	8848	8854	8859	1	1	2	2	3	3	4	5	5
7.7	8865	8871	8876	8882	8887	8893	8899	8904	8910	8915	1	1	2	2	3	3	4	4	5
7.8	8921	8927	8932	8938	8943	8949	8954	8960	8965	8971	1	1	2	2	3	3	4	4	5
7.9	8976	8982	8987	8993	8998	9004	9009	9015	9020	9025	1	1	2	2	3	3	4	4	5
8.0	9031	9036	9042	9047	9053	9058	9063	9069	9074	9079	1	1	2	2	3	3	4	4	5
8.1	9085	9090	9096	9101	9106	9112	9117	9122	9128	9133	1	1	2	2	3	3	4	4	5
8.2	9138	9143	9149	9154	9159	9165	9170	9175	9180	9186	1	1	2	2	3	3	4	4	5
8.3	9191	9196	9201	9206	9212	9217	9222	9227	9232	9238	1	1	2	2	3	3	4	4	5
8.4	9243	9248	9253	9258	9263	9269	9274	9279	9284	9289	1	1	2	2	3	3	4	4	5
8.5	9294	9299	9304	9309	9315	9320	9325	9330	9335	9340	1	1	2	2	3	3	4	4	5
8.6	9345	9350	9355	9360	9365	9370	9375	9380	9385	9390	1	1	2	2	3	3	4	4	5
8.7	9395	9400	9405	9410	9415	9420	9425	9430	9435	9440	0	1	1	2	2	3	3	4	4
8.8	9445	9450	9455	9460	9465	9469	9474	9479	9484	9489	0	1	1	2	2	3	3	4	4
8.9	9494	9499	9504	9509	9513	9518	9523	9528	9533	9538	0	1	1	2	2	3	3	4	4
9.0	9542	9547	9552	9557	9562	9566	9571	9576	9581	9586	0	1	1	2	2	3	3	4	4
9.1	9590	9595	9600	9605	9609	9614	9619	9624	9628	9633	0	1	1	2	2	3	3	4	4
9.2	9638	9643	9647	9652	9657	9661	9666	9671	9675	9680	0	1	1	2	2	3	3	4	4
9.3	9685	9689	9694	9699	9703	9708	9713	9717	9722	9727	0	1	1	2	2	3	3	4	4
9.4	9731	9736	9741	9745	9750	9754	9759	9763	9768	9773	0	1	1	2	2	3	3	4	4
9.5	9777	9782	9786	9791	9795	9800	9805	9809	9814	9818	0	1	1	2	2	3	3	4	4
9.6	9823	9827	9832	9836	9841	9845	9850	9854	9859	9863	0	1	1	2	2	3	3	4	4
9.7	9868	9872	9877	9881	9886	9890	9894	9899	9903	9908	0	1	1	2	2	3	3	4	4
9.8	9912	9917	9921	9926	9930	9934	9939	9943	9948	9952	0	1	1	2	2	3	3	4	4
9.9	9956	9961	9965	9969	9974	9978	9983	9987	9991	9996	0	1	1	2	2	3	3	3	4

GREEK ALPHABET

A, α	alpha	H, η	eta	N, ν	nu	T, τ	tau
B, β	beta	Θ, ϑ	theta	Ξ, ξ	xi	Υ, υ	upsilon
Γ, γ	gamma	I, ι	iota	O, o	omicron	Φ, φ	phi
Δ, δ	delta	K, κ	kappa	Π, π	pi	X, χ	chi
E, ϵ	epsilon	Λ, λ	lamba	P, ρ	rho	Ψ, ψ	psi
Z, ζ	zeta	M, μ	mu	Σ, σ	sigma	Ω, ω	omega

TRIGONOMETRIC SOLUTION OF TRIANGLES

Right-Angle Triangle

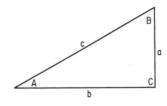

$$a^2 + b^2 = c^2 \qquad A + B = 90° \qquad \text{Area} = \tfrac{1}{2}ab$$
$$\sin A = a/c \qquad \cos A = b/c \qquad \tan A = a/b$$
$$\csc A = c/a \qquad \sec A = c/b \qquad \cot A = b/a$$
$$a = \sqrt{c^2 - b^2} = c \sin A = b \tan A = b/\cot A = c/\csc A$$
$$b = \sqrt{c^2 - a^2} = c \cos A = a/\tan A = a \cot A = c/\sec A$$
$$c = \sqrt{a^2 + b^2} = a/\sin A = b/\cos A = a \csc A = b \sec A$$
$$A = 90° - B \qquad B = 90° - A \qquad C = 90°$$

Oblique-Angle Triangle

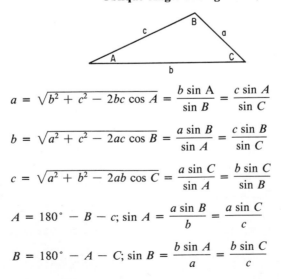

$$a = \sqrt{b^2 + c^2 - 2bc \cos A} = \frac{b \sin A}{\sin B} = \frac{c \sin A}{\sin C}$$

$$b = \sqrt{a^2 + c^2 - 2ac \cos B} = \frac{a \sin B}{\sin A} = \frac{c \sin B}{\sin C}$$

$$c = \sqrt{a^2 + b^2 - 2ab \cos C} = \frac{a \sin C}{\sin A} = \frac{b \sin C}{\sin B}$$

$$A = 180° - B - c;\ \sin A = \frac{a \sin B}{b} = \frac{a \sin C}{c}$$

$$B = 180° - A - C;\ \sin B = \frac{b \sin A}{a} = \frac{b \sin C}{c}$$

$$C = 180° - A - B; \sin C = \frac{c \sin A}{a} = \frac{c \sin B}{b}$$

$$\text{Area} = \tfrac{1}{2}ab \sin C = \frac{a^2 \sin B \sin C}{2 \sin A} = \frac{bc \sin A}{2} = \frac{ac \sin B}{2}$$

TRIGONOMETRIC FORMULAS*

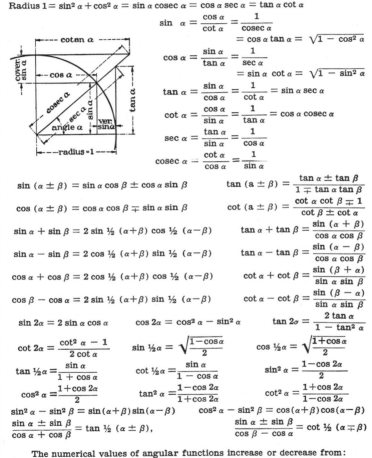

Radius $1 = \sin^2 \alpha + \cos^2 \alpha = \sin \alpha \operatorname{cosec} \alpha = \cos \alpha \sec \alpha = \tan \alpha \cot \alpha$

$$\sin \alpha = \frac{\cos \alpha}{\cot \alpha} = \frac{1}{\operatorname{cosec} \alpha}$$
$$= \cos \alpha \tan \alpha = \sqrt{1 - \cos^2 \alpha}$$

$$\cos \alpha = \frac{\sin \alpha}{\tan \alpha} = \frac{1}{\sec \alpha}$$
$$= \sin \alpha \cot \alpha = \sqrt{1 - \sin^2 \alpha}$$

$$\tan \alpha = \frac{\sin \alpha}{\cos \alpha} = \frac{1}{\cot \alpha} = \sin \alpha \sec \alpha$$

$$\cot \alpha = \frac{\cos \alpha}{\sin \alpha} = \frac{1}{\tan \alpha} = \cos \alpha \operatorname{cosec} \alpha$$

$$\sec \alpha = \frac{\tan \alpha}{\sin \alpha} = \frac{1}{\cos \alpha}$$

$$\operatorname{cosec} \alpha = \frac{\cot \alpha}{\cos \alpha} = \frac{1}{\sin \alpha}$$

$$\sin (\alpha \pm \beta) = \sin \alpha \cos \beta \pm \cos \alpha \sin \beta$$

$$\tan (a \pm \beta) = \frac{\tan \alpha \pm \tan \beta}{1 \mp \tan \alpha \tan \beta}$$

$$\cos (\alpha \pm \beta) = \cos \alpha \cos \beta \mp \sin \alpha \sin \beta$$

$$\cot (a \pm \beta) = \frac{\cot \alpha \cot \beta \mp 1}{\cot \beta \pm \cot \alpha}$$

$$\sin \alpha + \sin \beta = 2 \sin \tfrac{1}{2} (\alpha + \beta) \cos \tfrac{1}{2} (\alpha - \beta)$$

$$\tan \alpha + \tan \beta = \frac{\sin (\alpha + \beta)}{\cos \alpha \cos \beta}$$

$$\sin \alpha - \sin \beta = 2 \cos \tfrac{1}{2} (\alpha + \beta) \sin \tfrac{1}{2} (\alpha - \beta)$$

$$\tan \alpha - \tan \beta = \frac{\sin (\alpha - \beta)}{\cos \alpha \cos \beta}$$

$$\cos \alpha + \cos \beta = 2 \cos \tfrac{1}{2} (\alpha + \beta) \cos \tfrac{1}{2} (\alpha - \beta)$$

$$\cot \alpha + \cot \beta = \frac{\sin (\beta + \alpha)}{\sin \alpha \sin \beta}$$

$$\cos \beta - \cos \alpha = 2 \sin \tfrac{1}{2} (\alpha + \beta) \sin \tfrac{1}{2} (\alpha - \beta)$$

$$\cot \alpha - \cot \beta = \frac{\sin (\beta - \alpha)}{\sin \alpha \sin \beta}$$

$$\sin 2\alpha = 2 \sin \alpha \cos \alpha \qquad \cos 2\alpha = \cos^2 \alpha - \sin^2 \alpha \qquad \tan 2\alpha = \frac{2 \tan \alpha}{1 - \tan^2 \alpha}$$

$$\cot 2\alpha = \frac{\cot^2 \alpha - 1}{2 \cot \alpha} \qquad \sin \tfrac{1}{2}\alpha = \sqrt{\frac{1 - \cos \alpha}{2}} \qquad \cos \tfrac{1}{2}\alpha = \sqrt{\frac{1 + \cos \alpha}{2}}$$

$$\tan \tfrac{1}{2}\alpha = \frac{\sin \alpha}{1 + \cos \alpha} \qquad \cot \tfrac{1}{2}\alpha = \frac{\sin \alpha}{1 - \cos \alpha} \qquad \sin^2 \alpha = \frac{1 - \cos 2\alpha}{2}$$

$$\cos^2 \alpha = \frac{1 + \cos 2\alpha}{2} \qquad \tan^2 \alpha = \frac{1 - \cos 2\alpha}{1 + \cos 2\alpha} \qquad \cot^2 \alpha = \frac{1 + \cos 2\alpha}{1 - \cos 2\alpha}$$

$$\sin^2 \alpha - \sin^2 \beta = \sin (\alpha + \beta) \sin (\alpha - \beta) \qquad \cos^2 \alpha - \sin^2 \beta = \cos (\alpha + \beta) \cos (\alpha - \beta)$$

$$\frac{\sin \alpha \pm \sin \beta}{\cos \alpha + \cos \beta} = \tan \tfrac{1}{2} (\alpha \pm \beta), \qquad \frac{\sin \alpha \pm \sin \beta}{\cos \beta - \cos \alpha} = \cot \tfrac{1}{2} (\alpha \mp \beta)$$

The numerical values of angular functions increase or decrease from:

Function	In Quadrant I 0°—90°	In Quadrant II 90°—180°	In Quadrant III 180°—270°	In Quadrant IV 270°—360°
sin	0 to +1	+1 to 0	0 to −1	−1 to 0
cos	+1 to 0	0 to −1	−1 to 0	0 to +1
tan	0 to +∞	−∞ to 0	0 to +∞	−∞ to 0
cot	+∞ to 0	0 to −∞	+∞ to 0	0 to −∞
sec	+1 to +∞	−∞ to −1	−1 to −∞	+∞ to +1
cosec	+∞ to +1	+1 to +∞	−∞ to −1	−1 to −∞

° *Courtesy of* New Departure-Hyatt Bearings Division of General Motors Corp.

FIG. 1-1

TABLE 1-4 TRIGONOMETRIC FUNCTIONS

SINES

De-grees	Ra-dians	0′	10′	20′	30′	40′	50′	60′	
0	.0000	.0000	.0029	.0058	.0087	.0116	.0145	.0175	89
1	.0175	.0175	.0204	.0233	.0262	.0291	.0320	.0349	88
2	.0349	.0349	.0378	.0407	.0436	.0465	.0494	.0523	87
3	.0524	.0523	.0552	.0581	.0610	.0640	.0669	.0698	86
4	.0698	.0698	.0727	.0756	.0785	.0814	.0843	.0872	85
5	.0873	.0872	.0901	.0929	.0958	.0987	.1016	.1045	84
6	.1047	.1045	.1074	.1103	.1132	.1161	.1190	.1219	83
7	.1222	.1219	.1248	.1276	.1305	.1334	.1363	.1392	82
8	.1396	.1392	.1421	.1449	.1478	.1507	.1536	.1564	81
9	.1571	.1564	.1593	.1622	.1650	.1679	.1708	.1736	80
10	.1745	.1736	.1765	.1794	.1822	.1851	.1880	.1908	79
11	.1920	.1908	.1937	.1965	.1994	.2022	.2051	.2079	78
12	.2094	.2079	.2108	.2136	.2164	.2193	.2221	.2250	77
13	.2269	.2250	.2278	.2306	.2334	.2363	.2391	.2419	76
14	.2443	.2419	.2447	.2476	.2504	.2532	.2560	.2588	75
15	.2618	.2588	.2616	.2644	.2672	.2700	.2728	.2756	74
16	.2793	.2756	.2784	.2812	.2840	.2868	.2896	.2924	73
17	.2967	.2924	.2952	.2979	.3007	.3035	.3062	.3090	72
18	.3142	.3090	.3118	.3145	.3173	.3201	.3228	.3256	71
19	.3316	.3256	.3283	.3311	.3338	.3365	.3393	.3420	70
20	.3491	.3420	.3448	.3475	.3502	.3529	.3557	.3584	69
21	.3665	.3584	.3611	.3638	.3665	.3692	.3719	.3746	68
22	.3840	.3746	.3773	.3800	.3827	.3854	.3881	.3907	67
23	.4014	.3907	.3934	.3961	.3987	.4014	.4041	.4067	66
24	.4189	.4067	.4094	.4120	.4147	.4173	.4200	.4226	65
25	.4363	.4226	.4253	.4279	.4305	.4331	.4358	.4384	64
26	.4538	.4384	.4410	.4436	.4462	.4488	.4514	.4540	63
27	.4712	.4540	.4566	.4592	.4617	.4643	.4669	.4695	62
28	.4887	.4695	.4720	.4746	.4772	.4797	.4823	.4848	61
29	.5061	.4848	.4874	.4899	.4924	.4950	.4975	.5000	60
30	.5236	.5000	.5025	.5050	.5075	.5100	.5125	.5150	59
31	.5411	.5150	.5175	.5200	.5225	.5250	.5275	.5299	58
32	.5585	.5299	.5324	.5348	.5373	.5398	.5422	.5446	57
33	.5760	.5446	.5471	.5495	.5519	.5544	.5568	.5592	56
34	.5934	.5592	.5616	.5640	.5664	.5688	.5712	.5736	55
35	.6109	.5736	.5760	.5783	.5807	.5831	.5854	.5878	54
36	.6283	.5878	.5901	.5925	.5948	.5972	.5995	.6018	53
37	.6458	.6018	.6041	.6065	.6088	.6111	.6134	.6157	52
38	.6632	.6157	.6180	.6202	.6225	.6248	.6271	.6293	51
39	.6807	.6293	.6316	.6338	.6361	.6383	.6406	.6428	50
40	.6981	.6428	.6450	.6472	.6494	.6517	.6539	.6561	49
41	.7156	.6561	.6583	.6604	.6626	.6648	.6670	.6691	48
42	.7330	.6691	.6713	.6734	.6756	.6777	.6799	.6820	47
43	.7505	.6820	.6841	.6862	.6884	.6905	.6926	.6947	46
44	.7679	.6947	.6967	.6988	.7009	.7030	.7050	.7071	45
45	.7854	.7071							

| | | 60′ | 50′ | 40′ | 30′ | 20′ | 10′ | 0′ | De-grees |

COSINES

TABLE 1-4   TRIGONOMETRIC FUNCTIONS *(Continued)*

SINES

De-grees	Ra-dians	0'	10'	20'	30'	40'	50'	60'	
45	0.7854	.7071	.7092	.7112	.7133	.7153	.7173	.7193	44
46	0.8029	.7193	.7214	.7234	.7254	.7274	.7294	.7314	43
47	0.8203	.7314	.7333	.7353	.7373	.7392	.7412	.7431	42
48	0.8378	.7431	.7451	.7470	.7490	.7509	.7528	.7547	41
49	0.8552	.7547	.7566	.7585	.7604	.7623	.7642	.7660	40
50	0.8727	.7660	.7679	.7698	.7716	.7735	.7753	.7771	39
51	0.8901	.7771	.7790	.7808	.7826	.7844	.7862	.7880	38
52	0.9076	.7880	.7898	.7916	.7934	.7951	.7969	.7986	37
53	0.9250	.7986	.8004	.8021	.8039	.8056	.8073	.8090	36
54	0.9425	.8090	.8107	.8124	.8141	.8158	.8175	.8192	35
55	0.9599	.8192	.8208	.8225	.8241	.8258	.8274	.8290	34
56	0.9774	.8290	.8307	.8323	.8339	.8355	.8371	.8387	33
57	0.9948	.8387	.8403	.8418	.8434	.8450	.8465	.8480	32
58	1.0123	.8480	.8496	.8511	.8526	.8542	.8557	.8572	31
59	1.0297	.8572	.8587	.8601	.8616	.8631	.8646	.8660	30
60	1.0472	.8660	.8675	.8689	.8704	.8718	.8732	.8746	29
61	1.0647	.8746	.8760	.8774	.8788	.8802	.8816	.8829	28
62	1.0821	.8829	.8843	.8857	.8870	.8884	.8897	.8910	27
63	1.0996	.8910	.8923	.8936	.8949	.8962	.8975	.8988	26
64	1.1170	.8988	.9001	.9013	.9026	.9038	.9051	.9063	25
65	1.1345	.9063	.9075	.9088	.9100	.9112	.9124	.9135	24
66	1.1519	.9135	.9147	.9159	.9171	.9182	.9194	.9205	23
67	1.1694	.9205	.9216	.9228	.9239	.9250	.9261	.9272	22
68	1.1868	.9272	.9283	.9293	.9304	.9315	.9325	.9336	21
69	1.2043	.9336	.9346	.9356	.9367	.9377	.9387	.9397	20
70	1.2217	.9397	.9407	.9417	.9426	.9436	.9446	.9455	19
71	1.2392	.9455	.9465	.9474	.9483	.9492	.9502	.9511	18
72	1.2566	.9511	.9520	.9528	.9537	.9546	.9555	.9563	17
73	1.2741	.9563	.9572	.9580	.9588	.9596	.9605	.9613	16
74	1.2915	.9613	.9621	.9628	.9636	.9644	.9652	.9659	15
75	1.3090	.9659	.9667	.9674	.9681	.9689	.9696	.9703	14
76	1.3265	.9703	.9710	.9717	.9724	.9730	.9737	.0744	13
77	1.3439	.9744	.9750	.9757	.9763	.9769	.9775	.9781	12
78	1.3614	.9781	.9787	.9793	.9799	.9805	.9811	.9816	11
79	1.3788	.9816	.9822	.9827	.9833	.9838	.9843	.9848	10
80	1.3963	.9848	.9853	.9858	.9863	.9868	.9872	.9877	9
81	1.4137	.9877	.9881	.9886	.9890	.9894	.9899	.9903	8
82	1.4312	.9903	.9907	.9911	.9914	.9918	.9922	.9925	7
83	1.4486	.9925	.9929	.9932	.9936	.9939	.9942	.9945	6
84	1.4661	.9945	.9948	.9951	.9954	.9957	.9959	.9962	5
85	1.4835	.9962	.9964	.9967	.9969	.9971	.9974	.9976	4
86	1.5010	.9976	.9978	.9980	.9981	.9983	.9985	.9986	3
87	1.5184	.9986	.9988	.9989	.9990	.9992	.9993	.9994	2
88	1.5359	.9994	.9995	.9996	.9997	.9997	.9998	.9998	1
89	1.5533	.9998	.9999	.9999	.9999	1.0000	1.0000	1.0000	0
90	1.5708	1.0000							
		60'	50'	40'	30'	20'	10'	0'	De-grees

COSINES

TABLE 1-4 TRIGONOMETRIC FUNCTIONS *(Continued)*

TANGENTS

De-grees	Ra-dians	0'	10'	20'	30'	40'	50'	60'	
0	.0000	.0000	.0029	.0058	.0087	.0116	.0145	.0175	89
1	.0175	.0175	.0204	.0233	.0262	.0291	.0320	.0349	88
2	.0349	.0349	.0378	.0407	.0437	.0466	.0495	.0524	87
3	.0524	.0524	.0553	.0582	.0612	.0641	.0670	.0699	86
4	.0698	.0699	.0729	.0758	.0787	.0816	.0846	.0875	85
5	.0873	.0875	.0904	.0934	.0963	.0992	.1022	.1051	84
6	.1047	.1051	.1080	.1110	.1139	.1169	.1198	.1228	83
7	.1222	.1228	.1257	.1287	.1317	.1346	.1376	.1405	82
8	.1396	.1405	.1435	.1465	.1495	.1524	.1554	.1584	81
9	.1571	.1584	.1614	.1644	.1673	.1703	.1733	.1763	80
10	.1745	.1763	.1793	.1823	.1853	.1883	.1914	.1944	79
11	.1920	.1944	.1974	.2004	.2035	.2065	.2095	.2126	78
12	.2094	.2126	.2156	.2186	.2217	.2247	.2278	.2309	77
13	.2269	.2309	.2339	.2370	.2401	.2432	.2462	.2493	76
14	.2443	.2493	.2524	.2555	.2586	.2617	.2648	.2679	75
15	.2618	.2679	.2711	.2742	.2773	.2805	.2836	.2867	74
16	.2793	.2867	.2899	.2931	.2962	.2994	.3026	.3057	73
17	.2967	.3057	.3089	.3121	.3153	.3185	.3217	.3249	72
18	.3142	.3249	.3281	.3314	.3346	.3378	.3411	.3443	71
19	.3316	.3443	.3476	.3508	.3541	.3574	.3607	.3640	70
20	.3491	.3640	.3673	.3706	.3739	.3772	.3805	.3839	69
21	.3665	.3839	.3872	.3906	.3939	.3973	.4006	.4040	68
22	.3840	.4040	.4074	.4108	.4142	.4176	.4210	.4245	67
23	.4014	.4245	.4279	.4314	.4348	.4383	.4417	.4452	66
24	.4189	.4452	.4487	.4522	.4557	.4592	.4628	.4663	65
25	.4363	.4663	.4699	.4734	.4770	.4806	.4841	.4877	64
26	.4538	.4877	.4913	.4950	.4986	.5022	.5059	.5095	63
27	.4712	.5095	.5132	.5169	.5206	.5243	.5280	.5317	62
28	.4887	.5317	.5354	.5392	.5430	.5467	.5505	.5543	61
29	.5061	.5543	.5581	.5619	.5658	.5696	.5735	.5774	60
30	.5236	.5774	.5812	.5851	.5890	.5930	.5969	.6009	59
31	.5411	.6009	.6048	.6088	.6128	.6168	.6208	.6249	58
32	.5585	.6249	.6289	.6330	.6371	.6412	.6453	.6494	57
33	.5760	.6494	.6536	.6577	.6619	.6661	.6703	.6745	56
34	.5934	.6745	.6787	.6830	.6873	.6916	.6959	.7002	55
35	.6109	.7002	.7046	.7089	.7133	.7177	.7221	.7265	54
36	.6283	.7265	.7310	.7355	.7400	.7445	.7490	.7536	53
37	.6458	.7536	.7581	.7627	.7673	.7720	.7766	.7813	52
38	.6632	.7813	.7860	.7907	.7954	.8002	.8050	.8098	51
39	.6807	.8098	.8146	.8195	.8243	.8292	.8342	.8391	50
40	.6981	.8391	.8441	.8491	.8541	.8591	.8642	.8693	49
41	.7156	.8693	.8744	.8796	.8847	.8899	.8952	.9004	48
42	.7330	.9004	.9057	.9110	.9163	.9217	.9271	.9325	47
43	.7505	.9325	.9380	.9435	.9490	.9545	.9601	.9657	46
44	.7679	.9657	.9713	.9770	.9827	.9884	.9942	1.0000	45
45	.7854	1.0000							

| | | 60' | 50' | 40' | 30' | 20' | 10' | 0' | De-grees |

COTANGENTS

TABLE 1-4 TRIGONOMETRIC FUNCTIONS (Continued)

TANGENTS

De-grees	Ra-dians	0′	10′	20′	30′	40′	50′	60′	
45	0.7854	1.000	1.006	1.012	1.018	1.024	1.030	1.036	44
46	0.8029	1.036	1.042	1.048	1.054	1.060	1.066	1.072	43
47	0.8203	1.072	1.079	1.085	1.091	1.098	1.104	1.111	42
48	0.8378	1.111	1.117	1.124	1.130	1.137	1.144	1.150	41
49	0.8552	1.150	1.157	1.164	1.171	1.178	1.185	1.192	40
50	0.8727	1.192	1.199	1.206	1.213	1.220	1.228	1.235	39
51	0.8901	1.235	1.242	1.250	1.257	1.265	1.272	1.280	38
52	0.9076	1.280	1.288	1.295	1.303	1.311	1.319	1.327	37
53	0.9250	1.327	1.335	1.343	1.351	1.360	1.368	1.376	36
54	0.9425	1.376	1.385	1.393	1.402	1.411	1.419	1.428	35
55	0.9599	1.428	1.437	1.446	1.455	1.464	1.473	1.483	34
56	0.9774	1.483	1.492	1.501	1.511	1.520	1.530	1.540	33
57	0.9948	1.540	1.550	1.560	1.570	1.580	1.590	1.600	32
58	1.0123	1.600	1.611	1.621	1.632	1.643	1.653	1.664	31
59	1.0297	1.664	1.675	1.686	1.698	1.709	1.720	1.732	30
60	1.0472	1.732	1.744	1.756	1.767	1.780	1.792	1.804	29
61	1.0647	1.804	1.816	1.829	1.842	1.855	1.868	1.881	28
62	1.0821	1.881	1.894	1.907	1.921	1.935	1.949	1.963	27
63	1.0996	1.963	1.977	1.991	2.006	2.020	2.035	2.050	26
64	1.1170	2.050	2.066	2.081	2.097	2.112	2.128	2.145	25
65	1.1345	2.145	2.161	2.177	2.194	2.211	2.229	2.246	24
66	1.1519	2.246	2.264	2.282	2.300	2.318	2.337	2.356	23
67	1.1694	2.356	2.375	2.394	2.414	2.434	2.455	2.475	22
68	1.1868	2.475	2.496	2.517	2.539	2.560	2.583	2.605	21
69	1.2043	2.605	2.628	2.651	2.675	2.699	2.723	2.747	20
70	1.2217	2.747	2.773	2.798	2.824	2.850	2.877	2.904	19
71	1.2392	2.904	2.932	2.960	2.989	3.018	3.047	3.078	18
72	1.2566	3.078	3.108	3.140	3.172	3.204	3.237	3.271	17
73	1.2741	3.271	3.305	3.340	3.376	3.412	3.450	3.487	16
74	1.2915	3.487	3.526	3.566	3.606	3.647	3.689	3.732	15
75	1.3090	3.732	3.776	3.821	3.867	3.914	3.962	4.011	14
76	1.3265	4.011	4.061	4.113	4.165	4.219	4.275	4.331	13
77	1.3439	4.331	4.390	4.449	4.511	4.574	4.638	4.705	12
78	1.3614	4.705	4.773	4.843	4.915	4.989	5.066	5.145	11
79	1.3788	5.145	5.226	5.309	5.396	5.485	5.576	5.671	10
80	1.3963	5.671	5.769	5.871	5.976	6.084	6.197	6.314	9
81	1.4137	6.314	6.435	6.561	6.691	6.827	6.968	7.115	8
82	1.4312	7.115	7.269	7.429	7.596	7.770	7.953	8.144	7
83	1.4486	8.144	8.345	8.556	8.777	9.010	9.255	9.514	6
84	1.4661	9.514	9.788	10.078	10.385	10.712	11.059	11.430	5
85	1.4835	11.430	11.826	12.251	12.706	13.197	13.727	14.301	4
86	1.5010	14.301	14.924	15.605	16.350	17.169	18.075	19.081	3
87	1.5184	19.081	20.206	21.470	22.904	24.542	26.432	28.636	2
88	1.5359	28.636	31.242	34.368	38.188	42.964	49.104	57.290	1
89	1.5533	57.290	68.750	85.940	114.59	171.89	343.77	infinity	0
90	1.5708	infinity							
		60′	50′	40′	30′	20′	10′	0′	De-grees

COTANGENTS

CALCULUS

FORMULAS FOR DIFFERENTIATION $\left(D_x = \frac{d}{dx}\right)$

$D_x(c) = 0$

$D_x(x) = 1$

$D_x(u + v - w) = D_x(u) + D_x(v) - D_x(w)$

$D_x(cv) = c \cdot D_x(v)$

$D_x(uv) = u \cdot D_x(v) + v\, D_x(u)$

$D_x(v^n) = n \cdot v^{n-1}\, D_x(v)$

$D_x(\sqrt{v}) = D_x(v)/2\sqrt{v}$

$D_x\left(\dfrac{c}{v^n}\right) = -c \cdot n \cdot D_x(v)/v^{n+1}$

$D_x\left(\dfrac{u}{v}\right) = \dfrac{v \cdot D_x(u) - u \cdot D_x(v)}{v^2}$

$D_x\left(\dfrac{u}{c}\right) = \dfrac{1}{c} \cdot D_x(u)$

$D_x\left(\dfrac{c}{v}\right) = -c \cdot D_x(v)/v^2$

$D_x(\log_a v) = \log_a e \cdot \dfrac{D_x(v)}{v}$

$D_x(\log v) = \dfrac{D_x(v)}{v}$

$D_x(a^v) = a^v \log a\, D_x(v)$

$D_x(e^v) = e^v\, D_x(v)$

$D_x(u^v) = v \cdot u^{v-1}\, D_x(u) + u^v \log u \cdot D_x(v)$

$D_x(\sin v) = \cos v \cdot D_x(v)$

$D_x(\cos v) = -\sin v \cdot D_x(v)$

$D_x(\tan v) = \sec^2 v \cdot D_x(v)$

$D_x(\cot v) = -\csc^2 v \cdot D_x(v)$

$D_x(\sec v) = \sec v \cdot \tan v \cdot D_x(v)$

$D_x(\csc v) = -\csc v \cdot \cot v \cdot D_x(v)$

c = constant

FORMULAS FOR INTEGRATION

$\displaystyle \int dv = v + C$

$\displaystyle \int a\, dv = a \int dv$

$\displaystyle \int (du + dv - dw) = \int du + \int dv - \int dw$

$\displaystyle \int v^n\, dv = \frac{v^{n+1}}{n+1} + C$

$\displaystyle \int \frac{dv}{v^n} = \frac{1}{(1-n)v^{n-1}} + C$

$\displaystyle \int \frac{dv}{\sqrt{v}} = 2\sqrt{v} + C$

$\displaystyle \int \frac{dv}{v} = \log v + C = \log cv$

$\displaystyle \int a^v\, dv = \frac{a^v}{\log a} + C$

$\displaystyle \int e^v\, dv = e^v + C$

$\displaystyle \int \sin v \cdot dv = -\cos v + C$

$\displaystyle \int \cos v \cdot dv = \sin v + C$

$\displaystyle \int \sec^2 v \cdot dv = \tan v + C$

$\displaystyle \int \csc^2 v \cdot dv = -\cot v + C$

$\displaystyle \int \sec v \cdot \tan v \cdot dv = \sec v + C$

$\displaystyle \int \csc v \cdot \cot v \cdot dv = -\csc v + C$

$\displaystyle \int \tan v \cdot dv = \begin{cases} \log \sec v + C, \text{ or} \\ -\log \cos v + C \end{cases}$

$\displaystyle \int \cot v \cdot dv = \begin{cases} \log \sin v + C, \text{ or} \\ -\log \csc v + C \end{cases}$

$\displaystyle \int \sec v \cdot dv = \begin{cases} \log (\sec v + \tan v) + C, \text{ or} \\ \log \tan (v/2 + \pi/4) + C \end{cases}$

$\displaystyle \int \csc v \cdot dv = \begin{cases} \log (\csc v - \cot v) + C, \text{ or} \\ \log \tan v/2 + C \end{cases}$

$\displaystyle \int \frac{dv}{\sqrt{a^2 - v^2}} = \sin^{-1} \frac{v}{a} + C$

$\displaystyle \int \frac{dv}{v^2 + a^2} = \frac{1}{a} \tan^{-1} \frac{v}{a} + C$

$\displaystyle \int \frac{dv}{v\sqrt{v^2 - a^2}} = \frac{1}{a} \sec^{-1} \frac{v}{a} + C$

$\displaystyle \int \frac{dv}{v^2 - a^2} = \frac{1}{2a} \log \frac{v-a}{v+a} + C, \left[\text{or } \frac{a-v}{a+v}\right]$

$\displaystyle \int \frac{dv}{a^2 - v^2} = \frac{1}{2a} \log \frac{a+v}{a-v} + C, \left[\text{or } \frac{v+a}{v-a}\right]$

$\displaystyle \int \frac{dv}{\sqrt{v^2 \pm a^2}} = \log (v + \sqrt{v^2 \pm a^2}) + C$

$\displaystyle \int \sqrt{a^2 - v^2} \cdot dv = \frac{v}{2} \sqrt{a^2 - v^2} + \frac{a^2}{2} \sin^{-1} \frac{v}{a} + C$

$\displaystyle \int \sqrt{v^2 \pm a^2} \cdot dv = \frac{v}{2} \sqrt{v^2 \pm a^2} \pm \frac{a^2}{2} \log (v + \sqrt{v^2 \pm a^2}) + C$

FIG. 1-2

TABLE 1-5 DECIMAL EQUIVALENTS

1/64——	.015625	
1/32———	.03125	
3/64——	.046875	
1/16———————	.0625	
5/64——	.078125	
3/32———	.09375	
7/64——	.109375	
1/8——————	.125	
9/64——	.140625	
5/32———	.15625	
11/64——	.171875	
3/16— ————	.1875	
13/64——	.203125	
7/32———	.21875	
15/64——	.234375	
1/4——————	.250	
17/64——	.265625	
9/32———	.28125	
19/64——	.296875	
5/16—— ———	.3125	
21/64——	.328125	
11/32———	.34375	
23/64——	.359375	
3/8——————	.375	
25/64——	.390625	
13/32-———	.40625	
27/64——	421875	
7/16———————	.4375	
29/64——	.453125	
15/32———	.46875	
31/64——	.484375	
1/2———————	.500	

33/64——	.515625
17/32———	.53125
35/64——	.546875
9/16———————	.5625
37/64——	.578125
19/32———	.59375
39/64——	.609375
5/8——————	.625
41/64——	.640625
21/32———	.65625
43/64——	.671875
11/16— -———	.6875
45/64——	.703125
23/32———	.71875
47/64——	.734375
3/4——————	.750
49/64——	.765625
25/32———	.78125
51/64——	.796875
13/16———————	.8125
53/64——	.828125
27/32———	.84375
55/64——	.859375
7/8——————	.875
57/64——	.890625
29/32———	.90625
59/64——	.921875
15/16———————	.9375
61/64——	.953125
31/32———	.96875
63/64——	.984375
1 ———————	1.000

TABLE 1-6 CONVERSION TABLES: U.S. CUSTOMARY SYSTEM

To convert	Multiply by	To obtain	To convert	Multiply by	To obtain
		Units of length			
in	0.0833	ft	yd	36	in
in	0.0278	yd	yd	3	ft
in	0.0000158	mi	yd	0.000568	mi
ft	12	in	mi	63,360	in
ft	0.3333	yd	mi	5280	ft
ft	0.000189	mi	mi	1760	yd
ft	0.000165	mi (nautical)	mi (nautical)	6076	ft
		Units of area			
in^2	0.00694	ft^2	yd^2	9	ft^2
in^2	0.000772	yd^2	yd^2	0.0002066	acres
ft^2	144	in^2	acres	43,560	ft^2
ft^2	0.1111	yd^2	acres	4840	yd^2
ft^2	0.00002296	acres	yd^2	1296	in^2
		Units of volume			
in^3	0.00433	gal	ft^3	0.0370	yd^3
in^3	0.000579	ft^3	ft^3	0.0000230	acre·ft
in^3	0.0000214	yd^3	yd^3	46,656	in^3
gal	231	in^3	yd^3	202	gal
gal	0.1337	ft^3	yd^3	27	ft^3
gal	0.00495	. yd^3			
gal	0.00000307	acre·ft	acre·ft	325,800	gal
gal	0.0238	bbl (oil)	bbl (oil)	42	gal
gal	0.8327	imperial gal	imperial gal	1.2	gal
ft^3	1728	in^3	acre·ft	43,560	ft^3
ft^3	7.48	gal			
		Units of weight			
gr	0.00229	oz	lb	7000	gr
gr	0.0001429	lb	lb	16	oz
gr	0.0000000714	tons	lb	0.0005	tons
oz	438	gr	tons	14,000,000	gr
oz	0.0625	lb	tons	32,000	oz
oz	0.00003125	tons	tons	2000	lb
gal	8.3322	lb*	lb/h*	0.0020003	gal/min
lb	0.12002	gal*	gal/min	499.925	lb/h*
		Units of velocity			
ft/s	60	ft/min	ft/h	0.0002778	ft/s
ft/s	3600	ft/h	ft/h	0.01667	ft/min
ft/s	0.682	mi/h	ft/h	0.0001894	mi/h
ft/s	0.592	kn	kn	1.689	ft/s
ft/min	0.01667	ft/s	mi/h	1.467	ft/s
ft/min	60	ft/h	mi/h	88	ft/min
ft/min	0.01136	mi/h	mi/h	5280	ft/h

*Water at 68°F.

TABLE 1-6 CONVERSION TABLES: U.S. CUSTOMARY SYSTEM *(Continued)*

To convert	Multiply by	To obtain	To convert	Multiply by	To obtain
		Units of time			
Seconds	0.01667	Minutes	Days	24	Hours
Seconds	0.0002778	Hours	Days	0.0329	Months
Seconds	0.00001157	Days	Days	0.00274	Years
Minutes	60	Seconds	Months	2,628,000	Seconds
Minutes	0.01667	Hours	Months	43,800	Minutes
Minutes	0.000694	Days	Months	730	Hours
Hours	3600	Seconds	Months	30.42	Days
Hours	60	Minutes	Months	0.0833	Years
Hours	0.0417	Days	Years	31,536,000	Seconds
Hours	0.001370	Months	Years	525,600	Minutes
Hours	0.0001142	Years	Years	8760	Hours
Days	86,400	Seconds	Years	365	Days
Days	1440	Minutes	Years	12	Months
		Units of work, energy, and heat			
Btu	9340	in·lb	ft·lb	0.0000003766	kWh
Btu	778.3	ft·lb	ft·lb	0.000000505	hp·h
Btu	0.000293	kWh	kWh	3413	Btu
Btu	0.000393	hp·h	kWh	31,872,000	in·lb
in·lb	0.000107	Btu	kWh	2,656,000	ft·lb
in·lb	0.0833	ft·lb	kWh	1.341	hp·h
in·lb	0.00000003138	kWh	hp·h	2545	Btu
in·lb	0.0000000421	hp·h	hp·h	23,760,000	in·lb
ft·lb	0.001285	Btu	hp·h	1,980,000	ft·lb
ft·lb	12	in·lb	hp·h	0.7455	kWh
		Units of power			
kW	1.341	hp	ft·lb/s	0.0771	Btu/min
kW	738	ft·lb/s	ft·lb/min	0.00002260	kW
kW	44,260	ft·lb/min	ft·lb/min	0.0000303	hp
kW	0.948	Btu/s	ft·lb/min	0.01667	ft·lb/s
kW	56.9	Btu/min	ft·lb/min	0.00002141	Btu/s
kW	3413	Btu/h	ft·lb/min	0.001285	Btu/min
hp	0.7455	kW	Btu/s	1.055	kW
hp	550	ft·lb/s	Btu/s	1.415	hp
hp	33,000	ft·lb/min	Btu/s	778.3	ft·lb/s
hp	0.707	Btu/s	Btu/s	46,700	ft·lb/min
hp	42.41	Btu/min	Btu/s	60	Btu/min
hp	2545	Btu/h	Btu/min	0.01758	kW
ft·lb/s	0.001356	kW	Btu/min	0.02357	hp
ft·lb/s	0.001818	hp	Btu/min	12.97	ft·lb/s
ft·lb/s	60	ft·lb/min	Btu/min	778.3	ft·lb/min
ft·lb/s	0.001285	Btu/s	Btu/min	0.01667	Btu/s

TABLE 1-6   CONVERSION TABLES: U.S. CUSTOMARY
SYSTEM *(Continued)*

To convert	Multiply by	To obtain	To convert	Multiply by	To obtain
Units of pressure (water, mercury at 68°F)					
in water	0.0833	ft water	oz/ft²	0.01205	in water
in water	0.0736	in mercury	oz/ft²	0.001004	ft water
in water	82.98	oz/ft²	oz/ft²	0.000887	in mercury
in water	0.03602	lb/in² (psi)	oz/ft²	0.000434	lb/in² (psi)
in water	5.1869	lb/ft²	oz/ft²	0.0625	lb/ft²
ft water	12	in water	lb/in² (psi)	27.762	in water
ft water	0.8832	in mercury	lb/in² (psi)	2.314	ft water
ft water	995.8	oz/ft²	lb/in² (psi)	2.314/sp gr	ft (any liquid)
ft water	0.4322	lb/in² (psi)	lb/in² (psi)	2.043	in mercury
ft water	62.24	lb/ft²	lb/in² (psi)	230.4	oz/ft²
ft (any liquid)	0.4322 × sp gr	lb/in² (psi)	lb/in² (psi)	144	lb/ft²
in mercury	13.57	in water	lb/in² (psi)	0.06802	atm
in mercury	1.131	ft water	lb/ft²	0.1928	in water
in mercury	1128	oz/ft²	lb/ft²	0.01607	ft water
in mercury	0.4894	lb/in² (psi)	lb/ft²	0.01419	in mercury
in mercury	70.47	lb/ft²	lb/ft²	16	oz/ft²
in mercury	0.03342	atm	lb/ft²	0.00694	lb/in² (psi)
atm	29.92	in mercury	atm	14.7	lb/in² (psi)
Weight—time rates					
lb/s	60	lb/min	lb/h*	0.0020003	gal/min
lb/s	3600	lb/h	lb/h	0.0002778	lb/s
lb/s	86,400	lb/day	lb/h	0.01667	lb/min
lb/min	0.01667	lb/s	lb/h	24	lb/day
lb/min	60	lb/h	lb/day	0.00001157	lb/s
lb/min	1440	lb/day	lb/day	0.000694	lb/min
			lb/day	0.0417	lb/h
Volume—flow rates					
ft³/s	60	ft³/min	gal/s	8.022	ft³/min
ft³/s	3600	ft³/h	gal/s	481.3	ft³/h
ft³/s	7.48	gal/s	gal/s	60	gal/min
ft³/s	448.8	gal/min	gal/s	3600	gal/h
ft³/s	26,930	gal/h	gal/min	0.00223	ft³/s
ft³/s	646,317	gal/day	gal/min	0.1337	ft³/min
ft³/s	1.983	acre·ft/day	gal/min	8.022	ft³/h
ft³/min	0.01667	ft³/s	gal/min	0.01667	gal/s
ft³/min	60	ft³/h	gal/min	60	gal/h
ft³/min	0.1247	gal/s	gal/min	499.925	lb/h*
ft³/min	7.48	gal/min	gal/h	0.0000371	ft³/s
ft³/min	448.8	gal/h	gal/h	0.00223	ft³/min
ft³/h	0.0002778	ft³/s	gal/h	0.1337	ft³/h
ft³/h	0.01667	ft³/min	gal/h	0.0002778	gal/s
ft³/h	0.002078	gal/s	gal/h	0.01667	gal/min
ft³/h	0.1247	gal/min	bbl/min (oil)	42	gal/min
ft³/h	7.48	gal/h	bbl/day (oil)	0.0292	gal/min
gal/s	0.1337	ft³/s	acre·ft/day	0.5042	ft³/s

*Water at 68°F.

CONVERSION TABLES: METRIC (SI) SYSTEM OF UNITS*

The modernized metric system is called the international system of units (with the international abbreviation SI). This system of metric units was established for international use in 1954 by the General Conference of Weights and Measures on the International System of Units, and the name *international system* was adopted by the same body in 1960.

The SI units are classified into three categories: basic units, derived units, and supplementary units. The basic SI units, seven in number, are listed in Table 1-7.

The SI system was established by international agreement to provide a logical and interconnected framework for *all* measurements in science, industry, and commerce. The system is built upon the *base units* listed below. All other SI units are derived from these base units. Multiples and submultiples are expressed in a decimal system as in Table 1-8.

The seven basic units of the SI listed below are relatively easy to use and to convert to or from U.S. customary units. It is in the area of *derived units,* i.e.,

*Condensed from *McGraw-Hill Metrication Manual*, New York, 1972.

TABLE 1-7   Basic SI Units

Quantity	Unit
Length	meter (m)
Mass	kilogram (kg)
Time	second (s)
Electric current	ampere (A)
Temperature (thermodynamic)	kelvin (K)
Amount of substance	mole (mol)
Luminous intensity	candela (cd)

TABLE 1-8   Prefixes for SI Units

Multiple and submultiple		Prefix	Symbol
1,000,000,000,000 =	10^{12}	tera	T
1,000,000,000 =	10^{9}	giga	G
1,000,000 =	10^{6}	mega	M
1,000 =	10^{3}	kilo	k
100 =	10^{2}	hecto	h
10 =	10	deka	da
0.1 =	10^{-1}	deci	d
0.01 =	10^{-2}	centi	c
0.001 =	10^{-3}	milli	m
0.000 001 =	10^{-6}	micro	μ
0.000 000 001 =	10^{-9}	nano	n
0.000 000 000 001 =	10^{-12}	pico	p
0.000 000 000 000 001 =	10^{-15}	femto	f
0.000 000 000 000 000 001 =	10^{-18}	atto	a

TABLE 1-9 Derived Units of the International System

Quantity	Name of unit	Unit symbol or abbreviation, where differing from basic form	Unit expressed in terms of basic or supplementary units*
Area	square meter		m^2
Volume	cubic meter		m^3
Frequency	hertz, cycle per second†	Hz	s^{-1}
Density	kilogram per cubic meter		kg/m^3
Velocity	meter per second		m/s
Angular velocity	radian per second		rad/s
Acceleration	meter per second squared		m/s^2
Angular acceleration	radian per second squared		rad/s^2
Volumetric flow rate	cubic meter per second		m^3/s
Force	newton	N	$kg \cdot m/s^2$
Surface tension	newton per meter, joule per square meter	N/m, J/m^2	kg/s^2
Pressure	newton per square meter, pascal†	N/m^2, Pa†	$kg/m \cdot s^2$
Viscosity, dynamic	newton-second per square meter, poiseuille†	$N \cdot s/m^2$, Pl†	$kg/m \cdot s$
Viscosity, kinematic	meter squared per second		m^2/s
Work, torque, energy, quantity of heat	joule, newton-meter, watt-second	J, $N \cdot m$, $W \cdot s$	$kg \cdot m^2/s^2$
Power, heat flux	watt, joule per second	W, J/s	$kg \cdot m^2/s^3$
Heat flux density	watt per square meter	W/m^2	kg/s^3
Volumetric heat release rate	watt per cubic meter	W/m^3	$kg/m \cdot s^3$
Heat transfer coefficient	watt per square meter kelvin	$W/m^2 \cdot K$	$kg/s^3 \cdot K$
Heat capacity (specific)	joule per kilogram degree	$J/kg \cdot K$	$m^2/s^2 \cdot K$
Capacity rate	watt per kelvin	W/K	$kg \cdot m^2/s^3 \cdot K$
Thermal conductivity	watt per meter kelvin	$W/m \cdot K$, $\dfrac{Jm}{\cdot m^2 \cdot K}$	$kg \cdot m/s^3 \cdot K$

*Supplementary units are plane angle, radian (rad); solid angle, steradian (sr).
†Not used in all countries.

those units of measure which are derived from one or more of the basic units listed above, that readers unfamiliar with the SI metric system might meet difficulties. Further, certain units are expressed only in terms of other derived units.

One SI metric unit which initially might cause difficulty is the unit of force, or the *newton*. This unit enters a variety of calculations in civil, mechanical, and electrical engineering and is defined as the force which produces an acceleration of one meter per second squared when applied to a mass of one kilogram.

Table 1-9 includes the more common derived units used in SI.

Conversion Factors as Exact Numerical Multiples of SI Units

Table 1-10 expresses the definitions of various units of measure as exact numerical multiples of coherent SI units and provides multiplying factors for converting numbers and miscellaneous units to corresponding new numbers and SI units.

The first two digits of each numerical entry represent a power of 10. An asterisk follows each number which expresses an exact definition. For example, the entry "−02 2.54*" expresses the fact that 1 inch = 2.54 × 10⁻² meter, exactly, by definition. Most of the definitions are extracted from National Bureau of Standards documents. Numbers not followed by an asterisk are only approximate representations of definitions or are the results of physical measurements.

TABLE 1-10   Listing by Physical Quantity

To convert from	To	Multiply by
Acceleration		
foot/second^2	meter/second^2	−01 3.048*
free fall, standard	meter/second^2	+00 9.806 65*
gal (galileo)	meter/second^2	−02 1.00*
inch/second^2	meter/second^2	−02 2.54*
Area		
acre	meter^2	+03 4.046 856 422 4*
are	meter^2	+02 1.00*
barn	meter^2	−28 1.00*
circular mil	meter^2	−10 5.067 074 8
foot^2	meter^2	−02 9.290 304*
hectare	meter^2	+04 1.00*
inch^2	meter^2	−04 6.4516*
mile^2 (U.S. statute)	meter^2	+06 2.589 988 110 336*
section	meter^2	+06 2.589 988 110 336*
township	meter^2	+07 9.323 957 2
yard^2	meter^2	−01 8.361 273 6*
Density		
gram/ centimeter^3	kilogram/meter^3	+03 1.00*
lbm/inch^3	kilogram/meter^3	+04 2.767 990 5
lbm/foot^3	kilogram/meter^3	+01 1.601 846 3
slug/foot^3	kilogram/meter^3	+02 5.153 79
Energy		
British thermal unit (ISO/TC 12)	joule	+03 1.055 06
British thermal unit (International Steam Table)	joule	+03 1.055 04
British thermal unit (mean)	joule	+03 1.055 87
British thermal unit (thermochemical)	joule	+03 1.054 350 264 488
British thermal unit (39°F)	joule	+03 1.059 67
British thermal unit (60°F)	joule	+03 1.054 68
calorie (International Steam Table)	joule	+00 4.1868
calorie (mean)	joule	+00 4.190 02

TABLE 1-10 Listing by Physical Quantity *(Continued)*

To convert from	To	Multiply by
calorie (thermochemical)	joule	+00 4.184*
calorie (15°C)	joule	+00 4.185 80
calorie (20°C)	joule	+00 4.181 90
calorie (kilogram, International Steam Table)	joule	+03 4.1868
calorie (kilogram, mean)	joule	+03 4.190 02
calorie (kilogram, thermochemical)	joule	+03 4.184*
electron volt	joule	−19 1.602 10
erg	joule	−07 1.00*
foot-lbf	joule	+00 1.355 817 9
foot-poundal	joule	−02 4.214 011 0
joule (international of 1948)	joule	+00 1.000 165
kilocalorie (International Steam Table)	joule	+03 4.1868
kilocalorie (mean)	joule	+03 4.190 02
kilocalorie (thermochemical)	joule	+03 4.184*
kilowatthour	joule	+06 3.60*
kilowatthour (international of 1948)	joule	+06 3.600 59
ton (nuclear equivalent of TNT)	joule	+09 4.20
watthour	joule	+03 3.60*

Energy/area time		
Btu (thermochemical)/foot²-second	watt/meter²	+04 1.134 893 1
Btu (thermochemical)/foot²-minute	watt/meter²	+02 1.891 488 5
Btu (thermochemical)/foot²-hour	watt/meter²	+00 3.152 480 8
Btu (thermochemical)/inch²-second	watt/meter²	+06 1.634 246 2
calorie (thermochemical)/cm²-minute	watt/meter²	+02 6.973 333 3
erg/centimeter²-second	watt/meter²	−03 1.00*
watt/centimeter²	watt/meter²	+04 1.00*

Force		
dyne	newton	−05 1.00*
kilogram-force (kgf)	newton	+00 9806 65*
kilopond	newton	+00 9.806 65*
kip	newton	+03 4.448 221 615 260 5*
lbf (pound-force, avoirdupois)	newton	+00 4.448 221 615 260 5*
ounce-force (avoirdupois)	newton	−01 2.780 138 5
pound-force, lbf (avoirdupois)	newton	+00 4.448 221 615 260 5*
poundal	newton	−01 1.382 549 543 76*

Length		
angstrom	meter	−10 1.00*
astronomical unit	meter	+11 1.495 978 9
cable	meter	+02 2.194 56*
caliber	meter	−04 2.54*
chain (surveyor's or Gunter's)	meter	+01 2.011 68*
chain (engineer's)	meter	+01 3.048*
cubit	meter	−01 4.572*
fathom	meter	+00 1.8288*
fermi (femtometer)	meter	−15 1.00*
foot	meter	−01 3.048*
foot (U.S. survey)	meter	+00 1200/3937*
foot (U.S. survey)	meter	−01 3.048 006 096
furlong	meter	+02 2.011 68*

TABLE 1-10 Listing by Physical Quantity *(Continued)*

To convert from	To	Multiply by
hand	meter	−01 1.016*
inch	meter	−02 2.54*
league (U.K. nautical)	meter	+03 5.559 552*
league (international nautical)	meter	+03 5.556*
league (statute)	meter	+03 4.828 032*
light-year	meter	+15 9.460 55
link (engineer's)	meter	−01 3.048*
link (surveyor's or Gunter's)	meter	−01 2.011 68*
meter	wavelengths Kr 86	+06 1.650 763 73*
micrometer	meter	−06 1.00*
mil	meter	−05 2.54*
mile (U.S. statute)	meter	+03 1.609 344*
mile (U.K. nautical)	meter	+03 1.853 184*
mile (international nautical)	meter	+03 1.852*
mile (U.S. nautical)	meter	+03 1.852*
nautical mile (U.K.)	meter	+03 1.853 184*
nautical mile (international)	meter	+03 1.852*
nautical mile (U.S.)	meter	+03 1.852*
pace	meter	−01 7.62*
parsec	meter	+16 3.083 74
perch	meter	+00 5.0292*
pica (printer's)	meter	−03 4.217 517 6*
point (printer's)	meter	−04 3.514 598*
pole	meter	+00 5.0292*
rod	meter	⏐00 5.0292*
skein	meter	+02 1.097 28*
span	meter	−01 2.286*
statute mile (U.S.)	meter	+03 1.609 344*
yard	meter	−01 9.144*

	Mass	
carat (metric)	kilogram	−04 2.00*
dram (avoirdupois)	kilogram	−03 1.771 845 195 312 5*
dram (troy or apothecaries')	kilogram	−03 3.887 934 6*
grain	kilogram	−05 6.479 891*
gram	kilogram	−03 1.00*
hundredweight (long)	kilogram	+01 5.080 234 544*
hundredweight (short)	kilogram	+01 4.535 923 7*
kgf-second^2/meter (mass)	kilogram	+00 9.806 65*
kilogram mass	kilogram	+00 1.00*
lbm (pound mass, avoirdupois)	kilogram	−01 4.535 923 7*
ounce mass (avoirdupois)	kilogram	−02 2.834 952 312 5*
ounce mass (troy or apothecaries')	kilogram	−02 3.110 347 68*
pennyweight	kilogram	−03 1.555 173 84*
pound mass, lbm (avoirdupois)	kilogram	−01 4.535 923 7*
pound mass (troy or apothecaries')	kilogram	−01 3.732 417 216*
scruple (apothecaries')	kilogram	−03 1.295 978 2*
slug	kilogram	+01 1.459 390 29
ton (assay)	kilogram	−02 2.916 666 6
ton (long)	kilogram	+03 1.016 046 908 8*
ton (metric)	kilogram	+03 1.00*
ton (short, 2000 pounds)	kilogram	+02 9.071 847 4*
tonne	kilogram	+03 1.00*

TABLE 1-10   Listing by Physical Quantity *(Continued)*

To convert from	To	Multiply by
Power		
Btu (thermochemical)/second	watt	+03 1.054 350 264 488
Btu (thermochemical)/minute	watt	+01 1.757 250 4
calorie (thermochemical)/second	watt	+00 4.184*
calorie (thermochemical)/minute	watt	−02 6.973 333 3
foot-lbf/hour	watt	−04 3.766 161 0
foot-lbf/minute	watt	−02 2.259 696 6
foot-lbf/second	watt	+00 1.355 817 0
horsepower (550 foot-lbf/second)	watt	+02 7.456 998 7
horsepower (boiler)	watt	+03 9.809 50
horsepower (electric)	watt	+02 7.46*
horsepower (metric)	watt	+02 7.354 99
horsepower (U.K.)	watt	+02 7.457
horsepower (water)	watt	+02 7.460 43
kilocalorie (thermochemical)/minute	watt	+01 6.973 333 3
kilocalorie (thermochemical)/second	watt	+03 4.184*
watt (international of 1948)	watt	+00 1.000 165
Pressure		
atmosphere	newton/meter^2	+05 1.013 25*
bar	newton/meter^2	+05 1.00*
barye	newton/meter^2	−01 1.00*
centimeter of mercury (0°C)	newton/meter^2	+03 1.333 22
centimeter of water (4°C)	newton/meter^2	+01 9.806 38
dyne/centimeter^2	newton/meter^2	−01 1.00*
foot of water (39.2°F)	newton/meter^2	+03 2.988 98
inch of mercury (32°F)	newton/meter^2	+03 3.386 389
inch of mercury (60°F)	newton/meter^2	+03 3.376 85
inch of water (39.2°F)	newton/meter^2	+02 2.490 82
inch of water (60°F)	newton/meter^2	+02 2.4884
kgf/centimeter^2	newton/meter^2	+04 9.806 65*
kgf/meter^2	newton/meter^2	+00 9.806 65*
lbf/foot^2	newton/meter^2	+01 4.788 025 8
lbf/inch^2 (psi)	newton/meter^2	+03 6.894 757 2
millibar	newton/meter^2	+02 1.00*
millimeter of mercury (0°C)	newton/meter^2	+02 1.333 224
pascal	newton/meter^2	+00 1.00*
psi (lbf/inch^2)	newton/meter^2	+03 6.894 757 2
torr (0°C)	newton/meter^2	+02 1.333 22
Speed		
foot/hour	meter/second	−05 8.466 666 6
foot/minute	meter/second	−03 5.08*
foot/second	meter/second	−01 3.048*
inch/second	meter/second	−02 2.54*
kilometer/hour	meter/second	−01 2.777 777 8
knot (international)	meter/second	−01 5.144 444 444
mile/hour (U.S. statute)	meter/second	−01 4.4704*
mile/minute (U.S. statute)	meter/second	+01 2.682 24*
mile/second (U.S. statute)	meter/second	+03 1.609 344
Temperature		
Celsius	kelvin	$t_K = t_C + 273.15$
Fahrenheit	kelvin	$t_K = (5/9)(t_F + 459.67)$

TABLE 1-10   Listing by Physical Quantity *(Continued)*

To convert from	To	Multiply by
Fahrenheit	Celsius	$t_C = (5/9)(t_F - 32)$
Rankine	kelvin	$t_K = (5/9)t_R$

	Time	
day (mean solar)	second (mean solar)	+04 8.64*
day (sidereal)	second (mean solar)	+04 8.616 409 0
hour (mean solar)	second (mean solar)	+03 3.60*
hour (sidereal)	second (mean solar)	+03 3.590 170 4
minute (mean solar)	second (mean solar)	+01 6.00*
minute (sidereal)	second (mean solar)	+01 5.983 617 4
month (mean calendar)	second (mean solar)	+06 2.628*
second (ephemeris)	second	+00 1.000 000 000
second (mean solar)	second (ephemeris)	Consult *American Ephemeris and Nautical Almanac*
second (sidereal)	second (mean solar)	−01 9.972 695 7
year (calendar)	second (mean solar)	+07 3.1536*
year (sidereal)	second (mean solar)	+07 3.155 815 0
year (tropical)	second (mean solar)	+07 3.155 692 6
year 1900, tropical, January, day 0, hour 12	second (ephemeris)	+07 3.155 692 597 47*
year 1900, tropical, January, day 0, hour 12	second	+07 3.155 692 597 47

	Viscosity	
centistoke	meter^2/second	−06 1.00*
stoke	meter^2/second	−04 1.00*
foot^2/second	meter^2/second	−02 9.290 304*
centipoise	newton-second/meter^2	−03 1.00*
lbm/foot-second	newton-second/meter^2	+00 1.488 163 9
lbf-second/foot^2	newton-second/meter^2	+01 4.788 025 8
poise	newton-second/meter^2	−01 1.00*
poundal-second/foot^2	newton-second/meter^2	+00 1.488 163 9
slug/foot-second	newton-second/meter^2	+01 4.788 025 8
rhe	meter^2/newton-second	+01 1.00*

	Volume	
acre-foot	meter^3	+03 1.233 481 9
barrel (petroleum, 42 gallons)	meter^3	−01 1.589 873
board foot	meter^3	−03 2.359 737 216*
bushel (U.S.)	meter^3	−02 3.523 907 016 688*
cord	meter^3	+00 3.624 566 3
cup	meter^3	−04 2.365 882 365*
dram (U.S. fluid)	meter^3	−06 3.696 691 195 312 5*
fluid ounce (U.S.)	meter^3	−05 2.957 352 956 25*
foot^2	meter^3	−02 2.831 684 659 2*
gallon (U.K. liquid)	meter^3	−03 4.546 087
gallon (U.S. dry)	meter^3	−03 4.404 883 770 86*
gallon (U.S. liquid)	meter^3	−03 3.785 411 784*
gill (U.K.)	meter^3	−04 1.420 652
gill (U.S.)	meter^3	−04 1.182 941 2
hogshead (U.S.)	meter^3	−01 2.384 809 423 92*
inch^3	meter^3	−05 1.638 706 4*
liter	meter^3	−03 1.00*

TABLE 1-10   Listing by Physical Quantity *(Continued)*

To convert from	To	Multiply by
ounce (U.S. fluid)	meter^3	−05 2.957 352 956 25*
peck (U.S.)	meter^3	−03 8.809 767 541 72*
pint (U.S. dry)	meter^3	−04 5.506 104 713 575*
pint (U.S. liquid)	meter^3	−04 4.731 764 73*
quart (U.S. dry)	meter^3	−03 1.101 220 942 715*
quart (U.S. liquid)	meter^3	−04 9.463 529 5
stere	meter^3	+00 1.00*
tablespoon	meter^3	−05 1.478 676 478 125*
teaspoon	meter^3	−06 4.928 921 593 75*
ton (register)	meter^3	+00 2.831 684 659 2*
yard^3	meter^3	−01 7.645 548 579 84*

*Exact by definition.

TABLE 1-11 TEMPERATURE CONVERSION TABLE

Degrees Fahrenheit to degrees Celsius: $^\circ F = \%\,^\circ C + 32^\circ$ $^\circ C = \%(^\circ F - 32^\circ)$

F	C	F	C	F	C	F	C	F	C	F	C
−40	−40.00	+30	−1.11	+80	+26.67	+250	+121.11	+500	+260.00	+900	+482.22
−38	−38.89	31	−0.56	81	27.22	255	123.89	505	262.78	910	487.78
−36	−37.78	32	0.00	82	27.78	260	126.67	510	265.56	920	493.33
−34	−36.67	33	+0.56	83	28.33	265	129.44	515	268.33	930	498.89
−32	−35.56	34	1.11	84	28.89	270	132.22	520	271.11	940	504.44
−30	−34.44	35	1.67	85	29.44	275	135.00	525	273.89	950	510.00
−28	−33.33	36	2.22	86	30.00	280	137.78	530	276.67	960	515.56
−26	−32.22	37	2.78	87	30.56	285	140.55	535	279.44	970	521.11
−24	−31.11	38	3.33	88	31.11	290	143.33	540	282.22	980	526.67
−22	−30.00	39	3.89	89	31.67	295	146.11	545	285.00	990	532.22
−20	−28.89	40	4.44	90	32.22	300	148.89	550	287.78	1000	537.78
−18	−27.78	41	5.00	91	32.78	305	151.67	555	290.55	1050	565.56
−16	−26.67	42	5.56	92	33.33	310	154.44	560	293.33	1100	593.33
−14	−25.56	43	6.11	93	33.89	315	157.22	565	296.11	1150	612.11
−12	−24.44	44	6.67	94	34.44	320	160.00	570	298.89	1200	648.89
−10	−23.33	45	7.22	95	35.00	325	162.78	575	301.67	1250	676.67
− 8	−22.22	46	7.78	96	35.56	330	165.56	580	304.44	1300	704.44
− 6	−21.11	47	8.33	97	36.11	335	168.33	585	307.22	1350	732.22
− 4	−20.00	48	8.89	98	36.67	340	171.11	590	310.00	1400	760.00
− 2	−18.89	49	9.44	99	37.22	345	173.89	595	312.78	1450	787.78
0	−17.78	50	10.00	100	37.78	350	176.67	600	315.56	1500	815.56
+ 1	−17.22	51	10.56	105	40.55	355	179.44	610	321.11	1550	843.33
2	−16.67	52	11.11	110	43.33	360	182.22	620	326.67	1600	871.11
3	−16.11	53	11.67	115	46.11	365	185.00	630	332.22	1650	898.89
4	−15.56	54	12.22	120	48.89	370	187.78	640	337.78	1700	926.67
5	−15.00	55	12.78	125	51.67	375	190.55	650	343.33	1750	954.44
6	−14.44	56	13.33	130	54.44	380	193.33	660	348.89	1800	982.22
7	−13.89	57	13.89	135	57.22	385	196.11	670	354.44	1850	1010.00
8	−13.33	58	14.44	140	60.00	390	198.89	680	360.00	1900	1037.78
9	−12.78	59	15.00	145	62.78	395	201.67	690	365.56	1950	1065.56
10	−12.22	60	15.56	150	65.56	400	204.44	700	371.11	2000	1093.33
11	−11.67	61	16.11	155	68.33	405	207.22	710	376.67	2050	1121.11
12	−11.11	62	16.67	160	71.11	410	210.00	720	382.22	2100	1148.89
13	−10.56	63	17.22	165	73.89	415	212.78	730	387.78	2150	1176.67
14	−10.00	64	17.78	170	76.67	420	215.56	740	393.33	2200	1204.44
15	− 9.44	65	18.33	175	79.44	425	218.33	750	398.89	2250	1232.22
16	− 8.89	66	18.89	180	82.22	430	221.11	760	404.44	2300	1260.00
17	− 8.33	67	19.44	185	85.00	435	223.89	770	410.00	2350	1287.78
18	− 7.78	68	20.00	190	87.78	440	226.67	780	415.56	2400	1315.56
19	− 7.22	69	20.56	195	90.55	445	229.44	790	421.11	2450	1343.33
20	− 6.67	70	21.11	200	93.33	450	232.22	800	426.67	2500	1371.11
21	− 6.11	71	21.67	205	96.11	455	235.00	810	432.22	2550	1398.89
22	− 5.56	72	22.22	210	98.89	460	237.78	820	437.78	2600	1426.67
23	− 5.00	73	22.78	215	101.67	465	240.55	830	443.33	2650	1454.44
24	− 4.44	74	23.33	220	104.44	470	243.33	840	448.89	2700	1482.22
25	− 3.89	75	23.89	225	107.22	475	246.11	850	454.44	2750	1510.00
26	− 3.33	76	24.44	230	110.00	480	248.89	860	460.00	2800	1537.78
27	− 2.78	77	25.00	235	112.78	485	251.67	870	465.56	2850	1565.59
28	− 2.22	78	25.56	240	115.56	490	254.44	880	471.11	2900	1593.33
29	− 1.67	79	26.11	245	118.33	495	257.22	890	476.67	2950	1621.11

Mechanical and Physical Data

MECHANICAL PROPERTIES OF MATERIALS

All materials have properties which must be known to promote their proper use. These properties are essential to selection of the best material for a given member.
    Mechanical properties used by engineers are:

1. Ultimate tensile strength
2. Tensile yield strength
3. Elongation
4. Modulus of elasticity
5. Compressive strength
6. Shear strength
7. Endurance limit
8. Toughness
9. Transition temperature
10. Critical stress intensity factor

1. *Ultimate tensile strength* is defined as the maximum load per unit of original cross-sectional area sustained by a material during a tension test.

2. *Tensile yield strength* is defined as the stress corresponding to some permanent deformation from the modulus slope, e.g., 0.2 percent offset in the case of heat-treated alloy steels.

3. *Elongation* is defined as the amount of permanent extension in a ruptured tensile-test specimen; it is usually expressed as a percentage of the original gauge length. Elongation is usually taken as a measure of ductility.

4. *Modulus of elasticity* is the property of the material which indicates its rigidity. This property is the ratio of stress to the strain within the elastic range.

$$\frac{\text{Stress } \sigma}{\text{Strain } \epsilon} = \text{modulus of elasticity } E$$

On a stress-strain diagram, the modulus of elasticity is represented usually by the straight portion of the curve when the stress is directly proportional to the strain. The steeper the curve, the higher the modulus of elasticity and the stiffer the material. Any steel has a modulus of elasticity in tension of approximately 30 million lb/in^2 (psi). Other materials may vary according to the specific alloy. Cast iron, for example, has a modulus of elasticity in tension between 10 million and 25 million lb/in^2, depending on the grade.

5. *Compressive strength* is defined as the maximum compressive stress that a material is capable of developing on the basis of the original cross-sectional area. The general design practice is to assume that the compressive strength of a steel is equal to its tensile strength, although it is actually somewhat greater.

6. *Shear strength* is defined as the stress required to produce fracture in the plane of cross section, the conditions of loading being such that the directions of force and of resistance are parallel and opposite although their paths are offset a specified minimum amount. The ultimate shear strength is generally assumed to be three-fourths of the material's ultimate tensile strength.

7. *Endurance limit* is defined as the maximum stress to which the material can be subjected for an indefinite service life. Although standards vary for various types of members and different industries, it is common practice to accept the assumption that carrying a certain load for several million cycles of stress rever-

sals indicates that the load can be carried for an indefinite time. When a load or a member is constantly varying in value, is repeated by relatively high frequency, or constitutes a complete reversal of stresses with each operating cycle, the material's endurance limit must be substituted for the ultimate strength as called for by the design formulas. The geometry of the member, the presence of local areas of high stress concentration, and the condition of the material (pits, surface irregularities, and corrosion) have considerable influence on the real endurance limit.

8. *Toughness* is the ability of a metal to absorb energy and deform plastically before fracturing. It is usually measured by the energy absorbed in a notch impact test, but the area under the stress-strain curve in tensile testing is also a measure of toughness.

9. *Transition temperature* is (*a*) an arbitrarily defined temperature within the temperature range in which metal fracture characteristics determined usually by notched tests are changing rapidly such as from primarily fibrous (shear) to primarily crystalline (cleavage) fracture. Commonly used definitions are "transition temperature for 50 percent cleavage fracture," "10-ft · lb transition temperature," and "transition temperature for half maximum energy." (*b*) The term sometimes is also used to denote the arbitrarily defined temperature in a range in which ductility changes rapidly with temperature.

10. *Critical stress intensity factor, K_C or K_{IC}.* This material property is the value of stress intensity K at which unstable crack growth occurs. Stress intensity is a function of stress level and crack size. This property is part of the linear elastic fracture mechanics approach to failure analysis and is based on three major assumptions:

a. Cracks and similar flaws are inherently present in parts or specimens.

b. A crack is a flat, internal free surface in a linear elastic stress field—a purely elastic stress field in an isotropic continuum (featureless solid).

c. The quantity of stored energy released from a cracking specimen or part during rapid crack propagation is a basic material property independent of specimen or part size.

GLOSSARY

The following is a collection of metallurgical terms.

Age hardening. Process of aging that increases hardness and strength and ordinarily decreases ductility. Age hardening usually follows rapid cooling or cold working.

Alloy. Substance that has metallic properties and is composed of two or more chemical elements of which at least one is a metal.

Annealing. Process involving heating and cooling, usually applied to induce softening. The term also refers to treatments intended to alter mechanical or physical properties, produce a definite microstructure, or remove gases. When applicable, the following more specific terms should be used.

Black annealing   Isothermal annealing
Blue annealing   Malleableizing
Box annealing   Process annealing
Bright annealing   Spheroidizing
Full annealing   Stabilizing annealing
Graphitizing

Definitions of the above terms will not be given in this brief glossary. When applied to ferrous alloys, the term *annealing* without question implies full annealing. Any process of annealing will usually reduce stresses, but if the treatment is applied for the sole purpose of such relief, it should be designated as *stress-relieving*.

Austenitizing. Process of forming austenite by heating a ferrous alloy into the transformation range (partial austenitizing) or above the transformation range (complete austenitizing).

Carbonitriding. Process in which a ferrous alloy is case-hardened by first being heated in a gaseous atmosphere of such composition that the alloy absorbs carbon and nitrogen simultaneously and then being cooled at a rate that will produce desired properties.

Carburizing. Process that introduces carbon into a solid ferrous alloy by heating the metal in contact with a carbonaceous material—solid, liquid,or gas—to a temperature above the transformation range and holding at that temperature. Carburizing is generally followed by quenching to produce a hardened case.

Case hardening. Process of hardening a ferrous alloy so that the surface layer, or case, is made substantially harder than the interior, or core. Typical case-hardening processes are carburizing, cyaniding, carbonitriding, induction hardening, and flame hardening.

Charpy test. Pendulum type of impact test in which a specimen, supported at both ends as a simple beam, is broken by the impact of the falling pendulum. The energy absorbed in breaking the specimen, as determined by the decreased rise of the pendulum, is a measure of the impact strength of the metal.

Cold work. Plastic deformation at such temperatures and rates that substantial increases occur in the strength and hardness of the metal. Visible structural changes include changes in grain shape and, in some instances, mechanical twinning or banding.

Corrosion embrittlement. Embrittlement caused in certain alloys by exposure to a corrosive environment. Such material is usually susceptible to the intergranular type of corrosion attack.

Corrosion fatigue. Repeated cyclic stressing of a metal in a corrosive medium, resulting in more rapid deterioration of properties than would be encountered as a result of either cyclic stressing or corrosion alone.

Creep. Time-dependent strain occurring under stress. In steels, it usually occurs at elevated temperatures.

Cyaniding. Process of case-hardening a ferrous alloy by heating it in a molten cyanide, thus causing the alloy to absorb carbon and nitrogen simultaneously. Cyaniding is usually followed by quenching to produce a hard case.

Dezincification. Corrosion of an alloy containing zinc (usually brass), involving loss of zinc and a surface residue or deposit of one or more less active components (usually copper).

Dye-penetrant inspection. Method of nondestructive inspection for determining the existence and extent of discontinuities that are open to the surface of the part being inspected.

Endurance limit. Maximum stress to which a material can be subjected for an indefinite service life. Although standards vary for various types of members and for different industries, it is common practice to accept the assumption that carrying a certain load for several million cycles of stress reversals indicates that the load can be carried for an indefinite time. When a load on a member is constantly varying in value, is repeated by relatively high frequency, or constitutes a complete reversal of stresses with each operating cycle, the material's endurance limit must be substituted for the ultimate strength as called for by the design formulas. The geometry of the member, the presence of local areas of high stress concentration, and the condition of the material (pits, surface irregularities, and corrosion) have considerable influence on the real endurance limit.

Hardening. Any process for increasing the hardness of metal by suitable treatment, usually involving heating and cooling.

Heat treatment. Combination of heating and cooling operations timed and applied to a metal or an alloy in the solid state in a way that will produce desired properties. Heating for the sole purpose of hot working is excluded from the meaning of this definition.

Magnetic-particle inspection. Nondestructive method of inspection for determining the existence and extent of possible defects in magnetic materials. Finely divided magnetic particles applied to a magnetized part are attracted to and outline the pattern of any magnetic leakage fields created by discontinuities.

Mechanical properties. Those properties of a material that reveal the elastic and inelastic reaction when force is applied or that involve the relationship between stress and strain, for example, the modulus of elasticity, tensile strength, and fatigue limit. These properties have often been designated as *physical properties,* but the term *mechanical properties* is preferred.

Modulus of elasticity. Slope of the elastic portion of the stress-strain curve in mechanical testing. The unit stress is divided by the unit elongation. The tensile or compressive elastic modulus is called *Young's modulus;* the torsional elastic modulus is known as the *shear modulus* or *modulus of rigidity.*

Nitriding. Process of case hardening in which a ferrous alloy, usually of special composition, is heated in an atmosphere of ammonia or in contact with nitrogenous material to produce surface hardening by the absorption of nitrogen without quenching.

Normalizing. Process in which a ferrous alloy is heated to a suitable temperature above the transformation range and is subsequently cooled in still air to room temperature.

Physical metallurgy. Science concerned with the physical and mechanical characteristics of metals and alloys.

Physical properties. Those properties familiarly discussed in physics exclusive of those described under the heading "Mechanical properties," for example, density, electrical conductivity, and coefficient of thermal expansion. This term has often been used to describe mechanical properties, but this usage is not recommended.

Precipitation hardening. Process of hardening an alloy in which a constituent precipitates from supersaturated solid solution. This hardening usually consists of two steps: (1) *solution heat treatment* (as described below) at a high temperature that softens and prepares the part; (2) *age hardening* at a lower temperature that hardens and strengthens the part.

Proof stress. In a test, stress that will cause a specified permanent deformation in a material, usually 0.01 percent or less.

Proportional limit. Greatest stress that a material is capable of sustaining without a deviation from the law of proportionality of stress to strain.

Quenching. Process of rapid cooling from an elevated temperature by contact with liquids, gases, or solids.

Radiography. Nondestructive method of internal examination in which metal or other objects are exposed to a beam of x-ray or gamma radiation. Differences in thickness, density, or absorption caused by internal discontinuities are apparent in the shadow image on photographic film placed behind the object.

Reduction in area. Difference between the original cross-sectional area and that of the smallest area at the point of rupture, usually stated as a percentage of the original area; also called *contraction of area.*

Residual stress. Macroscopic stresses that are set up within a metal as the result of nonuniform plastic deformation. This deformation may be caused by cold working or by drastic gradients of temperature from quenching or welding.

Solution heat treatment. Process in which an alloy is heated to a suitable temperature, is held at this temperature long enough to allow a certain constituent to enter into solid solution, and is then cooled rapidly to hold the constituent in solution. The metal is left in a supersaturated, unstable state and may subsequently exhibit age hardening.

Stabilizing anneal. Treatment applied to austenitic stainless steels that contain titanium or columbium. It consists of heating to a temperature below that of a full anneal in order to precipitate the maximum amount of carbon as titanium carbide or columbium carbide. This eliminates precipitation at a lower temperature, which might reduce the resistance of the steel to corrosion.

Steel, alloy. Steel containing significant quantities of alloying elements (other than carbon and the commonly accepted amounts of manganese, silicon, sulfur, and phosphorus) added to effect changes in mechanical or physical properties.

Steel, carbon. Steel containing up to about 2 percent carbon and only residual quantities of other elements except those added for deoxidation, with silicon usually limited to 0.60 percent and manganese to about 1.65 percent. It is also termed *plain carbon steel, ordinary steel,* and *straight carbon steel.*

Steel, stainless. Not a single alloy composition but rather the name applied to a group of low-carbon alloy steels containing at least 10 percent chromium and to which other alloying elements, principally nickel, may be added for increased corrosion resistance and/or improved fabricating characteristics.

The three basic types of stainless steels are:

Group A: martensitic hardenable steels. These steels contain chromium and carbon as the principal alloying elements. They are the cutlery-type steels. They respond to heat treatment and can be hardened in a manner similar to that employed with the familiar alloy and tool steels to provide a wide range of mechanical properties. They are magnetic.

Group B: ferritic nonhardenable steels. This group comprises the chromium-iron alloys much used for decorator trim. Like Group A, they are magnetic; however, they do not respond to heat treatment and are normally used in the annealed state, in which they exhibit their maximum softness, ductility, and corrosion resistance. The mechanical properties of these steels can be increased to a small extent by cold working.

Group C: austenitic nonhardenable steels. The chromium-nickel alloys which form the basis of this group offer a greater degree of corrosion resistance than the steels of Groups A and B. They are strong, tough, and ductile, and although they cannot be hardened by heat treatment, they can be appreciably strengthened by cold working. These austenitic steels are nonmagnetic.

Stress-corrosion cracking. Failure of a part by cracking due to a combination of stress, whether residual or applied, and corrosive environment. Cracking may be intergranular or transgranular, depending on material and environment.

Stress relieving. Process of reducing residual stresses in a metal object by heating it to a suitable temperature and holding for a sufficient time. This treatment may be applied to relieve stresses induced by casting, quenching, normalizing, machining, cold working, or welding.

Temper. Condition produced in a metal or alloy by mechanical or thermal treatment and having characteristic structure and mechanical properties. A given alloy may be in the fully softened or annealed temper, or it may be cold-worked to the hard temper or further to spring temper. Intermediate tempers produced by cold working (rolling or drawing) are called *quarter hard, half hard,* and *three-quarters hard* and are determined by the amount of cold reduction and the resulting tensile properties. In addition to the annealed temper, conditions produced by thermal treatment are the solution-heat-treated temper and the heat-treated and artificially aged temper. Other tempers involve a combination of mechanical and thermal treatments and include the temper produced by cold working after heat treating and that produced by artificial aging of alloys that are as cast, as extruded, as forged, and heat-treated and worked.

Tempering. Process of reheating quench-hardened or normalized steel to a temperature below the transformation range and then cooling at any rate desired.

Tensile strength. Value obtained by dividing the maximum load observed during tensile straining by the specimen cross-sectional area before straining; also called *ultimate strength.*

Ultrasonic testing. Method of nondestructive inspection in which a beam of sound is transmitted into a material to examine it for possible discontinuities. Discontinuities in the material will reflect or dissipate sound. These sound reflections can be electronically monitored.

Wire wooling. Abrasive-wear failure of shafts by wire wooling has been observed under circumstances in which contact occurs between the shaft and a stationary part, resulting in removal by machining of wire shavings that resemble steel wool. This type of failure as been found on turbine and turbine-generator shafts made of 3 percent chromium–0.5 percent molybdenum steels, on 12 percent chromium stainless steels, or on 18 percent chromium–8 percent nickel stainless steels and on nonchromium steels in the presence of certain chloride-containing oils. Although the mechanisms of wire wooling are not clearly understood, it is known that wire wooling requires contact between a shaft and a bearing, either directly or through a buildup of deposits. If the deposit on a stationary part contains hard particles, fine slivers can be cut or spun off the shaft surface. As fine slivers or pieces come off, if the resultant friction and heat are sufficient, additional hard particles may be formed by reactions between iron and/or chromium and the oil or gas present. Sometimes scabs or solid chunks of laminated or compacted slivers and other deposits are formed.

Methods for prevention of wire wooling include:

1. Changing or coating the shaft material
2. Using a softer bearing or labyrinth material
3. Changing to a different oil containing polar olefin additives
4. Eliminating the deposits
5. Providing greater clearance

Yield point. In mild or medium-carbon steel, the stress at which a marked increase in deformation occurs without an increase in load. In other steels and in nonferrous metals this phenomenon is not observed.

Yield strength. Stress at which a material exhibits a specified limiting deviation from proportionality of stress to strain. An offset of 0.2 percent is used for many metals. Copper-base alloys often use 0.5 percent total elongation under load.

TABLE 2-1 CHEMICAL COMPOSITION, MECHANICAL PROPERTIES, AND PHYSICAL CONSTANTS OF SOME METALS AND ALLOYS[a]

Material	Approx chemical composition, %	Form and condition	Average mechanical properties				Average physical constants				
			Yield strength (0.2% offset), 1,000 psi	Tensile strength, 1,000 psi	Elongation, % in 2 in.	Brinell hardness	Density, lb/cu in.	Specific gravity	Thermal expansion coefficient (32–212°F), 10^6 in./(in.)(°F)	Tensile modulus of elasticity, 10^{-6} psi	Torsional modulus of elasticity, 10^{-6} psi
Monel alloy 410 (cast)	Ni 66, Cu 30.5, Fe 1.0, Mn 0.8, Si 1.6	As cast	35	75	40	150	0.312	8.63	9.2	23	
Monel alloy K-500 bars and forgings AMS-4676	Ni bal, Cu 29.5, Al 2.8, Fe 1.0, Ti 0.5, Mn 0.6, C 0.15, Si 0.15	Rod as hot-rolled	49	97	44	155	0.306	8.46	7.6[e]	26	9.5
		Hot-rolled aged	111	160	23.5	300					
Monel alloy 505 (cast)	Ni 64, Cu 29, Si 4, Fe 2, Mn 0.8	Casting annealed	75	115	10	225	0.302	8.36	8.9[d]	24	
		As cast or annealed and aged	110	135	2	340					
Nickel (cast)	Ni 95.6, Cu 0.5, Fe 0.5, Mn 0.8, Si 1.5, C 0.8	As cast	25	57	22	110	0.301	8.34	8.85	21.5	
Hastelloy B	Ni 61, Mo 28, Fe 5.0, Co 2.5, Cr 1.0, C 0.05, others 3	Cast or wrought	50	90	10	180	0.334	8.81	5.3[e]	27	
Aluminum alloy 2017 bar, rod, and wire ASTM B 211	Al bal, Cu 4.0, Mn 0.5, Mg 0.5	Bar annealed	10	26	22	45	0.101	2.79	12.7[g]	10.5	4.0
		Heat-treated	40	62	22	105			. . . .	10.5	4.0
Ni-Resist type 1 ASTM A 436	C 3 max, Si 2, Mn 1.25, Ni 15.5, Cr 2.0, Cu 6.5, Fe bal	As cast	. . . .	27	. . . .	150	0.264	7.30	10.4[c]	15.6	4.5
Cast gray iron ASTM A 48 C 1.30	C 3.4, Si 1.8, Mn 0.8, Fe bal	As cast	. . . .	32	. . . .	190	0.260	7.20	6.7	14	
Cast carbon steel	Fe bal, Mn 0.7, Si 0.4, C 0.3	Cast	40	72	26	140	0.283	7.84	6.7	30	
Carbon steel SAE 1020 ASTM A 285	Fe bal, Mn 0.45, Si 0.25, C 0.20	Annealed	38	65	30	130	0.284	7.86	6.7	30	
		Quenched and tempered at 1000°F	62	90	25	179					

TABLE 2-1 CHEMICAL COMPOSITION, MECHANICAL PROPERTIES, AND PHYSICAL CONSTANTS OF SOME METALS AND ALLOYS (Continued)

Material	Composition	Condition									
AISI 4340 low-alloy steel	Fe bal, Ni 1.75, Cr 0.80, Mo 0.25	Heat-treated	125	140	17	285	0.280	7.80	6.5	29	10.5
Stainless steel type 304	Fe bal, Cr 19, Ni 9, C 0.98, max	Annealed / Cold-rolled[a]	30 / 160	85 / 185	50 / 8	160 / 400	0.286	7.92	9.3	29	10.5
Stainless steel type 310	Fe bal, Cr 25, Ni 20, C 0.25 max	Annealed	40	100	50	165	0.285	7.90	8.5	29.5	11.1
Cast stainless steel type 316	Fe bal, Cr 19, Ni 9, Mo 3, C 0.1	Cast	44	80	50	150	0.286	7.92	8.9	29	
Stainless steel type 410	Fe bal, Cr 12.5, C 0.15 max	Annealed / Heat-treated	40 / 115	75 / 150	30 / 15	150 / 300	0.277	7.67	6.1	28	10
Stainless steel type 420	Fe bal, Cr 13, C 0.35	Annealed / Heat-treated	60 / 200	98 / 250	28 / 8	180 / 480	0.278	7.70	5.7	28	10
Stainless steel type 430	Fe bal, Cr 16, C 0.12 max	Annealed / Cold-rolled	40 / 95	70 / 110	35 / 10	165 / 225	0.275	7.61	6.0	29	10.5
Stainless steel 17-4 PH	Fe bal, Cr 17, Ni 4, Cu 4	Heat-treated	155	165	12	330	0.280	7.80	6.0	29	10.5
Stainless steel type 422	Cr 13, Ni 0.75, Va 0.30, W 1.00, Mo 1.00	Heat-treated	90	125	18	250	0.280	7.78	5.9	28	10.0
Copper CA110: sheet—ASTM B 152, rod—B 124, B 133, wire—B 1, B 2, B 3	Cu 99.9 min	Strip annealed / Spring temper	10[b] / 50[b]	32 / 55	45 / 4	42 / 107	0.322	8.91	9.4[g] / 9.8[h]	17	6.4
Phosphor bronze CA-524: sheet—ASTM B 103, rod—ASTM B 139, wire—ASTM B 159	Cu bal, Sn 10, P 0.2	Annealed / Spring temper	28[b] /	66 / 122	68 / 4	100 / 119	0.317	8.78	10.2[h]	16	6.0
G bronze (88-10-2) castings, ASTM B 143 A1,1A	Cu bal, Sn 10, Zn 2	As cast	21[b]	45	31	80	0.315	8.72	....	15	
Leaded red brass (85-5-5-5) castings, ASTM B 145	Cu 85, Zn 5, Pb 5, Sn 5	As cast	17[b]	35	25	60	0.317	8.75	10.2	13	

[a] Abridged with permission from *Properties of Some Metals and Alloys*, The International Nickel Co., Inc.
[b] 0.5 percent extension.
[c] 68 to 392°F.
[d] 70 to 1100°F.
[e] 70 to 200°F.
[f] Maximum for wrought alloy. Cast alloys have slightly different composition, notably higher carbon.
[g] 68 to 212°F.
[h] 68 to 572°F.

TABLE 2-2 HARDNESS—TENSILE-STRENGTH CONVERSION TABLE*

Brinell, 10-mm carbide ball, 3000-kg load		Diamond pyramid hardness no.	Rockwell		Shore	Approximate tensile strength, 1000 lb/in^2
Indentation diameter, mm	Hardness no.		C scale 150-kg Brale	B scale 100-kg $\frac{1}{16}$-in ball		
. . . .	. . . .	940	68	. . . .	97	
. . . .	767	880	66.4	. . . .	93	
2.25	745	840	65.3	. . . .	91	
2.30	712					
2.35	682	737	61.7	. . . .	84	
2.40	653	697	60	. . . .	81	
2.45	627	667	58.7	. . . .	79	
2.50	601	640	57.3	. . . .	77	
2.55	578	615	56	. . . .	75	
2.60	555	591	54.7	. . . .	73	298
2.65	534	569	53.5	. . . .	71	288
2.70	514	547	52.1	. . . .	70	274
2.75	495	528	51	. . . .	68	264
2.80	477	508	49.6	. . . .	66	252
2.85	461	491	48.5	. . . .	65	242
2.90	444	472	47.1	. . . .	63	230
2.95	429	455	45.7	. . . .	61	219
3.00	415	440	44.5	. . . .	59	212
3.05	401	425	43.1	. . . .	58	202
3.10	388	410	41.8	. . . .	56	193
3.15	375	396	40.4	. . . .	54	184
3.20	363	383	39.1	. . . .	52	177
3.25	352	372	37.9	(110)	51	171
3.30	341	360	36.6	(109)	50	164
3.35	331	350	35.5	(108.5)	48	159
3.40	321	339	34.3	(108)	47	154
3.45	311	328	33.1	(107.5)	46	149
3.50	302	319	32.1	(107)	45	146
3.55	293	309	30.9	(106)	43	141
3.60	285	301	29.9	(105.5)	. . . .	138
3.65	277	292	28.8	(104.5)	41	134
3.70	269	284	27.6	(104)	40	130
3.75	262	276	26.6	(103)	39	127
3.80	255	269	25.4	(102)	38	123
3.85	248	261	24.2	(101)	37	120
3.90	241	253	22.8	100	36	116
3.95	235	247	21.7	99	35	114
4.00	229	241	20.5	98.2	34	111
4.05	223	234	(18.8)	97.3		
4.10	217	228	(17.5)	96.4	33	105
4.20	207	218	(15.2)	94.6	32	100
4.30	197	207	(12.7)	92.8	30	95
4.40	187	196	(10.0)	90.7	. . . .	90

TABLE 2-2   HARDNESS—TENSILE-STRENGTH CONVERSION TABLE* (*Continued*)

Brinell, 10-mm carbide ball, 3000-kg load		Diamond pyramid hardness no.	Rockwell			Approximate tensile strength, 1000 lb/in²
Indentation diameter, mm	Hardness no.		C scale 150-kg Brale	B scale 100-kg ¹⁄₁₆-in ball	Shore	
4.50	179	188	(8.0)	89.0	27	87
4.60	170	178	(5.4)	86.8	26	83
4.70	163	171	(3.3)	85.0	25	79
4.80	156	163	(0.9)	82.9	. . . .	76
5.00	143	150	. . . .	78.7	22	71
5.20	131	137	. . . .	74.0	. . . .	65
5.40	121	127	. . . .	69.8	19	60
5.60	111	117	. . . .	65.7	15	56

*Approximate data from *Society of Automotive Engineers Handbook,* Warrendale, Pa., 1979. Values in parentheses are beyond normal range and are for information only.

	A (area)	I	c	k
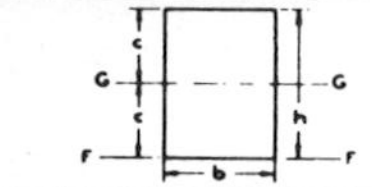	$A = bh$	$I_G = \dfrac{bh^3}{12}$ $I_F = \dfrac{bh^3}{3}$	$c = \dfrac{h}{2}$	$k_G = \dfrac{h}{\sqrt{12}}$ $k_F = \dfrac{h}{\sqrt{3}}$
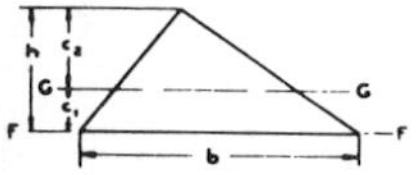	$A = \tfrac{1}{2}bh$	$I_G = \dfrac{bh^3}{36}$ $I_F = \dfrac{bh^3}{12}$	$c_1 = \dfrac{h}{3}$ $c_2 = \dfrac{2h}{3}$	$k_G = \dfrac{h}{\sqrt{18}}$ $k_F = \dfrac{h}{\sqrt{6}}$
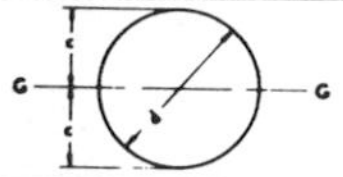	$A = \dfrac{\pi}{4}d^2$	$I_G = \dfrac{\pi d^4}{64}$	$c = \dfrac{d}{2}$	$k_G = \dfrac{d}{4}$
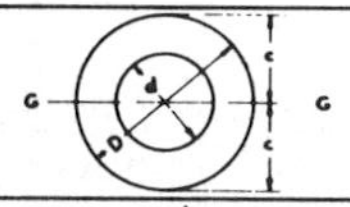	$A = \dfrac{\pi}{4}(D^2 - d^2)$	$I_G = \dfrac{\pi}{64}(D^4 - d^4)$	$c = \dfrac{D}{2}$	$k_G = \dfrac{\sqrt{D^2 + d^2}}{4}$
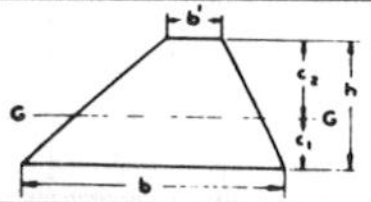	$A = \dfrac{h(b + b')}{2}$	$I_G = \dfrac{h^3(b^2 + 4bb' + b'^2)}{36(b + b')}$	$c_1 = \dfrac{h(b + 2b')}{3(b + b')}$ $c_2 = \dfrac{h(b' + 2b)}{3(b + b')}$	$k_G = \dfrac{h\sqrt{2(b^2 + 4bb' + b'^2)}}{6(b + b')}$
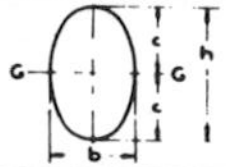	$A = \dfrac{\pi bh}{4}$	$I_G = \dfrac{\pi bh^3}{64}$	$c = \dfrac{h}{2}$	$k_G = \dfrac{h}{4}$
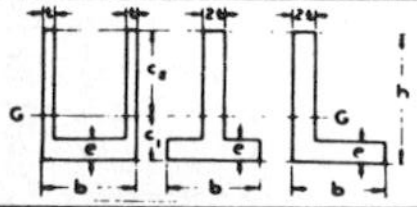	$A = 2ht + (b - 2t)e$	$I_G = 2th^3 + \dfrac{(b - 2t)e^3}{3} - Ac_1^2$	$c_1 = \dfrac{ht^2 + \frac{1}{2}e^2(b - 2t)}{A}$ $c_2 = h - c_1$	$k_G = \sqrt{\dfrac{I_G}{A}}$
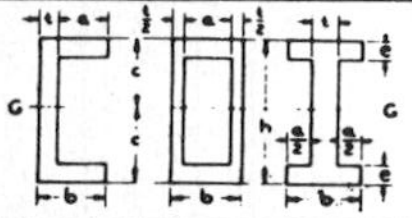	$A = bh - a(h - 2e)$	$I_G = \dfrac{bh^3 - a(h - 2e)^3}{12}$	$c = \dfrac{h}{2}$	$k_G = \sqrt{\dfrac{bh^3 - a(h - 2e)^3}{12A}}$

FIG. 2-1 The above table gives the properties of various sections where A = area, I = moment of inertia, c = distance from the center of gravity to the extreme point or edge, and k = radius of gyration. In the sections shown, the axis GG passes through the center of gravity, while the axis FF passes through one edge. If e is the distance between GG and any parallel axis BB, the moment of inertia of the section about the BB axis is $I_B = I_G = Ae^2$. In general, the radius of gyration about axis BB of a section is $k_B = \sqrt{I_B/A}$.

BEAM FORMULAS (UNIFORM SECTION)

Loading	Reaction R	Bending moment M	Deflection y
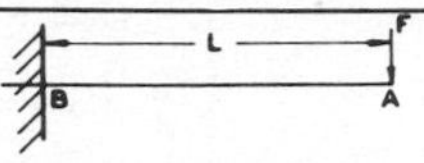	$R_B = F$	$M_B = FL$	$y_A = \dfrac{1}{3}\dfrac{FL^3}{EI}$
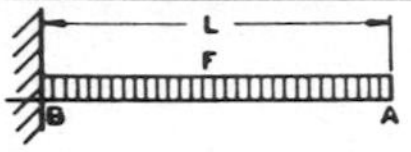	$R_B = F$	$M_B = \dfrac{FL}{2}$	$y_A = \dfrac{1}{8}\dfrac{FL^3}{EI}$
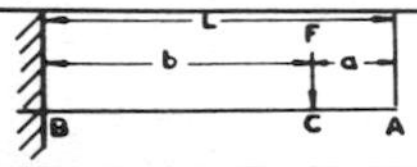	$R_B = F$	$M_B = Fb;\ M_C = 0$	$y_A = \dfrac{F(2L^3 - 3L^2 a + a^3)}{6\ EI}$
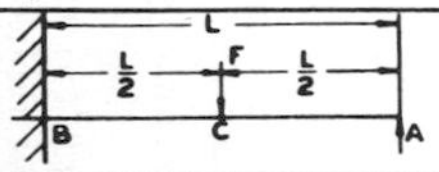	$R_A = \dfrac{5}{16}F;\ R_B = \dfrac{11}{16}F$	$M_B = \dfrac{3}{16}FL;\ M_C = \dfrac{5}{32}FL$	$y_C = 0.0093\dfrac{FL^3}{EI}$
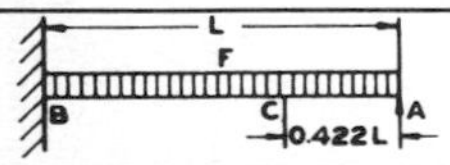	$R_A = \dfrac{3}{8}F;\ R_B = \dfrac{5}{8}F$	$M_B = \dfrac{1}{8}FL;\ M_C = 0.07FL$	$y_C = 0.0054\dfrac{FL^3}{EI}$
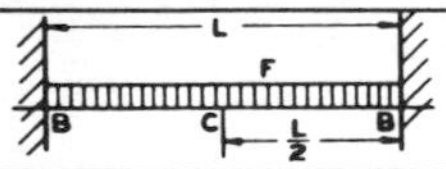	$R_B = \dfrac{F}{2}$	$M_B = \dfrac{FL}{12};\ M_C = \dfrac{FL}{24}$	$y_C = \dfrac{1}{384}\dfrac{FL^3}{EI}$
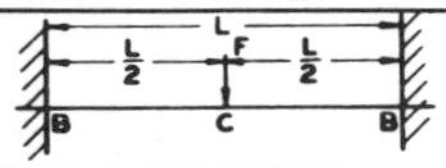	$R_B = \dfrac{F}{2}$	$M_B = M_C = \dfrac{FL}{8}$	$y_C = \dfrac{1}{192}\dfrac{FL^3}{EI}$
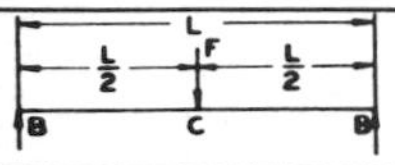	$R_B = \dfrac{F}{2}$	$M_C = \dfrac{FL}{4}$	$y_C = \dfrac{1}{48}\dfrac{FL^3}{EI}$
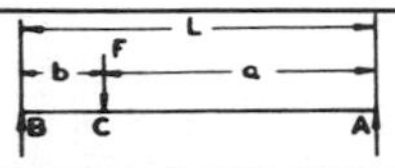	$R_A = \dfrac{Fb}{L};\ R_B = \dfrac{Fa}{L}$	$M_C = \dfrac{Fab}{L}$	$y_{max} = \dfrac{Fab(a + 2b)\sqrt{3a(a + 2b)}}{27EIL}$
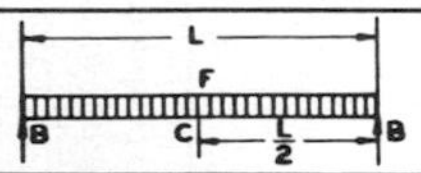	$R_B = \dfrac{F}{2}$	$M_C = \dfrac{1}{8}FL$	$y_C = \dfrac{5}{384}\dfrac{FL^3}{EI}$
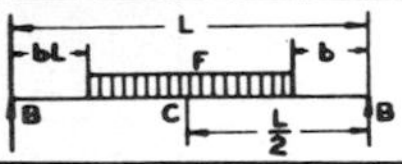	$R_B = \dfrac{F}{2}$	$M_C = \dfrac{F}{8}(2b + L)$	$y_C = \dfrac{(5 - 24b^2 + 16b^4)}{384\ (1 - 2b)}\dfrac{FL^3}{EI}$
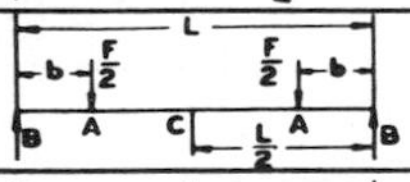	$R_B = \dfrac{F}{2}$	$M_A = M_C = \dfrac{Fb}{2}$	$y_C = \dfrac{Fb}{12EI}(\tfrac{3}{4}L^2 - b^2)$
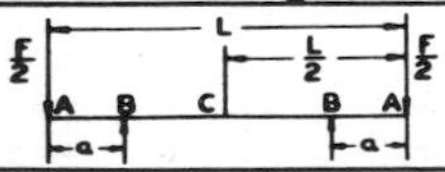	$R_B = \dfrac{F}{2}$	$M_B = M_C = \dfrac{Fa}{2}$	$y_A = \dfrac{Fa^2(3L - 4a)}{12\ EI}$ $y_C = \dfrac{Fa(L - 2a)^2}{16\ EI}$

FIG. 2-2 The units to be used in the above equations should be inches and pounds. The bending stress in the beam is given by the equation $s = Mc/I$, where s = stress, lb/in^2; M = bending moment, in·lb; c = distance from neutral axis or center of gravity to outer fiber, in; and I = moment of inertia, in^4 (see Fig. 2-1 for calculation of I and c).

SHAFT DESIGN

Shafts are designed on the basis of the torsional moment or torque which they must transmit. This torque may be found from the equation $T = 63,025 \text{ hp}/n$, where T = torque, in·lb, hp = transmitted horsepower, and n = shaft r/min. The shear stress developed for a given transmitted torque is given by the formula

$$s_s = \frac{16T}{\pi d^3}$$

where d = shaft diameter, in

The curve in Fig. 2-3 may be used to determine the shaft diameter required for a shear stress of 10,000 lb/in² for a given horsepower and speed. If it is desired to use some other design stress, the shaft diameter found in Fig. 2-3 may be corrected from the graph in Fig. 2-4.

Example: If 1000 hp is to be transmitted at a shaft speed of 1500 r/min, the shaft diameter as read from the chart in Fig. 2-3 for a shear stress of 10,000 lb/in² is 2.78

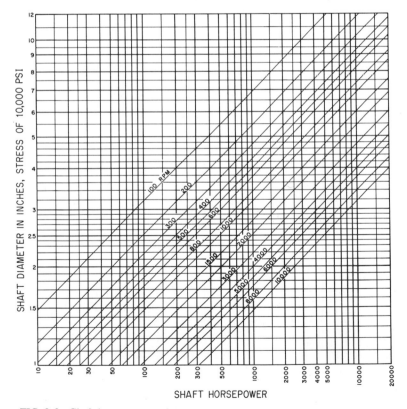

FIG. 2-3 Shaft-horsepower rating curve.

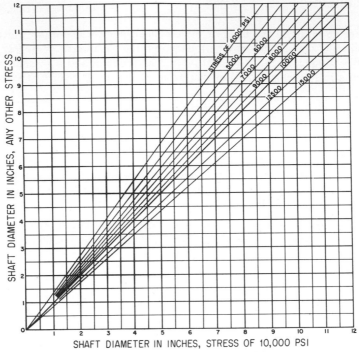

FIG. 2-4 Correction curve for shaft diameters at various stresses.

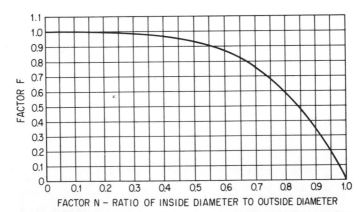

FIG. 2-5 Torsional conversion factor for hollow shafts. (1) Compute factor N. (2) Select F from curve. (3) Rating equals product of F and rating of solid shaft.

in. If a stress of 5000 lb/in^2 is desired, the correction graph in Fig. 2-4 shows that a shaft diameter of 3.5 in is required.

HOLLOW SHAFTS

Figure 2-5 can be used for hollow shafts by applying the factors and following the simple procedure indicated. This curve has been calculated by the formula

$$1 - \left(\frac{\text{inside diameter}}{\text{outside diameter}} \right)^4$$

EFFECT OF KEYWAYS

For standard keyways in which the width equals one-fourth of the shaft diameter and the depth is one-half of the width, the approximate effect on the torsional strength of a solid shaft is as follows:

No. keyways	Comparison with strength of shaft without keyways, %
1	85
2	78
3	73
4	70

BENDING MOMENTS OF SHAFTS

The maximum bending moment M at a for a shaft having an overhang load P at b, as shown in Fig. 2-6, neglecting the weight of the shaft, is

$$M = PL$$

where M = bending moment, in·lb
P = load, lb
L = distance from P to bearing support, in

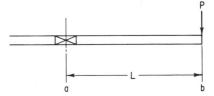

FIG. 2-6   Bending moment.

COMBINED TORSIONAL AND BENDING LOADS OF SHAFTS

For shafts which are subjected to bending as well as torsion, the following formula is recommended:

$$\text{Equivalent torsion} = \sqrt{T^2 + M^2}$$

where T = actual torsional moment
M = actual bending moment

Figure 2-7 simplifies this operation. These curves can be used for any range of loads by considering the bending and torsional moment graduations as units rather than as absolute figures and then pointing off the desired decimal places, just as is done in making calculations by means of a slide rule.

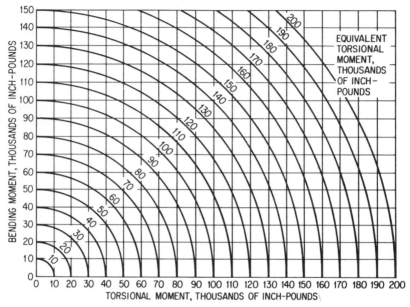

FIG. 2-7 Combined torsional and bending loads.

STEPPED SHAFTS

Too little attention has been paid in the past to the effect of shoulders in producing stress concentrations, often resulting in failures which could easily have been prevented simply by the use of a properly proportioned fillet.

Figure 2-8a, b, c, and d shows the proper fillet proportions based upon data presented by L. S. Jacobsen in *ASME Transactions* (vol. 47, p. 619, 1925).

SAFE SPEEDS OF SHAFTS

Figure 2-9 has been included to show the maximum safe speeds with regard to whipping for plain shafts of uniform diameter without other members and of various lengths between supports.

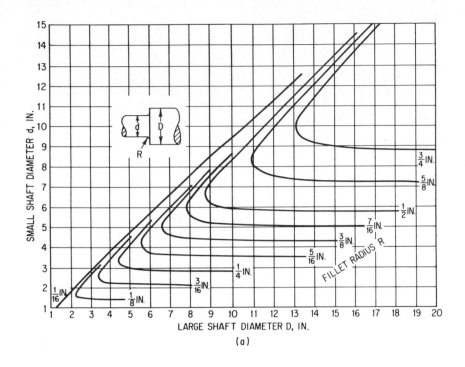

(a)

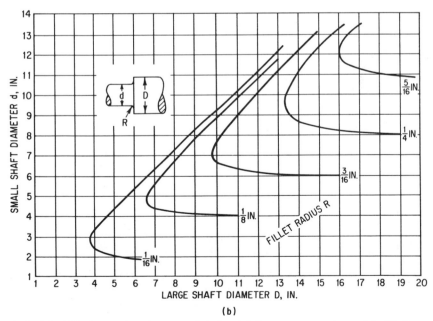

(b)

FIG. 2-8 Fillet proportions for stepped shafts. (*a*) Stress concentration = 1.5. (*b*) Stress concentration = 2.0.

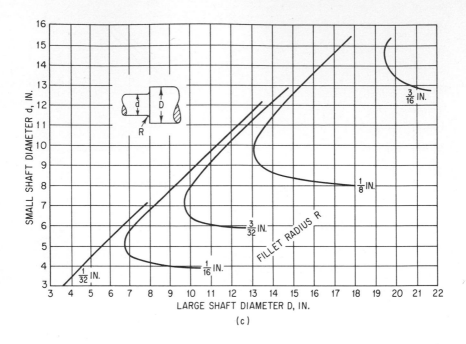

(c)

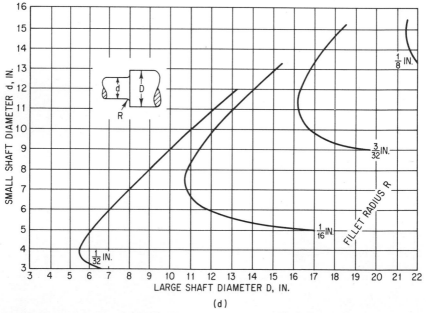

(d)

FIG. 2-8 *(Cont.)* Fillet proportions for stressed shafts. (*c*) Stress concentration = 2.5. (*d*) Stress concentration = 3.0.

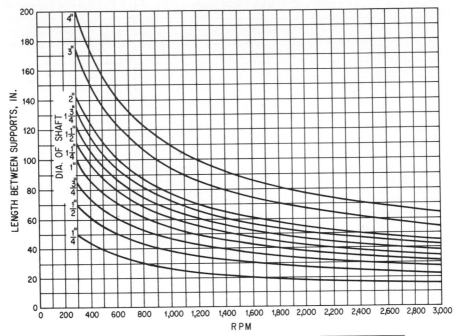

FIG. 2-9 Maximum safe speed of uniform-diameter shaft. $L = \sqrt{3,000,000 \times D/n}$, where D = diameter in inches and n = revolutions per minute. This formula permits operation up to approximately 63 percent of the critical speed.

TABLE 2-3   ALLOWABLE STRESSES IN TYPICAL STEELS

Because steady torsional loads cause failure in shear, shafts should be designed on the basis of the yield strength in shear, which is usually taken at 60 percent of the yield strength in tension. This table gives the average mechanical properties of some popular steels.

Material	Size rounds, in.	Tensile strength	Yield point	Elonga-tion, %	Reduc-tion area	Brinell hardness
SAE 1020 hot-rolled	1	65,000	40,000	30	55	130
SAE 1020 hot-rolled	6	60,000	35,000	30	40	120
SAE 1020 forged	12	55,000	30,000	20	30	110
SAE 1040 hot-rolled	1	94,000	58,000	27	52	187
SAE 1040 hot-rolled	6	84,000	46,000	19	30	160
SAE 1040 forged	12	82,000	44,000	16	28	160
SAE 1040 hot-rolled, water-quenched, tempered at 1200°F	1	100,000	70,000	27	60	200
SAE 1040 hot-rolled, water-quenched, tempered at 1200°F	6	82,000	52,000	25	48	160

TABLE 2-3   ALLOWABLE STRESSES IN TYPICAL STEELS   *(Continued)*

Material	Size rounds, in.	Tensile strength	Yield point	Elonga-tion, %	Reduc-tion area	Brinell hardness
SAE 1040 forged, water-quenched, tempered at 1200°F	12	78,000	44,000	23	44	155
SAE 2340 hot-rolled, oil-quenched, tempered at 1200°F	1	112,000	85,000	25	63	230
SAE 2340 hot-rolled, oil-quenched, tempered at 1200°F	6	104,000	75,000	27	58	210
SAE 2340 forged, oil-quenched, tempered at 1200°F*	12	100,000	70,000	21	48	200
SAE 2340 normalized, tempered at 1200°F	12	100,000	65,000	20	45	200
SAE 4140 hot-rolled, oil-quenched, tempered at 1200°F	1	145,000	125,000	17	56	293
SAE 4140 hot-rolled, oil-quenched, tempered at 1200°F	6	108,000	80,000	21	54	220
SAE 4140 forged, oil-quenched, tempered at 1200°F*	12	103,000	65,000	20	45	210
SAE 4140 forged, normalized, tempered at 1200°F	12	95,000	57,000	21	48	190
SAE 4340 hot-rolled, oil-quenched, tempered at 1200°F	1	150,000	130,000	20	58	302
SAE 4340 hot-rolled, oil-quenched, tempered at 1200°F	6	125,000	100,000	18	54	250
SAE 4340 forged, oil-quenched, tempered at 1200°F*	12	110,000	90,000	17	50	230
SAE 4340 forged, normalized, tempered at 1200°F	12	95,000	70,000	20	45	220

*It is not generally considered good practice to liquid-quench solid diameters larger than 10 in.

When reversing or shock loads are encountered, the shafts should be designed on the basis of the endurance limit in shear. This is taken as one-fourth of the ultimate tensile strength. These are the stresses where failure will begin, and to guard against breakdowns it is necessary to apply a factor of safety.

The usual factor of safety for ordinary service is 3, but when exceptionally heavy shocks are encountered or with smooth loads when an occasional shock occurs, larger factors should be used. It is difficult to recommend the precise factor to be used, as this depends upon the intensity and frequency of the shock. Here experience and good judgment play an important part.

It must also be borne in mind that when mechanical properties of steels are given, the specimens are usually heat-treated in 1-in diameter and then turned to a standard test diameter of 0.505 in for testing.

WR^2 CALCULATION

Figure 2-10 may be found useful in calculating the WR^2 for cylindrical bodies often required in determining the flywheel effect.

$$WR^2 = W \times R^2$$

where WR^2 is in lb·ft²
   W = weight, lb
   R^2 = radius of gyration squared from Fig. 2-10

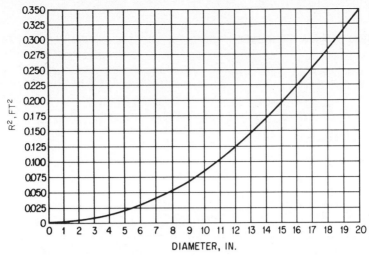

FIG. 2-10 Determination of WR^2 of cylindrical bodies. To determine the WR^2 for the flywheel effect, multiply the weight in pounds by the value taken from the curve. For members composed of various diameters, calculate the WR^2 of each diameter separately and take their sum, deducting for hollow sections. The curve follows the formula $R^2 = (\text{diameter in inches}/24)^2/2$ and gives R^2 in feet.

SQUARE AND FLAT KEYS

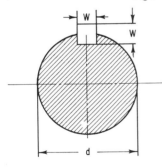

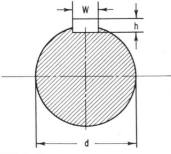

Dimensions of Standard Sizes in Inches

Shaft dia d (inclusive)	W	h
$7/16$–$9/16$	$1/8$	$3/32$
$9/16$–$7/8$	$3/16$	$1/8$
$7/8$–$1\,1/4$	$1/4$	$3/16$
$1\,1/4$–$1\,3/8$	$5/16$	$1/4$
$1\,3/8$–$1\,3/4$	$3/8$	$1/4$
$1\,3/4$–$2\,1/4$	$1/2$	$3/8$
$2\,1/4$–$2\,3/4$	$5/8$	$7/16$
$2\,3/4$–$3\,1/4$	$3/4$	$1/2$
$3\,1/4$–$3\,3/4$	$7/8$	$5/8$
$3\,3/4$–$4\,1/2$	1	$3/4$
$4\,1/2$–$5\,1/2$	$1\,1/4$	$7/8$
$5\,1/2$–$6\,1/2$	$1\,1/2$	1
$6\,1/2$–$7\,1/2$	$1\,3/4$	$1\,1/2$°
$7\,1/2$–9	2	$1\,1/2$

From ANSI Standard B17.1–1967.

° Some key standards show $1\,1/4$ in. Preferred size in $1\,1/2$ in.

FIG. 2-11

UNIFIED SCREW THREADS

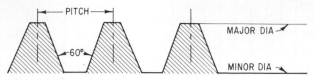

FIG. 2-12 Abridged from ANSI Standard B1.1-1974.

Size	Basic major diameter, in	Coarse-thread series			Fine-thread series		
		Threads per inch	Minor diameter, in	Area, minor diameter, in^2	Threads per inch	Minor diameter, in	Area, minor diameter, in^2
0	0.0600				80	0.0447	0.00151
1	0.0730	64	0.0538	0.00218	72	0.0560	0.00237
2	0.0860	56	0.0641	0.00310	64	0.0668	0.00339
3	0.0990	48	0.0734	0.00406	56	0.0771	0.00451
4	0.1120	40	0.0813	0.00496	48	0.0864	0.00566
5	0.1250	40	0.0943	0.00672	44	0.0971	0.00716
6	0.1380	32	0.0997	0.00745	40	0.1073	0.00874
8	0.1640	32	0.1257	0.01196	36	0.1299	0.01285
10	0.1900	24	0.1389	0.01450	32	0.1517	0.0175
12	0.2160	24	0.1649	0.0206	28	0.1722	0.0226
$^1/_4$	0.2500	20	0.1887	0.0269	28	0.2062	0.0326
$^5/_{16}$	0.3125	18	0.2443	0.0454	24	0.2614	0.0524
$^3/_8$	0.3750	16	0.2983	0.0678	24	0.3239	0.0809
$^7/_{16}$	0.4375	14	0.3499	0.0933	20	0.3762	0.1090
$^1/_2$	0.5000	13	0.4056	0.1257	20	0.4387	0.1486
$^9/_{16}$	0.5625	12	0.4603	0.162	18	0.4943	0.189
$^5/_8$	0.6250	11	0.5135	0.202	18	0.5568	0.240
$^3/_4$	0.7500	10	0.6273	0.302	16	0.6733	0.351
$^7/_8$	0.8750	9	0.7387	0.419	14	0.7874	0.480
1	1.0000	8	0.8466	0.551	12	0.8978	0.625
$1^1/_8$	1.1250	7	0.9497	0.693	12	1.0228	0.812
$1^1/_4$	1.2500	7	1.0747	0.890	12	1.1478	1.024
$1^3/_8$	1.3750	6	1.1705	1.054	12	1.2728	1.260
$1^1/_2$	1.5000	6	1.2955	1.294	12	1.3978	1.521
$1^3/_4$	1.7500	5	1.5046	1.74			
2	2.0000	$4^1/_2$	1.7274	2.30			
$2^1/_4$	2.2500	$4^1/_2$	1.9774	3.02			
$2^1/_2$	2.5000	4	2.1933	3.72			
$2^3/_4$	2.7500	4	2.4433	4.62			
3	3.0000	4	2.6933	5.62			
$3^1/_4$	3.2500	4	2.9433	6.72			
$3^1/_2$	3.5000	4	3.1933	7.92			
$3^3/_4$	3.7500	4	3.4433	9.21			
4	4.0000	4	3.6933	10.61			

TABLE 2-4   DIMENSIONS OF HEXAGON-HEAD CAP SCREWS AND NUTS IN INCHES

ANSI Standard

Size	Hexagon-head cap screws°			Hexagon nuts†		
	Width across flats	Width across corners (max)	Nominal height	Width across flats	Width across corners (max)	Nominal height
$1/4$	$7/16$	0.505	$5/32$	$7/16$	0.505	$7/32$
$5/16$	$1/2$	0.577	$13/64$	$1/2$	0.577	$17/64$
$3/8$	$9/16$	0.650	$15/64$	$9/16$	0.650	$21/64$
$7/16$	$5/8$	0.722	$9/32$	$11/16$	0.794	$3/8$
$1/2$	$3/4$	0.866	$5/16$	$3/4$	0.866	$7/16$
$9/16$	$13/16$	0.938	$23/64$	$7/8$	1.010	$31/64$
$5/8$	$15/16$	1.083	$25/64$	$15/16$	1.083	$35/64$
$3/4$	$1 1/8$	1.299	$15/32$	$1 1/8$	1.299	$41/64$
$7/8$	$1 5/16$	1.516	$35/64$	$1 5/16$	1.516	$3/4$
1	$1 1/2$	1.732	$39/64$	$1 1/2$	1.732	$55/64$
$1 1/8$	$1 11/16$	1.949	$11/16$	$1 11/16$	1.949	$31/32$
$1 1/4$	$1 7/8$	2.165	$25/32$	$1 7/8$	2.165	$1 1/16$
$1 3/8$	$2 1/16$	2.382	$27/32$	$2 1/16$	2.382	$1 11/64$
$1 1/2$	$2 1/4$	2.598	$15/16$	$2 1/4$	2.598	$1 9/32$
$1 3/4$	$2 5/8$	3.031	$1 3/32$			
2	3	3.464	$1 7/32$			
$2 1/4$	$3 3/8$	3.897	$1 3/8$			
$2 1/2$	$3 3/4$	4.330	$1 17/32$			
$2 3/4$	$4 1/8$	4.763	$1 11/16$			
3	$4 1/2$	5.196	$1 7/8$			

*From ANSI Standard B18.2.1–1972.
†From ANSI Standard B18.2.2–1972.

TABLE 2-5   DIMENSIONS OF HEXAGON-HEAD CAP SCREWS AND NUTS IN INCHES

ANSI Standard Heavy

Size	Hexagon-head cap screws°			Hexagon nuts†		
	Width across flats	Width across corners (max)	Nominal height	Width across flats	Width across corners (max)	Nominal height
$1/4$	. . . .	. . . .	. . . . .	$1/2$	0.577	$15/64$
$5/16$	. . . .	. . . .	. . . . .	$9/16$	0.650	$19/64$
$3/8$	. . . .	. . . .	. . . . .	$11/16$	0.794	$23/64$
$7/16$	. . . .	. . . .	. . . . .	$3/4$	0.866	$27/64$
$1/2$	$7/8$	1.010	$5/16$	$7/8$	1.010	$31/64$
$9/16$	. . . .	. . . .	. . . . .	$15/16$	1.083	$35/64$
$5/8$	$1 1/16$	1.227	$25/64$	$1 1/16$	1.227	$39/64$
$3/4$	$1 1/4$	1.443	$15/32$	$1 1/4$	1.443	$47/64$
$7/8$	$1 7/16$	1.660	$35/64$	$1 7/16$	1.660	$55/64$
1	$1 5/8$	1.876	$39/64$	$1 5/8$	1.876	$63/64$
$1 1/8$	$1 13/16$	2.093	$11/16$	$1 13/16$	2.093	$1 7/64$
$1 1/4$	2	2.309	$25/32$	2	2.309	$1 7/32$
$1 3/8$	$2 3/16$	2.526	$27/32$	$2 3/16$	2.526	$1 11/32$

*From ANSI Standard B18.2.1–1972.
†From ANSI Standard B18.2.2–1972.

TABLE 2-5   DIMENSIONS OF HEXAGON-HEAD CAP SCREWS AND NUTS IN INCHES (*Continued*)

ANSI Standard Heavy

Size	Hexagon-head cap screws°			Hexagon nuts†		
	Width across flats	Width across corners (max)	Nominal height	Width across flats	Width across corners (max)	Nominal height
$1\frac{1}{2}$	$2\frac{3}{8}$	2.742	$^{15}/_{16}$	$2\frac{3}{8}$	2.742	$1^{15}/_{32}$
$1\frac{5}{8}$	. . . .	. . . .	. . . . .	$2^{9}/_{16}$	2.959	$1^{19}/_{32}$
$1\frac{3}{4}$	$2\frac{3}{4}$	3.175	$1^{3}/_{32}$	$2\frac{3}{4}$	3.175	$1^{23}/_{32}$
$1\frac{7}{8}$	. . . .	. . . .	. . . . .	$2^{15}/_{16}$	3.392	$1^{27}/_{32}$
2	$3\frac{1}{8}$	3.608	$1^{7}/_{32}$	$3\frac{1}{8}$	3.608	$1^{31}/_{32}$
$2\frac{1}{4}$	$3\frac{1}{2}$	4.041	$1\frac{3}{8}$	$3\frac{1}{2}$	4.041	$2^{13}/_{64}$
$2\frac{1}{2}$	$3\frac{7}{8}$	4.474	$1^{17}/_{32}$	$3\frac{7}{8}$	4.474	$2^{29}/_{64}$
$2\frac{3}{4}$	$4\frac{1}{4}$	4.907	$1^{11}/_{16}$	$4\frac{1}{4}$	4.907	$2^{45}/_{64}$
3	$4\frac{5}{8}$	5.340	$1\frac{7}{8}$	$4\frac{5}{8}$	5.340	$2^{16}/_{64}$
$3\frac{1}{4}$	. . . .	. . . .	. . . . .	5	5.774	$3^{3}/_{16}$
$3\frac{1}{2}$	. . . .	. . . .	. . . . .	$5\frac{3}{8}$	6.207	$3^{7}/_{16}$
$3\frac{3}{4}$	. . . .	. . . .	. . . . .	$5\frac{3}{4}$	6.640	$3^{11}/_{16}$
4	. . . .	. . . .	. . . . .	$6\frac{1}{8}$	7.073	$3^{15}/_{16}$

*From ANSI Standard B18.2.1-1972.
†From ANSI Standard B18.2.2-1972.

TABLE 2-6   ANSI STANDARD PIPE THREADS

Dimensions in inches.

Nominal pipe size	Pipe outside diameter	Threads per inch	Length of effective threads	Length of hand-tight engagement	Total thread length	Tap drill in cast iron*
1/16	0.3125	27	0.26	0.16	0.39	0.246
1/8	0.405	27	0.26	0.16	0.39	0.332
1/4	0.540	18	0.40	0.23	0.60	7/16
3/8	0.675	18	0.41	0.24	0.60	9/16
1/2	0.840	14	0.53	0.32	0.78	45/64
3/4	1.050	14	0.55	0.34	0.79	29/32
1	1.315	11½	0.68	0.40	0.99	1 5/64
1¼	1.660	11½	0.71	0.42	1.01	1 31/64
1½	1.900	11½	0.72	0.42	1.03	1 47/64
2	2.375	11½	0.76	0.44	1.06	2 13/64
2½	2.875	8	1.14	0.68	1.57	2 5/8
3	3.500	8	1.20	0.77	1.63	
3½	4.000	8	1.25	0.82	1.68	
4	4.500	8	1.30	0.84	1.73	
5	5.563	8	1.41	0.94	1.84	
6	6.625	8	1.51	0.96	1.95	
8	8.625	8	1.71	1.06	2.15	
10	10.750	8	1.93	1.21	2.36	
12	12.750	8	2.13	1.36	2.56	
14 OD	14.000	8	2.25	1.56	2.68	
16 OD	16.000	8	2.45	1.81	2.88	
18 OD	18.000	8	2.65	2.00	3.08	
20 OD	20.000	8	2.85	2.13	3.28	
24 OD	24.000	8	3.25	2.38	3.68	

Table abridged from ANSI Standard B2.1–1968.
*ANSI Standard twist drill sizes, without use of reamer.

TABLE 2-7   PROPERTIES OF WELDED AND SEAMLESS STEEL PIPE

Size, nominal and outside dia, in.	Identification Schedule No.	Identification Standard, X-strong, XX-strong	Wall thickness, in.	ID, in.	Inside area, sq in.	Wt/ft, lb	Wt of water, lb/ft	External surface, sq ft/ft
1/8	40	STD	0.068	0.269	0.0568	0.244	0.025	0.106
(0.405)	80	XS	0.095	0.215	0.0364	0.314	0.016	
1/4	40	STD	0.088	0.364	0.1041	0.424	0.045	0.141
(0.540)	80	XS	0.119	0.302	0.0716	0.535	0.031	
3/8	40	STD	0.091	0.493	0.1910	0.567	0.083	0.177
(0.675)	80	XS	0.126	0.423	0.1405	0.738	0.061	
1/2	40	STD	0.109	0.622	0.3040	0.850	0.132	0.220
(0.840)	80	XS	0.147	0.546	0.2340	1.087	0.101	
	160	. . . . . .	0.188	0.464	0.1691	1.311	0.073	
		XXS	0.294	0.252	0.0499	1.714	0.022	
3/4	40	STD	0.113	0.824	0.5330	1.130	0.230	0.275
(1.050)	80	XS	0.154	0.742	0.4330	1.473	0.187	
	160	. . . . . .	0.219	0.612	0.2942	1.944	0.127	
		XXS	0.308	0.434	0.1479	2.440	0.063	
1	40	STD	0.133	1.049	0.8640	1.678	0.374	0.344
(1.315)	80	XS	0.179	0.957	0.7190	2.171	0.311	
	160	. . . . . .	0.250	0.815	0.5217	2.840	0.226	
		XXS	0.358	0.599	0.2818	3.659	0.122	
1 1/4	40	STD	0.140	1.380	1.495	2.272	0.647	0.434
(1.660)	80	XS	0.191	1.278	1.283	2.996	0.555	
	160	. . . . . .	0.250	1.160	1.057	3.764	0.457	
		XXS	0.382	0.896	0.630	5.214	0.273	
1 1/2	40	STD	0.145	1.610	2.036	2.717	0.882	0.497
(1.900)	80	XS	0.200	1.500	1.767	3.631	0.765	
	160	. . . . . .	0.281	1.338	1.406	4.858	0.610	
		XXS	0.400	1.100	0.950	6.408	0.412	
2	40	STD	0.154	2.067	3.355	3.65	1.45	0.622
(2.375)	80	XS	0.218	1.939	2.953	5.02	1.28	
	160	. . . . . .	0.344	1.687	2.235	7.46	0.97	
		XXS	0.436	1.503	1.774	9.03	0.77	
2 1/2	40	STD	0.203	2.469	4.788	5.79	2.07	0.753
(2.875)	80	XS	0.276	2.323	4.238	7.66	1.83	
	160	. . . . . .	0.375	2.125	3.547	10.01	1.54	
		XXS	0.552	1.771	2.464	13.70	1.07	
3	40	STD	0.216	3.068	7.393	7.58	3.20	0.916
(3.500)	80	XS	0.300	2.900	6.605	10.25	2.86	
	160	. . . . . .	0.438	2.624	5.407	14.31	2.34	
		XXS	0.600	2.300	4.155	18.58	1.80	
3 1/2	40	STD	0.226	3.548	9.886	9.11	4.28	1.047
(4.000)	80	XS	0.318	3.364	8.888	12.51	3.85	
4	40	STD	0.237	4.026	12.730	10.79	5.51	1.178
(4.500)	80	XS	0.337	3.826	11.497	14.98	4.98	
	120	. . . . . .	0.438	3.624	10.315	18.98	4.47	
	160	. . . . . .	0.531	3.438	9.283	22.52	4.02	
		XXS	0.674	3.152	7.803	27.54	3.38	
5	40	STD	0.258	5.047	20.006	14.62	8.66	1.456
(5.563)	80	XS	0.375	4.813	18.194	20.78	7.87	
	120	. . . . . .	0.500	4.563	16.353	27.04	7.08	
	160	. . . . . .	0.625	4.313	14.610	32.96	6.32	
		XXS	0.750	4.063	12.966	38.55	5.62	

TABLE 2-7   PROPERTIES OF WELDED AND SEAMLESS STEEL PIPE *(Continued)*

Size, nominal and outside dia, in.	Identification		Wall thickness, in.	ID, in.	Inside area, sq in.	Wt/ft, lb	Wt of water, lb/ft	External surface, sq ft/ft
	Schedule No.	Standard, X-strong, XX-strong						
6	40	STD	0.280	6.065	28.89	18.97	12.5	1.734
(6.625)	80	XS	0.432	5.761	26.07	28.57	11.3	
	120	. . . . . .	0.562	5.501	23.77	36.42	10.3	
	160	. . . . . .	0.719	5.187	21.13	45.34	9.2	
		XXS	0.864	4.897	18.83	53.16	8.1	
8	20	. . . . . .	0.250	8.125	51.8	22.36	22.5	2.258
(8.625)	30	. . . . . .	0.277	8.071	51.2	24.70	22.2	
	40	STD	0.322	7.981	50.0	28.55	21.6	
	60	. . . . . .	0.406	7.813	47.9	35.66	20.8	
	80	XS	0.500	7.625	45.7	43.39	19.8	
	100	. . . . . .	0.594	7.437	43.4	50.93	18.8	
	120	. . . . . .	0.719	7.187	40.6	60.69	17.6	
	140	. . . . . .	0.812	7.001	38.5	67.79	16.7	
		XXS	0.875	6.875	37.1	72.42	16.1	
	160	. . . . . .	0.906	6.813	36.5	74.71	15.8	
10	20	. . . . . .	0.250	10.250	82.5	28.04	35.9	2.814
(10.750)	30	. . . . . .	0.307	10.136	80.7	34.24	35.0	
	40	STD	0.365	10.020	78.9	40.48	34.1	
	60	XS	0.500	9.750	74.7	54.74	32.3	
	80	. . . . . .	0.594	9.562	71.8	64.40	31.1	
	100	. . . . . .	0.719	9.312	68.1	77.00	29.5	
	120	. . . . . .	0.844	9.062	64.5	89.27	27.9	
	140	XXS	1.000	8.750	60.1	104.13	26.1	
	160	. . . . . .	1.125	8.500	56.7	115.65	24.6	
12	20	. . . . . .	0.250	12.250	118.0	33.38	51.3	3.338
(12.750)	30	. . . . . .	0.330	12.090	114.8	43.77	49.7	
		STD	0.375	12.000	113.1	49.56	48.9	
	40	. . . . . .	0.406	11.938	111.9	53.56	48.5	
		XS	0.500	11.750	108.4	65.42	46.9	
	60	. . . . . .	0.562	11.626	106.2	73.22	46.0	
	80	. . . . . .	0.688	11.374	101.6	88.57	44.0	
	100	. . . . . .	0.844	11.062	96.1	107.29	41.6	
	120	XXS	1.000	10.750	90.8	125.49	39.3	
	140	. . . . . .	1.125	10.500	86.6	139.68	37.5	
	160	. . . . . .	1.312	10.126	80.5	160.33	34.9	
14	10	. . . . . .	0.250	13.500	143.0	36.71	62.1	3.665
(14.000)	20	. . . . . .	0.312	13.376	140.5	45.68	60.9	
	30	STD	0.375	13.250	137.9	54.57	59.7	
	40	. . . . . .	0.438	13.124	135.3	63.37	58.5	
		XS	0.500	13.000	132.7	72.09	57.4	
	60	. . . . . .	0.594	12.812	128.9	85.01	55.8	
	80	. . . . . .	0.750	12.500	122.7	106.13	53.2	
	100	. . . . . .	0.938	12.124	115.4	130.79	50.0	
	120	. . . . . .	1.094	11.812	109.6	150.76	47.5	
	140	. . . . . .	1.250	11.500	103.9	170.22	45.0	
	160	. . . . . .	1.406	11.188	98.3	189.15	42.6	
16	10	. . . . . .	0.250	15.500	188.7	42.05	81.7	4.189
(16.000)	20	. . . . . .	0.312	15.376	185.7	52.36	80.4	
	30	STD	0.375	15.250	182.6	62.58	79.1	
	40	XS	0.500	15.000	176.7	82.77	76.5	
	60	. . . . . .	0.656	14.688	169.4	107.54	73.4	
	80	. . . . . .	0.844	14.312	160.9	136.58	69.7	
	100	. . . . . .	1.031	13.938	152.6	164.86	66.0	
	120	. . . . . .	1.219	13.562	144.5	192.40	62.6	
	140	. . . . . .	1.438	13.124	135.3	223.57	58.6	
	160	. . . . . .	1.594	12.812	129.0	245.22	55.8	

TABLE 2-7 PROPERTIES OF WELDED AND SEAMLESS STEEL PIPE (Continued)

| Size, nominal and outside dia, in. | Identification | | Wall thickness, in. | ID, in. | Inside area, sq in. | Wt/ft, lb | Wt of water, lb/ft | External surface, sq ft/ft |
	Schedule No.	Standard, X-strong, XX-strong						
18	10	......	0.250	17.500	241.0	47.39	104.6	4.712
(18.000)	20	......	0.312	17.376	237.1	59.03	102.7	
		STD	0.375	17.250	233.7	70.59	101.2	
	30	......	0.438	17.124	229.5	82.06	99.5	
		XS	0.500	17.000	227.0	93.45	98.2	
	40	......	0.562	16.876	224.0	104.76	97.2	
	60	......	0.750	16.500	213.8	138.17	92.5	
	80	......	0.938	16.124	204.2	170.84	88.4	
	100	......	1.156	15.688	193.3	208.00	83.7	
	120	......	1.375	15.250	182.7	244.14	79.2	
	140	......	1.562	14.876	173.8	274.30	75.3	
	160	......	1.781	14.438	163.7	308.55	71.0	
20	10	......	0.250	19.500	299.0	52.73	130.0	5.236
(20.000)	20	STD	0.375	19.250	291.1	78.60	126.0	
	30	XS	0.500	19.000	283.5	104.13	122.8	
	40	......	0.594	18.812	277.9	123.06	120.4	
	60	......	0.812	18.376	265.2	166.50	114.9	
	80	......	1.031	17.938	252.7	208.92	109.4	
	100	......	1.281	17.438	238.8	256.15	103.4	
	120	......	1.500	17.000	227.0	296.37	98.3	
	140	......	1.750	16.500	213.8	341.10	92.6	
	160	......	1.969	16.062	202.6	379.14	87.8	
22	10	......	0.250	21.500	363.1	58.07	157.4	5.760
(22.000)	20	STD	0.375	21.250	354.7	86.61	153.7	
	30	XS	0.500	21.000	346.4	114.81	150.2	
	60	......	0.875	20.250	322.1	197.42	139.6	
	80	......	1.125	19.750	306.4	250.82	132.8	
	100	......	1.375	19.250	291.0	302.88	126.2	
	120	......	1.625	18.750	276.1	353.61	119.6	
	140	......	1.875	18.250	261.6	403.01	113.3	
	160	......	2.125	17.750	247.4	451.07	107.2	
24	10	......	0.250	23.500	435.0	63.41	187.9	6.283
(24.000)	20	STD	0.375	23.250	424.6	94.62	183.9	
		XS	0.500	23.000	416.0	125.49	180.0	
	30	......	0.562	22.876	411.0	140.80	178.0	
	40	......	0.688	22.624	402.0	171.17	174.1	
	60	......	0.969	22.062	382.3	238.29	165.6	
	80	......	1.219	21.562	365.2	296.53	158.2	
	100	......	1.531	20.938	344.3	367.45	149.3	
	120	......	1.812	20.376	326.1	429.50	141.4	
	140	......	2.062	19.876	310.3	483.24	134.4	
	160	......	2.344	19.312	292.9	542.09	126.9	
26	10	......	0.312	25.376	505.8	85.73	219.2	6.807
(26.000)		STD	0.375	25.250	500.7	102.63	217.1	
	20	XS	0.500	25.000	490.9	136.17	212.8	
28	10	......	0.312	27.376	588.6	92.41	255.0	7.330
(28.000)		STD	0.375	27.250	583.2	110.64	252.6	
	20	XS	0.500	27.000	572.6	146.85	248.0	
	30	......	0.625	26.750	562.0	182.73	243.4	
30	10	......	0.312	29.376	677.8	99.08	293.7	7.854
(30.000)		STD	0.375	29.250	672.0	118.65	291.2	
	20	XS	0.500	29.000	660.5	157.53	286.2	
	30	......	0.625	28.750	649.2	196.08	281.3	
32	10	......	0.312	31.376	773.2	105.76	335.2	8.378
(32.000)		STD	0.375	31.250	766.9	126.66	332.5	

TABLE 2-7 PROPERTIES OF WELDED AND SEAMLESS STEEL PIPE *(Continued)*

Size, nominal and outside dia, in.	Schedule No.	Standard, X-strong, XX-strong	Wall thickness, in.	ID, in.	Inside area, sq in.	Wt/ft, lb	Wt of water, lb/ft	External surface, sq ft/ft
	20	XS	0.500	31.000	754.7	168.21	327.2	
	30	......	0.625	30.750	742.5	209.43	321.9	
	40	......	0.688	30.624	736.6	229.92	319.0	
34	10	......	0.312	33.376	874.9	112.43	379.3	8.901
(34.000)		STD	0.375	33.250	868.3	134.67	376.2	
	20	XS	0.500	33.000	855.3	178.89	370.8	
	30	......	0.625	32.750	842.4	222.78	365.0	
	40	......	0.688	32.624	835.9	244.60	362.1	
36	10	......	0.312	35.376	982.9	119.11	426.1	9.425
(36.000)		STD	0.375	35.250	975.8	142.68	423.1	
	20	XS	0.500	35.000	962.1	189.57	417.1	
	30	......	0.625	34.750	948.3	236.13	411.1	
	40	......	0.750	34.500	934.7	282.36	405.3	

NOTE: Two systems of rating pipe wall thickness are utilized. The newer schedule numbers correspond to definite pressure-stress ratios and are expressed simply as follows:

$$\text{Schedule no.} = 1000 \times \frac{p}{s}$$

where p = internal pressure, psig
s = allowable fiber stress, lb/in^2

Properties of the traditional designation for pipe entitled *standard, extra strong,* and *double extra strong* are also shown in the tables.

TABLE 2-8 ANSI STANDARD STEEL PIPE FLANGES

All dimensions in inches.

Nominal pipe size	Flange OD	Flange thickness	Bolt circle dia	Bore weld neck socket weld‡	No. of bolts	Bolt dia
			150-lb Standard°			
1/2	3 1/2	7/16	2 3/8	0.62	4	1/2
3/4	3 7/8	1/2	2 3/4	0.82	4	1/2
1	4 1/4	9/16	3 1/8	1.05	4	1/2
1 1/4	4 5/8	5/8	3 1/2	1.38	4	1/2
1 1/2	5	11/16	3 7/8	1.61	4	1/2
2	6	3/4	4 3/4	2.07	4	5/8
2 1/2	7	7/8	5 1/2	2.47	4	5/8
3	7 1/2	15/16	6	3.07	4	5/8
3 1/2	8 1/2	15/16	7	3.55	8	5/8
4	9	15/16	7 1/2	4.03	8	5/8
5	10	15/16	8 1/2	5.05	8	3/4
6	11	1	9 1/2	6.07	8	3/4
8	13 1/2	1 1/8	11 3/4	7.98	8	3/4
10	16	1 3/16	14 1/4	10.02	12	7/8
12	19	1 1/4	17	12.00	12	7/8
14 OD	21	1 3/8	18 3/4	†	12	1
16 OD	23 1/2	1 7/16	21 1/4	†	16	1
18 OD	25	1 9/16	22 3/4	†	16	1 1/8
20 OD	27 1/2	1 11/16	25	†	20	1 1/8
24 OD	32	1 7/8	29 1/2	†	20	1 1/4

For footnotes, see end of table.

TABLE 2-8 ANSI STANDARD STEEL PIPE FLANGES
(Continued)

Nominal pipe size	Flange OD	Flange thickness	Bolt circle dia	Bore weld neck socket weld‡	No. of bolts	Bolt dia
			300-lb Standard°			
1/2	3³/₄	⁹/₁₆	2⁵/₈	0.62	4	1/2
3/4	4⁵/₈	5/8	3¹/₄	0.82	4	5/8
1	4⁷/₈	¹¹/₁₆	3¹/₂	1.05	4	5/8
1¹/₄	5¹/₄	3/4	3⁷/₈	1.38	4	5/8
1¹/₂	6¹/₈	¹³/₁₆	4¹/₂	1.61	4	3/4
2	6¹/₂	7/8	5	2.07	8	5/8
2¹/₂	7¹/₂	1	5⁷/₈	2.47	8	3/4
3	8¹/₄	1¹/₈	6⁵/₈	3.07	8	3/4
3¹/₂	9	1³/₁₆	7¹/₄	3.55	8	3/4
4	10	1¹/₄	7⁷/₈	4.03	8	3/4
5	11	1³/₈	9¹/₄	5.05	8	3/4
6	12¹/₂	1⁷/₁₆	10⁵/₈	6.07	12	3/4
8	15	1⁵/₈	13	7.98	12	7/8
10	17¹/₂	1⁷/₈	15¹/₄	10.02	16	1
12	20¹/₂	2	17³/₄	12.00	16	1¹/₈
14 OD	23	2¹/₈	20¹/₄	†	20	1¹/₈
16 OD	25¹/₂	2¹/₄	22¹/₂	†	20	1¹/₄
18 OD	28	2³/₈	24³/₄	†	24	1¹/₄
20 OD	30¹/₂	2¹/₂	27	†	24	1¹/₄
24 OD	36	2³/₄	32	†	24	1¹/₂
			400-lb Standard§			
1/2	3³/₄	⁹/₁₆	2⁵/₈	†	4	1/2
3/4	4⁵/₈	5/8	3¹/₄	†	4	5/8
1	4⁷/₈	¹¹/₁₆	3¹/₂	†	4	5/8
1¹/₄	5¹/₄	¹³/₁₆	3⁷/₈	†	4	5/8
1¹/₂	6¹/₈	7/8	4¹/₂	†	4	3/4
2	6¹/₂	1	5	†	8	5/8
2¹/₂	7¹/₂	1¹/₈	5⁷/₈	†	8	3/4
3	8¹/₄	1¹/₄	6⁵/₈	†	8	3/4
3¹/₂	9	1³/₈	7¹/₄	†	8	7/8
4	10	1³/₈	7⁷/₈	†	8	7/8
5	11	1¹/₂	9¹/₄	†	8	7/8
6	12¹/₂	1⁵/₈	10⁵/₈	†	12	7/8
8	15	1⁷/₈	13	†	12	1
10	17¹/₂	2¹/₈	15¹/₄	†	16	1¹/₈
12	20¹/₂	2¹/₄	17³/₄	†	16	1¹/₄
14 OD	23	2³/₈	20¹/₄	†	20	1¹/₄
16 OD	25¹/₂	2¹/₂	22¹/₂	†	20	1³/₈
18 OD	28	2⁵/₈	24³/₄	†	24	1³/₈
20 OD	30¹/₂	2³/₄	27	†	24	1¹/₂
24 OD	36	3	32	†	24	1³/₄

TABLE 2-8 ANSI STANDARD STEEL PIPE FLANGES
(Continued)

Nominal pipe size	Flange OD	Flange thickness	Bolt circle dia	Bore weld neck socket weld‡	No. of bolts	Bolt dia
			600-lb Standard§			
$1/2$	$3^3/4$	$9/16$	$2^5/8$	†	4	$1/2$
$3/4$	$4^5/8$	$5/8$	$3^1/4$	†	4	$5/8$
1	$4^7/8$	$11/16$	$3^1/2$	†	4	$5/8$
$1^1/4$	$5^1/4$	$13/16$	$3^7/8$	†	4	$5/8$
$1^1/2$	$6^1/8$	$7/8$	$4^1/2$	†	4	$3/4$
2	$6^1/2$	1	5	†	8	$5/8$
$2^1/2$	$7^1/2$	$1^1/8$	$5^7/8$	†	8	$3/4$
3	$8^1/4$	$1^1/4$	$6^5/8$	†	8	$3/4$
$3^1/2$	9	$1^3/8$	$7^1/4$	†	8	$7/8$
4	$10^3/4$	$1^1/2$	$8^1/2$	†	8	$7/8$
5	13	$1^3/4$	$10^1/2$	†	8	1
6	14	$1^7/8$	$11^1/2$	†	12	1
8	$16^1/2$	$2^3/16$	$13^3/4$	†	12	$1^1/8$
10	20	$2^1/2$	17	†	16	$1^1/4$
12	22	$2^5/8$	$19^1/4$	†	20	$1^1/4$
14 OD	$23^3/4$	$2^3/4$	$20^3/4$	†	20	$1^3/8$
16 OD	27	3	$23^3/4$	†	20	$1^1/2$
18 OD	$29^1/4$	$3^1/4$	$25^3/4$	†	20	$1^5/8$
20 OD	32	$3^1/2$	$28^1/2$	†	24	$1^5/8$
24 OD	37	4	33	†	24	$1^7/8$
			900-lb Standard§			
$1/2$	$4^3/4$	$7/8$	$3^1/4$	†	4	$3/4$
$3/4$	$5^1/8$	1	$3^1/2$	†	4	$3/4$
1	$5^7/8$	$1^1/8$	4	†	4	$7/8$
$1^1/4$	$6^1/4$	$1^1/8$	$4^3/8$	†	4	$7/8$
$1^1/2$	7	$1^1/4$	$4^7/8$	†	4	1
2	$8^1/2$	$1^1/2$	$6^1/2$	†	8	$7/8$
$2^1/2$	$9^5/8$	$1^5/8$	$7^1/2$	†	8	1
3	$9^1/2$	$1^1/2$	$7^1/2$	†	8	$7/8$
4	$11^1/2$	$1^3/4$	$9^1/4$	†	8	$1^1/8$
5	$13^3/4$	2	11	†	8	$1^1/4$
6	15	$2^3/16$	$12^1/2$	†	12	$1^1/8$
8	$18^1/2$	$2^1/2$	$15^1/2$	†	12	$1^3/8$
10	$21^1/2$	$2^3/4$	$18^1/2$	†	16	$1^3/8$
12	24	$3^1/8$	21	†	20	$1^3/8$
14 OD	$25^1/4$	$3^3/8$	22	†	20	$1^1/2$
16 OD	$27^3/4$	$3^1/2$	$24^1/4$	†	20	$1^5/8$
18 OD	31	4	27	†	20	$1^7/8$
20 OD	$33^3/4$	$4^1/4$	$29^1/2$	†	20	2
24 OD	41	$5^1/2$	$35^1/2$	†	20	$2^1/2$

TABLE 2-8 ANSI STANDARD STEEL PIPE FLANGES
(Continued)

Nominal pipe size	Flange OD	Flange thickness	Bolt circle dia	Bore weld neck socket weld	No. of bolts	Bolt dia
			1,500-lb Standard§			
$\frac{1}{2}$	$4\frac{3}{4}$	$\frac{7}{8}$	$3\frac{1}{4}$	†	4	$\frac{3}{4}$
$\frac{3}{4}$	$5\frac{1}{8}$	1	$3\frac{1}{2}$	†	4	$\frac{3}{4}$
1	$5\frac{7}{8}$	$1\frac{1}{8}$	4	†	4	$\frac{7}{8}$
$1\frac{1}{4}$	$6\frac{1}{4}$	$1\frac{1}{8}$	$4\frac{3}{8}$	†	4	$\frac{7}{8}$
$1\frac{1}{2}$	7	$1\frac{1}{4}$	$4\frac{7}{8}$	†	4	1
2	$8\frac{1}{2}$	$1\frac{1}{2}$	$6\frac{1}{2}$	†	8	$\frac{7}{8}$
$2\frac{1}{2}$	$9\frac{5}{8}$	$1\frac{5}{8}$	$7\frac{1}{2}$	†	8	1
3	$10\frac{1}{2}$	$1\frac{7}{8}$	8	†	8	$1\frac{1}{8}$
4	$12\frac{1}{4}$	$2\frac{1}{8}$	$9\frac{1}{2}$	†	8	$1\frac{1}{4}$
5	$14\frac{3}{4}$	$2\frac{7}{8}$	$11\frac{1}{2}$	†	8	$1\frac{1}{2}$
6	$15\frac{1}{2}$	$3\frac{1}{4}$	$12\frac{1}{2}$	†	12	$1\frac{3}{8}$
8	19	$3\frac{5}{8}$	$15\frac{1}{2}$	†	12	$1\frac{5}{8}$
10	23	$4\frac{1}{4}$	19	†	12	$1\frac{7}{8}$
12	$26\frac{1}{2}$	$4\frac{7}{8}$	$22\frac{1}{2}$	†	16	2
14 OD	$29\frac{1}{2}$	$5\frac{1}{4}$	25	†	16	$2\frac{1}{4}$
16 OD	$32\frac{1}{2}$	$5\frac{3}{4}$	$27\frac{3}{4}$	†	16	$2\frac{1}{2}$
18 OD	36	$6\frac{3}{8}$	$30\frac{1}{2}$	†	16	$2\frac{3}{4}$
20 OD	$38\frac{3}{4}$	7	$32\frac{3}{4}$	†	16	3
24 OD	46	8	39	†	16	$3\frac{1}{2}$
			2,500-lb Standard§			
$\frac{1}{2}$	$5\frac{1}{4}$	$1\frac{3}{16}$	$3\frac{1}{2}$	†	4	$\frac{3}{4}$
$\frac{3}{4}$	$5\frac{1}{2}$	$1\frac{1}{4}$	$3\frac{3}{4}$	†	4	$\frac{3}{4}$
1	$6\frac{1}{4}$	$1\frac{3}{8}$	$4\frac{1}{4}$	†	4	$\frac{7}{8}$
$1\frac{1}{4}$	$7\frac{1}{4}$	$1\frac{1}{2}$	$5\frac{1}{8}$	†	4	1
$1\frac{1}{2}$	8	$1\frac{3}{4}$	$5\frac{3}{4}$	†	4	$1\frac{1}{8}$
2	$9\frac{1}{4}$	2	$6\frac{3}{4}$	†	8	1
$2\frac{1}{2}$	$10\frac{1}{2}$	$2\frac{1}{4}$	$7\frac{3}{4}$	†	8	$1\frac{1}{8}$
3	12	$2\frac{5}{8}$	9	†	8	$1\frac{1}{4}$
4	14	3	$10\frac{3}{4}$	†	8	$1\frac{1}{2}$
5	$16\frac{1}{2}$	$3\frac{5}{8}$	$12\frac{3}{4}$	†	8	$1\frac{3}{4}$
6	19	$4\frac{1}{4}$	$14\frac{1}{2}$	†	8	2
8	$21\frac{3}{4}$	5	$17\frac{1}{4}$	†	12	2
10	$26\frac{1}{2}$	$6\frac{1}{2}$	$21\frac{1}{4}$	†	12	$2\frac{1}{2}$
12	30	$7\frac{1}{4}$	$24\frac{3}{8}$	†	12	$2\frac{3}{4}$

From ANSI Standard *Steel Pipe Flanges and Flanged Fittings*, B16.5–1977.
*Flange thickness includes $\frac{1}{16}$-in raised face.
†To be specified by purchaser.
‡Socket-weld flanges not available at 400-lb rating.
§Flange thickness does not include $\frac{1}{4}$-in raised face.

TABLE 2-9 ANSI STANDARD CAST-IRON PIPE FLANGES

All dimensions in inches.

Nominal pipe size	Flange OD	Flange thickness	Bolt circle dia	No. of bolts	Bolt dia
125-lb Standard					
1	$4\frac{1}{4}$	$\frac{7}{16}$	$3\frac{1}{8}$	4	$\frac{1}{2}$
$1\frac{1}{4}$	$4\frac{5}{8}$	$\frac{1}{2}$	$3\frac{1}{2}$	4	$\frac{1}{2}$
$1\frac{1}{2}$	5	$\frac{9}{16}$	$3\frac{7}{8}$	4	$\frac{1}{2}$
2	6	$\frac{5}{8}$	$4\frac{3}{4}$	4	$\frac{5}{8}$
$2\frac{1}{2}$	7	$\frac{11}{16}$	$5\frac{1}{2}$	4	$\frac{5}{8}$
3	$7\frac{1}{2}$	$\frac{3}{4}$	6	4	$\frac{5}{8}$
$3\frac{1}{2}$	$8\frac{1}{2}$	$\frac{13}{16}$	7	8	$\frac{5}{8}$
4	9	$\frac{15}{16}$	$7\frac{1}{2}$	8	$\frac{5}{8}$
5	10	$\frac{15}{16}$	$8\frac{1}{2}$	8	$\frac{3}{4}$
6	11	1	$9\frac{1}{2}$	8	$\frac{3}{4}$
8	$13\frac{1}{2}$	$1\frac{1}{8}$	$11\frac{3}{4}$	8	$\frac{3}{4}$
10	16	$1\frac{3}{16}$	$14\frac{1}{4}$	12	$\frac{7}{8}$
12	19	$1\frac{1}{4}$	17	12	$\frac{7}{8}$
14 OD	21	$1\frac{3}{8}$	$18\frac{3}{4}$	12	1
16 OD	$23\frac{1}{2}$	$1\frac{7}{16}$	$21\frac{1}{4}$	16	1
18 OD	25	$1\frac{9}{16}$	$22\frac{3}{4}$	16	$1\frac{1}{8}$
20 OD	$27\frac{1}{2}$	$1\frac{11}{16}$	25	20	$1\frac{1}{8}$
24 OD	32	$1\frac{7}{8}$	$29\frac{1}{2}$	20	$1\frac{1}{4}$
30 OD	$38\frac{3}{4}$	$2\frac{1}{8}$	36	28	$1\frac{1}{4}$
36 OD	46	$2\frac{3}{8}$	$42\frac{3}{4}$	32	$1\frac{1}{2}$
42 OD	53	$2\frac{5}{8}$	$49\frac{1}{2}$	36	$1\frac{1}{2}$
48 OD	$59\frac{1}{2}$	$2\frac{3}{4}$	56	44	$1\frac{1}{2}$
54 OD	$66\frac{1}{4}$	3	$62\frac{3}{4}$	44	$1\frac{3}{4}$
60 OD	73	$3\frac{1}{8}$	$69\frac{1}{4}$	52	$1\frac{3}{4}$
250-lb Standard°					
1	$4\frac{7}{8}$	$\frac{11}{16}$	$3\frac{1}{2}$	4	$\frac{5}{8}$
$1\frac{1}{4}$	$5\frac{1}{4}$	$\frac{3}{4}$	$3\frac{7}{8}$	4	$\frac{5}{8}$
$1\frac{1}{2}$	$6\frac{1}{8}$	$\frac{13}{16}$	$4\frac{1}{2}$	4	$\frac{3}{4}$
2	$6\frac{1}{2}$	$\frac{7}{8}$	5	8	$\frac{5}{8}$
$2\frac{1}{2}$	$7\frac{1}{2}$	1	$5\frac{7}{8}$	8	$\frac{3}{4}$
3	$8\frac{1}{4}$	$1\frac{1}{8}$	$6\frac{5}{8}$	8	$\frac{3}{4}$
$3\frac{1}{2}$	9	$1\frac{3}{16}$	$7\frac{1}{4}$	8	$\frac{3}{4}$
4	10	$1\frac{1}{4}$	$7\frac{7}{8}$	8	$\frac{3}{4}$
5	11	$1\frac{3}{8}$	$9\frac{1}{4}$	8	$\frac{3}{4}$
6	$12\frac{1}{2}$	$1\frac{7}{16}$	$10\frac{5}{8}$	12	$\frac{3}{4}$
8	15	$1\frac{5}{8}$	13	12	$\frac{7}{8}$
10	$17\frac{1}{2}$	$1\frac{7}{8}$	$15\frac{1}{4}$	16	1
12	$20\frac{1}{2}$	2	$17\frac{3}{4}$	16	$1\frac{1}{8}$
14 OD	23	$2\frac{1}{8}$	$20\frac{1}{4}$	20	$1\frac{1}{8}$
16 OD	$25\frac{1}{2}$	$2\frac{1}{4}$	$22\frac{1}{2}$	20	$1\frac{1}{4}$
18 OD	28	$2\frac{3}{8}$	$24\frac{3}{4}$	24	$1\frac{1}{4}$
20 OD	$30\frac{1}{2}$	$2\frac{1}{2}$	27	24	$1\frac{1}{4}$
24 OD	36	$2\frac{3}{4}$	32	24	$1\frac{1}{2}$
30 OD	43	3	$39\frac{1}{4}$	28	$1\frac{3}{4}$
36 OD	50	$3\frac{3}{8}$	46	32	2
42 OD	57	$3\frac{11}{16}$	$52\frac{3}{4}$	36	2
48 OD	65	4	$60\frac{3}{4}$	40	2

From ANSI Standard *Cast Iron Pipe Flanges and Flange Fittings*, B16.1-1975.

*Flange thickness includes $\frac{1}{16}$-in raised face.

TABLE 2-10 CAST-IRON PIPE DIMENSIONS

Pipe size, in.	OD, in.	Internal pressure, psi																				
		50			100			150			200			250			300			350		
		Thickness class	Wall thickness, in.	ID, in.	Thickness class	Wall thickness, in.	ID, in.	Thickness class	Wall thickness, in.	ID, in.	Thickness class	Wall thickness, in.	ID, in.	Thickness class	Wall thickness, in.	ID, in.	Thickness class	Wall thickness, in.	ID, in.	Thickness class	Wall thickness, in.	ID, in.
3	3.96	22	0.32	3.32	22	0.32	3.32	22	0.32	3.32	22	0.32	3.32	22	0.32	3.32	22	0.32	3.32	22	0.32	3.32
4	4.80	22	0.35	4.10	22	0.35	4.10	22	0.35	4.10	22	0.35	4.10	22	0.35	4.10	22	0.35	4.10	22	0.35	4.10
6	6.90	22	0.38	6.14	22	0.38	6.14	22	0.38	6.14	22	0.38	6.14	22	0.38	6.14	22	0.38	6.14	22	0.38	6.14
8	9.05	22	0.41	8.23	22	0.41	8.23	22	0.41	8.23	22	0.41	8.23	22	0.41	8.23	22	0.41	8.23	22	0.41	8.23
10	11.10	22	0.44	10.22	22	0.44	10.22	22	0.44	10.22	22	0.44	10.22	22	0.44	10.22	23	0.48	10.14	24	0.52	10.06
12	13.20	22	0.48	12.24	22	0.48	12.24	22	0.48	12.24	22	0.48	12.24	23	0.52	12.16	23	0.52	12.16	24	0.56	12.08
14	15.30	21	0.48	14.39	22	0.51	14.28	22	0.51	14.28	23	0.55	14.20	24	0.59	14.12	24	0.59	14.12	25	0.64	14.02
16	17.40	22	0.54	16.32	22	0.54	16.32	22	0.54	16.32	23	0.58	16.24	24	0.63	16.14	25	0.68	16.04	25	0.68	16.04
18	19.50	21	0.54	18.42	22	0.58	18.34	22	0.58	18.34	23	0.63	18.24	24	0.68	18.14	25	0.73	18.04	26	0.79	17.92
20	21.60	21	0.57	20.46	22	0.62	20.36	22	0.62	20.36	23	0.67	20.26	24	0.72	20.16	25	0.78	20.04	26	0.84	19.92
24	25.80	21	0.63	24.54	22	0.68	24.44	23	0.73	24.34	24	0.79	24.22	24	0.79	24.22	25	0.85	24.10	26	0.92	23.96
30	32.00	22	0.79	30.42	22	0.79	30.42	23	0.85	30.30	24	0.92	30.16	25	0.99	30.02	26	1.07	29.86	27	1.16	29.68
36	38.30	22	0.87	36.56	22	0.87	36.56	23	0.94	36.42	24	1.02	36.26	25	1.10	36.10	26	1.19	35.92	27	1.29	35.72
42	44.50	22	0.97	42.56	22	0.97	42.56	23	1.05	42.40	24	1.13	42.24	25	1.22	42.06	26	1.32	41.86	27	1.43	41.64
48	50.80	22	1.06	48.68	22	1.06	48.68	23	1.14	48.52	24	1.23	48.34	25	1.33	48.14	27	1.56	47.68	28	1.68	47.44

This table gives the dimensions of cast-iron pipe, suitable for pressures from 50 to 350 lb/in², taken from specifications ANSI A21.6–1975 (AWWA H1-67) and applying to the following conditions:

1. Pipe laid in flat-bottom trench, backfill tamped
2. Depth of cover, 5 ft
3. Iron strength of 18/40, or iron having a bursting strength of 18,000 lb/in² and a ring modulus of rupture of 40,000 lb/in², centrifugally cast

The thickness class given in the table is a manufacturer's designation for standard wall thickness.
For conditions of installation other than those outlined in the table, the subject specification must be consulted.

Fluids Engineering Data

TABLE 3-1 PROPERTIES OF WATER AT VARIOUS TEMPERATURES*

Temp, °F	Temp, °C	Specific volume, cu ft/lb	Specific weight, lb/cu ft	Specific gravity†	Vapor pressure, psia	Vapor pressure, in. Hg abs
32	0.0	0.01602	62.420	1.0016	0.0886	0.1804
33	0.6	0.01602	62.423	1.0017	0.0922	0.1878
34	1.1	0.01602	62.423	1.0017	0.0960	0.1955
35	1.7	0.01602	62.423	1.0017	0.0999	0.2034
36	2.2	0.01602	62.423	1.0017	0.1040	0.2118
37	2.8	0.01602	62.428	1.0018	0.1082	0.2203
38	3.3	0.01602	62.428	1.0018	0.1125	0.2290
39	3.9	0.01602	62.428	1.0018	0.1170	0.2382
40	4.4	0.01602	62.428	1.0018	0.1216	0.2476
41	5.0	0.01602	62.428	1.0018	0.1264	0.2574
42	5.6	0.01602	62.427	1.0018	0.1314	0.2675
43	6.1	0.01602	62.426	1.0017	0.1366	0.2781
44	6.7	0.01602	62.424	1.0017	0.1419	0.2889
45	7.2	0.01602	62.423	1.0017	0.1474	0.3001
46	7.8	0.01602	62.422	1.0017	0.1531	0.3117
47	8.3	0.01602	62.417	1.0016	0.1590	0.3237
48	8.9	0.01602	62.417	1.0016	0.1651	0.3361
49	9.4	0.01602	62.415	1.0016	0.1714	0.3490
50	10.0	0.01602	62.412	1.0015	0.1780	0.3624
51	10.6	0.01602	62.408	1.0014	0.1847	0.3760
52	11.1	0.01602	62.405	1.0014	0.1916	0.3900
53	11.7	0.01602	62.400	1.0013	0.1988	0.4048
54	12.2	0.01603	62.397	1.0013	0.2062	0.4196
55	12.8	0.01603	62.392	1.0012	0.2139	0.4355
56	13.3	0.01603	62.389	1.0011	0.2218	0.4516
57	13.9	0.01603	62.383	1.0010	0.2300	0.4682
58	14.4	0.01603	62.379	1.0010	0.2384	0.4853
59	15.0	0.01603	62.374	1.0009	0.2471	0.5031
60	15.6	0.01603	62.368	1.0008	0.2561	0.5214
62	16.7	0.01604	62.357	1.0006	0.2749	0.5517
64	17.8	0.01604	62.345	1.0004	0.2950	0.6006
66	18.9	0.01604	62.331	1.0002	0.3163	0.6439
68	20.0	0.01605	62.318	1.0000	0.3389	0.6900
70	21.1	0.01605	62.304	0.9998	0.3629	0.7388
75	23.9	0.01606	62.263	0.9991	0.4296	0.8747
80	26.7	0.01607	62.217	0.9984	0.5068	1.0318
85	29.4	0.01608	62.169	0.9976	0.5958	1.2130
90	32.2	0.01610	62.116	0.9968	0.6981	1.4213
95	35.0	0.01611	62.058	0.9958	0.8153	1.6600
100	37.8	0.01613	62.00	0.9949	0.9492	1.9325
110	43.3	0.01616	61.86	0.9927	1.2750	2.5959
120	48.9	0.01620	61.71	0.9903	1.6927	3.4463
130	54.4	0.01625	61.56	0.9878	2.2230	4.5260
140	60.0	0.01629	61.38	0.9850	2.8892	5.8824
150	65.6	0.01634	61.20	0.9821	3.7184	7.570
160	71.1	0.01640	61.01	0.9790	4.7414	9.653
170	76.7	0.01645	60.79	0.9755	5.9926	12.200
180	82.2	0.01651	60.57	0.9720	7.5110	15.292
190	87.8	0.01657	60.35	0.9684	9.3400	19.016
200	93.3	0.01664	60.13	0.9649	11.526	23.467

TABLE 3-1 PROPERTIES OF WATER AT VARIOUS TEMPERATURES *(Continued)*

Temp, °F	Temp, °C	Specific volume, cu ft/lb	Specific weight, lb/cu ft	Specific gravity†	Vapor pressure, psia	Vapor pressure, in. Hg abs
210	98.9	0.01670	59.88	0.9609	14.123	
220	104.4	0.01678	59.63	0.9569	17.186	
230	110.0	0.01685	59.38	0.9529	20.779	
240	115.6	0.01693	59.10	0.9484	24.968	
250	121.1	0.01701	58.82	0.9439	29.825	
260	126.7	0.01709	58.51	0.9389	35.427	
270	132.2	0.01718	58.24	0.9346	41.856	
280	137.8	0.01726	57.94	0.9297	49.200	
290	143.3	0.01736	57.64	0.9249	57.550	
300	148.9	0.01745	57.31	0.9196	67.005	
310	154.4	0.01755	56.98	0.9143	77.667	
320	160.0	0.01766	56.66	0.9092	89.643	
330	165.6	0.01776	56.31	0.9036	103.045	
340	171.1	0.01787	55.96	0.8980	117.992	
350	176.7	0.01799	55.59	0.8920	134.604	
360	182.2	0.01811	55.22	0.8861	153.010	
370	187.8	0.01823	54.85	0.8802	173.339	
380	193.3	0.01836	54.47	0.8741	195.729	
390	198.9	0.01850	54.05	0.8673	220.321	
400	204.4	0.01864	53.65	0.8609	247.259	
410	210.0	0.01878	53.25	0.8545	276.694	
420	215.6	0.01894	52.80	0.8473	308.780	
430	221.1	0.01909	52.36	0.8402	343.674	
440	226.7	0.01926	51.92	0.8332	381.54	
450	232.2	0.01943	51.5	0.826	422.55	
460	237.8	0.01961	51.0	0.818	466.87	
470	243.3	0.01980	50.5	0.810	514.67	
480	249.9	0.02000	50.0	0.802	566.15	
490	254.4	0.02021	49.5	0.794	621.48	
500	260.0	0.02043	49.0	0.786	680.86	
510	265.6	0.02067	48.3	0.775	744.47	
520	271.1	0.02091	47.8	0.767	812.53	
530	276.7	0.02118	47.2	0.757	885.23	
540	282.2	0.02146	46.5	0.746	962.79	
550	287.8	0.02176	45.9	0.737	1,045.43	
560	293.3	0.02207	45.2	0.725	1,133.38	
570	298.9	0.02242	44.6	0.716	1,226.88	
580	304.4	0.02279	43.9	0.704	1,326.17	
590	310.0	0.02319	43.1	0.692	1,431.5	
600	315.6	0.02364	42.4	0.680	1,543.2	
610	321.1	0.02412	41.5	0.666	1,661.6	
620	326.7	0.02466	40.5	0.650	1,786.9	
630	332.2	0.02526	39.5	0.634	1,919.5	
640	337.8	0.02595	38.5	0.618	2,059.9	
650	343.3	0.02674	37.3	0.599	2,208.4	
660	348.9	0.02768	36.0	0.578	2,365.7	
670	354.4	0.02884	34.5	0.554	2,532.2	
680	360.0	0.03037	32.8	0.526	2,708.6	
690	365.6	0.03256	30.5	0.489	2,895.7	
705.47	374.1	0.05078	19.9	0.319	3,208.2	

*Computed by permission from *1967 ASME Steam Tables,* American Society of Mechanical Engineers, New York, 1967.
†Referred to water at 68°F, weighing 62.318 lb/ft³.

TABLE 3-2 CONVERSION OF POUNDS PER SQUARE INCH TO HEAD IN FEET OF WATER AT 68°F, WEIGHING 62.318 LB/FT³*

Psi	Head in Feet	Psi	Head in Feet	Psi	Head in Feet	Psi	Head in Feet
		50	115.55	100	231.1	500	1156
1	2.311	51	117.86	105	242.7	525	1213
2	4.622	52	120.17	110	254.2	550	1271
3	6.933	53	122.48	115	265.8	575	1329
4	9.244	54	124.79	120	277.3	600	1387
5	11.555	55	127.11	125	288.9	625	1444
6	13.866	56	129.42	130	300.4	650	1502
7	16.177	57	131.73	135	312.0	675	1560
8	18.488	58	134.04	140	323.5	700	1618
9	20.799	59	136.35	145	335.1	725	1675
10	23.110	60	138.66	150	346.7	750	1733
11	25.421	61	140.97	155	358.2	775	1791
12	27.732	62	143.28	160	369.8	800	1849
13	30.043	63	145.59	165	381.3	825	1907
14	32.354	64	147.90	170	392.9	850	1964
15	34.665	65	150.22	175	404.4	875	2022
16	36.976	66	152.53	180	416.0	900	2080
17	39.287	67	154.84	185	427.5	925	2138
18	41.598	68	157.15	190	439.1	950	2196
19	43.909	69	159.46	195	450.6	975	2253
20	46.220	70	161.77	200	462.2	1000	2311
21	48.531	71	164.08	210	485.3	1050	2427
22	50.842	72	166.39	220	508.4	1100	2542
23	53.153	73	168.70	230	531.5	1150	2658
24	55.464	74	171.01	240	554.6	1200	2773
25	57.775	75	173.33	250	577.8	1250	2889
26	60.086	76	175.64	260	600.9	1300	3004
27	62.397	77	177.95	270	624.0	1350	3120
28	64.708	78	180.26	280	647.1	1400	3235
29	67.019	79	182.57	290	670.2	1450	3351
30	69.330	80	184.88	300	693.3	1500	3467
31	71.641	81	187.19	310	716.4	1550	3582
32	73.952	82	189.50	320	739.5	1600	3698
33	76.263	83	191.81	330	762.6	1650	3813
34	78.574	84	194.12	340	785.7	1700	3929
35	80.885	85	196.44	350	808.9	1750	4044
36	83.196	86	198.75	360	832.0	1800	4160
37	85.507	87	201.06	370	855.1	1850	4275
38	87.818	88	203.37	380	878.2	1900	4391
39	90.129	89	205.68	390	901.3	1950	4506
40	92.440	90	207.99	400	924.4	2000	4622
41	94.751	91	210.30	410	947.5	2050	4738
42	97.062	92	212.61	420	970.6	2100	4853
43	99.373	93	214.92	430	993.7	2150	4969
44	101.68	94	217.23	440	1017	2200	5084
45	104.00	95	219.55	450	1040	2250	5200
46	106.31	96	221.86	460	1063	2300	5315
47	108.62	97	224.17	470	1086	2350	5431
48	110.93	98	226.48	480	1109	2400	5546
49	113.24	99	228.79	490	1132	2450	5662

*At other temperatures correction must be made for specific gravity.

TABLE 3-3 CONVERSION OF INCHES OF MERCURY (AT 68°F)* TO FEET OF WATER (AT 68°F)*

1 inHg = 1.131 ft water

Inches of Mercury	INCHES OF MERCURY IN FRACTIONS							
	0	1/8	1/4	3/8	1/2	5/8	3/4	7/8
	FEET OF WATER							
0	0.000	0.141	0.283	0.424	0.566	0.707	0.848	0.990
1	1.131	1.272	1.414	1.555	1.697	1.838	1.979	2.121
2	2.262	2.403	2.545	2.686	2.828	2.969	3.110	3.252
3	3.393	3.534	3.676	3.817	3.959	4.100	4.241	4.383
4	4.524	4.665	4.807	4.948	5.090	5.231	5.372	5.514
5	5.655	5.796	5.938	6.079	6.221	6.362	6.503	6.645
6	6.786	6.927	7.069	7.210	7.352	7.493	7.634	7.776
7	7.917	8.058	8.200	8.341	8.483	8.624	8.765	8.907
8	9.048	9.189	9.331	9.472	9.614	9.755	9.896	10.038
9	10.179	10.320	10.462	10.603	10.745	10.886	11.027	11.169
10	11.310	11.451	11.593	11.734	11.876	12.017	12.158	12.300
11	12.441	12.582	12.724	12.865	13.007	13.148	13.289	13.431
12	13.572	13.713	13.855	13.996	14.138	14.279	14.420	14.562
13	14.703	14.844	14.986	15.127	15.269	15.410	15.551	15.693
14	15.834	15.975	16.117	16.258	16.400	16.541	16.682	16.824
15	16.965	17.106	17.248	17.389	17.531	17.672	17.813	17.955
16	18.096	18.237	18.379	18.520	18.662	18.803	18.944	19.086
17	19.227	19.368	19.510	19.651	19.793	19.934	20.075	20.217
18	20.358	20.499	20.641	20.782	20.924	21.065	21.206	21.348
19	21.489	21.630	21.772	21.913	22.055	22.196	22.337	22.479
20	22.620	22.761	22.903	23.044	23.186	23.327	23.468	23.610
21	23.751	23.892	24.034	24.175	24.317	24.458	24.599	24.741
22	24.882	25.023	25.165	25.306	25.448	25.589	25.730	25.872
23	26.013	26.154	26.296	26.437	26.579	26.720	26.861	27.003
24	27.144	27.285	27.427	27.568	27.710	27.851	27.992	28.134
25	28.275	28.416	28.558	28.699	28.841	28.982	29.123	29.265
26	29.406	29.547	29.689	29.830	29.972	30.113	30.254	30.396
27	30.537	30.678	30.820	30.961	31.103	31.244	31.385	31.527
28	31.668	31.809	31.951	32.092	32.234	32.375	32.516	32.658
29	32.799	32.940	33.082	33.223	33.365	33.506	33.647	33.789
30	33.930	34.071	34.213	34.354	34.496	34.637	34.778	34.920

Temp. °F	Correction *	Temp. °F	Correction *	Temp. °F	Correction *
32	1.00199	65	0.99997	100	1.00196
35	1.00160	70	1.00004	105	1.00251
40	1.00105	75	1.00018	110	1.00313
45	1.00063	80	1.00040	115	1.00380
50	1.00032	85	1.00069	120	1.00451
55	1.00011	90	1.00105	125	1.00529
60	1.00000	95	1.00148	130	1.00612

*Correct tabular values to other temperatures by multiplying by correction. To use this table with a mercury and water differential manometer, subtract the number of inches of mercury divided by 12 from the corrected tabular value to obtain the equivalent number of feet of water.

TABLE 3-4   VELOCITY HEAD $V^2/2g$ FOR VARIOUS VELOCITIES

VELOC-ITY fps	VELOCITY IN TENTHS									
	0	**0.1**	**0.2**	**0.3**	**0.4**	**0.5**	**0.6**	**0.7**	**0.8**	**0.9**
	VELOCITY HEAD (FEET)									
0	0	0	0	0	0	0	0	0	0.01	0.01
1	0.02	0.02	0.02	0.03	0.03	0.04	0.04	0.05	0.05	0.06
2	0.06	0.07	0.08	0.08	0.09	0.10	0.11	0.11	0.12	0.13
3	0.14	0.15	0.16	0.17	0.18	0.19	0.20	0.21	0.22	0.24
4	0.25	0.26	0.27	0.29	0.30	0.32	0.33	0.34	0.36	0.37
5	0.39	0.40	0.42	0.44	0.45	0.47	0.49	0.51	0.52	0.54
6	0.56	0.58	0.60	0.62	0.64	0.66	0.68	0.70	0.72	0.74
7	0.76	0.78	0.81	0.83	0.85	0.88	0.90	0.92	0.95	0.97
8	0.99	1.00	1.05	1.07	1.10	1.12	1.15	1.18	1.20	1.23
9	1.26	1.29	1.32	1.35	1.37	1.40	1.43	1.46	1.49	1.52
10	1.56	1.59	1.62	1.65	1.68	1.71	1.75	1.78	1.81	1.85
11	1.88	1.92	1.95	1.99	2.02	2.06	2.09	2.13	2.17	2.20
12	2.24	2.28	2.31	2.35	2.39	2.43	2.47	2.51	2.55	2.59
13	2.63	2.67	2.71	2.75	2.79	2.83	2.88	2.92	2.96	3.00
14	3.05	3.09	3.14	3.18	3.22	3.27	3.31	3.36	3.41	3.45
15	3.50	3.55	3.59	3.64	3.69	3.74	3.78	3.83	3.88	3.93
16	3.98	4.03	4.08	4.13	4.18	4.23	4.28	4.34	4.39	4.44
17	4.49	4.55	4.60	4.65	4.71	4.76	4.82	4.87	4.93	4.98
18	5.04	5.10	5.15	5.21	5.25	5.32	5.38	5.44	5.50	5.55
19	5.61	5.67	5.73	5.79	5.85	5.91	5.97	6.03	6.10	6.16
20	6.22	6.28	6.34	6.41	6.47	6.53	6.60	6.66	6.73	6.79
21	6.86	6.92	6.99	7.05	7.12	7.19	7.25	7.32	7.39	7.46
22	7.53	7.59	7.66	7.73	7.80	7.87	7.94	8.01	8.08	8.15
23	8.22	8.30	8.37	8.44	8.51	8.59	8.66	8.73	8.81	8.88
24	8.96	9.03	9.11	9.18	9.26	9.33	9.41	9.49	9.56	9.64
25	9.72	9.80	9.87	9.95	10.03	10.11	10.19	10.27	10.35	10.43
26	10.51	10.59	10.67	10.75	10.84	10.92	11.00	11.08	11.17	11.25
27	11.33	11.42	11.50	11.59	11.67	11.76	11.84	11.93	12.02	12.10
28	12.19	12.28	12.36	12.45	12.54	12.63	12.72	12.81	12.90	12.99
29	13.08	13.17	13.26	13.35	13.44	13.53	13.62	13.71	13.81	13.90
30	13.99	14.09	14.18	14.27	14.37	14.46	14.56	14.65	14.75	14.84

TABLE 3-5   CONVERSION OF INCHES OF MERCURY TO POUNDS PER SQUARE INCH

Multiply inches of mercury by conversion factor to obtain pounds per square inch.

Temperature °F	Conversion Factor	Temperature °F	Conversion Factor
0	0.49276	70	0.48929
5	0.49251	75	0.48905
10	0.49226	80	0.48880
15	0.49201	85	0.48855
20	0.49176	90	0.48831
25	0.49152	95	0.48806
30	0.49127	100	0.48782
35	0.49102	105	0.48757
40	0.49077	110	0.48733
45	0.49053	115	0.48708
50	0.49028	120	0.48684
55	0.49003	125	0.48660
60	0.48979	130	0.48635
65	0.48954	135	0.48611

VISCOSITY OF FLUIDS

Viscosity is that property of any fluid (liquid or gas) which tends to resist a shearing force. It is important to fluid flow because nearly all fluid motion is accompanied by shearing forces.

The two basic viscosity parameters are the *dynamic* (or *absolute*) *viscosity* μ, having the dimension force $\times$ time/(length)^2, and the *kinematic viscosity* ν, having the dimension (length)^2/time. The parameters are related through the mass density ρ of the fluid, such that $\nu = \mu/\rho = \mu g/\gamma$, where γ is the specific weight and g is the acceleration of gravity.

The unit of dynamic viscosity in U.S. customary measure is the *pound-second per square foot*, which is numerically identical with the *slug per foot-second*. The unit of dynamic viscosity in SI measure is the pascal second, which is one-tenth of a *poise*. Numerical values generally are expressed in *centipoises,* a unit which is one-hundredth of a poise. A unit called the *reyn,* equal to 1 *pound-second per square inch,* is used in lubrication problems.

The unit of kinematic viscosity in U.S. customary measure is the *square foot per second*. The unit of kinematic viscosity in SI measure is the *square meter per second,* which is ⅟₁₀,₀₀₀ of a *stoke*. Numerical values generally are expressed in *centistokes,* a unit which is one-hundredth of a stoke.

Widespread use of the Saybolt viscosimeter has led to the use of the time of efflux in seconds, for 60 cm^3 of liquid, as an arbitrary unit of kinematic viscosity. The term SSU (Seconds Saybolt Universal) refers to the smaller, and the term SSF (Seconds Saybolt Furol) to the larger, of two orifices with which the instrument may be equipped. (Other empirical measures of kinematic viscosity may be converted to basic units by Fig. 3-5 and Table 3-8.)

The dynamic viscosity of any fluid is a function of temperature and pressure. The dynamic viscosity of most liquids increases with an increase of pressure, but fortunately the changes may be neglected for the ranges of pressure usually encountered in engineering problems. The dynamic viscosity of gases is virtually

TABLE 3-6   Viscosities of Air and Water at 68°F and Atmospheric Pressure

Fluid	Dynamic viscosity μ		Kinematic viscosity ν		
	Poises	lb·s/ft^2	Stokes	ft^2/s	SSU
Air	178.9×10^{-6}	0.3735×10^{-6}	0.1463	157.4×10^{-6}	
Water	0.010087	21.067×10^{-6}	0.010105	10.877×10^{-6}	30.1

independent of pressure except at extremely high or low pressures. Pressure is very important in determining the kinematic viscosity of a gas because of its influence on the mass density. (The kinematic viscosities of many fluids are shown in Figs. 3-6 to 3-9 inclusive.)

Newtonian Materials

Newton deduced that the viscosity of a given liquid should be constant at any particular temperature and pressure and independent of the rate of shear, as illustrated in Fig. 3-1. In such *newtonian fluids*, shear stress is directly proportional to rate of shear. At temperatures above their cloud points most mineral oils are newtonian fluids.

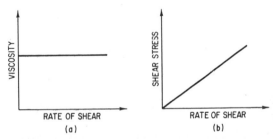

FIG. 3-1   Characteristics of newtonian liquids. (*a*) Viscosity is independent of rate of shear. (*b*) Shear stress is directly proportional to rate of shear.

Nonnewtonian Materials

The viscosities of some materials, such as greases and polymer-thickened mineral oils, are affected by shearing effects, and these materials are termed *nonnewtonian*. In other words, the viscosity of a nonnewtonian fluid will depend on the rate of shear at which it is measured. Since a nonnewtonian fluid can have an unlimited number of viscosity values (as the shear rate is varied), the term *apparent viscosity* is used to describe its viscous properties. Apparent viscosity is expressed in absolute units and is a measure of the resistance to flow at a given rate of shear. It has meaning only if the rate of shear used in the measurement is also given and is obtained experimentally by measuring and dividing the shear stress by the rate of shear. A *rheogram*, or *flow curve*, relating shear stress to rate of shear is frequently used to describe completely the viscous properties of a nonnewtonian material.

Nonnewtonian materials may be divided into five types: plastic, pseudo-plastic, dilatant, thixotropic, and rheopectic. Figure 3-2 presents characteristic rheograms in which shear stress (e.g., pressure in a steady-flow system) is plotted against rate of shear (e.g., flow velocity). The curves at the left in Fig. 3-3 illustrate how the apparent viscosities of nonnewtonian materials vary with changing rates of shear.

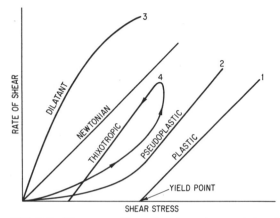

FIG. 3-2   Flow curves illustrating shear characteristics of various types of materials.

As illustrated in curve 1 of Fig. 3-2, a *plastic* material, such as a grease, putty, or molding clay, is characterized by a *yield point,* or *yield value.* This means that a definite minimum stress or force must be applied to the material before any flow takes place. From a rheological standpoint,* tomato catsup is a common example of a plastic material. If a bottle is shaken only gently, its contents may not flow out because the yield point has not been exceeded. However, if the bottle is struck or shaken more vigorously, the yield point is exceeded, the viscosity is reduced, and the catsup gushes forth.

While a pseudo-plastic fluid has no yield point, its apparent viscosity also decreases with increasing shear rates but stabilizes only at very high rates of shear. Many emulsions such as water-base fluids and resinous materials show this type of behavior.

Oppositely, the apparent viscosity of a dilatant fluid increases as the rate of shear increases. Such a fluid often solidifies at high rates of shear. Examples are pigment-vehicle suspensions such as paints and printing inks and some starches.

The three fluids described above—plastic, pseudo-plastic, and dilatant—are also known as time-independent nonnewtonian fluids, since their rheological, or flow, properties are independent of time. The rate of shear at any point in the fluid is a simple function of the shear stress at that point.

On the other hand, the flow properties of two other nonnewtonian materials—thixotropic and rheopectic—are dependent on time. The apparent viscosity of

*Rheology is the science treating of deformation and flow of matter.

these more complex fluids depends not only on the magnitude of the shear rate but also on the length of time during which shear has been applied, as illustrated in Fig. 3-3.

If a thixotropic fluid is subjected to a constant rate of shear for some time, its structure is gradually broken down and its apparent viscosity decreases to some

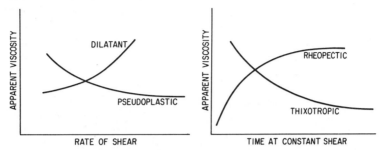

FIG. 3-3   Different types of nonnewtonian behavior.

minimum value. When the shear effect is removed and the fluid is at rest, the structure rebuilds gradually and apparent viscosity increases with time to the original value. This is called *reversible thixotropy*. If, however, upon removing the shear stress, a value less than the original viscosity is obtained with time, the phenomenon is known as *irreversible thixotropy*. Some oils containing high-molecular-weight polymers and mineral oils at temperatures below their cloud point show this latter effect.

During rotary drilling of deep oil wells, a very special *drilling mud* with thixotropic properties is pumped down the hollow drill stem to force cuttings back to the surface. As long as the mud is agitated by rotation of the drill stem and by pumping, it remains fluid and removes drilling debris. However, whenever drilling is stopped, the drilling mud solidifies to a gel, holds the cuttings in suspension, and thereby prevents them from settling and interfering with subsequent drilling.

Quicksand is also thixotropic, since it becomes more and more fluid when agitated; therefore persons caught in this water-and-sand mixture improve their chance of survival by remaining as motionless as possible.

If a rheopectic fluid is subjected to a constant rate of shear for a given period of time, its apparent viscosity increases to some maximum value. Upon cessation of shearing and resting for a time, its apparent viscosity decreases again.

Some greases are intentionally manufactured to have partial rheopectic properties, which facilitate pumping from a drum or central grease storage in which the grease is in a relatively fluid condition. Upon shearing in a bearing, however, the grease builds up to a higher apparent viscosity or consistency and stays in place. Such a grease does not have full rheopectic characteristics, however, since after shearing and resting, it still retains a higher consistency.

Since the viscosity of a nonnewtonian lubricant depends upon the rate of shear acting on it, the importance of measuring viscosity at various shear rates that will be encountered in the use of such a lubricant can be readily seen. In some machine elements, shear rates up to 3 million reciprocal seconds may be encountered, while in other applications only a few reciprocal seconds or a few tenths are the order of magnitude. In dispensing greases, shear rates as low as 0.1 reciprocal second are sometimes encountered, while leakage from housings during periods of shutdown involves an even lower range.

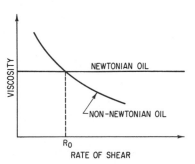

FIG. 3-4   Viscosity versus rate of shear for newtonian and nonnewtonian oils.

As illustrated by Fig. 3-4, the determination of viscosity of a nonnewtonian liquid at only one shear rate is not usually sufficient. Incorrect conclusions would be drawn and application difficulties would be invited if the viscosities of a newtonian and a nonnewtonian oil were measured at some specific shear rate R_0 where the two curves happened to cross each other. While both oils have the same apparent viscosity at this one point, the remainders of their viscosity-shear curves are entirely different.

TABLE 3-7   VISCOSITY CONVERSION FACTORS

Multiply	by	to obtain
poises	100	centipoises
pound-seconds/sq ft	47,880.1	centipoises
reyns	6.89473×10^6	centipoises
centipoises	2.08855×10^{-5}	pound-seconds/sq ft
centipoises	1.45038×10^{-7}	reyns
stokes	100	centistokes
sq ft/second	92,903.4	centistokes
centistokes	1.07639×10^{-5}	sq ft/second
sq ft/second	$1488.16 \times \gamma$ in pounds/cu ft	centipoises
centipoises	$6.71970 \times 10^{-4}/\gamma$ in pounds/cu ft	sq ft/second

Saybolt Viscosimeter Conversion Formulas

μ, in centistokes $= 0.226 \times SSU - 195/SSU$ for $SSU \leq 100$

μ, in centistokes $= 0.220 \times SSU - 135/SSU$ for $SSU > 100$

μ, in centistokes $= 2.24 \times SSF - 184/SSF$ for $25 \leq SSF \leq 40$

μ, in centistokes $= 2.16 \times SSF - 60/SSF$ for $SSF > 40$

VISCOSITY CONVERSION CHART

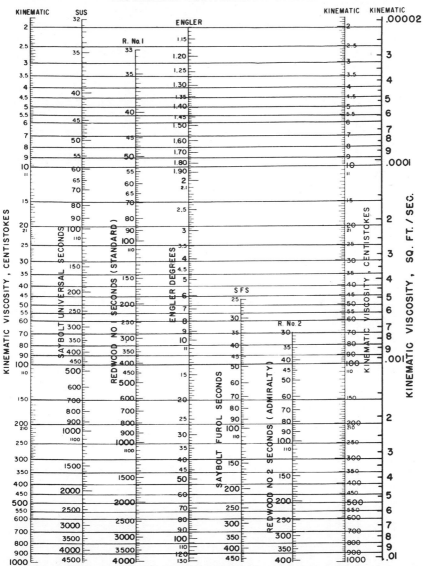

FIG. 3-5 Viscosities at the *same temperature* on all scales are equivalent. To extend the range of only the kinematic, Saybolt Universal, Redwood No. 1, and Engler scales, multiply by 10 the viscosities on these scales between 100 and 1000 cSt on the kinematic scale and the corresponding viscosities on the other three scales. For further extension, multiply these scales as above by 100 or a higher power of 10. Example: 1500 cSt = 150 × 10 cSt = 695 × 10 SSU = 6950 SSU. *(Courtesy of Texaco, Inc.)*

TABLE 3-8 VISCOSITY CONVERSION TABLE

This table gives a comparison of various viscosity ratings, so that if the viscosity is given in terms other than Saybolt Universal, it can be translated quickly by following horizontally to the Saybolt Universal column.

Seconds Saybolt Universal SSU	Kinematic viscosity, centistokes	Seconds Saybolt Furol SSF	Seconds Redwood 1 (Standard)	Seconds Redwood 2 (Admiralty)	Degrees Engler	Degrees Barbey	Seconds Parlin Cup No. 7	Seconds Parlin Cup No. 10	Seconds Parlin Cup No. 15	Seconds Parlin Cup No. 20	Seconds Ford Cup No. 3	Seconds Ford Cup No. 4
31	1.00	....	29	....	1.00	6,200						
35	2.56	....	32.1	....	1.16	2,420						
40	4.30	....	36.2	5.10	1.31	1,440						
50	7.40	....	44.3	5.83	1.58	838						
60	10.3	....	52.3	6.77	1.88	618						
70	13.1	12.95	60.9	7.60	2.17	483						
80	15.7	13.70	69.2	8.44	2.45	404						
90	18.2	14.44	77.6	9.30	2.73	348						
100	20.6	15.24	85.6	10.12	3.02	307						
150	32.1	19.30	128	14.48	4.48	195						
200	43.2	23.5	170	18.90	5.92	144	40					
250	54.0	28.0	212	23.45	7.35	114	46					
300	65.0	32.5	254	28.0	8.79	95	52.5	15	6.0	3.0	30	20
400	87.60	41.9	338	37.1	11.70	70.8	66	21	7.2	3.2	42	28
500	110.0	51.6	423	46.2	14.60	56.4	79	25	7.8	3.4	50	34
600	132	61.4	508	55.4	17.50	47.0	92	30	8.5	3.6	58	40
700	154	71.1	592	64.6	20.45	40.3	106	35	9.0	3.9	67	45
800	176	81.0	677	73.8	23.35	35.2	120	39	9.8	4.1	74	50
900	198	91.0	762	83.0	26.30	31.3	135	41	10.7	4.3	82	57
1,000	220	100.7	896	92.1	29.20	28.2	149	43	11.5	4.5	90	62
1,500	330	150	1,270	138.2	43.80	18.7	. . . .	65	15.2	6.3	132	90
2,000	440	200	1,690	184.2	58.40	14.1	. . . .	86	19.5	7.5	172	118
2,500	550	250	2,120	230	73.0	11.3	. . . .	108	24	9	218	147
3,000	660	300	2,540	276	87.60	9.4	. . . .	129	28.5	11	258	172
4,000	880	400	3,380	368	117.0	7.05	. . . .	172	37	14	337	230
5,000	1,100	500	4,230	461	146	5.64	. . . .	215	47	18	425	290
6,000	1,320	600	5,080	553	175	4.70	. . . .	258	57	22	520	350
7,000	1,540	700	5,920	645	204.5	4.03	. . . .	300	67	25	600	410

8,000	1,760	800	6,770	737	233.5	3.52	. . . . .	344	76	29	. . . . .	680	465
9,000	1,980	900	7,620	829	263	3.13	. . . . .	387	86	32	. . . . .	780	520
10,000	2,200	1,000	8,460	921	292	2.82	. . . . .	430	96	35	. . . . .	850	575
15,000	3,300	1,500	13,700	. . . . .	438	2.50	. . . . .	650	147	53	. . . . .	1,280	860
20,000	4,400	2,000	18,400	. . . . .	584	1.40	. . . . .	860	203	70	. . . . .	1,715	1,150

Reprinted from *Hydraulic Institute Engineering Data Book*, 1st ed., Cleveland, 1979.

*Kinematic viscosity (in centistokes) = $\dfrac{\text{absolute viscosity (in centipoises)}}{\text{specific gravity}}$. Above 300 SSU, use the following approximate conversion: SSU = centistokes × 4.635. Above the range of this table and within the range of the viscosimeter, multiply their rating by the following factors to convert to SSU:

Viscosimeter	Factor
Saybolt Furol	10
Redwood Standard. . . .	1.095
Redwood Admiralty. . .	10.87
Engler, degrees	34.5

Viscosimeter	Factor
Parlin cup No. 15	98.2
Parlin cup No. 20	187.0
Ford cup No. 4	17.4

3-15

VISCOSITY OF FLUIDS

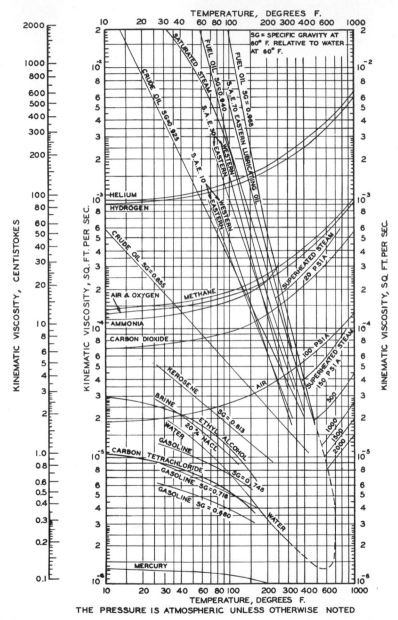

FIG. 3-6

EFFECT OF TEMPERATURE ON FUEL-OIL VISCOSITY

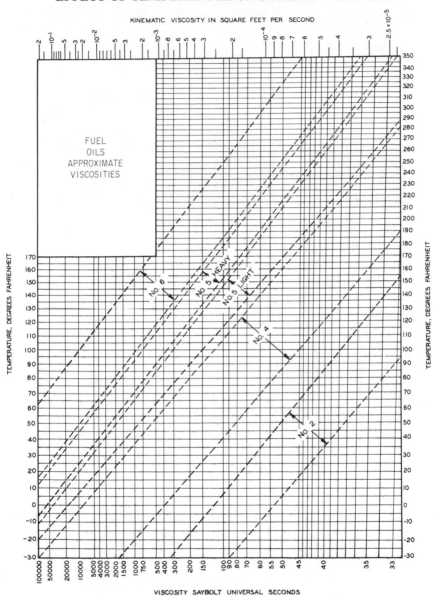

FIG. 3-7 *(Courtesy of Texaco, Inc.)*

EFFECT OF TEMPERATURE ON TURBINE-OIL VISCOSITY

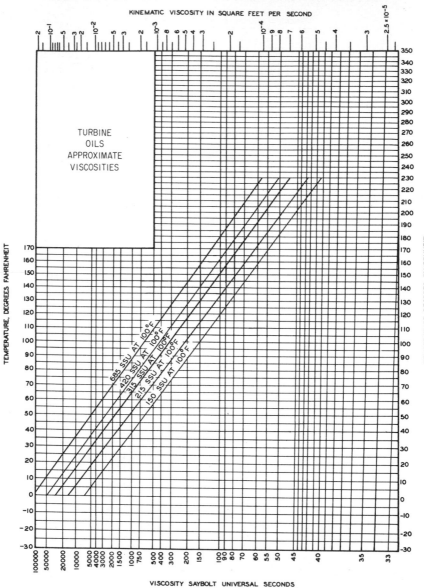

FIG. 3-8 *(Courtesy of Texaco, Inc.)*

CRANKCASE-OIL CLASSIFICATION

SAE Recommended Practice*

The SAE viscosity numbers constitute a classification for crankcase lubricating oils in terms of viscosity only. Other factors of oil character or quality are not considered.

TABLE 3-9   Viscosity Values

SAE viscosity No.	Viscosity range, SSU			
	At 0°F		At 210°F	
	Min	Max	Min	Max
5W	. . . .	6,000		
10W	6,000°	Less than 12,000		
20W	12,000†	48,000		
20	. . . .	. . . . . .	45	Less than 58
30	. . . .	. . . . . .	58	Less than 70
40	. . . .	. . . . . .	70	Less than 85
50	. . . .	. . . . . .	85	110

NOTE: SAE crankcase oils should not be used in lubrication systems for which turbine oils are recommended

*Minimum viscosity at 0°F can be waived, provided viscosity at 210°F is not below 40 SSU.

†Minimum viscosity at 0°F can be waived, provided viscosity at 210°F is not below 45 SSU.

Viscosity numbers without an additional symbol are based on the viscosity at 210°F. Viscosity numbers with the additional symbol W are based on the viscosity at 0°F.

A multiviscosity numbered oil is one whose 0°F viscosity falls within the prescribed range of one of the W-number classifications and whose 210°F viscosity falls within the prescribed range of one of the non-W-number classifications.

For the effect of temperature on SAE crankcase-oil viscosity, see Fig. 3-9.

FRICTION LOSSES IN PIPES

Pipe Friction

The resistance to the incompressible flow of any fluid in any pipe may be computed from the equation

$$h_f = f \frac{L}{D} \frac{V^2}{2g} \tag{3-1}$$

*Taken from the *Society of Automotive Engineers Handbook,* Warrendale, Pa., 1969, by permission.

EFFECT OF TEMPERATURE ON SAE CRANKCASE-OIL VISCOSITY

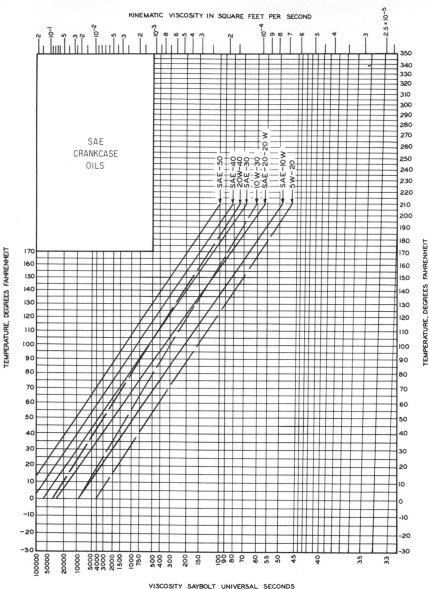

FIG. 3-9   *(Courtesy of Texaco, Inc.)*

where h_f = frictional resistance, ft of fluid

L = length of pipe, ft, including the centerline length of all valves and fittings plus the allowance from Table 3-12

D = average internal diameter of pipe, ft (For noncircular cross sections, replace D by 4 times the cross-sectional area in square feet divided by the wetted perimeter in feet.)

V = average velocity in pipe, ft/s

g = acceleration due to gravity = 32.17 ft/s² at sea level and approximately 45° latitude

f = friction factor to be taken from Fig. 3-10 according to the diameter, kind of pipe, and the *Reynolds number*

The Reynolds number R is a dimensionless flow parameter defined by

$$R = \frac{VD}{v} \qquad (3\text{-}2)$$

where V and D are as defined above

v = kinematic viscosity, ft²/s

The flow in most commercial pipe systems may be assumed to be turbulent for all values of Reynolds number greater than 3000. A critical state of flow usually occurs between $R = 2000$ and $R = 3000$ in which the frictional resistance is difficult to predict. For all values of the Reynolds number less than 2000, the flow is laminar and the friction factor for any kind and size of pipe is given by

$$f = \frac{64}{R} \qquad R \lessgtr 2000 \text{ only} \qquad (3\text{-}3)$$

Figure 3-10 may be used to compute the friction loss due to the turbulent flow of any fluid, liquid or gas, in pipes of any material.

If the fluid is fresh water at 60°F or atmospheric air at 60°F, the two respective scales at the bottom may be used in lieu of the Reynolds number, the scale reading being the product of the average velocity in feet per second and the internal diameter in inches (VD'').

Tables for Water

Tables 3-10 and 3-11 show the average velocity and the friction loss per 100 ft of new clean pipe for fresh water at 60°F. Table 3-10 is for Schedule 40 steel or wrought-iron pipe, ANSI Specification B36.10. Table 3-11 is for asphalt-dipped cast-iron pipe. Asphalt-dipped pipe should not be confused with pipes lined with asphalt bitumen, for which the friction losses are considerably smaller. The allowance to be made for valves and fittings is given in Table 3-12.

Charts for Viscous Fluids

Figure 3-12 is useful to determine the friction losses for various fluids flowing through Schedule 40 new steel or wrought-iron pipes, ANSI Specification B36.10. Fluid viscosities are given in Seconds Saybolt Universal (SSU) and centistokes.

FRICTION FACTORS FOR ANY KIND AND SIZE OF PIPE

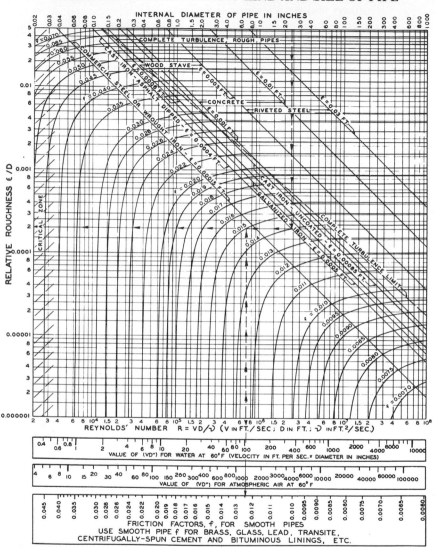

FIG. 3-10 Example: Given a 24-in-inside-diameter cast-iron pipe handling 6500 gal/min of water, from Table 3-11 find $V = 4.61$ ft/s, $VD'' = 4.61 \times 24 = 110.6$. This is equivalent to a Reynolds number $R = 7 \times 10^5$. Find $f = 0.015$ and $\epsilon/D = 0.0002$ by following dotted lines shown on chart. From Eq. (3-1) find $h_f = 0.246$ ft/100 ft. Given any smooth pipe such as brass with $R = 7 \times 10^5$, find $f = 0.0124$.

Friction-loss moduli for *laminar flow* are shown by the 45° lines in the upper-left-hand portion of each chart. Moduli for *turbulent flow* are shown by the steeper curves in the lower-right-hand portion.

The horizontal scale at the bottom of each chart shows the rate of flow in gallons per minute and, at the top, the corresponding average velocity in the pipe in feet per second. The vertical scales of friction-loss modulus are converted to pressure loss in pounds per square inch or to head loss in feet of liquid as described on the charts. The allowance to be made for valves and fittings is given in Table 3-12.

Construction of Charts and Tables

The friction factor f has been determined by the Colebrook equation

$$\frac{1}{\sqrt{f}} = -2 \log_{10}\left(\frac{\epsilon}{3.7D} + \frac{2.51}{R\sqrt{f}}\right) \tag{3-4}$$

where ϵ = absolute roughness of pipe wall, ft
ϵ/D = relative roughness of pipe wall, dimensionless
R = Reynolds number defined by Eq. (3-2)

Friction factors determined from Eq. (3-4) are within 10 percent of measured values for most new pipes. Insufficient data are available to establish reliable values of ϵ/D for old or deteriorated pipes. Equations (3-1) to (3-4), together with appropriate values of ϵ, have been used to construct all the charts and tables of fluid friction given in this subsection. The tables are based on $\nu = 0.00001216$ ft^2/s, equivalent to 1.130 cSt, which is the value for pure fresh water at 60°F. The absolute roughness $\epsilon = 0.00015$ ft has been used for new steel or wrought-iron pipe, and $\epsilon = 0.0004$ ft has been used for asphalt-dipped cast-iron pipe.

TABLE 3-10 FRICTION OF WATER IN NEW STEEL OR WROUGHT-IRON PIPE—SCHEDULE 40°

Friction is in feet of water at 60°F per 100 ft of pipe.

gpm	Velocity fps	Friction ft./100 ft.	gpm	Velocity fps	Friction ft./100 ft.	gpm	Velocity fps	Friction ft./100 ft.
	1-Inch Pipe (1.049" I.D.)			1½-Inch Pipe (1.610" I.D.)			2½-Inch Pipe (2.469" I.D.)	
2	0.742	0.379	4	0.630	0.164	8	0.536	0.0712
3	1.114	0.772	5	0.788	0.242	10	0.670	0.105
4	1.48	1.295	6	0.946	0.333	12	0.804	0.145
5	1.86	1.93	8	1.26	0.558	14	0.938	0.191
6	2.23	2.68	10	1.58	0.829	16	1.07	0.243
8	2.97	4.54	15	2.36	1.74	18	1.21	0.300
10	3.71	6.86	20	3.15	2.94	20	1.34	0.362
12	4.45	9.62	25	3.94	4.46	22	1.47	0.430
14	5.20	12.8	30	4.73	6.26	24	1.61	0.502
16	5.94	16.5	35	5.52	8.37	26	1.74	0.580
18	6.68	20.6	40	6.30	10.79	28	1.88	0.663
20	7.42	25.1	45	7.10	13.5	30	2.01	0.753
22	8.17	30.2	50	7.88	16.4	35	2.35	1.00
24	8.91	35.6	60	9.46	23.2	40	2.68	1.28
26	9.65	41.6	70	11.03	31.3	45	3.02	1.60
28	10.39	47.9	80	12.6	40.5	50	3.35	1.94
30	11.1	54.6	90	14.2	51.0	55	3.69	2.32
35	13.0	73.3	100	15.8	62.2	60	4.02	2.72
40	14.8	95.0	125	19.7	95.5	65	4.36	3.16
45	16.7	119.	150	23.6	137.	70	4.69	3.63
	1¼-Inch Pipe (1.380" I.D.)			2-Inch Pipe (2.067" I.D.)		75	5.03	4.13
						80	5.36	4.66
						85	5.70	5.22
4	0.858	0.342	5	0.478	0.0731	90	6.03	5.82
5	1.073	0.508	10	0.956	0.248	95	6.37	6.45
6	1.29	0.704	15	1.44	0.513	100	6.70	7.11
7	1.50	0.930	20	1.91	0.868	110	7.37	8.51
8	1.72	1.18	25	2.38	1.30	120	8.04	10.0
10	2.15	1.77	30	2.87	1.82	130	8.71	11.7
12	2.57	2.48	35	3.35	2.42	140	9.38	13.5
14	3.00	3.28	40	3.82	3.10	150	10.05	15.4
16	3.43	4.20	45	4.30	3.85	160	10.7	17.4
18	3.86	5.22	50	4.78	4.67	170	11.4	19.6
20	4.29	6.34	60	5.74	6.59	180	12.1	21.9
25	5.36	9.61	70	6.69	8.86	190	12.7	24.2
30	6.44	13.6	80	7.65	11.4	200	13.4	26.7
35	7.51	18.2	90	8.60	14.2	220	14.7	32.2
40	8.58	23.5	100	9.56	17.4	240	16.1	38.1
45	9.65	29.4	125	11.96	26.8	260	17.4	44.5
50	10.7	36.0	150	14.3	38.0	280	18.8	51.3
60	12.9	51.0	175	16.8	51.2	300	20.1	58.5
70	15.0	68.8	200	19.1	66.3	350	23.5	79.2
80	17.2	89.2	225	21.5	83.7	400	26.8	103.

NOTE: No allowance has been made for age, differences in diameter, or any abnormal condition of interior surface. Any factor of safety must be estimated from the local conditions and the requirements of each particular installation.

For an explanation of this table see subsection "Friction Losses in Pipes."

TABLE 3-11 FRICTION OF WATER IN NEW ASPHALT-DIPPED CAST-IRON PIPE*

Friction is in feet of water at 60°F per 100 ft of pipe.

gpm	Velocity fps	Friction ft./100 ft.	gpm	Velocity fps	Friction ft./100 ft.	gpm	Velocity fps	Friction ft./100 ft.
3-Inch Pipe			4-Inch Pipe			6-Inch Pipe		
10	0.454	0.0435	20	0.511	0.03700	70	0.794	0.0496
15	0.681	0.0900	30	0.766	0.0770	80	0.908	0.0635
20	0.908	0.1510	40	1.02	0.131	90	1.02	0.0789
25	1.13	0.2280	50	1.28	0.199	100	1.13	0.0958
30	1.36	0.320	60	1.53	0.278	120	1.36	0.130
35	1.59	0.427	70	1.79	0.370	140	1.59	0.178
40	1.82	0.549	80	2.04	0.476	160	1.82	0.229
45	2.04	0.683	90	2.30	0.594	180	2.04	0.282
50	2.27	0.830	100	2.55	0.725	200	2.27	0.346
55	2.50	0.993	110	2.81	0.869	220	2.50	0.415
60	2.72	1.170	120	3.06	1.03	240	2.72	0.490
65	2.95	1.36	130	3.32	1.19	260	2.95	0.570
70	3.18	1.56	140	3.57	1.38	280	3.18	0.655
75	3.40	1.78	150	3.83	1.58	300	3.40	0.745
80	3.63	2.02	160	4.08	1.78	320	3.63	0.846
85	3.86	2.28	170	4.34	2.00	340	3.86	0.952
90	4.08	2.55	180	4.60	2.24	360	4.08	1.06
95	4.31	2.82	190	4.85	2.49	380	4.31	1.18
100	4.54	3.10	200	5.11	2.74	400	4.54	1.30
110	4.99	3.73	220	5.62	3.28	420	4.76	1.43
120	5.45	4.40	240	6.13	3.88	440	4.99	1.57
130	5.90	5.13	260	6.64	4.54	460	5.22	1.71
140	6.35	5.93	280	7.15	5.25	480	5.45	1.86
150	6.81	6.80	300	7.66	6.03	500	5.67	2.02
160	7.26	7.71	320	8.17	6.87	550	6.24	2.42
170	7.72	8.70	340	8.68	7.75	600	6.81	2.84
180	8.17	9.73	360	9.19	8.68	650	7.37	3.33
190	8.62	10.80	380	9.70	9.66	700	7.94	3.87
200	9.08	11.9	400	10.2	10.7	750	8.51	4.45
220	9.98	14.3	420	10.7	11.7	800	9.08	5.06
240	10.9	17.0	440	11.2	12.8	850	9.64	5.69
260	11.8	19.8	460	11.7	14.0	900	10.2	6.34
280	12.7	22.8	480	12.3	15.3	950	10.8	7.02
300	13.6	26.1	500	12.8	16.6	1,000	11.3	7.73
320	14.5	29.7	550	14.0	19.9	1,100	12.5	9.80
340	15.4	33.6	600	15.3	23.6	1,200	13.6	11.2
360	16.3	37.8	650	16.6	27.7	1,300	14.7	13.0
380	17.2	42.2	700	17.9	32.1	1,400	15.9	15.1
400	18.2	46.8	750	19.1	36.7	1,500	17.0	17.4
420	19.1	51.5	800	20.4	41.6	1,600	18.2	19.8
440	20.0	56.4	850	21.7	46.8	1,700	19.3	22.3
460	20.9	61.5	900	23.0	52.3	1,800	20.4	24.8
480	21.8	66.8	950	24.3	58.1	1,900	21.6	27.6
500	22.7	72.3	1,000	25.5	64.2	2,000	22.7	30.5

NOTE: No allowance has been made for age, differences in diameter, or any abnormal son-dition of interior surface. Any factor of safety must be estimated from the local conditions and the requirements of each particular installation.

*For an explanation of this table see subsection "Friction Losses in Pipes."

TABLE 3-11 FRICTION OF WATER IN NEW ASPHALT-DIPPED CAST-IRON PIPE *(Continued)*

Friction is in feet of water at 60°F per 100 ft of pipe.

gpm	Velocity fps	Friction ft./100 ft.	gpm	Velocity fps	Friction ft./100 ft.	gpm	Velocity fps	Friction ft./100 ft.
8-Inch Pipe			10-Inch Pipe			12-Inch Pipe		
200	1.28	0.0828	200	0.817	0.0276	300	0.851	0.0236
220	1.40	0.0989	220	0.899	0.0329	350	0.993	0.0316
240	1.53	0.1163	240	0.980	0.0387	400	1.13	0.0404
260	1.66	0.135	260	1.06	0.0449	450	1.28	0.0500
280	1.79	0.155	280	1.14	0.0514	500	1.42	0.0604
300	1.91	0.176	300	1.23	0.0583	550	1.56	0.0718
320	2.04	0.198	350	1.43	0.0778	600	1.70	0.0845
340	2.17	0.222	400	1.63	0.0990	650	1.84	0.0990
360	2.30	0.248	450	1.84	0.1235	700	1.99	0.115
380	2.43	0.275	500	2.04	0.151	750	2.13	0.131
400	2.55	0.304	550	2.25	0.181	800	2.27	0.148
450	2.87	0.380	600	2.45	0.214	850	2.41	0.166
500	3.19	0.464	650	2.66	0.250	900	2.55	0.184
550	3.51	0.557	700	2.86	0.288	950	2.69	0.203
600	3.83	0.658	750	3.06	0.328	1,000	2.84	0.224
650	4.15	0.767	800	3.27	0.370	1,100	3.12	0.272
700	4.47	0.884	850	3.47	0.415	1,200	3.40	0.321
750	4.79	1.01	900	3.68	0.462	1,300	3.69	0.372
800	5.11	1.14	950	3.88	0.512	1,400	3.97	0.428
850	5.42	1.29	1,000	4.09	0.565	1,500	4.26	0.488
900	5.74	1.44	1,100	4.49	0.680	1,600	4.54	0.552
950	6.06	1.60	1,200	4.90	0.805	1,700	4.82	0.621
1,000	6.38	1.76	1,300	5.31	0.945	1,800	5.11	0.695
1,100	7.02	2.14	1,400	5.72	1.09	1,900	5.39	0.774
1,200	7.66	2.53	1,500	6.13	1.25	2,000	5.67	0.858
1,300	8.30	2.94	1,600	6.54	1.42	2,200	6.24	1.03
1,400	8.93	3.40	1,700	6.94	1.60	2,400	6.81	1.22
1,500	9.57	3.91	1,800	7.35	1.78	2,600	7.38	1.43
1,600	10.2	4.45	1,900	7.76	1.97	2,800	7.94	1.65
1,700	10.8	5.00	2,000	8.17	2.17	3,000	8.51	1.88
1,800	11.5	5.58	2,200	8.99	2.64	3,500	9.93	2.55
1,900	12.1	6.19	2,400	9.80	3.12	4,000	11.3	3.31
2,000	12.8	6.84	2,600	10.6	3.63	4,500	12.8	4.18
2,200	14.0	8.26	2,800	11.4	4.18	5,000	14.2	5.13
2,400	15.3	9.80	3,000	12.3	4.79	5,500	15.6	6.17
2,600	16.6	11.47	3,200	13.1	5.47	6,000	17.0	7.30
2,800	17.9	13.3	3,400	13.9	6.18	6,500	18.4	8.55
3,000	19.1	15.2	3,600	14.7	6.91	7,000	19.9	9.92
3,200	20.4	17.3	3,800	15.5	7.68	7,500	21.3	11.4
3,400	21.7	19.5	4,000	16.3	8.50	8,000	22.7	13.0
3,600	23.0	21.9	4,500	18.4	10.7	8,500	24.1	14.7
3,800	24.3	24.4	5,000	20.4	13.2	9,000	25.5	16.4
4,000	25.5	27.0	5,500	22.5	15.9	9,500	26.9	18.2
4,500	28.7	34.0	6,000	24.5	18.9	10,000	28.4	20.2

NOTE: No allowance has been made for age, differences in diameter, or any abnormal condition of interior surface. Any factor of safety must be estimated from the local conditions and the requirements of each particular installation.

TABLE 3-11 FRICTION OF WATER IN NEW ASPHALT-DIPPED CAST-IRON PIPE *(Continued)*

Friction is in feet of water at 60°F per 100 ft of pipe.

gpm	Velocity fps	Friction ft./100 ft.	gpm	Velocity fps	Friction ft./100 ft.	gpm	Velocity fps	Friction ft./100 ft.
14-Inch Pipe			**16-Inch Pipe**			**18-Inch Pipe**		
400	0.834	0.0190	500	0.798	0.0148	700	0.883	0.0154
500	1.04	0.0284	600	0.957	0.0207	800	1.01	0.0197
600	1.25	0.0400	700	1.12	0.0276	900	1.13	0.0245
700	1.46	0.0533	800	1.28	0.0354	1,000	1.26	0.0298
800	1.67	0.0686	900	1.44	0.0441	1,100	1.39	0.0357
900	1.88	0.0859	1,000	1.60	0.0537	1,200	1.51	0.0420
1,000	2.08	0.1050	1,100	1.76	0.0642	1,300	1.64	0.0488
1,100	2.29	0.1256	1,200	1.91	0.0760	1,400	1.77	0.0560
1,200	2.50	0.148	1,300	2.07	0.0878	1,500	1.89	0.0639
1,300	2.71	0.172	1,400	2.23	0.101	1,600	2.02	0.0723
1,400	2.92	0.198	1,500	2.39	0.115	1,700	2.14	0.0812
1,500	3.13	0.225	1,600	2.55	0.130	1,800	2.27	0.0905
1,600	3.33	0.254	1,700	2.71	0.146	1,900	2.40	0.1001
1,700	3.54	0.285	1,800	2.87	0.163	2,000	2.52	0.110
1,800	3.75	0.318	1,900	3.03	0.180	2,500	3.15	0.170
1,900	3.96	0.353	2,000	3.19	0.200	3,000	3.78	0.240
2,000	4.17	0.391	2,200	3.51	0.240	3,500	4.41	0.320
2,200	4.59	0.470	2,400	3.83	0.282	4,000	5.04	0.415
2,400	5.00	0.554	2,600	4.15	0.329	4,500	5.67	0.525
2,600	5.42	0.648	2,800	4.47	0.379	5,000	6.30	0.645
2,800	5.84	0.749	3,000	4.79	0.433	5,500	6.93	0.775
3,000	6.25	0.858	3,200	5.11	0.490	6,000	7.56	0.920
3,200	6.67	0.971	3,400	5.43	0.553	6,500	8.20	1.07
3,400	7.09	1.095	3,600	5.74	0.618	7,000	8.83	1.24
3,600	7.50	1.22	3,800	6.06	0.684	7,500	9.46	1.42
3,800	7.92	1.36	4,000	6.38	0.754	8,000	10.09	1.61
4,000	8.34	1.50	4,500	7.18	0.948	8,500	10.7	1.81
4,500	9.38	1.89	5,000	7.98	1.17	9,000	11.3	2.02
5,000	10.4	2.31	5,500	8.78	1.41	10,000	12.6	2.48
5,500	11.5	2.79	6,000	9.57	1.66	11,000	13.9	3.01
6,000	12.5	3.31	7,000	11.2	2.26	12,000	15.1	3.56
6,500	13.5	3.88	8,000	12.8	2.96	13,000	16.4	4.19
7,000	14.6	4.50	9,000	14.4	3.73	14,000	17.7	4.85
7,500	15.6	5.16	10,000	16.0	4.57	15,000	18.9	5.56
8,000	16.7	5.87	11,000	17.6	5.50	16,000	20.2	6.31
8,500	17.7	6.60	12,000	19.1	6.52	17,000	21.4	7.11
9,000	18.8	7.42	13,000	20.7	7.63	18,000	22.7	7.95
9,500	19.8	8.27	14,000	22.3	8.81	19,000	24.0	8.85
10,000	20.8	9.15	15,000	23.9	10.1	20,000	25.2	9.79
11,000	22.9	11.05	16,000	25.5	11.5	21,000	26.5	10.78
12,000	25.0	13.0	17,000	27.1	13.0	22,000	27.7	11.9
13,000	27.1	15.2	18,000	28.7	14.6	23,000	29.0	12.9
14,000	29.2	17.6	19,000	30.3	16.3	24,000	30.3	14.1
15,000	31.3	20.2	20,000	31.9	18.1	25,000	31.5	15.3

NOTE: No allowance has been made for age, differences in diameter, or any abnormal condition of interior surface. Any factor of safety must be estimated from the local conditions and the requirements of each particular installation.

TABLE 3-11 FRICTION OF WATER IN NEW ASPHALT-DIPPED CAST-IRON PIPE *(Continued)*

Friction is in feet of water at 60°F per 100 ft of pipe.

gpm	Velocity fps	Friction ft./100 ft.	gpm	Velocity fps	Friction ft./100 ft.	gpm	Velocity fps	Friction ft./100 ft.
20-Inch Pipe			24-Inch Pipe			30-Inch Pipe		
800	0.817	0.0117	1,400	0.993	0.0135	2,000	0.908	0.00876
900	0.919	0.0146	1,600	1.135	0.0173	2,500	1.13	0.0132
1,000	1.02	0.0177	1,800	1.276	0.0216	3,000	1.36	0.0186
1,100	1.12	0.0212	2,000	1.42	0.0262	3,500	1.59	0.0248
1,200	1.23	0.0249	2,200	1.56	0.0313	4,000	1.82	0.0320
1,300	1.33	0.0289	2,400	1.70	0.0369	4,500	2.04	0.0400
1,400	1.43	0.0332	2,600	1.84	0.0429	5,000	2.27	0.0488
1,500	1.53	0.0378	2,800	1.99	0.0494	5,500	2.50	0.0585
1,600	1.63	0.0427	3,000	2.13	0.0563	6,000	2.72	0.0690
1,700	1.74	0.0478	3,500	2.48	0.0759	6,500	2.95	0.0803
1,800	1.84	0.0533	4,000	2.84	0.098	7,000	3.18	0.0923
1,900	1.94	0.0590	4,500	3.19	0.122	8,000	3.63	0.119
2,000	2.04	0.0650	5,000	3.55	0.149	9,000	4.08	0.149
2,500	2.55	0.0998	5,500	3.90	0.179	10,000	4.54	0.183
3,000	3.06	0.140	6,000	4.26	0.211	12,000	5.45	0.260
3,500	3.57	0.188	6,500	4.61	0.246	14,000	6.35	0.351
4,000	4.08	0.243	7,000	4.96	0.284	16,000	7.26	0.455
4,500	4.59	0.306	7,500	5.32	0.326	18,000	8.17	0.572
5,000	5.11	0.376	8,000	5.67	0.369	20,000	9.08	0.703
6,000	6.13	0.533	8,500	6.03	0.416	22,000	9.98	0.850
7,000	7.15	0.721	9,000	6.38	0.464	24,000	10.89	1.006
8,000	8.17	0.935	10,000	7.09	0.571	26,000	11.80	1.18
9,000	9.19	1.18	11,000	7.80	0.688	28,000	12.7	1.36
10,000	10.2	1.45	12,000	8.51	0.817	30,000	13.6	1.57
11,000	11.2	1.74	13,000	9.22	0.952	32,000	14.5	1.78
12,000	12.3	2.07	14,000	9.93	1.11	34,000	15.4	2.01
13,000	13.3	2.43	15,000	10.64	1.26	36,000	16.3	2.25
14,000	14.3	2.80	16,000	11.35	1.43	38,000	17.2	2.50
15,000	15.3	3.22	17,000	12.06	1.61	40,000	18.2	2.77
16,000	16.3	3.66	18,000	12.76	1.80	42,000	19.1	3.05
17,000	17.4	4.11	19,000	13.5	2.00	44,000	20.0	3.34
18,000	18.4	4.60	20,000	14.2	2.21	46,000	20.9	3.65
19,000	19.4	5.12	22,000	15.6	2.67	48,000	21.8	3.97
20,000	20.4	5.66	24,000	17.0	3.16	50,000	22.7	4.30
21,000	21.4	6.24	26,000	18.4	3.71	52,000	23.6	4.66
22,000	22.5	6.84	28,000	19.9	4.32	54,000	24.5	5.02
23,000	23.5	7.47	30,000	21.3	4.97	56,000	25.4	5.40
24,000	24.5	8.13	32,000	22.7	5.65	58,000	26.3	5.78
25,000	25.5	8.83	34,000	24.1	6.35	60,000	27.2	6.18
26,000	26.5	9.54	36,000	25.5	7.10	62,000	28.1	6.60
27,000	27.6	10.3	38,000	26.9	7.90	64,000	29.0	7.03
28,000	28.6	11.1	40,000	28.4	8.75	66,000	30.0	7.47
29,000	29.6	11.9	42,000	29.8	9.63	68,000	30.9	7.92
30,000	30.6	12.7	44,000	31.2	10.5	70,000	31.8	8.39

NOTE: No allowance has been made for age, differences in diameter, or any abnormal condition of interior surface. Any factor of safety must be estimated from the local conditions and the requirements of each particular installation.

TABLE 3-11 FRICTION OF WATER IN NEW ASPHALT-DIPPED CAST-IRON PIPE (Continued)

Friction is in feet of water at 60°F per 100 ft of pipe.

gpm	Velocity fps	Friction ft./100 ft.	gpm	Velocity fps	Friction ft./100 ft.	gpm	Velocity fps	Friction ft./100 ft.
	36-Inch Pipe			42-Inch Pipe			48-Inch Pipe	
3,000	0.946	0.00751	4,000	0.926	0.00602	5,000	0.887	0.00474
3,500	1.103	0.0101	5,000	1.16	0.00915	6,000	1.064	0.00667
4,000	1.26	0.0129	6,000	1.39	0.0128	7,000	1.24	0.00890
4,500	1.41	0.0161	7,000	1.62	0.0172	8,000	1.42	0.0114
5,000	1.58	0.0196	8,000	1.85	0.0222	9,000	1.60	0.0142
5,500	1.73	0.0234	9,000	2.08	0.0276	10,000	1.77	0.0173
6,000	1.89	0.0276	10,000	2.32	0.0337	11,000	1.95	0.0207
7,000	2.21	0.0369	11,000	2.55	0.0405	12,000	2.13	0.0244
8,000	2.52	0.0475	12,000	2.78	0.0478	13,000	2.30	0.0284
9,000	2.84	0.0593	13,000	3.01	0.0557	14,000	2.48	0.0327
10,000	3.15	0.0724	14,000	3.24	0.0641	15,000	2.66	0.0373
11,000	3.47	0.0868	15,000	3.47	0.0732	16,000	2.84	0.0422
12,000	3.78	0.103	16,000	3.71	0.0829	17,000	3.01	0.0474
13,000	4.10	0.120	17,000	3.94	0.0931	18,000	3.19	0.0529
14,000	4.41	0.139	18,000	4.17	0.104	19,000	3.37	0.0587
15,000	4.73	0.159	19,000	4.40	0.115	20,000	3.55	0.0648
16,000	5.04	0.180	20,000	4.63	0.127	22,000	3.90	0.0778
17,000	5.36	0.203	22,000	5.09	0.153	24,000	4.26	0.0920
18,000	5.67	0.227	24,000	5.56	0.181	26,000	4.61	0.1073
19,000	5.99	0.252	26,000	6.02	0.211	28,000	4.96	0.124
20,000	6.30	0.279	28,000	6.48	0.244	30,000	5.32	0.142
22,000	6.93	0.335	30,000	6.95	0.279	35,000	6.21	0.192
24,000	7.56	0.397	32,000	7.41	0.317	40,000	7.09	0.248
26,000	8.19	0.465	34,000	7.87	0.357	45,000	7.98	0.314
28,000	8.83	0.538	36,000	8.34	0.399	50,000	8.87	0.384
30,000	9.46	0.617	38,000	8.80	0.444	55,000	9.75	0.463
32,000	10.09	0.699	40,000	9.26	0.490	60,000	10.64	0.548
34,000	10.72	0.787	45,000	10.42	0.620	65,000	11.5	0.641
36,000	11.3	0.880	50,000	11.6	0.760	70,000	12.4	0.742
38,000	12.0	0.979	55,000	12.7	0.918	75,000	13.3	0.850
40,000	12.6	1.08	60,000	13.9	1.09	80,000	14.2	0.966
45,000	14.1	1.36	65,000	15.1	1.28	85,000	15.1	1.09
50,000	15.8	1.68	70,000	16.2	1.48	90,000	16.0	1.22
55,000	17.3	2.02	75,000	17.4	1.70	95,000	16.8	1.35
60,000	18.9	2.40	80,000	18.5	1.92	100,000	17.7	1.50
65,000	20.5	2.81	85,000	19.7	2.17	110,000	19.5	2.805
70,000	22.1	3.25	90,000	20.8	2.44	120,000	21.3	2.15
75,000	23.6	3.72	95,000	22.0	2.71	130,000	23.0	2.52
80,000	25.2	4.23	100,000	23.2	2.98	140,000	24.8	2.92
85,000	26.8	4.77	110,000	25.5	3.63	150,000	26.6	3.34
90,000	28.4	5.35	120,000	27.8	4.30	160,000	28.4	3.80
95,000	29.9	5.96	130,000	30.1	5.03	170,000	30.1	4.29
100,000	31.5	6.60	140,000	32.4	5.82	180,000	31.9	4.82

NOTE: No allowance has been made for age, differences in diameter, or any abnormal condition of interor surface. Any factor of safety must be estimated from the local conditions and the requirements of each particular installation.

TABLE 3-11 FRICTION OF WATER IN NEW ASPHALT-DIPPED CAST-IRON PIPE *(Continued)*

Friction is in feet of water at 60°F per 100 ft of pipe.

gpm	Velocity fps	Friction ft./100 ft.	gpm	Velocity fps	Friction ft./100 ft.	gpm	Velocity fps	Friction ft./100 ft.
54-Inch Pipe			60-Inch Pipe			72-Inch Pipe		
7,000	0.981	0.00499	8,000	0.908	0.00382	12,000	0.947	0.00329
8,000	1.121	0.00642	9,000	1.021	0.00476	13,000	1.025	0.00383
9,000	1.26	0.00800	10,000	1.13	0.00579	14,000	1.104	0.00438
10,000	1.40	0.00970	11,000	1.25	0.00691	15,000	1.18	0.00498
11,000	1.54	0.0116	12,000	1.36	0.00813	16,000	1.26	0.00563
12,000	1.68	0.0137	13,000	1.48	0.00945	17,000	1.34	0.00631
13,000	1.82	0.0159	14,000	1.59	0.01082	18,000	1.42	0.00703
14,000	1.96	0.0182	15,000	1.70	0.0123	19,000	1.50	0.00780
15,000	2.10	0.0207	16,000	1.82	0.0140	20,000	1.58	0.00859
16,000	2.24	0.0234	17,000	1.93	0.0156	25,000	1.97	0.0131
17,000	2.38	0.0263	18,000	2.04	0.0174	30,000	2.37	0.0184
18,000	2.52	0.0294	19,000	2.16	0.0195	35,000	2.76	0.0248
19,000	2.66	0.0326	20,000	2.27	0.0212	40,000	3.16	0.0320
20,000	2.80	0.0360	25,000	2.84	0.0325	45,000	3.55	0.0402
25,000	3.50	0.0550	30,000	3.40	0.0460	50,000	3.94	0.0493
30,000	4.20	0.0782	35,000	3.97	0.0618	55,000	4.34	0.0592
35,000	4.90	0.106	40,000	4.54	0.0800	60,000	4.73	0.0700
40,000	5.60	0.137	45,000	5.11	0.100	65,000	5.13	0.0816
45,000	6.30	0.172	50,000	5.67	0.124	70,000	5.52	0.0940
50,000	7.00	0.211	55,000	6.24	0.149	75,000	5.92	0.1074
55,000	7.70	0.254	60,000	6.81	0.176	80,000	6.31	0.122
60,000	8.40	0.301	70,000	7.94	0.237	90,000	7.10	0.154
65,000	9.11	0.352	80,000	9.08	0.307	100,000	7.89	0.189
70,000	9.81	0.408	90,000	10.21	0.387	110,000	8.68	0.228
75,000	10.51	0.467	100,000	11.3	0.478	120,000	9.47	0.271
80,000	11.21	0.530	110,000	12.5	0.578	130,000	10.25	0.317
85,000	11.91	0.597	120,000	13.6	0.688	140,000	11.04	0.365
90,000	12.6	0.668	130,000	14.8	0.805	150,000	11.8	0.418
95,000	13.3	0.743	140,000	15.9	0.930	160,000	12.6	0.475
100,000	14.0	0.820	150,000	17.0	1.062	170,000	13.4	0.534
110,000	15.4	0.982	160,000	18.2	1.20	180,000	14.2	0.598
120,000	16.8	1.18	170,000	19.3	1.36	190,000	15.0	0.665
130,000	18.2	1.38	180,000	20.4	1.52	200,000	15.8	0.736
140,000	19.6	1.59	190,000	21.6	1.69	220,000	17.4	0.887
150,000	21.0	1.82	200,000	22.7	1.87	240,000	18.9	1.050
160,000	22.4	2.07	210,000	23.8	2.06	260,000	20.5	1.237
170,000	23.8	2.34	220,000	25.0	2.26	280,000	22.1	1.43
180,000	25.2	2.62	230,000	26.1	2.47	300,000	23.7	1.64
190,000	26.6	2.91	240,000	27.2	2.69	320,000	25.2	1.86
200,000	28.0	3.22	250,000	28.4	2.92	340,000	26.8	2.10
210,000	29.4	3.55	260,000	29.5	3.16	360,000	28.4	2.35
220,000	30.8	3.89	270,000	30.6	3.40	380,000	30.0	2.62
230,000	32.2	4.25	280,000	31.8	3.66	400,000	31.6	2.90

NOTE: No allowance has been made for age, differences in diameter, or any abnormal condition of interior surface. Any factor of safety must be estimated from the local conditions and the requirements of each particular installaton.

RESISTANCE COEFFICIENTS FOR INCREASERS AND DIFFUSERS

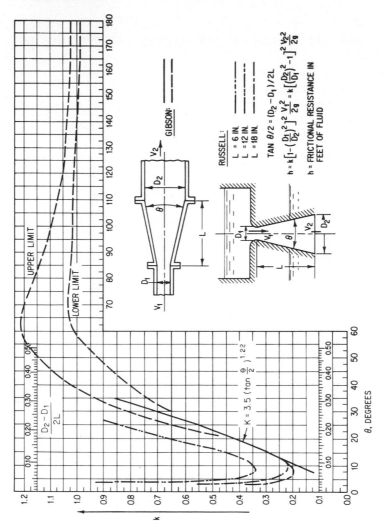

RESISTANCE COEFFICIENTS FOR REDUCERS

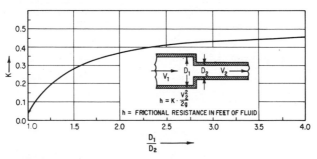

FIG. 3-11

TABLE 3-12 RESISTANCE OF VALVES AND FITTINGS TO FLOW

Equivalent Length of New Straight Pipe

Equivalent length is in *feet* and applies to *turbulent flow* only.

FITTINGS			¼	⅜	½	¾	1	1¼	1½	2	2½	3	4
REGULAR 90° ELBOW	SCREWED	STEEL	2.3	3.1	3.6	4.4	5.2	6.6	7.4	8.5	9.3	11	13
		C. I.										9.0	11
	FLANGED	STEEL			0.9	1.2	1.6	2.1	2.4	3.1	3.6	4.4	5.9
		C. I.										3.6	4.8
LONG RADIUS 90° ELBOW	SCREWED	STEEL	1.5	2.0	2.2	2.3	2.7	3.2	3.4	3.6	3.6	4.0	4.6
		C. I.										3.3	3.7
	FLANGED	STEEL			1.1	1.3	1.6	2.0	2.3	2.7	2.9	3.4	4.2
		C. I.										2.8	3.4
REGULAR 45° ELBOW	SCREWED	STEEL	0.3	0.5	0.7	0.9	1.3	1.7	2.1	2.7	3.2	4.0	5.5
		C. I.										3.3	4.5
	FLANGED	STEEL			0.5	0.6	0.8	1.1	1.3	1.7	2.0	2.6	3.5
		C. I.										2.1	2.9
TEE WITH LINE FLOW	SCREWED	STEEL	0.8	1.2	1.7	2.4	3.2	4.6	5.6	7.7	9.3	12	17
		C. I.										9.9	14
	FLANGED	STEEL			0.7	0.8	1.0	1.3	1.5	1.8	1.9	2.2	2.8
		C. I.										1.9	2.2
TEE WITH BRANCH FLOW	SCREWED	STEEL	2.4	3.5	4.2	5.3	6.6	8.7	9.9	12	13	17	21
		C. I.										14	17
	FLANGED	STEEL			2.0	2.6	3.3	4.4	5.2	6.6	7.5	9.4	12
		C. I.										7.7	10
180° RETURN BEND	SCREWED	STEEL	2.3	3.1	3.6	4.4	5.2	6.6	7.4	8.5	9.3	11	13
		C. I.										9.0	11
	REGULAR FLANGED	STEEL			0.9	1.2	1.6	2.1	2.4	3.1	3.6	4.4	5.9
		C. I.										3.6	4.8
	LONG RADIUS FLANGED	STEEL			1.1	1.3	1.6	2.0	2.3	2.7	2.9	3.4	4.2
		C. I.										2.8	3.4

TABLE 3-12   RESISTANCE OF VALVES AND FITTINGS TO FLOW (Continued)

Equivalent Length of New Straight Pipe

Equivalent length is in *feet* and applies to *turbulent flow* only.

NOMINAL PIPE SIZE IN INCHES																	
5	6	8	10	12	14	16	18	20	24	30	36	42	48	54	60	72	84
7.3	8.9	12	14	17	18	21	23	25	31	40	50	58	65	72	79	94	108
	7.2	9.8	12	15	17	19	22	24	28	36	43	50	57	64	69	82	95
5.0	5.7	7.0	7.7	9.0	9.4	10	11	12	14	16	19	21	23	25	28	31	34
	4.7	5.7	6.8	7.8	8.6	9.6	11	11	13	14	16	19	22	22	24	27	30
4.5	5.6	7.7	9.0	11	13	15	16	18	22								
	4.5	6.3	8.1	9.7	12	13	15	17	20								
3.3	3.8	4.7	5.2	6.0	6.4	7.2	7.6	8.2	9.7	12	14	16	18	19	21	24	27
	3.1	3.9	4.6	5.2	5.9	6.5	7.2	7.7	8.9	11	12	14	16	17	18	21	24
15	18	24	28	34	37	43	47	52	62	78	96	108	123	137	150	178	202
	15	20	25	30	35	39	44	49	57	70	82	95	110	120	131	155	178
7.3	8.9	12	14	17	18	21	23	25	30								
	7.2	9.8	12	15	17	19	22	24	28								
5.0	5.7	7.0	7.7	9.0	9.4	10	11	12	14								
	4.7	5.7	6.8	7.8	8.6	9.6	11	11	13								

TABLE 3-12 RESISTANCE OF VALVES AND FITTINGS TO FLOW *(Continued)*

Equivalent Length of New Straight Pipe

Equivalent length is in *feet* and applies to *turbulent flow* only.

FITTINGS			¼	⅜	½	¾	1	1¼	1½	2	2½	3	4
			NOMINAL PIPE SIZE IN INCHES										
GLOBE VALVE	SCREWED	STEEL	21	22	22	24	29	37	42	54	62	79	110
		C. I.										65	86
	FLANGED	STEEL			38	40	45	54	59	70	77	94	120
		C. I.										77	99
GATE VALVE	SCREWED	STEEL	0.3	0.5	0.6	0.7	0.8	1.1	1.2	1.5	1.7	1.9	2.5
		C. I.										1.6	2.0
	FLANGED	STEEL								2.8	2.7	2.8	2.9
		C. I.										2.3	2.4
ANGLE VALVE	SCREWED	STEEL	13	15	15	15	17	18	18	18	18	18	18
		C. I.										15	15
	FLANGED	STEEL			15	15	17	18	18	21	22	28	38
		C. I.										23	31
SWING CHECK VALVE	SCREWED	STEEL	7.2	7.3	8.0	8.8	11	13	15	19	22	27	38
		C. I.										22	31
	FLANGED	STEEL			3.8	5.3	7.2	10	12	17	21	27	38
		C. I.										22	31
COUPLING OR UNION	SCREWED	STEEL	.14	.18	.21	.24	.29	.36	.39	.45	.47	.53	.65
		C. I.										.44	.52
BELL MOUTH INLET		STEEL	.04	.07	.10	.13	.18	.26	.31	.43	.52	.67	.95
		C. I.										.55	.77
SQUARE EDGE INLET		STEEL	.44	.68	.96	1.3	1.8	2.6	3.1	4.3	5.2	6.7	9.5
		C. I.										5.5	7.7
RE-ENTRANT PIPE		STEEL	0.9	1.4	1.9	2.6	3.6	5.1	6.2	8.5	10	13	19
		C. I.										11	15
SUDDEN ENLARGE-MENT		LOSS OF HEAD $= h = (V_1 - V_2)^2/2g$ FEET OF FLUID;											

TABLE 3-12 RESISTANCE OF VALVES AND FITTINGS TO FLOW *(Continued)*

Equivalent Length of New Straight Pipe

Equivalent length is in *feet* and applies to *turbulent flow* only.

NOMINAL PIPE SIZE IN INCHES																	
5	6	8	10	12	14	16	18	20	24	30	36	42	48	54	60	72	84
150	190	260	310	390													
	150	210	270	330													
3.1	3.2	3.2	3.2	3.2	3.2	3.2	3.2	3.2	3.3	3.5	3.5	3.7	3.7	3.6	3.7	3.9	3.8
	2.6	2.7	2.8	2.9	2.9	3.0	3.0	3.0	3.0	3.1	3.0	3.2	3.2	3.2	3.2	3.5	3.4
50	63	90	110	140	160	190	210	240	300								
	52	74	98	120	150	170	200	230	280								
50	63	90	110	140													
	52	74	98	120													
1.3	1.6	2.3	2.7	3.5	4.0	4.7	5.3	6.1	7.7	10	13	16	19	21	24	30	36
	1.3	1.9	2.4	3.0	3.6	4.3	5.0	5.7	7.1	9.2	11	14	16	19	21	26	32
13	16	23	27	35	40	47	53	61	77	103	133	159	186	214	242	301	360
	13	19	24	30	36	43	50	57	71	92	114	139	162	188	212	263	318
25	32	45	55	70	80	95	110	120	154	206	265	318	372	428	484	602	720
	26	37	49	61	73	86	100	110	142	184	227	278	324	376	424	526	635

IF $V_2 = O$, $h = V_1^2/2g$ FEET OF FLUID.

Where V_1 is velocity in smaller pipe and V_2 is the velocity in larger pipe.

FRICTION LOSS FOR VISCOUS FLUIDS

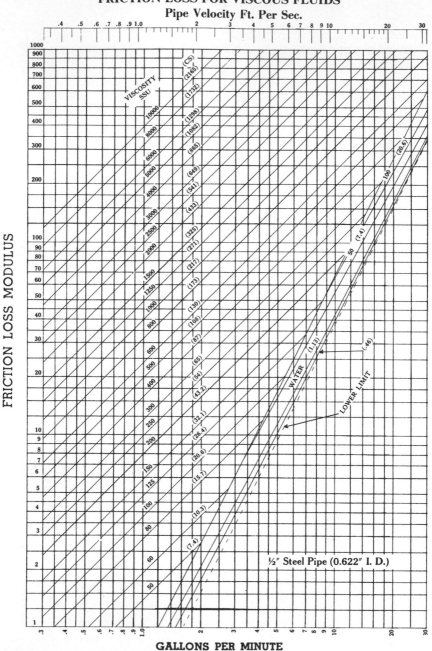

FIG. 3-12

Friction-Loss Modulus for 100 Feet of Pipe

Loss, lb/in^2 = modulus $\times$ specific gravity

Loss, ft of liquid = modulus $\times$ 2.31

FRICTION LOSS FOR VISCOUS FLUIDS *(Continued)*

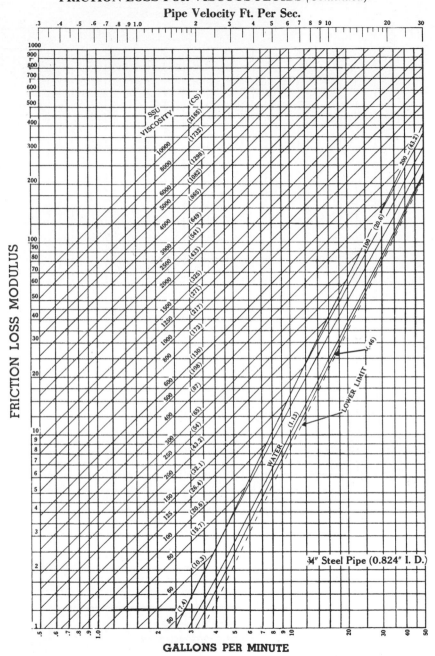

FIG. 3-12 *(Cont.)*

Friction-Loss Modulus for 100 Feet of Pipe

Loss, lb/in² = modulus × specific gravity

Loss, ft of liquid = modulus × 2.31

FRICTION LOSS FOR VISCOUS FLUIDS *(Continued)*

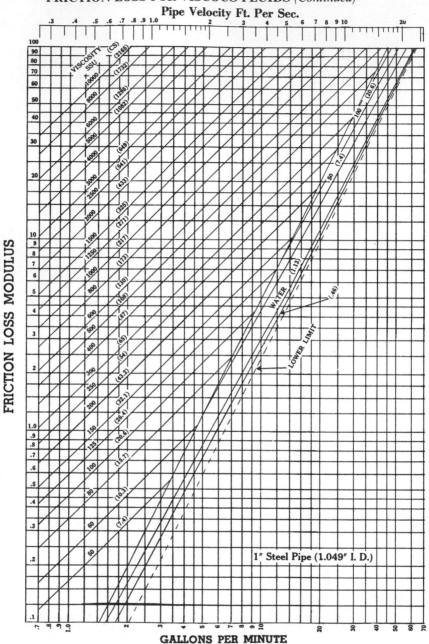

FIG. 3-12 *(Cont.)*

Friction-Loss Modulus for 100 Feet of Pipe

Loss, lb/in^2 = modulus × specific gravity

Loss, ft of liquid = modulus × 2.31

3-38

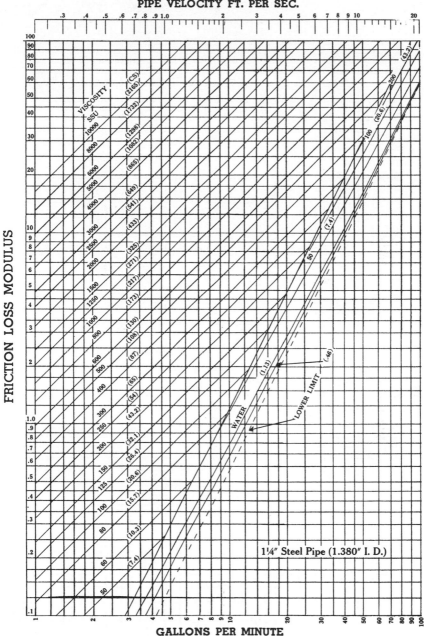

FIG. 3-12 *(Cont.)*

Friction-Loss Modulus for 100 Feet of Pipe

Loss, lb/in^2 = modulus $\times$ specific gravity

Loss, ft of liquid = modulus $\times$ 2.31

FRICTION LOSS FOR VISCOUS FLUIDS *(Continued)*
PIPE VELOCITY FT. PER SEC.

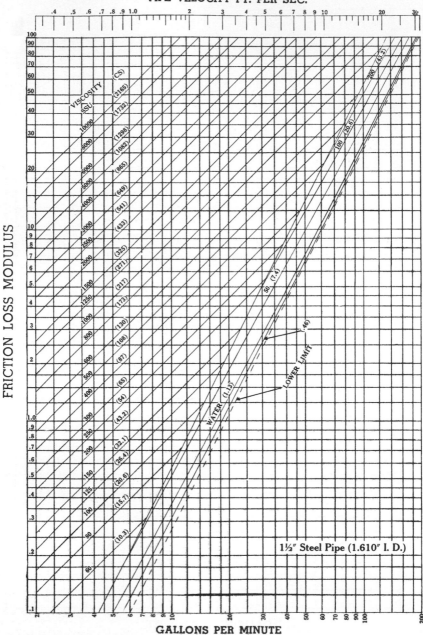

FIG. 3-12 *(Cont.)*

Friction-Loss Modulus for 100 Feet of Pipe

Loss, lb/in² = modulus × specific gravity

Loss, ft of liquid = modulus × 2.31

FRICTION LOSS FOR VISCOUS FLUIDS *(Continued)*
PIPE VELOCITY FT. PER SEC.

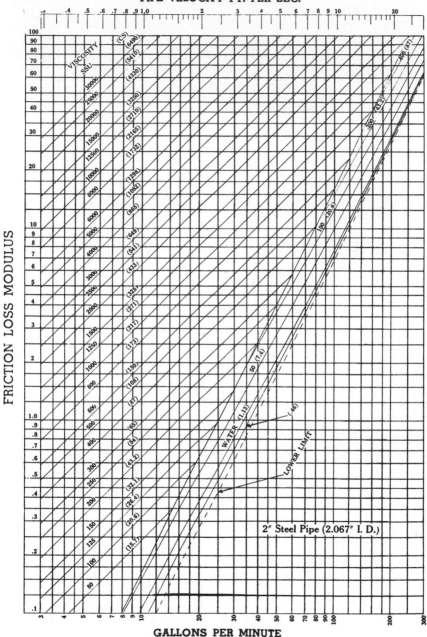

FIG. 3-12 *(Cont.)*

Friction-Loss Modulus for 100 Feet of Pipe

Loss, lb/in^2 = modulus $\times$ specific gravity

Loss, ft of liquid = modulus $\times$ 2.31

FRICTION LOSS FOR VISCOUS FLUIDS *(Continued)*
PIPE VELOCITY FT. PER SEC.

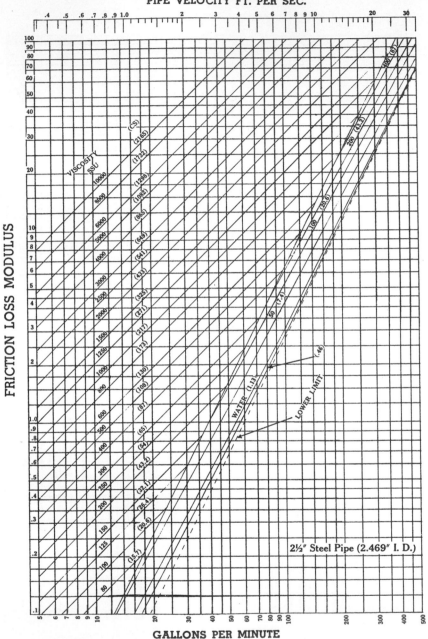

FIG. 3-12 *(Cont.)*

Friction-Loss Modulus for 100 Feet of Pipe

Loss, lb/in^2 = modulus $\times$ specific gravity

Loss, ft of liquid = modulus $\times$ 2.31

FRICTION LOSS FOR VISCOUS FLUIDS *(Continued)*

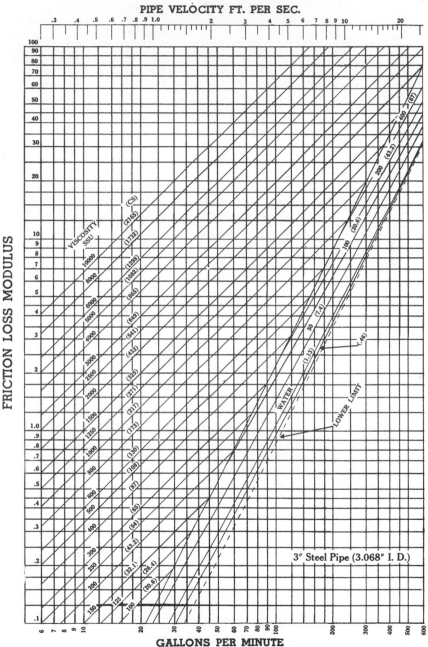

FIG. 3-12 *(Cont.)*

Friction-Loss Modulus for 100 Feet of Pipe

Loss, lb/in^2 = modulus $\times$ specific gravity

Loss, ft of liquid = modulus $\times$ 2.31

FRICTION LOSS FOR VISCOUS FLUIDS *(Continued)*

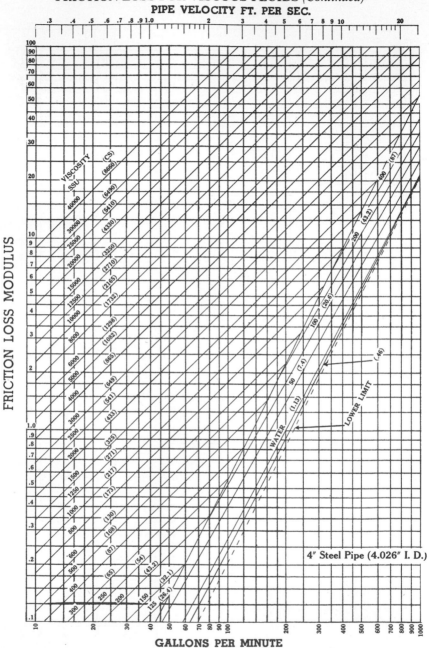

FIG. 3-12 *(Cont.)*

Friction-Loss Modulus for 100 Feet of Pipe

Loss, lb/in^2 = modulus × specific gravity

Loss, ft of liquid = modulus × 2.31

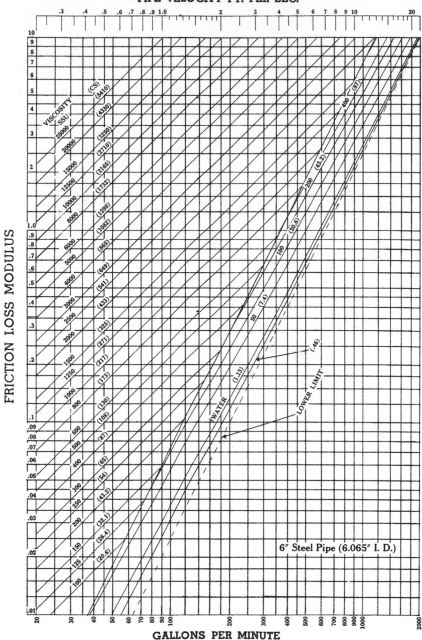

FRICTION LOSS FOR VISCOUS FLUIDS *(Continued)*
PIPE VELOCITY FT. PER SEC.

FRICTION LOSS MODULUS

GALLONS PER MINUTE

6″ Steel Pipe (6.065″ I. D.)

FIG. 3-12 *(Cont.)*

Friction-Loss Modulus for 100 Feet of Pipe

Loss, lb/in^2 = modulus $\times$ specific gravity

Loss, ft of liquid = modulus $\times$ 2.31

FRICTION LOSS FOR VISCOUS FLUIDS *(Continued)*

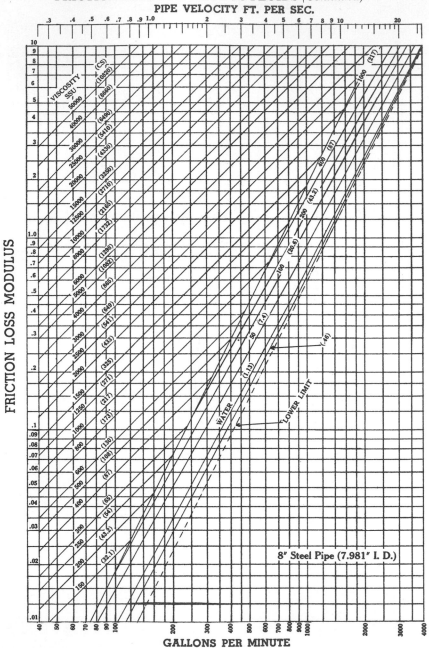

FIG. 3-12 *(Cont.)*

Friction-Loss Modulus for 100 Feet of Pipe

Loss, lb/in^2 = modulus × specific gravity

Loss, ft of liquid = modulus × 2.31

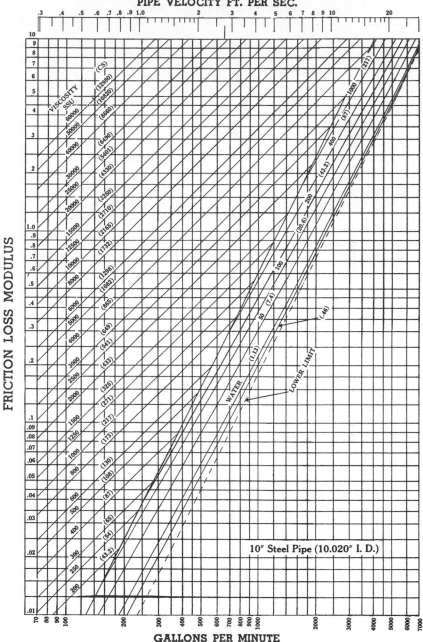

FIG. 3-12 *(Cont.)*

Friction-Loss Modulus for 100 Feet of Pipe

Loss, lb/in^2 = modulus $\times$ specific gravity

Loss, ft of liquid = modulus $\times$ 2.31

FRICTION LOSS FOR VISCOUS FLUIDS *(Continued)*

PIPE VELOCITY FT. PER SEC.

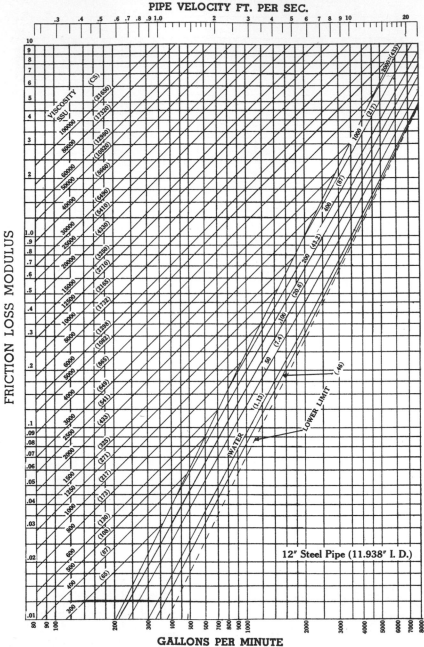

FIG. 3-12 *(Cont.)*

Friction-Loss Modulus for 100 Feet of Pipe

Loss, lb/in² = modulus × specific gravity

Loss, ft of liquid = modulus × 2.31

TABLE 3-13   PROPERTIES OF SATURATED STEAM

Abs pressure, psi	Temp, °F	Specific volume		Enthalpy			Entropy			Internal energy, evap
		Liquid	Vapor	Liquid	Evap	Vapor	Liquid	Evap	Vapor	
1.0	101.74	0.01614	333.60	69.73	1,036.1	1,105.8	0.1326	1.8455	1.9781	976.4
1.2	107.91	0.01616	280.96	75.90	1,032.6	1,108.5	0.1436	1.8192	1.9628	970.2
1.4	113.26	0.01618	243.02	81.23	1,029.5	1,110.7	0.1529	1.7969	1.9498	966.6
1.6	117.98	0.01620	214.33	85.95	1,026.8	1,112.7	0.1611	1.7775	1.9386	963.4
1.8	122.22	0.01621	191.85	90.18	1,024.3	1,114.5	0.1684	1.7604	1.9288	960.4
2.0	126.07	0.01623	173.76	94.03	1,022.1	1,116.2	0.1750	1.7450	1.9200	957.8
2.2	129.61	0.01624	158.87	97.57	1,020.1	1,117.6	0.1810	1.7311	1.9121	955.4
2.4	132.88	0.01626	146.40	100.84	1,018.2	1,119.0	0.1865	1.7183	1.9048	953.2
2.6	135.93	0.01627	135.80	103.88	1,016.4	1,120.3	0.1917	1.7065	1.8982	951.0
2.8	138.78	0.01629	126.67	106.73	1,014.7	1,121.5	0.1964	1.6956	1.8921	949.1
3.0	141.47	0.01630	118.73	109.42	1,013.2	1,122.6	0.2009	1.6854	1.8864	947.3
4.0	152.96	0.01636	90.64	120.92	1,006.4	1,127.3	0.2199	1.6428	1.8626	939.3
5.0	162.24	0.01640	73.53	130.20	1,000.9	1,131.1	0.2349	1.6094	1.8443	932.9
6.0	170.05	0.01645	61.98	138.03	996.2	1,134.2	0.2474	1.5820	1.8294	927.4
7.0	176.84	0.01649	53.65	144.83	992.1	1,136.9	0.2581	1.5587	1.8168	922.6
8.0	182.86	0.01653	47.34	150.87	988.5	1,139.3	0.2676	1.5384	1.8060	918.4
9.0	188.27	0.01656	42.40	156.30	985.1	1,141.4	0.2760	1.5204	1.7964	914.5
10	193.21	0.01659	38.42	161.26	982.1	1,143.3	0.2836	1.5043	1.7879	911.1
11	197.75	0.01662	35.14	165.82	979.3	1,145.1	0.2906	1.4896	1.7802	907.8
12	201.96	0.01665	32.39	170.05	976.6	1,146.7	0.2970	1.4763	1.7731	904.8
13	205.88	0.01668	30.05	174.00	974.2	1,148.2	0.3029	1.4638	1.7667	901.9
14	209.56	0.01670	28.04	177.71	971.9	1,149.5	0.3085	1.4522	1.7607	899.2
14.696	212.00	0.01672	26.79	180.17	970.3	1,150.5	0.3121	1.4447	1.7568	897.5
15	213.03	0.01673	26.29	181.21	969.7	1,150.9	0.3137	1.4415	1.7552	896.7
16	216.32	0.01675	24.75	184.52	967.6	1,152.1	0.3186	1.4314	1.7500	984.3
17	219.44	0.01677	23.38	187.66	965.6	1,153.2	0.3232	1.4219	1.7451	892.1
18	222.41	0.01679	22.17	190.66	963.7	1,154.3	0.3276	1.4129	1.7405	889.9
19	225.24	0.01681	21.07	193.52	961.8	1,155.3	0.3318	1.4043	1.7361	887.7
20	227.96	0.01683	20.087	196.27	960.1	1,156.3	0.3358	1.3962	1.7320	885.8
21	230.57	0.01685	19.190	198.90	958.4	1,157.3	0.3396	1.3885	1.7281	883.9
22	233.07	0.01687	18.373	201.44	956.7	1,158.1	0.3433	1.3811	1.7244	882.0
23	235.49	0.01689	17.624	203.88	955.1	1,159.0	0.3468	1.3740	1.7208	880.2
24	237.82	0.01691	16.936	206.24	953.6	1,159.8	0.3502	1.3672	1.7174	878.4
25	240.07	0.01693	16.301	208.5	952.1	1,160.6	0.3535	1.3607	1.7141	876.8
26	242.25	0.01694	15.714	210.7	950.6	1,161.4	0.3566	1.3544	1.7110	875.2
27	244.36	0.01696	15.168	212.9	949.2	1,162.1	0.3596	1.3483	1.7080	873.5
28	246.41	0.01698	14.661	214.9	947.9	1,162.8	0.3626	1.3425	1.7050	871.9
29	248.40	0.01699	14.181	217.0	946.5	1,163.5	0.3654	1.3368	1.7022	870.5
30	250.34	0.01701	13.744	218.9	945.2	1,164.1	0.3682	1.3313	1.6995	869.1
31	252.22	0.01702	13.328	220.8	943.9	1,164.8	0.3709	1.3260	1.6969	867.6
32	254.05	0.01704	12.938	222.7	942.7	1,165.4	0.3735	1.3209	1.6944	866.2
33	255.84	0.01705	12.570	224.5	941.5	1,166.0	0.3760	1.3159	1.6919	864.8
34	257.58	0.01707	12.223	226.3	940.3	1,166.6	0.3785	1.3110	1.6895	863.5
35	259.29	0.01708	11.896	228.0	939.1	1,167.1	0.3809	1.3063	1.6872	862.2
36	260.95	0.01709	11.586	229.7	938.0	1,167.7	0.3833	1.3017	1.6849	860.9
37	262.58	0.01711	11.292	231.4	936.9	1,168.2	0.3856	1.2972	1.6827	659.6
38	264.17	0.01712	11.014	233.0	935.8	1,168.8	0.3878	1.2928	1.6806	858.4
39	265.72	0.01714	10.749	234.6	934.7	1,169.3	0.3900	1.2885	1.6785	857.2
40	267.25	0.01715	10.496	236.1	933.6	1,169.8	0.3921	1.2844	1.6765	856.1
41	268.74	0.01716	10.256	237.7	932.6	1,170.2	0.3942	1.2803	1.6745	854.9
42	270.21	0.01718	10.027	239.2	931.5	1,170.7	0.3962	1.2763	1.6726	853.8
43	271.65	0.01719	9.808	240.6	930.5	1,171.2	0.3983	1.2724	1.6707	852.6
44	273.06	0.01720	9.599	242.1	929.5	1,171.6	0.4002	1.2686	1.6689	851.6
45	274.44	0.01721	9.399	243.5	928.6	1,172.0	0.4021	1.2649	1.6671	850.5
46	275.80	0.01723	9.207	244.9	927.6	1,172.5	0.4040	1.2613	1.6653	849.4
47	277.14	0.01724	9.023	246.2	926.6	1,172.9	0.4059	1.2577	1.6636	848.3
48	278.45	0.01725	8.846	247.6	925.7	1,173.3	0.4077	1.2542	1.6619	847.3
49	279.74	0.01726	8.677	248.9	924.8	1,173.7	0.4095	1.2507	1.6602	846.2

TABLE 3-13 PROPERTIES OF SATURATED STEAM
(Continued)

Abs pressure, psi	Temp, °F	Specific volume		Enthalpy			Entropy			Internal energy, evap
		Liquid	Vapor	Liquid	Evap	Vapor	Liquid	Evap	Vapor	
50	281.02	0.01727	8.514	250.2	923.9	1,174.1	0.4112	1.2474	1.6586	845.2
51	282.27	0.01728	8.357	251.5	923.0	1,174.5	0.4130	1.2441	1.6570	844.3
52	283.50	0.01730	8.206	252.8	922.1	1,174.9	0.4147	1.2408	1.6555	843.3
53	284.71	0.01731	8.060	254.0	921.2	1,175.2	0.4163	1.2376	1.6539	842.4
54	285.90	0.01732	7.920	255.2	920.4	1,175.6	0.4180	1.2345	1.6524	841.4
55	287.08	0.01733	7.785	256.4	919.5	1,175.9	0.4196	1.2314	1.6510	840.5
56	288.24	0.01734	7.654	257.6	918.7	1,176.3	0.4212	1.2284	1.6495	839.6
57	289.38	0.01735	7.528	258.8	917.8	1,176.6	0.4227	1.2254	1.6481	838.6
58	290.50	0.01736	7.406	259.9	917.0	1,177.0	0.4243	1.2224	1.6467	837.7
59	291.62	0.01737	7.288	261.1	916.2	1,177.3	0.4258	1.2196	1.6453	836.8
60	292.71	0.01738	7.174	262.2	915.4	1,177.6	0.4273	1.2167	1.6440	836.0
61	293.79	0.01739	7.063	263.3	914.6	1,177.9	0.4287	1.2139	1.6427	835.1
62	294.86	0.01740	6.956	264.4	913.8	1,178.2	0.4302	1.2112	1.6413	834.2
63	295.91	0.01741	6.852	265.5	913.0	1,178.6	0.4316	1.2084	1.6401	833.4
64	296.95	0.01742	6.751	266.6	912.3	1,178.8	0.4330	1.2058	1.6388	832.5
65	297.98	0.01743	6.653	267.6	911.5	1,179.1	0.4344	1.2031	1.6375	831.7
66	298.99	0.01744	6.558	268.7	910.8	1,179.4	0.4358	1.2005	1.6363	830.8
67	299.99	0.01745	6.466	269.7	910.0	1,179.7	0.4372	1.1979	1.6351	830.1
68	300.99	0.01746	6.376	270.7	909.3	1,180.0	0.4385	1.1954	1.6339	829.3
69	301.96	0.01747	6.290	271.7	908.5	1,180.3	0.4398	1.1929	1.6327	828.5
70	302.93	0.01748	6.205	272.7	907.8	1,180.6	0.4411	1.1905	1.6316	827.7
71	303.89	0.01749	6.123	273.7	907.1	1,180.8	0.4424	1.1880	1.6304	826.9
72	304.83	0.01750	6.042	274.7	906.4	1,181.1	0.4437	1.1856	1.6293	826.1
73	305.77	0.01751	5.964	275.7	905.7	1,181.3	0.4449	1.1833	1.6282	825.4
74	306.69	0.01752	5.888	276.6	905.0	1,181.6	0.4462	1.1809	1.6271	824.6
75	307.61	0.01753	5.814	277.6	904.3	1,181.9	0.4474	1.1786	1.6260	823.9
76	308.51	0.01754	5.742	278.5	903.6	1,182.1	0.4486	1.1763	1.6250	823.1
77	309.41	0.01755	5.672	279.4	902.9	1,182.4	0.4498	1.1741	1.6239	822.3
78	310.29	0.01756	5.603	280.3	902.3	1,182.6	0.4510	1.1718	1.6229	821.6
79	311.17	0.01756	5.536	281.3	901.6	1,182.8	0.4522	1.1696	1.6218	820.9
80	312.04	0.01757	5.471	282.1	900.9	1,183.1	0.4534	1.1675	1.6207	820.2
81	312.90	0.01758	5.407	283.0	900.3	1,183.3	0.4545	1.1653	1.6197	819.5
82	313.75	0.01759	5.345	283.9	899.6	1,183.5	0.4556	1.1632	1.6187	818.7
83	314.60	0.01760	5.284	284.8	899.0	1,183.8	0.4568	1.1611	1.6177	818.1
84	315.43	0.01761	5.225	285.7	898.3	1,184.0	0.4579	1.1590	1.6168	817.4
85	316.26	0.01762	5.167	286.5	897.7	1,184.2	0.4590	1.1569	1.6158	816.7
86	317.08	0.01762	5.110	287.4	897.0	1,184.4	0.4601	1.1549	1.6149	816.0
87	317.89	0.01763	5.055	288.2	896.4	1,184.6	0.4611	1.1529	1.6139	815.4
88	318.69	0.01764	5.000	289.0	895.8	1,184.8	0.4622	1.1509	1.6130	814.6
89	319.49	0.01765	4.947	289.9	895.2	1,185.0	0.4633	1.1489	1.6121	814.0
90	320.28	0.01766	4.895	290.7	894.6	1,185.3	0.4643	1.1470	1.6113	813.3
91	321.06	0.01767	4.844	291.5	893.9	1,185.5	0.4654	1.1450	1.6104	812.7
92	321.84	0.01768	4.795	292.3	893.3	1,185.7	0.4664	1.1431	1.6095	812.0
93	322.61	0.01768	4.746	293.18	892.7	1,185.9	0.4674	1.1412	1.6086	811.4
94	323.37	0.01769	4.698	293.9	892.1	1,186.0	0.4684	1.1393	1.6078	810.7
95	324.13	0.01770	4.651	294.7	891.5	1,186.2	0.4694	1.1375	1.6069	810.1
96	324.88	0.01771	4.606	295.5	891.0	1,186.4	0.4704	1.1356	1.6061	809.4
97	325.63	0.01772	4.560	296.3	890.4	1,186.6	0.4714	1.1338	1.6052	808.9
98	326.36	0.01772	4.517	297.0	889.8	1,186.8	0.4724	1.1320	1.6044	808.2
99	327.10	0.01773	4.473	297.8	889.2	1,187.0	0.4733	1.1302	1.6036	807.5
100	327.82	0.01774	4.431	298.5	888.6	1,187.2	0.4743	1.1284	1.6027	807.0
102	329.26	0.01776	4.349	300.0	887.5	1,187.5	0.4762	1.1249	1.6011	805.7
104	330.67	0.01777	4.270	301.5	886.4	1,187.9	0.4780	1.1215	1.5995	804.5
106	332.06	0.01779	4.193	303.0	885.2	1,188.2	0.4799	1.1181	1.5980	803.4
108	333.44	0.01780	4.120	304.4	884.1	1,188.5	0.4817	1.1148	1.5965	802.2
110	334.79	0.01782	4.048	305.8	883.1	1,188.9	0.4834	1.1115	1.5950	801.1
112	336.12	0.01783	3.980	307.2	882.0	1,189.2	0.4852	1.1083	1.5935	799.9
114	337.43	0.01785	3.914	308.6	880.9	1,189.5	0.4869	1.1052	1.5921	798.7
116	338.73	0.01786	3.850	309.9	879.9	1,189.8	0.4886	1.1021	1.5906	797.7
118	340.01	0.01787	3.788	311.3	878.8	1,190.1	0.4903	1.0990	1.5892	796.5
120	341.27	0.01789	3.728	312.6	877.8	1,190.4	0.4919	1.0960	1.5879	795.4
122	342.51	0.01790	3.670	313.9	876.8	1,190.7	0.4935	1.0930	1.5865	794.3

TABLE 3-13 PROPERTIES OF SATURATED STEAM
(Continued)

Abs pressure, psi	Temp, °F	Specific volume		Enthalpy			Entropy			Internal energy, evap
		Liquid	Vapor	Liquid	Evap	Vapor	Liquid	Evap	Vapor	
124	343.74	0.01792	3.613	315.2	875.8	1,190.9	0.4951	1.0901	1.5852	793.2
126	344.95	0.01793	3.559	316.4	874.8	1,191.2	0.4967	1.0872	1.5839	792.2
128	346.15	0.01794	3.506	317.7	873.8	1,191.5	0.4982	1.0843	1.5826	791.1
130	347.33	0.01796	3.454	319.0	872.8	1,191.7	0.4998	1.0815	1.5813	790.1
132	348.50	0.01797	3.405	320.2	871.3	1,192.0	0.5013	1.0788	1.5800	789.1
134	349.65	0.01799	3.356	321.4	870.8	1,192.2	0.5028	1.0760	1.5788	788.1
136	350.79	0.01800	3.309	322.6	869.9	1,192.5	0.5043	1.0733	1.5776	787.1
138	351.92	0.01801	3.263	323.8	869.9	1,192.7	0.5057	1.0707	1.5764	786.1
140	353.04	0.01803	3.219	325.0	868.0	1,193.0	0.5071	1.0681	1.5752	785.1
142	354.14	0.01804	3.176	326.1	867.1	1,193.2	0.5086	1.0655	1.5740	784.1
144	355.23	0.01805	3.134	327.3	866.2	1,193.4	0.5100	1.0629	1.5729	783.1
146	356.31	0.01806	3.093	328.4	865.2	1,193.6	0.5114	1.0604	1.5717	782.2
148	357.38	0.01808	3.053	329.5	864.3	1,193.9	0.5127	1.0579	1.5706	781.3
150	358.43	0.01809	3.014	330.6	863.4	1,194.1	0.5141	1.0554	1.5695	780.3
152	359.48	0.01810	2.976	331.8	862.5	1,194.3	0.5154	1.0530	1.5684	779.4
154	360.51	0.01812	2.939	332.8	861.6	1,194.5	0.5168	1.0506	1.5673	778.4
156	361.53	0.01813	2.903	333.9	860.8	1,194.7	0.5181	1.0482	1.5662	777.5
158	362.55	0.01814	2.868	335.0	859.9	1,194.9	0.5194	1.0458	1.5652	776.5
160	363.55	0.01815	2.834	336.1	859.0	1,195.1	0.5206	1.0435	1.5641	775.7
162	364.54	0.01817	2.800	337.1	858.2	1,195.3	0.5219	1.0412	1.5631	774.7
164	365.53	0.01818	2.767	338.2	857.3	1,195.5	0.5232	1.0389	1.5621	773.9
166	366.50	0.01819	2.736	339.2	856.5	1,195.7	0.5244	1.0367	1.5611	773.0
168	367.47	0.01820	2.704	340.2	855.6	1,195.8	0.5256	1.0344	1.5601	772.1
170	368.42	0.01821	2.674	341.2	854.8	1,196.0	0.5269	1.0322	1.5591	771.2
172	369.37	0.01823	2.644	342.2	853.9	1,196.2	0.5281	1.0300	1.5581	770.3
174	370.31	0.01824	2.615	343.2	853.1	1,196.4	0.5293	1.0279	1.5571	769.5
176	371.24	0.01825	2.586	344.2	852.3	1,196.5	0.5305	1.0257	1.5562	768.7
178	372.16	0.01826	2.558	345.2	851.5	1,196.7	0.5316	1.0236	1.5552	767.8
180	373.08	0.01827	2.531	346.2	850.7	1,196.9	0.5328	1.0215	1.5543	766.9
182	373.98	0.01828	2.504	347.2	849.9	1,197.0	0.5330	1.0194	1.5534	766.2
184	374.88	0.01830	2.478	348.1	849.1	1,197.2	0.5351	1.0174	1.5525	765.3
186	375.77	0.01831	2.453	349.1	848.3	1,197.3	0.5362	1.0153	1.5516	764.5
188	376.65	0.01832	2.428	350.0	847.5	1,197.5	0.5373	1.0133	1.5507	763.6
190	377.53	0.01833	2.403	350.9	846.7	1,197.6	0.5384	1.0113	1.5498	762.8
192	378.40	0.01834	2.379	351.9	845.9	1,197.8	0.5395	1.0094	1.5489	762.0
194	379.26	0.01835	2.355	352.8	845.1	1,197.9	0.5406	1.0074	1.5480	761.3
196	380.12	0.01836	2.332	353.7	844.4	1,198.1	0.5417	1.0054	1.5471	760.5
198	380.96	0.01838	2.310	354.6	843.6	1,198.2	0.5428	1.0035	1.5463	759.7
200	381.80	0.01839	2.287	355.5	842.8	1,198.3	0.5438	1.0016	1.5454	758.9
205	383.88	0.01841	2.233	357.7	840.9	1,198.7	0.5465	0.9969	1.5434	756.9
210	385.91	0.01844	2.182	359.9	839.1	1,199.0	0.5490	0.9923	1.5413	755.0
215	387.91	0.01847	2.133	362.1	837.2	1,199.3	0.5515	0.9878	1.5393	753.1
220	389.88	0.01850	2.086	364.2	835.4	1,199.6	0.5540	0.9834	1.5374	751.2
225	391.80	0.01852	2.0414	366.2	833.6	1,199.9	0.5564	0.9790	1.5354	749.4
230	393.70	0.01855	1.9985	368.3	831.8	1,200.1	0.5588	0.9748	1.5336	747.6
235	395.56	0.01857	1.9572	370.3	830.1	1,200.4	0.5611	0.9706	1.5317	745.8
240	397.39	0.01860	1.9177	372.3	828.4	1,200.6	0.5634	0.9665	1.5299	744.1
245	399.19	0.01863	1.8797	374.2	826.6	1,200.9	0.5657	0.9625	1.5281	742.2
250	400.97	0.01865	1.8432	376.1	825.0	1,201.1	0.5679	0.9585	1.5264	740.5
260	404.44	0.01870	1.7742	379.9	821.6	1,201.5	0.5722	0.9508	1.5230	737.2
270	407.80	0.01875	1.7101	383.6	818.3	1,201.9	0.5764	0.9433	1.5197	733.9
280	411.07	0.01880	1.6505	387.1	815.1	1,202.3	0.5805	0.9361	1.5166	730.6
290	414.25	0.01885	1.5948	390.6	812.0	1,202.6	0.5844	0.9291	1.5135	727.4
300	417.35	0.01889	1.5427	394.0	808.9	1,202.9	0.5882	0.9223	1.5105	724.3
320	423.31	0.01899	1.4480	400.5	802.9	1,203.4	0.5956	0.9092	1.5048	718.3
340	428.99	0.01908	1.3640	406.8	797.0	1,203.8	0.6026	0.8969	1.4994	712.4
360	434.41	0.01917	1.2891	412.8	791.3	1,204.1	0.6092	0.8851	1.4943	706.8
380	439.61	0.01925	1.2218	418.6	785.8	1,204.4	0.6156	0.8738	1.4894	701.3
400	444.60	0.01934	1.1610	424.2	780.4	1,204.6	0.6217	0.8630	1.4847	696.0
420	449.40	0.01942	1.1057	429.6	775.2	1,204.7	0.6276	0.8527	1.4802	690.8
440	454.03	0.01950	1.0557	434.8	770.0	1,204.8	0.6332	0.8427	1.4759	685.6
460	458.50	0.01959	1.0092	439.8	765.0	1,204.8	0.6387	0.8331	1.4718	680.7
480	462.82	0.01967	0.9668	444.7	760.9	1,204.8	0.6439	0.8238	1.4677	675.9

TABLE 3-13 PROPERTIES OF SATURATED STEAM
(Continued)

Abs pressure, psi	Temp, °F	Specific volume		Enthalpy			Entropy			Internal energy
		Liquid	Vapor	Liquid	Evap	Vapor	Liquid	Evap	Vapor	Vapor
500	467.01	0.0198	0.9276	449.5	755.1	1,204.7	0.6490	0.8148	1.4639	1,118.8
520	471.07	0.0198	0.8914	454.2	750.4	1,204.5	0.6540	0.8062	1.4601	1,118.8
540	475.01	0.0199	0.8577	458.7	745.7	1,204.4	0.6587	0.7977	1.4565	1,118.7
560	478.84	0.0200	0.8264	463.1	741.0	1,204.2	0.6634	0.7895	1.4529	1,118.5
580	482.57	0.0201	0.7971	467.5	736.5	1,203.9	0.6679	0.7816	1.4495	1,118.4
600	486.20	0.0201	0.7698	471.7	732.0	1,203.7	0.6723	0.7738	1.4461	1,118.2
620	489.74	0.0202	0.7441	475.8	727.5	1,203.4	0.6766	0.7662	1.4428	1,118.0
640	493.19	0.0203	0.7199	479.9	723.1	1,203.0	0.6808	0.7588	1.4396	1,117.8
660	496.57	0.0204	0.6972	483.9	718.8	1,202.7	0.6849	0.7516	1.4365	1,117.5
680	499.86	0.0204	0.6758	487.8	714.5	1,202.3	0.6889	0.7446	1.4334	1,117.2
700	503.08	0.0205	0.6556	491.6	710.2	1,201.8	0.6928	0.7377	1.4304	1,116.9
720	506.23	0.0206	0.6364	495.4	706.0	1,201.4	0.6966	0.7309	1.4275	1,116.6
740	509.32	0.0206	0.6182	499.1	701.9	1,200.9	0.7003	0.7243	1.4246	1,116.3
760	512.34	0.0207	0.6010	502.7	697.7	1,200.4	0.7040	0.7178	1.4218	1,115.9
780	515.30	0.0208	0.5846	506.3	693.6	1,199.9	0.7076	0.7114	1.4190	1,115.5
800	518.21	0.0209	0.5690	509.8	689.6	1,199.4	0.7111	0.7051	1.4163	1,115.2
820	521.06	0.0209	0.5541	513.3	685.5	1,198.8	0.7146	0.6990	1.4136	1,114.8
840	523.86	0.0210	0.5399	516.7	681.5	1,198.2	0.7180	0.6929	1.4109	1,114.3
860	526.60	0.0211	0.5263	520.1	677.6	1,197.7	0.7214	0.6869	1.4083	1,113.9
880	529.30	0.0212	0.5133	523.4	673.6	1,197.0	0.7247	0.6811	1.4057	1,113.4
900	531.95	0.0212	0.5009	526.7	669.7	1,196.4	0.7279	0.6753	1.4032	1,113.0
920	534.56	0.0213	0.4890	530.0	665.8	1,195.7	0.7311	0.6696	1.4007	1,112.5
940	537.13	0.0214	0.4776	533.2	661.9	1,195.1	0.7342	0.6640	1.3982	1,112.0
960	539.65	0.0214	0.4666	536.3	658.0	1,194.4	0.7373	0.6584	1.3958	1,111.5
980	542.14	0.0215	0.4561	539.5	654.2	1,193.7	0.7404	0.6530	1.3934	1,111.0
1,000	544.58	0.0216	0.4460	542.6	650.4	1,192.9	0.7434	0.6476	1.3910	1,110.4
1,050	550.53	0.0218	0.4222	550.1	640.9	1,191.0	0.7507	0.6344	1.3851	1,109.0
1,100	556.28	0.0220	0.4006	557.5	631.5	1,189.1	0.7578	0.6216	1.3794	1,107.4
1,150	561.82	0.0221	0.3807	564.8	622.2	1,187.0	0.7647	0.6091	1.3738	1,106.7
1,200	567.19	0.0223	0.3624	571.9	613.0	1,184.8	0.7714	0.5966	1.3683	1,104.0
1,250	572.38	0.0225	0.3456	578.8	603.8	1,182.6	0.7780	0.5850	1.3630	1,102.6
1,300	577.42	0.0227	0.3299	585.6	594.6	1,180.2	0.7843	0.5733	1.3577	1,000.9
1,350	582.35	0.0229	0.3148	592.1	584.0	1,176.1	0.7902	0.5604	1.3506	1,099.0
1,400	587.07	0.0231	0.3018	598.8	576.5	1,175.3	0.7966	0.5507	1.3474	1,097.1
1,450	591.73	0.0233	0.2884	605.2	565.5	1,170.7	0.8023	0.5379	1.3402	1,095.2
1,500	596.20	0.0235	0.2772	611.7	558.4	1,170.1	0.8085	0.5288	1.3373	1,093.1
1,600	604.87	0.0239	0.2554	624.2	540.3	1,164.5	0.8199	0.5076	1.3274	1,088.9
1,700	613.13	0.0243	0.2361	636.5	522.2	1,158.6	0.8309	0.4867	1.3176	1,084.4
1,800	621.02	0.0247	0.2186	648.5	503.8	1,153.3	0.8417	0.4662	1.3079	1,079.5
1,900	628.56	0.0252	0.2028	660.4	485.2	1,145.6	0.8522	0.4459	1.2981	1,074.3
2,000	635.80	0.0256	0.1883	672.1	466.2	1,138.3	0.8625	0.4256	1.2881	1,068.6
2,200	649.45	0.0267	0.1627	695.5	426.7	1,122.2	0.8828	0.3848	1.2676	1,055.9
2,400	662.11	0.0279	0.1408	719.0	384.8	1,103.7	0.9031	0.3430	1.2460	1,041.2
2,600	673.91	0.0294	0.1211	744.5	337.6	1,082.0	0.9247	0.2977	1.2225	1,023.8
2,800	684.96	0.0313	0.1031	770.7	285.1	1,055.8	0.9468	0.2491	1.1958	1,002.4
3,000	695.33	0.0343	0.0850	801.8	218.4	1,020.3	0.9728	0.1891	1.1619	973.1
3,200	705.08	0.0447	0.0566	875.5	56.1	931.6	1.0351	0.0482	1.0832	898.1
3,208.2	705.47	0.0508	0.0508	906.0	0.0	906.0	1.0612	0.0000	1.0612	875.9

Computed by permission from *1967 ASME Steam Tables,* American Society of Mechanical Engineers, New York, 1967.

TABLE 3-14   STEAM TABLE FOR USE IN CONDENSER CALCULATIONS

Temp, °F t	Abs pressure p		Specific volume sat vapor v_g	Enthalpy			Entropy	
	psi	in. Hg		Sat liquid h_f	Evap h_{fg}	Sat vapor H_g	Sat liquid s_f	Sat vapor s_g
50	0.17796	0.3623	1,704.8	18.05	1,065.3	1,083.4	0.0361	2.1262
52	0.19165	0.3901	1,589.2	20.06	1,064.2	1,084.2	0.0400	2.1197
54	0.20625	0.4199	1,482.2	22.06	1,063.1	1,085.1	0.0439	2.1134
56	0.22183	0.4516	1,383.6	24.06	1,061.9	1,086.0	0.0478	2.1070
58	0.23843	0.4854	1,292.2	26.06	1,060.8	1,086.9	0.0516	2.1008
60	0.25611	0.5214	1,207.6	28.06	1,059.7	1,087.7	0.0555	2.0946
62	0.27494	0.5597	1,129.2	30.06	1,058.5	1,088.6	0.0595	2.0885
64	0.29497	0.6005	1,056.5	32.06	1,057.4	1,089.5	0.0632	2.0824
66	0.31626	0.6439	989.1	34.06	1,056.3	1,090.4	0.0670	2.0764
68	0.33889	0.6899	926.5	36.05	1,055.2	1,091.2	0.0708	2.0704
70	0.36292	0.7389	868.4	38.05	1,054.0	1,092.1	0.0745	2.0645
72	0.38844	0.7908	814.3	40.05	1,052.9	1,093.0	0.0783	2.0587
74	0.41550	0.8459	764.1	42.05	1,051.8	1,093.8	0.0821	2.0529
76	0.44420	0.9044	717.4	44.04	1,050.7	1,094.7	0.0858	2.0472
78	0.47461	0.9663	673.9	46.04	1,049.5	1,095.6	0.0895	2.0415
80	0.50683	1.0319	633.3	48.04	1,048.4	1,096.4	0.0932	2.0359
82	0.54093	1.1013	595.6	50.03	1,047.3	1,097.3	0.0969	2.0303
84	0.57702	1.1748	560.3	52.03	1,046.1	1,098.2	0.1006	2.0248
86	0.61518	1.2525	527.5	54.03	1,045.0	1,099.0	0.1043	2.0193
88	0.65551	1.3346	496.8	56.02	1,043.9	1,099.9	0.1079	2.0139
90	0.69813	1.4214	468.1	58.02	1,042.7	1,100.8	0.1115	2.0086
92	0.74313	1.5130	441.3	60.01	1,041.6	1,101.6	0.1152	2.0033
94	0.79062	1.6097	416.3	62.01	1,040.5	1,102.5	0.1188	1.9980
96	0.84072	1.7117	392.9	64.01	1,039.3	1,103.3	0.1224	1.9928
98	0.89356	1.8193	370.9	66.00	1,038.2	1,104.2	0.1260	1.9876
100	0.94924	1.9326	350.4	68.00	1,037.1	1,105.1	0.1295	1.9825
102	1.00789	2.0520	331.1	70.00	1,035.9	1,105.9	0.1331	1.9775
104	1.06965	2.1778	313.1	71.99	1,034.8	1,106.8	0.1366	1.9725
106	1.1347	2.3102	296.2	73.99	1,033.6	1,107.6	0.1402	1.9675
108	1.2030	2.4493	280.3	75.98	1,032.5	1,108.5	0.1437	1.9626
110	1.2750	2.5956	265.4	77.98	1,031.4	1,109.3	0.1472	1.9577
112	1.3505	2.7496	251.4	79.98	1,030.2	1,110.2	0.1507	1.9528
114	1.4299	2.9112	238.2	81.97	1,029.1	1,111.0	0.1542	1.9480
116	1.5133	3.0810	225.8	83.97	1,027.9	1,111.9	0.1577	1.9433
118	1.6009	3.2594	214.2	85.97	1,026.8	1,112.7	0.1611	1.9386
120	1.6927	3.4463	203.26	87.97	1,025.6	1,113.6	0.1646	1.9339
122	1.7891	3.6423	192.95	89.96	1,024.5	1,114.4	0.1680	1.9293
124	1.8901	3.8482	183.24	91.96	1,023.3	1,115.3	0.1715	1.9247
126	1.9959	4.0636	174.09	93.96	1,022.2	1,116.1	0.1749	1.9202
128	2.1068	4.2894	165.47	95.96	1,021.0	1,117.0	0.1783	1.9157

Computed by permission from *1967 ASME Steam Tables*, American Society of Mechanical Engineers, New York, 1967.

TABLE 3-15 SUPERHEATED-STEAM TABLES

v = specific volume, ft³/lb; h = enthalpy, Btu/lb; s = entropy

Pressure, psi (saturation temp, °F)		Temp of steam, °F								
		340	380	420	460	500	550	600	650	700
20 (227.96)	v	23.59	24.82	26.04	27.25	28.46	29.96	31.47	32.97	34.46
	h	1,210.6	1,229.7	1,248.7	1,267.8	1,286.9	1,310.8	1,334.9	1,359.1	1,383.5
	s	1.8052	1.8285	1.8506	1.8718	1.8921	1.9164	1.9397	1.9621	1.9836
40 (267.25)	v	11.679	12.311	12.934	13.552	14.165	14.927	15.685	16.441	17.195
	h	1,206.8	1,226.6	1,246.2	1,265.6	1,285.0	1,309.3	1,333.6	1,358.0	1,382.5
	s	1.7250	1.7492	1.7720	1.7936	1.8143	1.8389	1.8624	1.8849	1.9065
60 (292.71)	v	7.705	8.140	8.566	8.985	9.400	9.914	10.425	10.932	11.438
	h	1,202.8	1,223.3	1,243.5	1,263.4	1,283.2	1,307.7	1,332.3	1,356.8	1,381.5
	s	1.6764	1.7015	1.7250	1.7471	1.7681	1.7931	1.8168	1.8395	1.8612
80 (312.03)	v	5.715	6.053	6.381	6.702	7.018	7.408	7.794	8.178	8.560
	h	1,198.6	1,220.0	1,240.8	1,261.2	1,281.3	1,306.2	1,330.9	1,355.7	1,380.5
	s	1.6405	1.6667	1.6909	1.7136	1.7349	1.7602	1.7842	1.8070	1.8289
100 (327.81)	v	4.519	4.799	5.068	5.331	5.588	5.904	6.216	6.525	6.833
	h	1,194.2	1,216.5	1,238.0	1,258.9	1,279.3	1,304.6	1,329.6	1,354.5	1,379.5
	s	1.6116	1.6389	1.6638	1.6870	1.7088	1.7344	1.7586	1.7816	1.8036
120 (341.25)	v	. . . .	3.962	4.193	4.416	4.634	4.901	5.163	5.424	5.681
	h	. . . .	1,212.9	1,235.1	1,256.5	1,277.4	1,302.9	1,328.2	1,353.3	1,378.4
	s	. . . .	1.6154	1.6412	1.6650	1.6872	1.7132	1.7376	1.7608	1.7829
140 (353.02)	v	. . . .	3.363	3.567	3.763	3.953	4.184	4.412	4.636	4.858
	h	. . . .	1,209.2	1,232.1	1,254.1	1,275.3	1,301.3	1,326.8	1,352.2	1,377.4
	s	. . . .	1.5948	1.6215	1.6459	1.6686	1.6949	1.7196	1.7430	1.7652
160 (363.53)	v	. . . .	2.913	3.097	3.272	3.441	3.647	3.848	4.046	4.242
	h	. . . .	1,205.3	1,229.1	1,251.6	1,273.3	1,299.6	1,325.4	1,351.0	1,376.4
	s	. . . .	1.5764	1.6041	1.6291	1.6522	1.6790	1.7039	1.7275	1.7499
180 (373.06)	v	. . . .	2.562	2.730	2.890	3.043	3.229	3.409	3.587	3.762
	h	. . . .	1,201.3	1,225.9	1,249.0	1,271.2	1,297.9	1,324.0	1,349.8	1,375.3
	s	. . . .	1.5596	1.5882	1.6140	1.6376	1.6647	1.6900	1.7137	1.7362

200 (381.79)	v	. . . .	. . . .	2.437	2.584	2.725	2.894	3.058	3.219	3.378
	h	. . . .	. . . .	1,222.6	1,246.4	1,269.0	1,296.2	1,322.6	1,348.6	1,374.3
	s	. . . .	. . . .	1.5737	1.6001	1.6242	1.6518	1.6773	1.7013	1.7239
220 (389.86)	v	. . . .	. . . .	2.196	2.334	2.464	2.620	2.771	2.919	3.064
	h	. . . .	. . . .	1,219.3	1,243.7	1,266.9	1,294.5	1,321.2	1,347.3	1,373.2
	s	. . . .	. . . .	1.5601	1.5873	1.6120	1.6400	1.6658	1.6900	1.7128
260 (404.42)	v	. . . .	. . . .	1.8246	1.9471	2.062	2.198	2.329	2.456	2.581
	h	. . . .	. . . .	1,212.2	1,238.2	1,262.4	1,290.9	1,318.2	1,344.9	1,371.1
	s	. . . .	. . . .	1.5353	1.5642	1.5899	1.6189	1.6453	1.6699	1.6930
300 (417.33)	v	. . . .	. . . .	1.5506	1.6627	1.7665	1.8833	2.004	2.117	2.226
	h	. . . .	. . . .	1,204.8	1,232.3	1,257.7	1,287.2	1,315.2	1,342.4	1,368.9
	s	. . . .	. . . .	1.5127	1.5433	1.5703	1.6003	1.6274	1.6524	1.6758
350 (431.72)	v	. . . .	. . . .	. . . .	1.3973	1.4913	1.6002	1.7028	1.8013	1.8970
	h	. . . .	. . . .	. . . .	1,224.7	1,251.5	1,282.4	1,311.4	1,339.2	1,366.2
	s	. . . .	. . . .	. . . .	1.5197	1.5483	1.5797	1.6077	1.6333	1.6571
400 (444.59)	v	. . . .	. . . .	. . . .	1.1970	1.2841	1.3836	1.4763	1.5646	1.6499
	h	. . . .	. . . .	. . . .	1,216.5	1,245.1	1,277.5	1,307.4	1,335.9	1,363.4
	s	. . . .	. . . .	. . . .	1.4978	1.5282	1.5611	1.5901	1.6163	1.6406

TABLE 3-15 SUPERHEATED-STEAM TABLES (Continued)

		Temp of steam, °F								
Pressure, psi (saturation temp, °F)		500	550	600	650	700	750	800	900	1000
450 (456.28)	v	1.1231	1.2154	1.3005	1.3810	1.4584	1.5337	1.6074	1.7516	1.8928
	h	1,238.4	1,272.0	1,302.8	1,331.9	1,359.9	1,387.3	1,414.3	1,467.7	1,521.0
	s	1.5095	1.5437	1.5735	1.6003	1.6250	1.6481	1.6699	1.7108	1.7486
500 (467.01)	v	0.9919	1.0791	1.1584	1.2327	1.3037	1.3725	1.4397	1.5708	1.6992
	h	1,231.2	1,267.0	1,299.1	1,329.1	1,357.7	1,385.4	1,412.7	1,466.6	1,520.3
	s	1.4921	1.5284	1.5595	1.5871	1.6123	1.6357	1.6578	1.6990	1.7371
550 (476.94)	v	0.8852	0.9686	1.0431	1.1124	1.1783	1.2419	1.3038	1.4241	1.5414
	h	1,223.7	1,261.2	1,294.3	1,324.9	1,354.0	1,382.3	1,409.9	1,464.3	1,518.2
	s	1.4751	1.5131	1.5451	1.5734	1.5991	1.6228	1.6452	1.6868	1.7250
600 (486.21)	v	0.7944	0.8746	0.9456	1.0109	1.0726	1.1318	1.1892	1.3008	1.4093
	h	1,215.9	1,255.6	1,290.3	1,322.0	1,351.8	1,380.4	1,408.3	1,463.0	1,517.4
	s	1.4590	1.4993	1.5329	1.5621	1.5884	1.6125	1.6351	1.6769	1.7155
700 (503.10)	v	. . . .	0.7271	0.7928	0.8520	0.9072	0.9596	1.0102	1.1078	1.2023
	h	. . . .	1,243.4	1,281.0	1,314.6	1,345.6	1,375.2	1,403.7	1,459.4	1,514.4
	s	. . . .	1.4726	1.5090	1.5509	1.5673	1.5923	1.6154	1.6580	1.6970
800 (518.23)	v	. . . .	0.6151	0.6774	0.7323	0.7828	0.8303	0.8759	0.9631	1.0470
	h	. . . .	1,230.1	1,271.1	1,306.8	1,339.3	1,369.8	1,399.1	1,455.8	1,511.4
	s	. . . .	1.4472	1.4869	1.5198	1.5484	1.5742	1.5980	1.6413	1.6807
900 (531.98)	v	. . . .	0.5263	0.5869	0.6388	0.6858	0.7296	0.7713	0.8504	0.9262
	h	. . . .	1,215.5	1,260.6	1,298.6	1,332.7	1,364.3	1,394.4	1,452.2	1,508.5
	s	. . . .	1.4223	1.4659	1.5010	1.5311	1.5578	1.5822	1.6263	1.6662
1,000 (544.61)	v	. . . .	0.4535	0.5137	0.5636	0.6080	0.6489	0.6875	0.7603	0.8295
	h	. . . .	1,199.3	1,249.3	1,290.1	1,325.9	1,358.7	1,389.6	1,448.5	1,505.4
	s	. . . .	1.3973	1.4457	1.4833	1.5149	1.5426	1.5677	1.6126	1.6530
1,100 (556.31)	v	. . . .	. . . .	0.4531	0.5017	0.5440	0.5826	0.6188	0.6865	0.7505
	h	. . . .	. . . .	1,237.3	1,281.2	1,318.8	1,352.0	1,384.7	1,444.7	1,502.4
	s	. . . .	. . . .	1.4259	1.4664	1.4996	1.5284	1.5542	1.6000	1.6410
1,200 (567.22)	v	. . . .	. . . .	0.4016	0.4497	0.4905	0.5273	0.5615	0.6250	0.6845
	h	. . . .	. . . .	1,224.2	1,271.8	1,311.5	1,346.9	1,379.7	1,440.9	1,499.4
	s	. . . .	. . . .	1.4061	1.4501	1.4851	1.5150	1.5415	1.5883	1.6298

P (T sat)								
1,400 (587.10)	v	0.3176	0.3667	0.4059	0.4400	0.4712	0.5282	0.5809
	h	1,194.1	1,251.4	1,296.1	1,334.5	1,369.3	1,433.2	1,493.2
	s	1.3652	1.4181	1.4575	1.4900	1.5182	1.5670	1.6096
1,600 (604.90)	v		0.3026	0.3415	0.3741	0.4032	0.4555	0.5031
	h		1,228.3	1,279.4	1,321.4	1,358.5	1,425.2	1,486.9
	s		1.3861	1.4312	1.4667	1.4968	1.5478	1.5916
1,800 (621.03)	v		0.2505	0.2906	0.3223	0.3500	0.3988	0.4426
	h		1,201.2	1,261.1	1,307.4	1,347.2	1,417.1	1,480.6
	s		1.3526	1.405	1.4446	1.4768	1.5302	1.5753
2,000 (635.82)	v		0.2056	0.2488	0.2805	0.3072	0.3534	0.3942
	h		1,168.3	1,240.9	1,292.6	1,335.4	1,408.7	1,474.1
	s		1.3154	1.3794	1.4231	1.4578	1.5138	1.5603
2,200 (649.46)	v		0.1636	0.2134	0.2458	0.2720	0.3161	0.3545
	h		1,123.9	1,218.0	1,276.8	1,323.1	1,400.0	1,467.6
	s		1.2691	1.3523	1.4020	1.4395	1.4984	1.5463

Computed by permission from *1967 ASME Steam Tables*, American Society of Mechanical Engineers, New York, 1967.

TABLE 3-16   PROPERTIES OF THE LOWER ATMOSPHERE

Altitude, ft	Pressure, psia	Pressure, in. Hg at 32°F	Specific wt, lb/cu ft	Temp, °F
0	14.696	29.921	0.07648	59.0
100	14.64	29.81	0.0763	58.6
200	14.59	29.71	0.0760	58.3
300	14.54	29.60	0.0758	57.9
400	14.48	29.49	0.0756	57.6
500	14.43	29.38	0.0754	57.2
600	14.38	29.28	0.0751	56.9
700	14.33	29.17	0.0749	56.5
800	14.28	29.07	0.0747	56.1
900	14.22	28.96	0.0745	55.8
1,000	14.17	28.86	0.0743	55.4
1,100	14.12	28.75	0.0740	55.1
1,200	14.07	28.65	0.0738	54.7
1,300	14.02	28.54	0.0736	54.4
1,400	13.97	28.44	0.0734	54.0
1,500	13.92	28.33	0.0732	53.7
1,600	13.87	28.23	0.0730	53.3
1,700	13.82	28.13	0.0727	52.9
1,800	13.76	28.02	0.0725	52.6
1,900	13.71	27.92	0.0723	52.2
2,000	13.66	27.82	0.0721	51.9
2,200	13.56	27.62	0.0717	51.2
2,400	13.47	27.42	0.0712	50.4
2,600	13.37	27.21	0.0708	49.7
2,800	13.27	27.02	0.0704	49.0
3,000	13.17	26.82	0.0700	48.3
3,500	12.93	26.33	0.0689	46.5
4,000	12.69	25.84	0.0679	44.7
4,500	12.46	25.37	0.0669	43.0
5,000	12.23	24.90	0.0659	41.2
6,000	11.78	23.98	0.0639	37.6
7,000	11.34	23.09	0.0620	34.0
8,000	10.92	22.22	0.0601	30.5
9,000	10.50	21.39	0.0583	26.9
10,000	10.11	20.58	0.0565	23.3
12,000	9.346	19.03	0.0530	16.2
14,000	8.633	17.58	0.0497	9.1
15,000	8.293	16.89	0.0481	5.5
16,000	7.965	16.21	0.0466	1.9
18,000	7.339	14.94	0.0436	− 5.2
20,000	6.753	13.75	0.0407	−12.3
25,000	5.453	11.10	0.0343	−30.2
30,000	4.364	8.885	0.0286	−48.0
35,000	3.458	7.041	0.0237	−65.8
40,000	2.720	5.538	0.0188	−69.7
45,000	2.139	4.355	0.0148	−69.7
50,000	1.682	3.425	0.0116	−69.7
55,000	1.323	2.693	0.00915	−69.7
60,000	1.040	2.118	0.00720	−69.7
65,000	0.8180	1.665	0.00566	−69.7

Data from *NASA Standard Atmosphere*, 1976.

SONIC VELOCITY

Because the speed of rotating machinery such as turbines and centrifugal compressors is continually being increased, the velocity of sound in the fluid becomes increasingly important. If the velocity of the fluid equals or exceeds that of sound in the fluid, shock waves are set up, and the actions which take place are vastly different from those when the fluid velocities are below that of sound.

The ratio of the fluid velocity to that of sound in the fluid is called the *Mach number*. For a Mach number of 1 the fluid velocity equals that of sound, for a Mach number of 2 the fluid velocity is twice that of sound, and so on.

The velocity of sound in air at 68°F is 1126 ft/s. The value for any gas may be calculated from the equation $V = \sqrt{kgp/\gamma}$ in feet per second, where k is the ratio of the specific heats, g is the acceleration due to gravity $= 32.2$ ft/s^2, p is the absolute pressure of the gas in pounds per square foot, and γ is the specific weight of the gas in pounds per cubic foot. The equation may also be written as follows:

$$V = \sqrt{kgpv} = \sqrt{kgRT}$$

where v is the specific volume in cubic feet per pound, R is the gas constant, and T is the absolute temperature in °F.

If a supercompressibility factor is used in the calculation, the equation for sonic velocity changes to

$$V = \sqrt{kgZRT}$$

where Z is the compressibility factor.

An analysis of the equations shows that the sonic velocity is independent of the pressure of the gas but is directly proportional to the square root of the absolute temperature.

TABLE 3-17 AVERAGE CALORIC VALUES OF FUELS

This table gives the average caloric value of various fuels for use in heat-balance calculations. For more accurate values of individual fuels obtained from different sources, consult the references given.

Fuels	High heat value, Btu/lb	Gases	High heat value per cu ft of dry gas at 60°F and 30 in. Hg pressure	Avg specific gravity (air = 1.0 weighing 0.0750 lb/cu ft)
Solid fuels:	. . . .	Natural gas	1,000–1,100	0.60
Wood (with 12% moisture) . .	. . . .	Producer gas	150–160	0.86
Oak	15,000	Blast-furnace gas	90–92	1.02
Pine, yellow	13,000	Coke-oven gas	510–575	0.40
		Carbureted water gas . . .	500–550	0.60
Coal (as delivered):	. . . .	Coal gas	480–530	0.50
Anthracite	12,000–14,000			
Bituminous A	12,000–14,000			
Bituminous C	9,000–11,000			
Coke	12,000–14,000			
Liquid fuels:				
Petroleums:				
Crude oil	18,000–19,500			
Gasoline	20,750			
Kerosene	19,800			
Fuel oil	18,500–19,300			

For references see *ASTM Standards on Coal and Coke*, ASTM Symposium on Industrial Fuels, 1936; L. S. Marks, *Standard Handbook for Mechanical Engineers*, 7th ed., ed. by Theodore Baumeister et al., McGraw-Hill Book Company, New York, 1967.

Basic Engineering Principles

MECHANICS

Nomenclature

Symbol	Name	Definition	Dimensions
a	(linear) acceleration	$\dfrac{dv}{dt}$	in/s², ft/s²
a, b, c	vectors	. . . .	according to use
A	area	. . . .	in², ft²
E	work done	$F \times S$	in·lb, ft·lb
E_K	kinetic energy	$\dfrac{mv^2}{2}$	in·lb, ft·lb
F	force	fundamental	lb
g	acceleration of gravity	. . . .	32.16 ft/s², 386 in/s²
h, H	hydrostatic head	$\dfrac{p}{\gamma}$	in, ft (of liquid)
hp	horsepower	550 ft·lb/s	hp
I	area moment of inertia	$\int y^2\, dA$	in⁴
J	polar mass moment of inertia	$\int r^2\, dm$	in·lb·s², ft·lb·s²
k	area radius of gyration	$I = k^2 A$	in, ft
k_0	polar radius of gyration	$J = k_0^2 m$	in, ft
K	spring rate (or modulus)	$\dfrac{F}{S}$	lb/in
kW	kilowatt	737.6 ft·lb/s	kW
m	mass	$\dfrac{W}{g}$	lb·s²/in, lb·s²/ft
M	moment, torque	$F \times r$	in·lb, ft·lb
N	rotating speed	revolutions per minute	r/min
N_0	initial rotating speed	. . . .	r/min

Symbol	Name	Definition	Dimensions
p	hydrostatic pressure	$\dfrac{F}{A}$	lb/in² (psi), lb/ft²
P	power	$\dfrac{FS}{T}, \omega M$	ft·lb/s, hp
q	flow rate	. . . .	ft³/s
r, R	radius	. . . .	in, ft
$\vec{R}$	vector force	. . . .	lb
S	displacement	. . . .	in, ft
S_0	initial displacement	. . . .	in, ft
t	time, as a variable	fundamental	s
T	period of time	. . . .	s
v	velocity	$\dfrac{ds}{dt}$	in/s, ft/s
v_0	initial velocity	. . . .	in/s, ft/s
V	volume	. . . .	in³, ft³
w	weight flow	. . . .	lb/s
W	weight	fundamental	lb
$x, X; y, Y; z, Z$	spatial coordinates	. . . .	in, ft
α	angular acceleration	$\dfrac{d\omega}{dt}$	rad/s², 1/s²
γ	specific weight	$\dfrac{W}{V}$	lb/in³, lb/ft³
θ	angle of rotation	. . . .	rad, in/in
μ	coefficient of friction		
ω	angular velocity	$\dfrac{d\theta}{dt}$	rad/s, 1/s
ω_0	initial angular velocity	. . . .	rad/s, 1/s

NOTE: Every mechanical relationship must be written in consistent dimensions unless expressly noted otherwise. Also, the two sides of an equation must have the same dimensions.

Example: In the formula

$$E_K = \frac{m}{2} v^2$$

m must be in lb· s²/in, v in in/s; then E_K will be obtained in in·lb; or m is in lb·s²/ft, v in ft/s; then E_K is in ft·lb.

Fundamental Laws of Engineering Mechanics

1. In a closed system, the sum of all masses remains constant throughout any process.

2. In a closed system, the sum of all energies remains constant throughout any process. (But one form of energy may be transformed into another.)

3. The momentum of a closed system remains constant throughout any mechanical process.

4. In a closed system, the sum of all energies at the end of any process is always at a lower potential than at the beginning. In any transformation of energy some *usable* energy is lost.

5. Of all conceivable modes of energy transformation under a set of conditions, the actually occurring process requires the least amount of work.

Basic Mechanical Relations (Kinematics)

Uniform motion

Linear motion	*Rotary motion*
$S = vt$	$\theta = \omega t$
$v = \text{const}$	$\omega = \dfrac{\pi N}{30} = \text{const}$
$a = 0$	$\alpha = 0$

Uniform acceleration; initial velocity and displacement zero

Linear motion	*Rotary motion*
$S = \dfrac{a}{2} t^2 = \dfrac{v}{2} t$	$\theta = \dfrac{\alpha}{2} t^2 = \dfrac{\omega}{2} t$
$v = at = \sqrt{2aS}$	$\omega = \alpha t = \sqrt{2\alpha\theta}$

Free fall

$$S = \frac{g}{2} t^2$$

Example: How long does an object take to reach the ground in free fall if it has been released from a height of 100 ft?

$$t = \sqrt{\frac{2S}{g}} = \sqrt{\frac{200}{32.16}} = 2.5 \text{ s}$$

Uniform acceleration or deceleration

Linear motion	*Rotary motion*
Initial velocity v_0	Initial angular velocity ω_0
$S = v_0 t \pm \dfrac{a}{2} t^2$	$\theta = \omega_0 t \pm \dfrac{\alpha}{2} t^2$
$v = v_0 \pm at$	$\omega = \omega_0 \pm \alpha t$

NOTE: Use + sign for acceleration, − sign for deceleration.

Action of a Force or Moment on a Mass (Dynamics)

No resistance. The force produces only acceleration or deceleration of the mass.

Linear motion   *Rotary motion*

$$F = ma = \frac{W}{g}a \qquad\qquad M = J\alpha$$

$$v = v_0 \pm \frac{F}{m}t \qquad\qquad \omega = \omega_0 \pm \frac{M}{J}t$$

$$S = v_0 t \pm \frac{F}{2m}t^2 \qquad\qquad \theta = \omega_0 t \pm \frac{M}{2J}t^2$$

$$E = FS \qquad\qquad E = M\theta$$

$$E_K = \frac{m}{2}(v^2 - v_0^2) \qquad\qquad E_K = \frac{J}{2}(\omega^2 - \omega_0^2)$$

$$E_K = \frac{m}{2}v^2 \ (v_0 = 0) \qquad\qquad E_K = \frac{J}{2}\omega^2 = \frac{J\pi^2 N^2}{1800} \ (\omega_0 = 0)$$

Example 1: If a force equal to its weight ($F = W$) acts on a free body, what will be its acceleration?

Answer: $\quad a = \dfrac{F}{m} = \dfrac{W}{W/g} = g$

Example 2: If a retarding moment (torque) of 1200 ft·lb is applied to a turbine rotor of $J = 300$ in·lb·s² rotating at 3600 r/min, what is the time to standstill ($\omega = 0$)?

Answer: $\quad t = (\omega_0 - \omega)\dfrac{J}{M} = \left(\dfrac{3600\pi}{30} - 0\right)\dfrac{300}{1200 \times 12} = 7.86 \text{ s}$

NOTE: Use $+$ sign for acceleration, $-$ sign for deceleration.

Constant resistance F_R, M_R = const. The force overcomes the resistance, and the excess force results in acceleration.

Linear motion   *Rotary motion*

$$F = ma + F_R \qquad\qquad M = J\alpha + M_R$$

$$v = v_0 \pm \frac{F - F_R}{m}t \qquad\qquad \omega = \omega_0 \pm \frac{M - M_R}{J}$$

$$S = v_0 t \pm \frac{F - F_R}{2m}t^2 \qquad\qquad \theta = \omega_0 t \pm \frac{M - M_R}{2J}t^2$$

$$E = FS \qquad\qquad E = M\theta$$

$$E_K = \frac{m}{2}v^2 \qquad\qquad E_K = \frac{J}{2}\omega^2 = \frac{J\pi^2 N^2}{1800}$$

Example: How far will a weight of 100 lb travel from standstill in 60 s against a constant friction force of 25 lb if it is pulled by a force of 40 lb (Fig. 4-1)?

Answer: $S = \dfrac{40 - 25}{2(100/32.16)} \times 60^2 = 8680$ ft

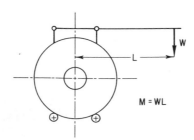

40LB

100 LB

25 LB   **FIG. 4-1**

No excess force; no acceleration

Linear motion	Rotary motion

$v = v_0 = \text{const}$ $\omega = \omega_0 = \dfrac{\pi N}{30} = \text{const}$

$S = v_0 t$ $\theta = \omega_0 t = \dfrac{\pi N}{30} t$

$E = F v_0 t$ $E = M\theta$

$E_K = \dfrac{m}{2} v_0^2$ $E_K = \dfrac{J}{2} \omega_0^2 = \dfrac{\pi^2}{1800} J N^2$

$P = F v_0 (\text{ft} \cdot \text{lb/s, in} \cdot \text{lb/s})$ $P = M \omega_0 (\text{ft} \cdot \text{lb/s, in} \cdot \text{lb/s})$

or $P = \dfrac{F(\text{lb}) v_0(\text{ft/s})}{550}$ (hp) $P = \dfrac{M(\text{in} \cdot \text{lb}) N(\text{r/min})}{63,025}$ (hp)

Example 1: What is the torque at the coupling that the shaft of a 2500-kW 1200-r/min steam turbine must carry?

Answer: $M = 1.34 \times 2500 \times \dfrac{63,025}{1200} = 175,945$ in $\cdot$ lb

Example 2: What is the power absorbed by a water brake rotating at 3000 r/min and registering a moment of 5040 in $\cdot$ lb (Fig. 4-2)?

Answer: $P = \dfrac{5040 \times 3000}{63,025} = 240$ hp

W

L

M = WL

FIG. 4-2

Viscous resistance; dragging a body through water, oil, air, etc.

Linear motion	*Rotary motion*
$F_R = Cv^2$	$M_R = C'\omega^2$
$F = ma + Cv^2$	$M = J\alpha + C'\omega^2$

Expressions for the calculation of velocities, displacements, and energy are beyond the scope of this section; however, the case of $M = 0$ (no external torque), $\omega_0 = \pi N_0/30$ is of special interest because it can be used to determine the mechanical resistance of a rotating machine.

Rotary motion

$$C' = \frac{30J}{\pi T}\left(\frac{1}{N_T} - \frac{1}{N_0}\right) \quad \text{in} \cdot \text{lb} \cdot \text{s}^2$$

$$M_R = \frac{\pi J N^2}{30T}\left(\frac{1}{N_T} - \frac{1}{N_0}\right) \quad \text{in} \cdot \text{lb}$$

$$P_R = \frac{JN^3}{601,843\,T}\left(\frac{1}{N_T} - \frac{1}{N_0}\right) \quad \text{hp}$$

where N_T is the speed in r/min after time T.

Example: A turbine running at 7200 r/min develops 3000 hp. After power is shut off, the turbine slows down to 3600 r/min in 6 min. What is the mechanical efficiency of the turbine-and-load combination? The WR^2 of turbine and connected load is 75,000 lb·in².

Answer: $\quad P_R \dfrac{75,000/386 \times 7200^3}{601,843 \times 6 \times 60}\left(\dfrac{1}{3600} - \dfrac{1}{7200}\right) = 46.5 \text{ hp}$

$$\text{Mechanical efficiency} = \frac{3000 - 46.5}{3000} = 0.984$$

Dynamic Forces and Moments

Gravitational attraction of masses

Magnitude

$$F = 3.44 \times 10^{-8}\frac{m_1 m_2}{r^2} \quad \text{lb}$$

(m_1, m_2 are lb·s²/ft, r is ft)

Electrodynamic attraction of two parallel conductors

Magnitude

$$F = 4.50 \times 10^{-8}\frac{I_1 I_2}{r} \quad \text{lb/ft}$$

(I_1, I_2 are amperes, r is ft)

Vertical impact $v = \sqrt{2gh}$; *the energy of the fall is completely converted into distortion of both bodies and of the foundation K (Fig. 4-3).*

Magnitude (plastic impact, spring depressed initially)

$$F = W + W' + \sqrt{W^2 + \left(\frac{2W^2Kh}{W + W'}\right)}$$

Magnitude (energy conserved, spring relaxed initially)

$$F = (W + W') + \sqrt{(W + W')^2 + 2WKh} \quad \text{if } W' = 0$$
$$F = W + \sqrt{W^2 + 2WKh}$$

Horizontal impact; the energy of the impact is wholly absorbed by the elastic distortions of the impacting bodies (Fig. 4-4).

$$F = v \sqrt{\frac{W}{g} K}$$

Example: An automobile of 2300 lb, traveling at 60 mi/h, strikes a guardrail; the elasticity of the guardrail is expressed by $K = 0.05 \times 10^6$ lb/in. What is the force F_a felt by the driver, who weighs 180 lb and is rigidly attached to the vehicle?

$$F = \frac{2300}{32.16} a = \frac{60 \times 5280}{3600} \sqrt{\frac{2300}{32.16} \times 0.05 \times 12 \times 10^6} = 577,600 \text{ lb}$$

$$F_a = \frac{180}{2300} \times F = 45,160 \text{ lb}$$

Torsional impact

$$M = \frac{\pi N}{30} \sqrt{JK}$$

Example: A turbine running at 1200 r/min seizes in a journal bearing and stops suddenly. The WR^2 of its rotor and connected load are 350,000 in^2·lb. What is the shock torque in the shaft next to the failed bearing? $K = 15 \times 10^6$ in·lb/rad.

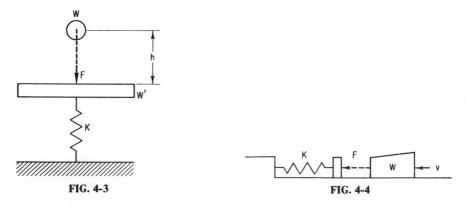

FIG. 4-3   **FIG. 4-4**

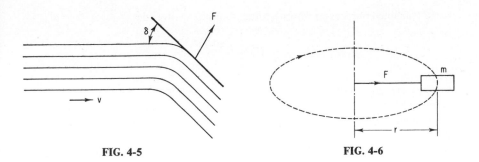

FIG. 4-5   FIG. 4-6

Answer: $M = \dfrac{\pi \times 1200}{30} \dfrac{350{,}000}{386} \times 15 \times 10^6 = 14{,}695{,}000 \text{ in} \cdot \text{lb}$

Hydraulic impact of a flowing fluid (water, air, etc.) on a plate under an angle δ to the direction of flow (Fig. 4-5)

$$F = 2\,\frac{w}{g}\,v\,\sin\left(\frac{\delta}{2}\right)$$

where w is the flow in lb/s.

Sudden load (as a weight placed on a table, without drop; or sudden pressure increase)

$$F = 2(F_{\text{static}})$$

Example: A pump casing, which is filled with water but not under pressure, is suddenly connected to a main at 2400 psig. What is the instantaneous pressure rise that all parts of the casing must withstand?

Answer: $P = 2 \times 2400 = 4800$ psig

Centrifugal force of a mass rotating about an axis (Fig. 4-6)

$$F = \frac{mv^2}{r} = \frac{W}{g} \times \frac{v^2}{r} = \frac{W}{g}\,r\left(\frac{\pi N}{30}\right)^2$$

where r is the distance from the axis to the center of gravity of m.

Example: A turbine disk has an unbalance of 1.5 oz at a radius of 14 in. What is the resulting unbalancing force on the shaft at 3600 r/min?

Answer: $F = \dfrac{1.5}{16 \times 386} \times 14 \times \left(\dfrac{\pi \times 3600}{30}\right)^2 = 483$ lb

Moments of Inertia of Bodies

Polar (mass) moment of inertia of a body about a given axis

$$J = r_1^2 m_1 + r_2^2 m_2 + r_3^2 m_3 + \cdots$$

For a (turbine-compressor-pump) rotor about its centerline,

$$J = \frac{1}{g} \Sigma W r^2$$

Definition of WR^2.

The expression WR^2 as used in industry for defining the moment of inertia of a symmetrical body about a given axis (turbine rotors, etc.) is the product of the rotor weight W in pounds times the polar radius of gyration squared, k_0^2, in feet squared.

For some simple bodies the radius of gyration and the corresponding WR^2 are given below:

	k_0, ft	WR^2, lb·ft²
Uniform disk	$\dfrac{D_0}{24\sqrt{2}} \cong \dfrac{D_0}{34}$	$W\dfrac{D_0^2}{1152}$
Ring	$\dfrac{\sqrt{D_0^2 + D_1^2}}{24\sqrt{2}} \cong \dfrac{\sqrt{D_0^2 + D_1^2}}{34}$	$W\dfrac{D_0^2 + D_1^2}{1152}$
Solid cone	$\dfrac{D_0}{24\sqrt{10/3}} \cong \dfrac{D_0}{44}$	$W\dfrac{D_0^2}{1920}$

where D_0 = OD, in
D_1 = ID, in

For a composite system, as, for instance, a motor-driven compressor connected through a speedup gear, the equivalent WR^2 as referred to the motor shaft will be

$$WR^2_{\text{motor equivalent}} = WR^2_{\text{motor and gear wheel}} + \left(\frac{N_{\text{compressor}}}{N_{\text{motor}}}\right)^2 WR^2_{\text{compressor and pinion}}$$

Combining and Resolving Vectors (Statics of Solids)

Displacements, velocities, forces, and moments can be represented as vectors, that is, by a straight arrow in a specified direction and of a length equal to the scaled value of the magnitude it represents. Addition and subtraction of vectors may be handled either graphically or analytically, depending upon convenience.

The vector of a moment is perpendicular to the plane of the moment and is oriented so that the moment turns clockwise when sighting in the direction of the arrow. A moment vector may be moved parallel to itself without changing the system.

In this subsection we shall deal only with vectors in the same plane.

Addition of two vectors through one point (Fig. 4-7)

$$(\overline{a + b}) = \sqrt{a^2 + b^2 + 2ab \cos \alpha}$$

$$\sin \beta = \frac{b \sin \alpha}{\sqrt{a^2 + b^2 + 2ab \cos \alpha}}$$

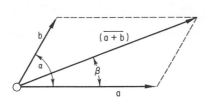

FIG. 4-7

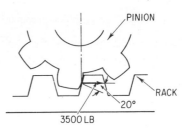

FIG. 4-8

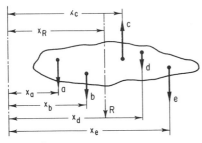

FIG. 4-9

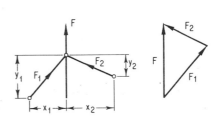

FIG. 4-10

Example: What is the tangential friction force on a pinion of 20° tooth form when driving a rack with 3500-lb pitch-line force (Fig. 4-8)? The friction coefficient = 0.05.

Answer: $F = 3500 \cos 20° \times 0.05 = 164.5$ lb

Addition of several parallel forces acting on a mass (Fig. 4-9)

$$R = a + b - c + d + e$$

$$X_R = \frac{1}{R}(ax_a + bx_b - cx_c + dx_d + ex_e)$$

Resolution of a force into two forces through two given points and intersecting in a third point (Fig. 4-10)

Graphically: Construct a vector diagram by drawing lines through the end points of vector F and parallel to F_1 and F_2. The sides of the triangle yield the values of F_1 and F_2 to the same scale as the scale of F.

Example: A turbine casing of 24,000 lb is lifted by two chains 4 ft long from a crane hook; the chains are attached to two eyebolts which are 54 in apart and one of which is 11 in higher from the horizontal joint than the other. What must be the rating of the chains if we assume that the center of gravity of the load is directly under C in Fig. 4-11?

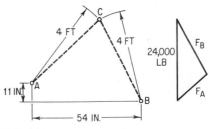

FIG. 4-11

Answer: The graphic approach is the most convenient.

1. Lay out the positions of eyebolts A and B to some scale.
2. To the same scale draw two circles 4 ft in diameter out of the eyebolts and intersect them; this fixes the position of crane hook C.
3. Draw the force polygon to a force scale so that the directions of the components are parallel to AC and BC.
4. Scale F_A and F_B. F_A scales 10,200 lb. F_B scales 18,500 lb. The rating of the chains should be at least 18,500 lb.

Hydrostatics

Force due to the weight of a column of liquid; hydrostatic head (Fig. 4-12)

$$F = AH\gamma$$

$$p = \frac{F}{A} = H\gamma$$

Note that pressure depends only upon the height of the column of liquid and its density but not upon its shape. Therefore, the head H is frequently used to express pressure.

Lateral pressure force (Fig. 4-13)

$$p = H\gamma$$

Hydrostatic pressure at a point is the same in all directions.

Hydrodynamics

Hydrodynamic reaction (Fig. 4-14)

$$F_x = \frac{w}{g}(v_2 \sin \alpha_2 - v_1 \sin \alpha_1)$$

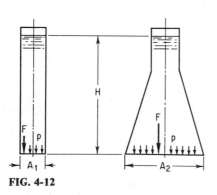

FIG. 4-12

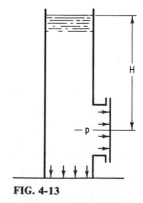

FIG. 4-13

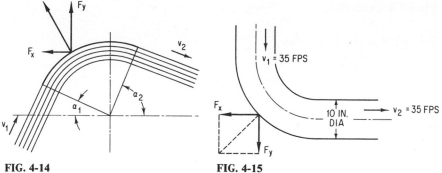

FIG. 4-14   **FIG. 4-15**

$$F_y = \frac{w}{g} (v_1 \cos \alpha_1 + v_2 \cos \alpha_2)$$

Example: What are the reactions along the two legs of a 10-in-diameter 90° elbow through which water is flowing with a velocity of 35 ft/s (Fig. 4-15)?

$$w = \left(\frac{10}{12}\right)^2 \frac{\pi}{4} \times 35 \times 62.4 = 1348 \text{ lb/s}$$

$$F_x = \frac{w}{g} (v_2 \sin 90° - v_1 \sin 0°) = \frac{w}{g} v_2$$

$$= \frac{1348}{32.2} \times 35 = 1470 \text{ lb}$$

$$F_y = \frac{w}{g} (v_1 \cos 0 + v_2 \cos 90°) = \frac{w}{g} v_1$$

$$= \frac{1348}{32.2} \times 35 = 1470 \text{ lb}$$

Forced vortex (fluid has same angular velocity as rotor) (Fig. 4-16)

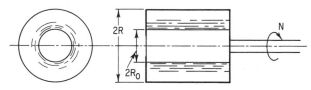

FIG. 4-16

Centrifugal pressure

$$p_R = \frac{\gamma}{2g} \left(\frac{\pi N}{30}\right)^2 (R^2 - R_0^2)$$

HEAT TRANSFER

Introduction and Definitions

Heat transfer can be defined as the transmission of energy from one region to another as a result of a temperature difference between them. There are three modes of heat transfer: conduction, convection, and radiation.

Conduction is the transfer of heat from a region of higher temperature to a region of lower temperature within a medium or between different media in direct physical contact. The energy is transmitted by direct molecular communication without appreciable displacement of the molecule.

Convection is a process which involves heat transport by the combined action of heat conduction, energy storage, and mixing motion.

Radiation is the transfer of heat in the form of an electromagnetic wave. All bodies above absolute-zero temperature radiate energy. Radiation incident on a body may be absorbed, reflected, and transmitted.

Conduction

The basic equation for conduction is the Fourier law (1828), which can be stated for the one-dimensional case in the steady state as follows:

$$q_k = -kA \frac{dT}{dx} \tag{4-1}$$

where q_k = rate of heat flow, Btu/h

k = thermal conductivity of material through which heat flows, Btu/$(h \cdot ft \cdot {}^\circ F)$

A = area of section through which heat flows by conduction, to be measured perpendicular to the direction of heat flow, ft^2

$-\dfrac{dT}{dx}$ = rate of change of temperature T with respect to distance x in the direction of heat flow. The negative sign is due to the fact that the heat flows in the direction of decreasing temperature. (${}^\circ F/ft$)

The value of k varies somewhat with temperature, but in most engineering applications the mean constant value is used and defined by

$$k_{\text{mean}} = \frac{1}{T_1 - T_2} \int_{T_2}^{T_1} k(T) \, dt \tag{4-2}$$

Over a moderate range, k varies linearly with T, and hence k_{mean} may be evaluated at the arithmetic mean of T_1 and T_2. Some commonly used values of k are given in Table 4-1.

For the case of heat flow through a plain wall of thickness L and cross-sectional area $\overline{A}$,

$$q_k = \frac{\overline{A}}{L} k_m (T_{\text{hot}} - T_{\text{cold}}) \tag{4-3}$$

TABLE 4-1 Thermal Conductivity k of Insulating Materials

Values of k are to be regarded as rough average values for the temperature range indicated.

Material	Bulk density, lb/ft³	Temp, °F	k Btu/ (h·ft·°F)	Material	Bulk density, lb/ft³	Temp, °F	k Btu/ (h·ft·°F)
Asbestos board, compressed asbestos and cement . . .	123	86	0.225	Sand, dry	94.8	68	0.188
Asbestos millboard	60.5	86	0.070	Sawdust, dry	13.4	68	0.042
Asbestos wool	25	212	0.058	Soil, dry	. . .	68	0.075
Concrete, sand, and gravel .	142	75	1.05	Soil, dry, including stones . .	127	68	0.30
Concrete, cinder	97	75	0.41	Snow	7–31	32	0.34–1.3
Cork, granulated	5.4	23	0.028	Wool, pure	5.6	86	0.021
Cotton wool	5.0	100	0.035	Woods, ovendry, across			
Earth plus 42% water,				grain:°			
frozen	108	0	0.62	Douglas fir	29	85	0.063
Glass, pyrex	139	200	0.59	Fir, white	26	85	0.069
Gypsum board	51	99	0.062	Maple, sugar	43	85	0.094
Ice	57.5	. . .	1.26	Oak, red	42	85	0.099
Mica	122	. . .	0.25	Pine, southern yellow . .	35	85	0.078
Rubber, hard	74.3	100	0.092	Pine, white	25	85	0.060
Rubber, soft, vulcanized . .	68.6	86	0.08	Redwood	25	85	0.062
				Spruce	21	85	0.052

Thermal Conductivity k, Specific Heat c, Density ρ, and Thermal diffusivity a of Metals and Alloys

Material	k, Btu/(hr)(ft)(°F)				c, Btu/(lb$_m$)(°F), 32°F	ρ, lb$_m$/cu ft, 32°F	a, sq ft/hr, 32°F
	32°F	212°F	572°F	932°F			
Metals:							
Aluminum	117	119	133	155	0.208	169	3.33
Copper, pure	224	218	212	207	0.091	558	4.42
Gold	169	170	. . .	. . .	0.030	1203	4.68
Iron, pure	35.8	36.6	. . .	. . .	0.104	491	0.70
Lead	20.1	19	18	. . .	0.030	705	0.95
Magnesium	91	92	. . .	. . .	0.232	109	3.60
Nickel	34.5	34	32	. . .	0.103	555	0.60
Silver	242	238	. . .	. . .	0.056	655	6.6
Tin	36	34	. . .	. . .	0.054	456	1.46
Zinc	65	64	59	. . .	0.091	446	1.60
Alloys:							
Admiralty metal	65	64					
Brass, 70 Cu, 30 Zn	56	60	66	. . .	0.092	532	1.14
Bronze, 75 Cu, 25 Sn . . .	15	. . .	. . .	. . .	0.082	540	0.34
Cast iron:							
Plain	33	31.8	27.7	24.8	0.11	474	0.63
Alloy	30	28.3	27	. . .	0.10	455	0.66
Steel, mild 0.1% C	26.5	26	25	22	0.11	490	0.49
18–8 stainless steel:							
Type 304	8.0	9.4	10.9	12.4	0.11	488	0.15
Type 347	8.0	9.3	11.0	12.8	0.11	488	0.15

NOTE: The thermal conductivity of different materials varies greatly. For metals and alloys k is high, while for certain insulating materials, such as glass wool, cork, and kapok it is very low. In general, k varies with the temperature, but in the case of metals the variation is relatively small. With most other substances, k increases with rising temperatures, but in the case of many crystalline materials the reverse is true.

°With heat flow parallel to the grain, k may be 2 to 3 times that with heat flow perpendicular to the grain.

For one-dimensional heat flow in other geometries, Eq. (4-3) can be applied as well if the area $\overline{A}$ is modified. For a hollow cylinder, heat flows in the radial direction

$$\overline{A} = (A_i - A_o)/\ln(A_i/A_o) \tag{4-4}$$

and for a hollow sphere,

$$\overline{A} = \sqrt{A_i A_o} \tag{4-5}$$

where A_i and A_o are inner and outer surface areas of the cylinder or sphere. In general, the mean surface area $\overline{A}$ for one-dimensional heat conduction can be obtained by integration, either numerically or graphically, by using the equation

$$\frac{1}{\overline{A}} = \frac{1}{L} \int_{L_1}^{L_2} \frac{dx}{A(x)} \tag{4-6}$$

Convection and Conduction

Convection is defined as the heat transfer between a surface and a fluid. The rate of heat transfer is determined by the equation established by Sir Isaac Newton in 1701, as follows:

$$q = hA\,\Delta T \tag{4-7}$$

where q = time rate of heat transfer by convection, Btu/h
  h = average convective heat-transfer film coefficient or surface coefficient of heat transfer, Btu/$(h \cdot ft^2 \cdot {}^\circ F)$
  A = heat-transfer area, ft^2
  ΔT = temperature difference between surface and fluid at a location away from the boundary layer, $^\circ F$

Convection is in itself a rather complex process. The coefficient h of a system depends on the geometry of the surface, the velocity of the fluid, the physical properties of the fluid, and the level and magnitude of the temperature difference ΔT. Because of the often wide variations in these parameters, one must distinguish between the local surface coefficient and the *average* surface coefficient, which is in most cases sufficient for engineering calculations and applications.

In most practical applications, heat is transmitted from one fluid to another through a composite wall as shown in Fig. 4-17. The rate of heat transfer, which involves both conduction and convection, in Btu per hour through a surface area of A square feet is

$$q = UA(T_{hot} - T_{cold})$$

where UA is the overall conductance and is defined as

$$\frac{1}{UA} = \frac{1}{h_i A_i} + \frac{1_1}{k_1 A_1} + \frac{1_2}{k_2 A_2} + \frac{1_3}{k_3 A_3} + \frac{1}{h_0 A_0} + \frac{R_s}{A_s} \tag{4-8}$$

in which R_s is the fouling resistance due to the presence of a scale deposit on the surface. Resistances for R_s are given in Table 4-2.

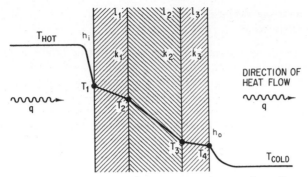

FIG. 4-17 Steady-state conduction through composite wall.

TABLE 4-2 Heat Transfer Resistances R_s for Scale Deposits from Water,* for Use in Eq. (4-8) [(h·ft²·°F)/Btu]

Temperature of heating medium	Up to 240°F		240–400°F†	
Temperature of water	125°F or less		Over 125°F	
	Water velocity, ft/s		Water velocity, ft/s	
Types of water	3 and less	Over 3	3 and less	Over 3
Seawater	0.0005	0.0005	0.001	0.001
Brackish water	0.002	0.001	0.003	0.002
Cooling tower and artificial spray pond				
Treated makeup	0.001	0.001	0.002	0.002
Untreated	0.003	0.003	0.005	0.004
City or well water (such as Great Lakes)	0.001	0.001	0.002	0.002
River water				
Minimum	0.002	0.001	0.003	0.002
Average	0.003	0.002	0.004	0.003
Muddy or silty	0.003	0.002	0.004	0.003
Hard (over 15 gr/gal)	0.003	0.003	0.005	0.005
Engine jacket	0.001	0.001	0.001	0.001
Distilled or closed-cycle condensate	0.0005	0.0005	0.0005	0.0005
Treated boiler feedwater	0.001	0.0005	0.001	0.001
Boiler blowdown	0.002	0.002	0.002	0.002

*1978 by Tubular Exchanger Manufacturers Association. All rights reserved.
†Ratings in cols. 3 and 4 are based on a temperature of the heating medium of 240–400°F. If the heating-medium temperature is over 400°F and the cooling medium is known to scale, these ratings should be modified accordingly.

Film Coefficients for Various Types of Heat Transfer by Convection

The mechanism of heat transfer by convection can be briefly explained by the concept of boundary layer introduced by Ludwig Prandtl in 1904.

When a fluid flows along a surface, the velocity increases from zero at the surface to its maximum value of the main stream within a relatively short distance. This short transition region is called the *boundary layer*. The thickness of the boundary layer has been defined as the distance from the surface at which the local velocity reaches 99 percent of the main velocity. Except for liquid metal, the thermal conductivity for fluid is small; therefore, the heat transfer depends largely on the mixing motion of the fluid particles. In the boundary-layer region, the mixing motion of fluid particles is relatively small and the heat is transmitted mainly by conduction. Hence, the thickness of this boundary layer becomes the dominant factor in determining the overall heat transfer between the surface and the fluid. The *film coefficient* is defined as the rate of heat transfer per unit area and unit temperature drop between the surface and the fluid. With the forced flow of gases or water, a condition usually encountered in practice, the flow is turbulent and the boundary-layer thickness can be greatly reduced by increasing the velocity of the fluid. As a result, the film coefficient can be greatly increased. For a given velocity and fluid, the boundary-layer thickness as well as the film coefficient depends upon the direction of fluid flow relative to the heating surface. With forced flow of high-viscosity liquids such as oil, laminar motion may prevail, and the film coefficient depends on thermal conductivity, specific heat, and the Reynolds number, since the boundary layer may extend to the entire flow region. With free or natural convection, in which the motion of the fluid is induced by the difference in specific gravity due to the temperature difference, the film coefficient for a given fluid and arrangement of surface depends upon physical properties such as viscosity, density, thermal conductivity, coefficient of expansion, and temperature difference.

For the simplest case of turbulent flow over a flat plate and inside a tube, the film coefficient can be determined analytically by applying the Reynolds analogy, which assumes that heat and momentum are transferred by a similar process in turbulent flow. For most practical cases, the film coefficient is established empirically with the aid of dimensionless groups. The dimensionless groups are as follows:

Reynolds number. The ratio of the inertia force to the viscous force, $Re = \rho \overline{V} D / \mu$, where ρ is the density of the fluid, $\overline{V}$ is the mean velocity, μ is the dynamic viscosity of the fluid, and D is the characteristic length of the system.

Prandtl number. The ratio of the kinematic viscosity μ/ρ and the thermal diffusivity $k/\rho C_p$, which relates the velocity distribution to the temperature distribution, $Pr = \mu C_p / k$. For $Pr = 1$, the temperature and velocity boundary layers in a flow field have the same thickness.

Nusselt number. The ratio of the temperature gradient of the fluid immediately in contact with the surface to a reference temperature gradient between the surface and the bulk-flow field, $Nu = hL/k$, where h is the film coefficient and L is the characteristic length of heat-transfer surface.

Grashof number. The ratio of buoyant to viscous force. The Grashof number is defined as $Gr = \rho^2 g \beta L^3 \Delta T / \mu^2$, where β is the coefficient of expansion.

In geometrically similar systems having the same Prandtl number Pr and Reynolds number Re, the temperature distribution will be similar and the numerical values of the Nusselt numbers for forced convection will be equal. For free or natural convection, the Nusselt numbers are the same when the values of Gr and Pr are the same in both systems. The correlation of the Nusselt number for a few simple cases will be listed below. For more complicated cases, the correlation can be found in the References and Bibliography at the end of this subsection.

Natural Convection Over horizontal cylinders with diameter D,

$$\overline{Nu}_D = 0.53(Gr_D Pr)^{1/4}, \text{ for } Pr > 0.5 \text{ and } 10^3 < Gr < 10^9 \qquad (4\text{-}9)$$

where $\overline{Nu}$ is the mean value of Nu over the heat-transfer surface.

For liquid metal in laminar flow over a horizontal cylinder

$$\overline{Nu}_D = 0.53 \left(\frac{Pr^2}{0.952 + Pr} \right) Gr_D^{0.25} \qquad (4\text{-}10)$$

For vertical plates and cylinders,

$$\overline{Nu}_D = C \, (Gr_D \cdot Pr)^n \qquad (4\text{-}11)$$

where

$C = 0.13$ $n = 1/3$ for $3.5 \times 10^7 < Gr \cdot Pr < 10^{12}$
$C = 0.55$ $n = 1/4$ for $10^4 < Gr \cdot Pr < 3.5 \times 10^7$

For a heated square plate facing upward or a cooled plate facing downward with L being the length of the side of the square,

$$Nu_L = 0.14(Gr_L Pr)^{1/3}$$

$$2 \times 10^7 < Gr < 10^{10} \quad \text{(turbulent range)} \qquad (4\text{-}12)$$

and

$$Nu_L = 0.54(Gr_L Pr)^{1/4}$$

$$10^5 < Gr < 10^7 \quad \text{(laminar range)} \qquad (4\text{-}13)$$

For heated plates facing downward or cooled plates facing upward, in the laminar range,

$$Nu_L = 0.27(Gr_L Pr)^{1/4}$$

$$3 \times 10^5 < Gr_L < 3 \times 10^{10} \qquad (4\text{-}14)$$

Forced Convection Flow over plane surfaces,

$$\overline{Nu}_L = 0.036 \, Pr^{1/3} Re_L^{0.8} \qquad Re_L > 5 \times 10^5 \qquad (4\text{-}15)$$

Flow normal to a single tube or wire,

$$\overline{Nu} = 1.1 \, CRe_D^n Pr^{0.31} \qquad (4\text{-}16)$$

TABLE 4-3 Coefficients for
Calculation of Average Heat-Transfer
Coefficient for a Circular Cylinder in a
Gas Flowing Normal to Its Axis

Re_D	C	n
0.4–4	0.891	0.350
4–40	0.821	0.385
40–4000	0.615	0.466
4000–40,000	0.174	0.618
40,000–400,000	0.0239	0.805

where C and n are empirical constants whose numerical values vary with the Reynolds number as shown in Table 4-3.

Fluid flow across banks of tubes that are 10 or more rows deep with Re > 6000,

$$Nu = C(Re)^{0.6}Pr^{1/3} \qquad (4\text{-}17)$$

where $C = 0.33$ for staggered tubes and $C = 0.26$ for in-line tubes.

For banks of tubes with fewer than 10 rows, the average film coefficient can be estimated by using the ratio tabulated in Table 4-4.

Turbulent flow inside clean tubes,

$$Nu = 0.023\,Re^{0.8}Pr^{0.4}, 0.5 < Pr < 120, 2300 < Re < 10^7 \qquad (4\text{-}18)$$

For an L/D (ratio of tube length to inside diameter) less than 60, multiply the right-hand side of Eq. (4-18) by $[1 + (D/L)^{0.7}]$.

Turbulent flow of liquid metal inside clean tubes with $0.005 < Pr < 0.05$, Re > 10,000,

$$Nu = 6.7 + 0.0041(Re \cdot Pr)^{0.793}e^{41.8Pr} \qquad (4\text{-}19)$$

For turbulent flow in annuli, use Eq. (4-18) with D taken as the clearance (inches), and base the heat-transfer area on the area actually heated or cooled.

For water in coiled pipes, multiply h_m for the straight pipe by the term $[1 + (3.5D_i/D_c)]$, where D_i is the inside diameter of the pipe and D_c is that of the coil.

Extended surfaces or fins are often used to increase the rate of heat transfer or cooling. The fin efficiency is the ratio of the heat transferred across the fin surface to the heat which would be transferred if the entire surface were at base temperature. Graphs of fin efficiency and heat-transfer coefficient for finned tubes of various types can be found in Gardner.[1]

TABLE 4-4 Ratio h_m for N Rows to h_m for 10 Rows Deep

N	1	2	4	6	8	10
Staggered tubes	0.68	0.75	0.89	0.95	0.98	1.0
In-line tubes	0.64	0.80	0.90	0.94	0.98	1.0

Heat Transfer inside Heat Exchanger

The basic equation for any steadily operated heat exchanger is $q = UA/\Delta T$ (Eq. 4-7), where U is defined as in Eq. (4-8). ΔT is the temperature difference across the heat-transfer surface. The proper value of ΔT can be obtained by integration. If we assume a constant U, a constant mass rate of flow, no change in phase, a constant specific heat, and an insulated exchanger, the integrated result of ΔT is

$$\Delta T = [(\Delta t)_1 - (\Delta t)_2]/\ln [(\Delta t)_1/(\Delta t)_2] = \text{LMTD}$$

which is valid for both parallel or countercurrent flow. ΔT is called the logarithmic mean overall temperature difference and is often designated by LMTD. The $(\Delta t)_1$ and $(\Delta t)_2$ are the terminal temperature differences between the hot and cold fluids.

For the more complex cases of multipass and cross-flow heat exchangers, the true mean temperature difference ΔT (or LMTD) is modified by a correction factor. This correction factor can be found in standard references.[2]

Heat Transfer by Radiation

Radiation is a process of transmission of energy by electromagnetic waves. All bodies with temperatures above absolute zero convert a part of their internal energy into electromagnetic waves. Radiation impinging on a body may be absorbed, reflected, and transmitted.

A *blackbody* is defined as a body which absorbs all radiation incident upon it or as a radiator which emits the maximum possible amount of thermal radiation at all wavelengths at any specific temperature. A blackbody does not reflect or transmit the radiation.

Kirchhoff's law states that at thermal equilibrium the ratio of the emissive power of a surface to its absorptivity is the same for all bodies, from which it can be established that the emissivity and absorptivity of a body are the same. The emissivity ϵ is defined as the ratio of emissive power of an actual surface to that of a blackbody. The absorptivity α is the fraction of incident radiation that is absorbed. For a blackbody, both emissivity and absorptivity are 1. Bodies with emissivity and absorptivity less than 1 are called graybodies. A graybody is a good approximation to a real body. However, it is worth noting that the emissivity ϵ varies with wavelength; Kirchhoff's law holds only at a given temperature or at a given wavelength. In practice, a body usually emits the bulk of its radiation at wavelengths which differ from those at which it receives radiation. Therefore, the total value (the integrated value over the whole wavelength spectrum) of ϵ and the total value of α are not necessarily the same.

The assumption of a graybody is equivalent to assuming that α and ϵ are constant over the entire wave spectrum and that their integrated values are identical.

The energy radiated by a blackbody is governed by the Stefan-Boltzmann law

$$E_b = \sigma T^4 \tag{4-20}$$

where E_b = radiated energy, $Btu/h \cdot ft^2$
σ = Stefan-Boltzmann constant
$\quad$ = $0.1717 \times 10^{-8}\ Btu/(h \cdot ft^2 \cdot °R^4)$
T = surface temperature, °R (Rankine)

For a graybody, the radiated energy is

$$E_g = \epsilon\sigma T^4 \tag{4-21}$$

where ϵ is the emissivity of the body and the subscript g denotes the graybody radiation. Representative values of the normal total emissivity of various surfaces are given in Table 4-5.

Radiation between Black Surfaces

In practice, radiation heat exchange involves two or more finite surfaces. Since they are normally located in open space, only a fraction of the total emission from each of the radiating surfaces will reach and be absorbed by the others. This fraction is called the *view factor* or *shape factor*.

Let F_{12} be the fraction of the total radiant energy leaving A_1 which is intercepted by A_2, and similarly F_{21} be the fraction of energy reaching A_1 from A_2. Then the energy transfer from A_1 to A_2 is

$$q_{1 \to 2} = A_1 F_{12}\sigma T_1^4$$

and from A_2 to A_1 is

$$q_{2 \to 1} = A_2 F_{21}\sigma T_2^4$$

The net energy change is

$$\Delta q_{1 \rightleftharpoons 2} = A_1 F_{12}\sigma T_1^4 - A_2 F_{21}\sigma T_2^4$$

If $T_1 = T_2$, $\Delta q_{1 \rightleftharpoons 2} = 0$, and it is necessary that $A_1 F_{12} = A_2 F_{21}$, which is the reciprocity theorem. The net rate of heat-flow radiation between any two black-bodies may then be written as

$$q_{1 \rightleftharpoons 2} = A_1 F_{12}\sigma(T_1^4 - T_2^4) = A_2 F_{21}\sigma(T_1^4 - T_2^4) \tag{4-22}$$

It is generally very tedious to determine the shape factors F since this requires evaluating a double integral over the entire surface area of the involved bodies. However, the shape factors of some simple and elementary surfaces have been evaluated and are available in graphical form.[2] It is also of practical interest to note that these basic data can be extended by simple addition and subtraction of the shape factors to permit the evaluation of a shape factor for other geometric arrangements which can be established from these elementary cases.[3,4]

Blackbody Enclosure The net radiation from surface A_i in a blackbody enclosure consisting of several surfaces is given by

$$q_i = \sum_{k=1}^{n} A_i F_{ik}(E_{bi} - E_{bk}) \tag{4-23}$$

where $E_{bi} = \sigma T_i^4$ = blackbody emission power for ith surface
$\quad E_{bk} = \sigma T_k^4$ = emission power for the kth surface

TABLE 4-5 Total Emissivities for Various Surfaces

Surface	t, °F	Emissivity
Metals and Their Oxides		
Aluminum:		
Highly polished plate, 98.3% pure..................	440–1070	0.039–0.057
Polished plate...............................	73	0.040
Rough plate................................	78	0.055
Oxidized at 1110°F..........................	390–1110	0.11–0.19
Al-surfaced roofing..........................	100	0.216
Calorized surfaces, heated at 1110°F:		
Copper..................................	390–1110	0.18–0.19
Steel....................................	390–1110	0.52–0.57
Brass:		
Highly polished:		
73.2 Cu, 26.7 Zn.........................	476–674	0.028–0.031
62.4 Cu, 36.8 Zn, 0.4 Pb, 0.3 Al...........	494–710	0.033–0.037
82.9 Cu, 17.0 Zn.........................	530	0.030
Polished................................	100–600	0.096
Rolled plate, natural surface..................	72	0.06
Rolled plate, rubbed with coarse emery...........	72	0.20
Dull plate...............................	120–660	0.22
Oxidized by heating at 1110°F................	390–1110	0.61–0.59
Chromium (see nickel alloys for Ni-Cr steels).........	100–1000	0.08–0.26
Copper:		
Carefully polished electrolytic.....................	176	0.018
Plate, heated long time, covered with thick oxide layer......	77	0.78
Plate heated at 1110°F..........................	390–1110	0.57
Cuprous oxide................................	1470–2010	0.66–0.54
Molten copper...............................	1970–2330	0.16–0.13
Iron and steel:		
Metallic surfaces (or very thin oxide layer):		
Electrolytic iron, highly polished..................	350–440	0.052–0.064
Polished iron.............................	800–1880	0.144–0.377
Oxidized surfaces:		
Iron plate, pickled, then rusted red................	68	0.612
Iron plate, pickled, then completely rusted............	67	0.685
Rolled sheet steel..........................	70	0.657
Oxidized iron.............................	212	0.736
Cast iron, oxidized at 1100°F....................	390–1110	0.64–0.78
Steel, oxidized at 1100°F......................	390–1110	0.79
Smooth oxidized electrolytic iron.................	260–980	0.78–0.82
Iron oxide...............................	930–2190	0.85–0.89
Rough ingot iron...........................	1700–2040	0.87–0.95
Wrought iron, dull oxidized....................	70–680	0.94
Steel plate, rough..........................	100–700	0.94–0.97
High-temp alloy steels (see nickel alloys)		
Molten metal:		
Cast iron.................................	2370–2550	0.29
Mild steel................................	2910–3270	0.28
Lead:		
Pure (99.96%), unoxidized......................	260–440	0.057–0.075
Gray oxidized.............................	75	0.281
Oxidized at 390°F..........................	390	0.63
Molybdenum filament..........................	1340–4700	0.096–0.292
Monel metal, oxidized at 1110°F..................	390–1110	0.41–0.46
Nickel:		
Electroplated on polished iron, then polished............	74	0.045
Technically pure (98.9 Ni + Mn), polished.............	440–710	0.07–0.087

TABLE 4-5 Total Emissivities for Various Surfaces *(Continued)*

Surface	t, °F	Emissivity
Nickel (continued):		
Electroplated on pickled iron, not polished	68	0.11
Wire	368–1844	0.096–0.186
Plate, oxidized by heating at 1110°F.	390–1110	0.37–0.48
Nickel oxide.	1200–2290	0.59–0.86
Nickel alloys:		
Chromnickel.	125–1894	0.64–0.76
Nickelin (18–32 Ni, 55–68 Cu, 20 Zn), gray oxidized	70	0.262
Tin, bright tinned iron sheet	76	0.043 and 0.064
Tungsten:		
Filament, aged	80–6000	0.032–0.35
Filament.	6000	0.39
Zinc:		
Commercial, 99.1%, polished.	440–620	0.045–0.053
Oxidized by heating at 750°F	750	0.11
Galvanized sheet iron, fairly bright	82	0.228
Galvanized sheet iron, gray oxidized	75	0.276

Refractories, Building Materials, Paints, and Miscellaneous

Surface	t, °F	Emissivity
Asbestos:		
Board.	74	0.96
Paper.	100–700	0.93–0.945
Brick:		
Red, rough, but no gross irregularities	70	0.93
Silica, unglazed, rough	1832	0.80
Silica, glazed, rough	2012	0.85
Grog brick, glazed	2012	0.75
See also refractory materials		
Carbon:		
T carbon (Gebr. Siemens) 0.9% ash. This started with emissivity		
at 260°F of 0.72, but on heating changed to values given	260–1160	0.81–0.79
Carbon filament.	1900–2560	0.526
Enamel, white fused, on iron	66	0.897
Glass, smooth	72	0.937
Gypsum, 0.02 in. thick on smooth or blackened plate.	70	0.903
Marble, light gray, polished.	72	0.931
Oak, planed	70	0.895
Oil layers on polished nickel (lub. oil):		
Polished surface, alone	68	0.045
+0.001-in. oil	. . . . . .	0.27
+0.002-in. oil	. . . . . .	0.46
+0.005-in. oil	. . . . . .	0.72
∞ thick oil layer.	. . . . . .	0.82
Oil layers on aluminum foil (linseed oil):		
Aluminum foil.	212	0.087
+ 1 coat oil.	212	0.561
+ 2 coats oil	212	0.574
Paints, lacquers, varnishes:		
Snow-white enamel varnish on rough iron plate	73	0.906
Black shiny lacquer, sprayed on iron	76	0.875
Oil paints, 16 different, all colors	212	0.92–0.96
Aluminum paints and lacquers:		
10% Al, 22% lacquer body, on rough or smooth surface	212	0.52
Paper, thin:		
Pasted on tinned iron plate.	66	0.924
Pasted on rough iron plate	66	0.929
Pasted on black lacquered plate	66	0.944

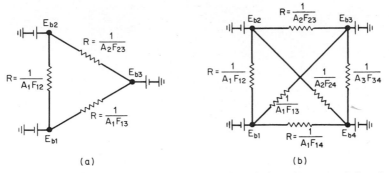

FIG. 4-18 Equivalent networks for radiation in blackbody enclosures consisting of three and four surfaces.

If it is assumed that none of the radiation emitted by a surface can strike directly on the surface itself, it can be concluded that

$$\sum_{k=1}^{n} A_i F_{ik} = A_i \qquad \text{or} \qquad \sum_{k=1}^{n} F_{ik} = 1 \qquad (4\text{-}24)$$

Equation (4-24) provides a means to find the shape factor for the last surface, which may have complicated geometry.

Analogy between radiation and electric current. If the blackbody emissive power E_b is considered to act as a potential and the shape factor $A_i F_{ik}$ as the conductance between two nodes with potentials E_{bi} and E_{bk}, then the resulting net flow of q_i is analogous to the flow of electric current in a similar network. As examples, networks for blackbody enclosures consisting of three and four heat-transfer surfaces are shown in Fig. 4-18. The network approach provides the simplest tool for solving the more practical problems, such as radiant heat transfer inside a graybody enclosure.

Blackbody Enclosure with Reradiating Surfaces A reradiating surface is a surface which diffusely reflects and emits radiation at the same rate at which it receives radiation. Figure 4-19 is a simplified sketch of a furnace which encloses a source T_1, a sink at T_2, and a reradiating wall at T_R.

The equivalent network of the heat transfer is constructed as shown in Fig. 4-20, and the net heat flow between T_1 and T_2 according to the network theory is

$$q_{1 \to 2} = A_1 \overline{F}_{12}(E_{b1} - E_{b2}) = A_1 \overline{F}_{12}\sigma(T_1^4 - T_2^4) \qquad (4\text{-}25)$$

where

$$\overline{F}_{12} = \frac{A_2 - A_1 F_{12}^2}{A_1 + A_2 - 2A_1 F_{12}} \qquad (4\text{-}26)$$

Enclosures Containing Gray Surfaces For gray surfaces, the amount of radiation will generally be expressed by the *radiocity*. The radiocity can be defined as the sum of the radiation emitted, reflected, and transmitted. By assuming an opaque body which transmits no radiation, the total amount of the radiation from a gray surface is

$$J = \rho G + \epsilon E_b \qquad (4\text{-}27)$$

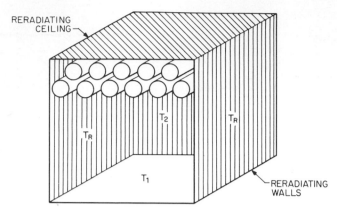

FIG. 4-19 Simplified sketch of a furnace.

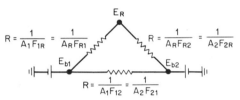

FIG. 4-20 Equivalent network for radiation between two blackbodies in a reradiating enclosure.

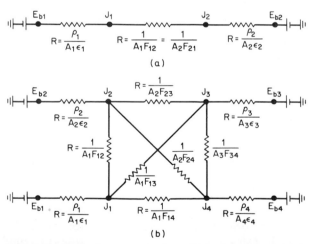

FIG. 4-21 Equivalent networks for radiation in gray enclosures. (a) Two graybody surfaces. (b) Four graybody surfaces.

where J = radiocity, Btu/h·ft^2

G = irradiation or radiation per unit time incident on unit surface area, Btu/h·ft^2

E_b = blackbody emissive power, Btu/h·ft^2

ρ = reflectivity

ϵ = emissivity

The rate of heat exchange of this graybody is

$$q_{\text{net}} = A(J - G) \quad \text{or} \quad q_{\text{net}} = \frac{\epsilon}{\rho} A(E_b - J)$$

For a graybody, the actual emissive power becomes the radiocity J; thus the shape factor applied to the emissive power E_b for the blackbody can be applied to J for the graybody. Accordingly, the radiation heat-transfer network for a graybody can be constructed as shown in Fig. 4-21 a and b for two and four gray surfaces respectively.

The net heat transfer between E_{b1} and E_{b2} in Fig. 4-21 a is

$$q_{1 \rightleftarrows 2} = A_1 \mathcal{F}_{12} \sigma (T_1^4 - T_2^4) \tag{4-28}$$

where $\mathcal{F}_{12}$ is defined as the graybody shape factor.

$$\mathcal{F}_{12} = \frac{1}{\dfrac{\rho_1}{\epsilon_1} + \dfrac{1}{F_{12}} + \left(\dfrac{A_1 \rho_2}{A_2 \epsilon_2}\right)} \tag{4-29}$$

For two parallel and infinite plates, Eq. (4-29) reduces to

$$\mathcal{F}_{12} = 1 \Big/ \left(\frac{1}{\epsilon_1} + \frac{1}{\epsilon_2} - 1\right)$$

For a small body in black surroundings with $A_2 \gg A_1$, Eq. (4-29) reduces to

$$\mathcal{F}_{12} = \epsilon_1$$

Inside most of the practical furnaces, it is adequate to divide the enclosure into two source-sink areas A_1 and A_2 and a nonflux area by lumping together areas of all refractory walls as A_R. The equivalent network of the radiant-heat transfer between the walls is similar to that shown in Fig. 4-20 except that the grounded node points E_{b1} and E_{b2} are replaced by those shown in Fig. 4-21. Accordingly, the graybody shape factor between A_1 and A_2 (or between E_{b1} and E_{b2}) is

$$\mathcal{F}_{12} = \frac{1}{\left(\dfrac{1}{\epsilon_1} - 1\right) + \dfrac{A_1}{A_2}\left(\dfrac{1}{\epsilon_2} - 1\right) + \dfrac{1}{\overline{F}_{12}}} \tag{4-30}$$

where $\overline{F}_{12}$ is defined as in Eq. (26).

For a gray enclosure consisting of more than two source-sink surfaces, the net heat flux can be determined by analyzing the heat balance at each node point (J_i in Fig. 4-21). At steady state, the net heat flow across the node point must be zero. Therefore, there are a total of n simultaneous equations to solve for n unknowns ($J_i, i = 1, n$) when the conditions on the surfaces are specified.

Nonluminous Gases   Many of the common gases and gas mixtures such as oxygen, nitrogen, hydrogen, and dry air have symmetrical molecules and are practically transparent to thermal radiation. They neither emit nor absorb appreciable amounts of radiant energy at temperatures of practical interest. However, other gases such as H_2O, CO_2, SO_2, CO, and various hydrocarbons and alcohols exhibit quite a different behavior and are important in the calculation of furnaces and heat exchangers of various kinds.

Whereas solids radiate at all wavelengths over the entire spectrum, these gases emit and absorb radiation only between narrow regions of wavelengths or bands. Radiation entering a gas volume is not absorbed within a small distance from the surface, as is the case in most solids. Instead, the intensity decreases slowly as the radiation passes through the gas. The emittance and absorptivity of a gas thus depend on its volume, pressure, and shape and the surface area. The rate of absorption of a monochromatic radiation through a gas medium obeys Beer's law, which states that

$$dI_\lambda = -a_\lambda c I_\lambda dx \tag{4-31}$$

where I_λ = intensity of radiation
  a_λ = absorption coefficient function of both temperature and wavelength
  c = concentration of absorbing molecules
  dx = path of radiation

If c is uniform along the path and $I_{\lambda 0}$ is the intensity at the entering surface, Eq. (4-31) becomes

$$I_\lambda = I_{\lambda 0} e^{-a_\lambda c L} \tag{4-32}$$

The monochromatic absorptivity of a gas is defined as

$$\alpha_\lambda = (I_{\lambda 0} - I_\lambda)/I_{\lambda 0}$$

The monochromatic emissivity of a heated gas body can be defined in a similar equation.

At a particular temperature, c is proportional to p_g, the partial pressure of the radiating gas. Then the summation over all significant wavelengths of α_λ leads to the overall radiation absorptivity and emissivity of a gas, which are functions of pressure, temperature, and a characteristic length of the gas body. For engineering calculations, a series of charts were developed by Hottel[2] on the basis of experimental data for evaluating the total emissivity of various gases at 1 atm. The value of ϵ as a function of $P_g L$ and gas temperature T_g from their graphs is multiplied by the correction factor C_g, which allows for departure from 1-atm pressure. The graphs apply strictly to a system in which hemispherical gas mass of radius L radiates to an element of surface located at the center of the hemisphere. The length L is defined as the effective beam length, which is the radius itself for hemisphere. For other simple shapes, the equivalent beam length is the product of the characteristic dimension of the shape and the corresponding constant as

TABLE 4-6 Mean Beam Lengths for Volume Radiation

Shape	Characteristic dimension D	$\dfrac{L_o}{D}$	$\dfrac{L_M}{D}$
Sphere	Diameter	0.67	0.63
Infinite cylinder	Diameter	1	0.94
Semi-infinite cylinder radiating to:			
Center of base	Diameter	1	0.90
Entire base	Diameter	0.81	0.65
Right-circle cylinder (height = diameter) radiating to:			
Center of base	Diameter	0.76	0.71
Whole surface	Diameter	0.67	0.60
Right-circle cylinder (height = 0.5 diameter) radiating to:			
End	Diameter	0.47	0.43
Side	Diameter	0.52	0.46
Total surface	Diameter	0.50	0.45
Right-circle cylinder (height = 2 × diameter) radiating to:			
End	Diameter	0.73	0.60
Side	Diameter	0.82	0.76
Total surface	Diameter	0.80	0.73
Infinite cylinder (half-circle cross section) radiating to spot on middle of flat side	Radius		1.26
Rectangular parallelepipeds			
1:1:1 (cube)	Edge	0.67	0.60
1:1:4, radiating to:			
1 × 4 face	Shortest edge	0.90	0.82
1 × 1 face	Shortest edge	0.86	0.71
Whole surface	Shortest edge	0.89	0.81
1:2:6, radiating to:			
2 × 6 face	Shortest edge	1.18	
1 × 6 face	Shortest edge	1.24	
1 × 2 face	Shortest edge	1.18	
Whole surface	Shortest edge	1.2	
Infinite parallel planes	Clearance	2.00	1.76
Space outside infinite bank of tubes, centers on equilateral triangles; tube diameter = clearance	Clearance	3.4	2.8
Same, except that tube diameter = 0.5 clearance	Clearance	4.45	3.8
Same, except that tube centers on squares, diameter = clearance	Clearance	4.1	3.5

listed in Table 4-6, in which L_M is the average mean beam length and L_o equals 4 times the ratio of gas volume to bounding area. More information on the effective beam length for parallel or perpendicular plates can be found in Rohsenow and Harnett.[3]

For an irregularly shaped gas mass, the equivalent beam length L can be roughly taken as 3.4 × volume/surface area.

The absorptivity of a gray gas is equal to its emissivity. For a real gas, if the surface is at T_1 and the gas at T_g, the absorptivity of the gas can be evaluated by the empirical equation

$$\alpha_{g1} = C_g \left(\frac{T_g}{T_1}\right)^n \epsilon_g \tag{4-33}$$

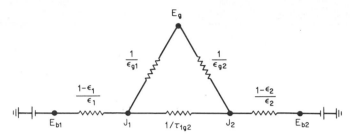

FIG. 4-22 Equivalent network for gas between two infinite-parallel plates.

where n = 0.65 for CO_2 and 0.45 for water vapor

C_g = pressure correction for gas emissivity

ϵ_g = gas emissivity evaluated at temperature T_1 and $(P_gL)_{\text{equivalent}}$ = $P_gL(T_1/T_g)$ from the graphs in McAdams[2]

When water vapor and CO_2 appear together, the emissivity can be estimated by adding the emissivities of the two constituents and subtracting a $\Delta\epsilon$ from the sum. This correction is due to the overlap of the absorption band of these two gases in the same wavelength. $\Delta\epsilon$ can also be found in McAdams.[2]

The approximate equation of radiant interchange between a gas and its bounding gray surface is

$$q/A = \sigma(\epsilon_g T_g^4 - \alpha_{g1} T_1^4)\left(\frac{\epsilon_1 + 1}{2}\right) \tag{4-34}$$

with $\epsilon_1 > 0.7$.

Gray Gas inside a Gray Enclosure   This problem can be solved by using the network concept with a slight modification; i.e., the shape factor between the surfaces must be multiplied by the transmissivity of the gray gas. The transmissivity τ_g is defined as $(1 - \alpha_g)$, where α_g is the absorptivity. For gray gas, $\alpha_g = \epsilon_g$.

Accordingly, Fig. 4-22 is the equivalent network for the case of the radiating heat transfer between two infinite plates with nonluminous gas in between. The overall shape factor between surface 1 and surface 2 is

$$\frac{1}{\mathscr{F}_{12}} = \frac{(1 - \epsilon_1)}{\epsilon_1} + \frac{1}{\tau_{1g2} + [1/[(1/\epsilon_{g1}) + (1/\epsilon_{g2})]]} + \frac{(1 - \epsilon_2)}{\epsilon_2} \tag{4-35}$$

The simplest mathematical model of a furnace chamber contains only a gray-gas zone, a heat or sink surface A_1, and a refractory area A_R. Figure 4-23 is the equivalent network diagram. The overall shape factor between the gas and A_1 is

$$\frac{1}{\mathscr{F}_{12}} = \frac{(1 - \epsilon_1)}{\epsilon_1} + \frac{1}{\tau_{1g2} + [1/[(1/\epsilon_{g1}) + (1/\epsilon_{g2})]]} + \frac{(1 - \epsilon_2)}{\epsilon_2} \tag{4-35}$$

For a black enclosure $\epsilon_1 = 1$,

$$\overline{F}_{1g} = \epsilon_{g1} + \frac{A_R/A_1}{(1/\epsilon_{gR}) + (1/F_{R1}\tau_{1gR})} \tag{4-36}$$

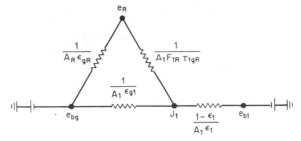

FIG. 4-23 Equivalent network for a gray gas in an enclosure of a gray hot surface and a refractory surface.

and
$$\frac{1}{\mathcal{F}_{1g}} = \frac{(1 - \epsilon_1)}{\epsilon_1} + \frac{1}{F_{1g}}$$
(4-37)

If we assume that $\epsilon_{g1} = \epsilon_{gR}$, Eq. (4-36) is simplified to give

$$\overline{F}_{1g} = \epsilon_g\left(1 + \frac{A_R/A_1}{1 + \epsilon_g/(1 - \epsilon_g)F_{R1}}\right)$$
(4-38)

For a "speckled furnace" model with surface A_1 and A_R so intimately mixed (this justifies the assumption that ϵ_{g1} and ϵ_{gR} are alike) that the view factors F_{11}, F_{1R}, F_{R1}, and F_{RR} are proportional to its own surface area, i.e., $F_{11} = F_{R1} = A_1/A_T = C$ and $F_{RR} = F_{1R} = (1 - C)$ with A_T being the total surface, Eq. (4-38) can be further simplified, and $1/\overline{F}_{1g} = 1 - C(1 - 1/\epsilon_g)$. For a furnace with $F_{11} = 0$, i.e., the surface A_1 cannot see itself, then $F_{1R} = 1$; Eq. (4-38) reduces to $1/\overline{F}_{1g} = 1 + C(1 - \epsilon_g)^2/\epsilon_g(1 - C\epsilon_g)$. If the refractory wall of a furnace can be completely segregated in a single plane A_R so that $F_{R1} = 1$, Eq. (4-38) becomes $1/\overline{F}_{1g} = 1 + (1 - \epsilon_g)[C - \epsilon_g(1 - C)]/\epsilon_g[1 - \epsilon_g(1 - C)]$. The simplest mathematical model of the furnace can be applied to a wide range of furnace types with substantially correct prediction of the relation among the dominant variables.

Luminous Flames and Clouds of Particles Powdered-coal or atomized-oil flames, soot from the thermal decomposition of hydrocarbons, and dust particles in flames all radiate as clouds of particles. Powdered-coal flames contain particles with an average size of around 0.001 in (25 μm). The luminous-soot particles have a dimension in the size range of 0.005 to 0.06 μm. Flames of heavy oil contain both solid particles of a diameter range between 50 and 200 μm and the luminous soot. Hence, the particles of powdered-coal and heavy-oil flames are essentially opaque to radiation, while the luminous flames due to soot interact with thermal radiation like semitransparent or scattering bodies. Consequently, the two kinds of luminosity obey different optical laws. However, the theory to determine the radiation properties of the luminous particles has not been completed or is inaccurate. Since there is still no standard textbook available, the reader is advised to consult the most recent papers for details of the development. For engineering calculations, it is suggested in Baumeister[4] that, for flames in the size range of 1 to 10 m, the emissivity of luminous soot ranges from 0.15 to 0.75. A crude allowance for soot in moderately luminous flames is to add 0.1 to the nonluminous-gas emissivity. For clouds of opaque large particles, emissivity may be calculated according to $\epsilon = 1 - e^{-cAL}$, where c is the number concentration of the particles, A is the projected area of a particle, and L is as defined in Table 4-6. With gas containing all three components, i.e., nonluminous gas, luminous soot, and opaque particles, the combined emissivity can be estimated by $[1 - (1 - \epsilon_g)(1 - \epsilon_s)(1 - \epsilon_p)]$, where ϵ_g, ϵ_s, ϵ_p are the separate emissivities due to the three components calculated as though no other emitter were present.

REFERENCES

1. K. A. Gardner, "Efficiency of Extended Surfaces," *Transactions of the American Society of Mechanical Engineers,* vol. 67, pp. 621–631, 1945.
2. W. H. McAdams, *Heat Transmission,* 3d ed., McGraw-Hill Book Company, New York, 1954.
3. W. M. Rohsenow and J. P. Harnett, *Handbook of Heat Transfer,* McGraw-Hill Book Company, New York, 1973.
4. T. Baumeister, *Marks' Standard Handbook for Mechanical Engineers,* 8th ed., McGraw-Hill Book Company, New York, 1978.

BIBLIOGRAPHY

Kays, W. M., and A. L. London: *Compact Heat Exchangers,* 2d ed., McGraw-Hill Book Company, New York, 1973.
Kreith, F.: *Principles of Heat Transfer,* 3d ed., Harper & Row Publishers, Inc., New York, 1973.
Rohsenow, Warren M., and Harry Y. Choi: *Heat, Mass and Momentum Transfer,* © 1961, pp. 192, 193, 351–352, 361, 363, 364. Reprinted by permission of Prentice-Hall, Inc., Englewood Cliffs, N.J.

THERMODYNAMICS

General Energy Equation for an Open System

The general energy equation for a fluid moving through a system represents the energy balance for a steady-state device. It ties together the various forms of energy and is based on the fact that mechanical work and heat are equivalent. *Steady state* means that the fluid going to and from the device is moving across any section continuously and at a constant rate. It is also assumed that thermal equilibrium exists and that there is no accumulation or diminution of energy within the device; in other words, the law of conservation of energy holds. All parts are at constant operating temperatures, and heat-transfer rates to and from are constant. One additional assumption is that there is no accumulation or diminution of fluid within the device; thus the law of the continuity of mass prevails.

The general energy-system diagram for a steady-state machine that could be a boiler, engine, turbine, compressor, pump, nozzle, or throttle valve is shown in Fig. 4-24, and the general energy equation is written as follows:

$$gZ_1 + \frac{v_1^2}{2} + \frac{p_1}{\rho_1} + u_1 = gZ_2 + \frac{v_2^2}{2} + \frac{p_2}{\rho_2} + u_2 + w_{1-2} - q_{1-2}$$

Each term must be taken in a consistent system of units, and in engineering the British thermal unit is mostly used. One pound of fluid is assumed. The various parts of the general energy equation are identified as follows:

Z = elevation above a datum line

$v^2/2$ = kinetic energy of fluid in approaching or leaving the device

p/ρ = flow work due to the motion of substance moving toward and away from the device

u = internal energy due to molecular activity in the fluid as measured above a convenient datum state

w_{1-2} = work done by the fluid between states 1 and 2

q_{1-2} = transferred heat added to the system between states 1 and 2

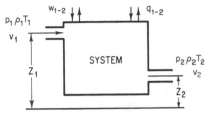

FIG. 4-24 General energy-system diagram: v = velocity; p = pressure; ρ = density; w = work done on or by fluid; q = heat added or subtracted.

Subscripts 1 and 2 refer to entering (or initial) and leaving (or final) conditions.

In most heat engines the change in potential energy from Z_1 to Z_2 is negligibly small. The internal-energy change depends on the nature of the substance, and the transferred heat and work depend as well upon the kind of process which the substance is undergoing.

According to the sign convention used, w and q, as solved in the equation, obey the following:

Positive w means that net work is done by the fluid.

Negative w means that net work is done on the fluid.

Positive q means that net heat is added to the fluid.

Negative q means that net heat is taken from the fluid.

This equation may also be applied to reciprocating engines even though admission of the working fluid is intermittent and not steady.

The energy q represents the amount of heat passing into the fluid. It is not a quantity that exists by virtue of a property or a characteristic state of the fluid such as velocity, pressure, or volume. Transferred heat is energy in transition as opposed to stored energy. In a similar sense, work is energy in transition. It may occur because of a change in stored energy, but it is not inherent in the fluid.

Laws of Thermodynamics

First Law of Thermodynamics　This is the law of conservation of energy that states that energy can be neither created nor destroyed. In equation form for an open system, it can be written as follows:

$$\Delta E = Q - W + \sum_i (h_i + e_{xi}) m_i$$

where ΔE = change in energy content of system

   Q = heat transferred to system

   W = work transferred from system

   h_i = enthalpy per unit mass

   e_{xi} = extrinsic energy, dependent on the frame of reference

   m_i = mass

The summation term represents energy convected into or out of the system by mass m_i. For a fluid system, e_{xi} = kinetic energy + potential energy = $V^2/2 + gZ$, where V is fluid velocity and gZ potential energy due to elevation.

In cases in which changes in kinetic energy and elevation of fluid stream can be neglected and the flow is adiabatic, the above equation reduces to the simple form

$$W \cong m(h_{in} - h_{out})$$

Thus work from a system approximately equals the decrease in enthalpy. This holds for the steady-state expansion in a turbine, for example.

Second Law of Thermodynamics   This law states that it is impossible without doing work to bring about any change or series of changes resulting in transfer of energy as heat from a low to a high temperature. In other words, heat will not by itself flow from low to high temperatures. It is advantageous, especially for direct quantitative-loss analysis in processes and systems, to express this law in the form of an equation as follows:

$$\Delta S = \frac{Q}{T} + I + \sum_i S_i$$

where ΔS = change in entropy of system

   $\frac{Q}{T} = \sum_i \frac{Q_i}{T_i}$ = sum of heat transferred over system boundaries Q_i

   T_i = local temperature at boundary

   I = irreversibility (≥ 0) (For a reversible process or cycle $I = 0$, and for an irreversible process $I > 0$.)

   $\sum_i S_i$ = entropy flow into and out of system with mass flow m_i

For example, a steady-flow adiabatic expansion through a turbine would reduce the above equation to the following form:

$$I = -\sum_i m_i S_i = -m(S_{in} - S_{out})$$

$$= m(S_{out} - S_{in})$$

The second law in the above equation form can be used for qualitative examination of all power plant processes regardless of the fluids used or the specific cycles employed.

Gas Laws of Thermodynamics   These laws represent fundamental rules for the behavior of gases. An ideal gas is defined as a gas which obeys equation of state:

$$pv = RT$$

where p = pressure
v = specific volume
R = gas constant
T = absolute temperature

When the specific volume is expressed in units of volume per pound mass–mole of gas, then the gas constant is the universal gas constant equal to 1545.3 ft·lbf/ (lbm·mol·°R).

Behavior of real gases often deviates from that given by the perfect-gas equation of state. One equation of state which accounts for this deviation is

$$pv = TRZ$$

where Z is the compressibility factor which, for a given gas, depends on temperature and pressure. (This equation is discussed in Sec. 9.) Another very useful equation of state of the vapor phase of many substances at pressures less than the critical pressure is the Beattie-Bridgman equation:

$$p = \frac{RT(1 - \epsilon)}{v^2}(v + B) - \frac{A}{v^2}$$

where $A = A_0(1 - a/v)$, $B = B_0(1 - b/v)$, $\epsilon = c/vT^3$, and A_0, a, B_0, b, and c are constants which are different for different gases. The specific volume is per mole of the gas, and R denotes the universal gas constant.

The Beattie-Bridgman equation can be used for a mixture of gases. In such case

$$A_0 = [x_i \sqrt{A_{0i}} + x_j \sqrt{A_{0j}} + \cdots]^2$$

$$a = x_i a_i + x_j a_j + \cdots$$

$$B_0 = x_i B_{0i} + x_j B_{0j} + \cdots$$

$$b = x_i b_i + x_j b_j + \cdots$$

$$c = x_i c_i + x_j c_j + \cdots$$

where x_i, x_j, etc., represent the mole fractions of substances i, j, etc., forming the mixture. The symbols A_0, a, B_0, b, and c with subscripts i, j, etc., represent the constants appearing in the Beattie-Bridgman equation for pure substances i, j, etc. This method results in a good representation of the behavior of mixtures of gases provided that the density of each constituent is less than one-half of the density of that constituent in its critical state.

Still another useful equation of state, used mainly for mixtures of hydrocarbons, is the Benedict-Webb-Rubin (B-W-R) equation. In this equation the compressibility factor Z is expressed as a function of temperature T and specific volume v. It has the form

$$\frac{Pv}{RT} = Z = 1 + \left(B_0 - \frac{A_0}{RT} - \frac{C_0}{RT^3}\right)(1/v) + \left(b - \frac{a}{RT}\right)(1/v^2)$$

$$+ (a\alpha/RT)(1/v^5) + (c/RT^3)[(1 + \gamma v^{-2})/v^{-2}]\, e^{-\gamma v^{-2}}$$

The constants A_0, B_0, C_0, a, b, c, α, and γ, tabulated for hydrocarbons CO_2, N_2, and SO_2, can be found in Reid and Sherwood,[1] which also gives formulas for calculation of the B-W-R constants for mixtures from the pure-component values.

Basic Heat-Engine Cycles: Steam

General   The essential elements for a thermodynamic cycle are (1) a working fluid or substance for receiving and rejecting heat and to do work, (2) a source of heat by means of which heat can be added to the working fluid, (3) a heat sink to which heat can be rejected by the working fluid, and (4) the engine itself. During every cycle operating in a steady state, the working substance goes through the same series of events or processes and always returns to its initial condition.

Thermal efficiency of a cycle, in its simplest form as shown in Fig. 4-25, is the output divided by the input. In the case of a power cycle, the output is represented by the produced power or net work; the input consists of heat added to the working substance from an external source of heat. The thermal efficiency can be written as follows:

$$\eta = \frac{W}{Q_A} = \frac{Q_A - Q_R}{Q_A} = \frac{\Sigma Q}{Q_A}$$

where W = net work, $W_{out} - W_{in}$
Q_A = heat added from external source
Q_R = heat rejected by the cycle

If the cycle is performed entirely within an engine, the above thermal efficiency represents that of the engine itself.

For any heat engine utilizing steam as a fluid, the Mollier enthalpy-entropy diagram is useful in following the changes in state (see Fig. 5-1).

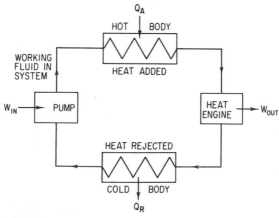

FIG. 4-25   Simple power cycle.

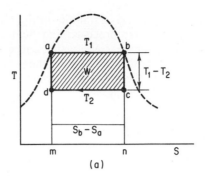

(a)

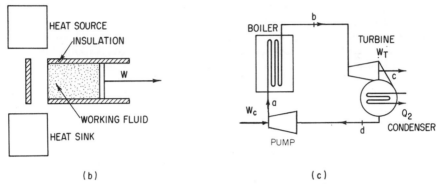

(b)   (c)

FIG. 4-26   The Carnot cycle. (*a*) *TS* diagram. (*b*) Nonflow system. (*c*) Flow system.

The Carnot Cycle   The three important contributions by the Frenchman Sadi Carnot (1824) were the concept of reversibility, the concept of the cycle, and the definition of a heat engine producing maximum work when operating cyclically between two fixed reservoirs at given temperatures. The knowledge of the ideal cycle in each application is useful and often essential in judging the performance of the actual cycle.

The components of the Carnot cycle are two isothermal and two isentropic processes as shown in Fig. 4-26*a* for a two-phase substance and Fig. 4-27 for an ideal gas. The cycle shown in Fig. 4-26*a* corresponds to a steam power plant, Fig. 4-26*c*. The Carnot cycle can be constructed for a flow system as well as for a nonflow system, Fig. 4-26*b*.

It can be shown that work derived from this cycle is represented by the enclosed area in the *TS* plane. The thermal efficiency is

$$\eta = \frac{T_1 - T_2}{T_1}$$

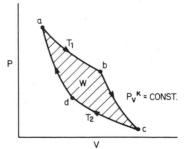

FIG. 4-27   The Carnot cycle: *pv* diagram, ideal gas.

In practice the Carnot cycle faces irreversibilities in the form of finite temperature differences during heat-transfer processes and fluid friction during work-
transfer processes. Also, a two-phase compression causes difficulties in many
applications. In overcoming these difficulties and others, the cost of the needed
equipment increases rapidly, and therefore other cycles provide a far more practical approach as shown in the following.

The Rankine Cycle   William Rankine (1859) established a sound thermodynamic basis for steam-power-plant practice by replacing the Carnot two-phase
compression with a simple liquid pump. This closely reproduced the actual case
in which the condensation process, inherent in the heat rejection, continues until
the saturated-liquid state is reached. The Rankine steam-power-plant cycle and
corresponding state changes are shown in Fig. 4-28. The volumes of liquid and
the temperature rise are indicated in the diagrams.

The Rankine cycle is a modification of the Carnot cycle which corresponds
much more closely to the sequence of states assumed by the working substance
in a steam power plant. The compression of liquid is performed by pumps.

The *regenerative* version of the Rankine cycle is a modification aimed at
increasing the thermal efficiency of the basic cycle. This efficiency increase is

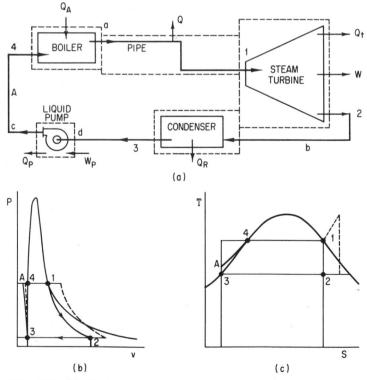

FIG. 4-28   The Rankine cycle. (*a*) Rankine steam-power-plant cycle. Q_A =
added heat; Q_R = rejected heat. (*b*) *pv* diagram. (*c*) *TS* diagram.

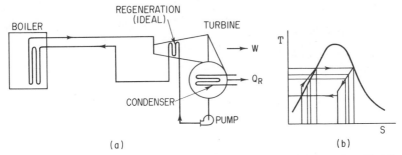

FIG. 4-29   Regenerative Rankine cycle. (*a*) Reheat after partial expansion. (*b*) *TS* diagram.

obtained by raising the mean temperature at which heat is added in the cycle. This is accomplished by raising the temperature at which the liquid water enters the boiler. In this way the regenerative cycle significantly decreases the irreversible heat flow that takes place if the condensed liquid is pumped directly into the boiler. The ideal regenerative Rankine cycle is shown in Fig. 4-29. In an actual cycle the heating of the liquid water is accomplished by means of steam bled from the turbine. This heating process is known as *regenerative feedwater heating*. It is accomplished in feedwater heaters, which can be of either the open or the closed type.

Using *reheat,* the Rankine-cycle thermal efficiency can be still further improved and the moisture content reduced to more acceptable levels in the last several stages of the turbine. Erosion damage by water droplets to rotating blading is limited by this means. Reheating is done by interrupting the expansion process and removing the vapor for reheat on the outside at constant pressure. It is thereafter returned for continued expansion down to condenser pressure. Reheating can be done in a special section of the boiler, in a separately fired heat exchanger, or even in a steam-to-steam heat exchanger. In a large power plant reheat permits an improvement of some 5 percentage points in thermal efficiency. Normally only one stage or a maximum of two stages of reheat is encountered in large power plant installations.

The *supercritical-pressure* steam power cycle has been made possible by advances in metallugy. Optimum cycle economy is obtained at a steam temperature of 1200°F and a pressure of 5000 psia. Both the regenerative cycle and the reheat cycle are effectively used for maximum thermal efficiency, which has reached values better than 40 percent. The supercritical cycle puts very severe demands on all cycle components but perhaps most on the boiler-feed pump, which must pump the heated water against extremely high heads. To do this successfully requires very special design features and experience with high-pressure water pumps with shaft input of up to 40,000 shp and more.

Many modifications to the original supercritical-pressure steam power cycle have already appeared, and changes are expected to continue when more actual experience in these power plants has been gained.

In the *back-pressure* power cycle steam is admitted into a turbine at a suitable initial pressure and emerges from there usually in a superheated state. A desu-

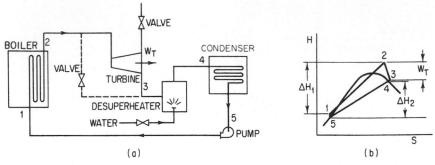

FIG. 4-30   Back-pressure steam power cycle. (*a*) Flow diagram. (*b*) *HS* diagram.

perheater, including a pressure water spray, is used for constant-temperature control. This can be achieved automatically. Saturated steam then enters the condenser and is completely condensed. The steam demands for power, process work, or heating can vary widely, and controls of pressure must be included in the cycle. Also, additional fresh steam can be supplied to the desuperheater when needed. In some versions the heating is done by hot water under high pressure, which is supplied by steam contained in an expansion drum.

The basic back-pressure steam power cycle is shown in Fig. 4-30.

The Nuclear Cycle   Most of the principles used in nuclear cycles are those also used in connection with the conventional steam power plant, but there are major differences in their application. For example, there is relative freedom in the selection of initial steam pressure for a conventional boiler plant but not in the case of a nuclear plant.

The *gas-cooled* nuclear plant uses single-pressure or dual-pressure steam cycles. In the first (Fig. 4-31), steam evaporation takes place at constant pressure. The dual-pressure steam cycle is essentially a high-pressure boiler in series with a low-pressure boiler. The high-pressure steam enters the turbine steam path at the beginning; the low-pressure steam, at a point farther down in the expansion.

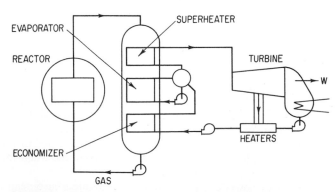

FIG. 4-31   Single-pressure nuclear steam cycle (gas-cooled).

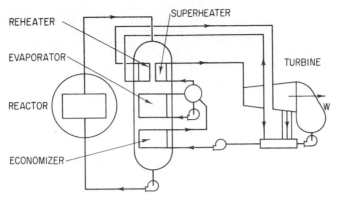

FIG. 4-32   Reheat nuclear steam cycle (gas-cooled).

The temperature of the gas leaving the high-pressure boiler can be much higher than could be tolerated on the grounds of blower power requirements. The low-pressure boiler is now given the task of cooling the gas to the required blower temperature. Therefore, substantial amounts of steam can be produced at 4 to 5 times the pressure otherwise obtainable. There are now two pinch points apart from those at the outlets of the superheaters. There are, again, definite relationships between the gas temperatures, the temperature approaches, the feed temperature, and the obtainable steam pressures. Also, there is a relation between these parameters and the relative proportions of high-pressure and low-pressure steam. The advantage of the dual-pressure steam cycle is likely to decrease when the simple single-pressure cycle attains increasingly higher steam pressures.

As was the case with conventional power plants, reheating is beneficially used in nuclear steam cycles. Steam is generated in the heat exchanger at a single pressure, expanded in part of the turbine, returned to the heat exchanger, resuperheated, and then expanded in the remainder of the turbine, as shown in Fig. 4-32. The optimum reheat pressure is about one-fourth of the initial steam pressure. A tolerable moisture content at turbine exhaust is taken at 13 percent. The steam pressure obtainable in the reheat cycle, for any given inlet- and outlet-gas temperatures, is lower than it is in the nonreheat cycle. This and the pressure losses in pipes between turbine and reheater are two reasons why expected advantages from application of reheat to the nuclear cycle do not fully materialize. By adopting the supercritical-pressure concept and utilizing the once-through boiler having no steam drum, further improvements to the nuclear cycle are to be expected, and overall net thermal efficiencies of 40 percent and beyond can be contemplated.

The *water-cooled*, water-monitored nuclear power plant is becoming popular because of the comparatively high density and good heat-transfer properties of the coolant, making heat removal from the reactor core relatively simple. Coolant circulators are used to raise the mean temperature of the coolant. However, available temperatures are limited by the pressure-temperature characteristics of saturated water. A margin is needed to suppress any tendency for boiling in the reactor core.

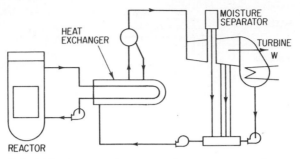

FIG. 4-33   Pressurized-water-cooled nuclear steam cycle.

In the pressurized-water reactor, cycle water is circulated through the reactor under high pressure and then through one or several heat exchangers which produce saturated steam. Moisture extraction and separation are used along the turbine steam path to control wetness of the steam at turbine exhaust. The amount of coolant circulating through the reactor is selected as high as possible for attaining the highest mean coolant temperature, leading to highest-quality steam being produced by the heat exchangers. The pressurized-water-reactor steam cycle is shown in Fig. 4-33.

Separate superheating using fossil fuels can be added to the above cycle, thus eliminating the difficulties caused by wetness of the steam and improving the net output of the cycle by about 45 percent. When the attainable steam temperatures in nuclear power plants become higher, as they have done in gas-cooled reactors, the attraction of separate superheating diminishes.

The boiling-water reactor allows steam to form in the reactor core itself. The steam-water mixture is then separated within the reactor or in separate steam-water drums. The saturated steam goes directly to the turbine, as shown in Fig. 4-34. Steam pressures as high as 6000 psia are used, and relatively good cycle efficiencies of 36 percent and above can be obtained. The level of auxiliary power

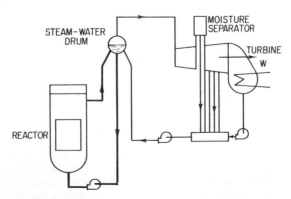

FIG. 4-34   Boiling-water-cooled nuclear steam cycle.

consumption, including the feed pump, becomes increasingly larger at the higher pressures and may well limit the selection of pressure level for the cycle.

The dual-pressure concept can be combined with the boiling-water cycle to combat the voidage in the reactor core. Coolant temperature at the inlet to the reactors is lower after passing through a heat exchanger of the pressurized-water-reactor type. In this system low-pressure steam is generated and taken to the turbine at an appropriate point of its steam path.

The *liquid-metal-cooled* reactor presents difficulties in heat transfer because of the violent reaction between suitable metals, such as molten sodium and potassium, and water. Any possible leakage must be eliminated; therefore, barrier liquids such as mercury are used, and the cycle becomes complicated.

The above brief review of the various steam cycles for nuclear-power-plant application shows the various possibilities available for the designer. It can be said that the nuclear steam cycle, as such, is here to stay, and only drastic improvements in attainable coolant temperatures from the nuclear reactor, based on some new technology, can change the status of steam as the basic fluid in the nuclear power cycle. However, closed gas-turbine nuclear-reactor cycles, with helium as the working medium, have shown promise in the solution of heat-transfer power problems without the use of water or steam.

Basic Heat-Engine Cycles: Combustion

Internal Combustion: Reciprocating Principle   In the internal-combustion type of heat engine the working fluid consists of the products of combustion of the fuel-air mixture itself. The two main advantages of the reciprocating internal-combustion engine are (1) the absence of heat exchangers in the working fluid stream, leading to simplicity of the engine; and (2) the fact that all its parts can work at temperatures well below the maximum cyclic temperature, which allows very high cyclic temperatures to be used, leading to high cyclic efficiencies. A third advantage is the low weight-horsepower ratio, which allows suitable applications in which no other engine system can compete.

The automotive engine represents a common highly developed internal-combustion engine based on the reciprocating principle. The basis of its operation is the indicated pv diagram composed of an expansion and a compression of the working fluid, as shown in Fig. 4-35. By means of the diagram, the mean effective pressure (mep) and the indicated horsepower (ihp) can be determined. The common processes of the cycle can be reviewed as follows:

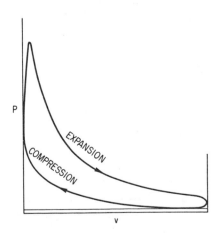

FIG. 4-35   Indicator pv diagram for internal-combustion engine.

Adiabatic process, one in which no heat is transferred

Isentropic process, one of constant entropy

Polytropic process, one internally reversible which conforms to the relation $p_1 v_1^n$ $= p_2 v_2^n$, where n is the polytropic exponent, which in practice has a value near to k, the specific heat ratio c_p/c_v.

Work can be expressed for all three processes with suitable formulas.

Internal-combustion engines fall basically into two classifications: the spark-ignition (SI) engine and the compression-ignition (CI) engine. The former represents lightweight and higher-revolutions-per-minute applications, the latter heavier and slow-speed applications. As the name indicates, the basic difference is in method of igniting the compressed combustible fuel-air mixture. The thermodynamic cycles applicable are as follows:

The Otto cycle by Nikolaus Otto (1876) represents the ideal spark-ignition condition, in which the combustion process takes place instantaneously at top dead center to give a constant-volume combustion of the fuel. This ideal cycle is shown in Fig. 4-36. On the basis of this principle, the reciprocating engine can have a four-stroke cycle (suction, compression, expansion, and exhaust) or a two-stroke cycle (scavenging by blowing air into the cylinder).

The actual cycle in an SI engine differs from the ideal fuel-air cycle because of leakage, incomplete combustion, progressive burning, time and heat losses, and exhaust losses due to opening the exhaust valve before bottom dead center.

The *diesel cycle* by Rudolf Diesel (1893) was originally developed to operate on coal as fuel. The present four-stroke-cycle diesel engine takes in only air during the suction stroke, and the liquid fuel is injected at the end of the compression stroke. Injection continues at such a rate that burning proceeds at constant pressure. The ideal diesel cycle is shown in Fig. 4-37. The diesel engine can also be built on the two-cycle principle, and many low-speed marine diesel engines are of this variety.

The actual diesel cycle differs from the ideal because (1) there is always a delay between the start of injection and the appearance of a flame or measurable pressure rise due to combustion, (2) most of the fuel tends to burn during the period of the very rapid rise in pressure, (3) the rise is followed by relatively slow combustion during the expansion stroke as the remaining unburned fuel finds the

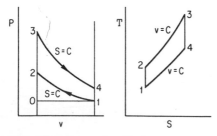

FIG. 4-36   Otto cycle—ideal.

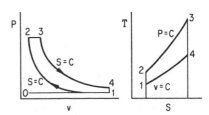

FIG. 4-37   Diesel cycle—ideal.

necessary oxygen, and (4) the crank angles for these three periods vary with design and operating conditions.

At any given engine revolutions per minute the crank angles are subject to a certain amount of control by means of injection timing, fuel-spray characteristics, and fuel composition. The actual diesel (compression-ignition) cycle differs from the ideal fuel-air cycle because of leakage, heat losses, time losses, and exhaust losses, as shown in Fig. 4-38.

The *Brayton cycle* was originally invented for a reciprocating engine (1906), but the cycle today is applied mainly to gas-turbine installations. The ideal Brayton cycle is shown in Fig. 4-39. A Carnot cycle operating through the same temperature limits T_3 and T_1 has a greater efficiency $(T_3 - T_1)/T_3$. The thermal efficiency for the Brayton cycle is $(T_3 - T_4)/T_3$. It increases as T_2 increases and as T_4 decreases.

The efficiency of an actual Brayton cycle is the actual net work delivered, divided by the actual energy chargeable against the cycle. In the case of a gas-turbine application, each of the three or more pieces of equipment must be evaluated separately for maximum efficiency. The fluid friction takes an

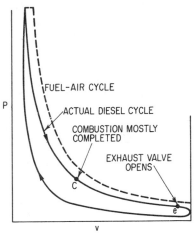

FIG. 4-38   Actual diesel cycle versus fuel-air cycle.

increasingly significant part in such a cycle. The actual Brayton-cycle gas-turbine diagram with its four parts is shown in Fig. 4-40. By using entropies for the various state points, the separate irreversible-loss effects of the cycle can be conveniently evaluated and examined independently with superposition of loss to determine the overall effect on the actual cycle.

Other gas-turbine cycles and turbochargers are discussed in Sec. 5 under "Gas Turbines, Turbochargers, and Expanders."

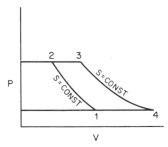

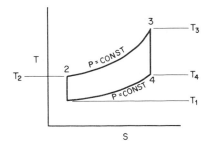

FIG. 4-39   Brayton cycle—ideal.

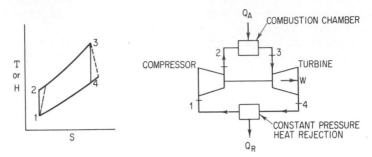

FIG. 4-40 Basic gas-turbine Brayton cycle.

Measuring Engine Performance

Heat-engine performance is measured after a steady-state condition has been reached, and *heat balance* constitutes an important tool in this procedure. This is a tabulation based on test results obtained and includes all pertinent thermodynamic parameters. Basically, the energy supplied to the system in one form or other must equal the energy being taken away. For mechanical work the joule equivalent should be used. The conventional dimension in heat-balance investigations is Btu per pound mass of fluid.

The heat-balance tabulation itemizes each energy transfer. The heat drops, both adiabatic and estimated, for the various losses are enumerated. Experimental results as well as previous tests and theoretical calculations are made available for loss calculations. When the energy tabulation has been completed, the efficiency of the heat engine can be obtained by a simple division. Also, other parameters such as heat rate and specific fuel consumption are then available from heat-balance data compiled similarly by simple mathematical means.

For *reciprocating engines* special measures must be made to obtain a yardstick for their performance. The indicator card is the most important measuring device and represents a picture record of the variation of pressure and volume of the working substance in a cylinder as the piston reciprocates. The record obtained by an indicator gives the indicated mean effective pressure for the engine. The indicated horsepower can then be calculated. On the basis of this information, a heat balance can be set up for any reciprocating heat engine. As measured on a dynamometer, the brake or shaft horsepower can be determined and brake mean pressure calculated. The mechanical efficiency is defined as the ratio of brake horsepower over indicated horsepower.

For *gas turbines* the setting up of a heat balance is difficult and hinges on accurate measurements in various locations of the open or closed loop. Many of the temperatures are highly elevated and, because of high fluid velocities, difficult to measure accurately. Also, factors such as combustion efficiency can often be only roughly estimated.

The two most important performance parameters for gas turbines are the specific fuel consumption and thermodynamic efficiency. In addition, such parameters as engine weight per horsepower or space occupied per horsepower are important, depending on the particular application.

REFERENCES

1. R. C. Reid and T. K. Sherwood, *The Properties of Gases and Liquids*, 2d ed., McGraw-Hill Book Company, New York, 1966.

BIBLIOGRAPHY

Doolittle, J. S., and A. H. Zerban: *Engineering Thermodynamics*, 3d ed., International Textbook Company, Scranton, Pa., 1964.
Faires, V. M.: *Thermodynamics*, 4th ed., The Macmillan Company, New York, 1962.
Jones, J. B., and G. A. Hawkins: *Engineering Thermodynamics*, John Wiley & Sons, Inc., New York, 1960.
Keenan, J. H.: *Thermodynamics*, John Wiley & Sons, Inc., New York, 1941.
Lee, J. F., and F. W. Sears: *Thermodynamics*, 2d ed., Addison-Wesley Publishing Company, Inc., Reading, Mass., 1963.
Sawyer, J. W.: *Gas Turbine Engineering Handbook*, Gas Turbine Publications, Inc., Stamford, Conn., 1966.
Taylor, C. F.: *The Internal Combustion Engine in Theory and Practice*, 2d ed., The M.I.T. Press, Cambridge, Mass., 1966.

FLUID MECHANICS

Introduction

This section deals with selected topics in the field of fluid dynamics; the descriptions have been kept simple. Other topics of fluid mechanics, such as hydrostatics, hydrodynamics, and pipe friction, have been treated in other sections. The various fields are related to each other.

It is interesting to note that Bernoulli's simple theorem has evolved into the general energy equation of thermodynamics, of which Bernoulli's theorem is now a special case. It may also be interesting to discover that the thrust of a jet engine or a rocket motor leads back to the same formula as the axial force on a section of reducing pipe.

Bernoulli's Theorem

The theorem of Daniel Bernoulli (1738) is a most important law of fluid dynamics for many engineering applications. In its original form, the law expresses the conservation of mechanical energy in an ideal (frictionless, nonviscous) and incompressible fluid in steady motion. Mechanical energy appears as kinetic energy, flow work, and potential energy or energy of elevation. These are interchangeable, and their sum must remain constant.

Bernoulli's theorem is applicable to fluid-flow problems in which changes of energy other than mechanical are considered to be negligible. These include many problems in liquid flow and low-speed aerodynamics.

Consider the motion of a small particle in a flowing ideal fluid. It generally

follows a three-dimensional curve called a streamline. If the shape of the streamlines and the velocity at any point do not change with time, the fluid is in steady
motion.

Consider the motion of a somewhat larger particle of a fluid in steady motion.
Its boundary is a surface of streamlines called a stream tube. No fluid crosses the
surface, and no mechanical work is exchanged with an adjacent stream tube.

Consider the stream tube to be sufficiently thin so that constant conditions
prevail over any cross section. Generally, the area, the fluid velocity, the pressure,
and the elevation will change from cross section to cross section. Such a stream
tube between area 1 and area 2 is shown in Fig. 4-41.

The flow in the stream tube is constant. The same amount of fluid enters
through area 1 as leaves through area 2 in any time interval, which may be
expressed by the following continuity
equation:

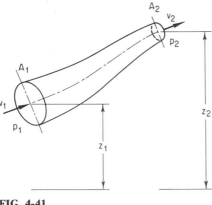

$$w = A_1 v_1 \gamma_1 = A_2 v_2 \gamma_2$$

where w = weight flow, lb/s
A = cross-sectional area, ft^2
v = velocity, ft/s
γ = specific weight, lb/ft^3
Av = volume flow, ft^3/s

If the specific weight remains constant, the velocity is inversely proportional to the area. This condition is
approximately met by the flow of a liquid through a pipe, the pipe being represented by a stream tube.

FIG. 4-41

Bernoulli's theorem expresses that, for any point along a stream tube as
described above, the total mechanical energy or the sum of the mechanical-energy
components must remain constant. This may be written as the following equation:

$$\frac{v^2}{2g} + \frac{p}{\gamma} + z = \text{const}$$

where g = acceleration, 32.174 ft/s^2
p = pressure, lb/ft^2
z = elevation, ft

and the terms are expressed as head, ft (unit mass).

The equation is used only to compare the conditions at different points, such
as the conditions at areas 1 and 2 of the stream-tube section shown above. The
absolute magnitude of the total mechanical energy is of no significance, and the
elevation may be referenced to any arbitrary datum.

Example: Consider the section of a pipe between area 1 and area 2 as shown by
the stream tube above. What is the fluid pressure at area 2 under the absence of losses
(cross-sectional velocities uniform)? The Bernoulli theorem tells that

$$\frac{v_1^2}{2g} + \frac{p_1}{\gamma_1} + z_1 = \frac{v_2^2}{2g} + \frac{p_2}{\gamma_2} + z_2$$

or

$$\frac{p_2}{\gamma_2} = \frac{p_1}{\gamma_1} - \Delta \frac{v^2}{2g} - \Delta z$$

where $\Delta \dfrac{v^2}{2g} = \dfrac{v_2^2}{2g} - \dfrac{v_1^2}{2g} = $ increase in velocity head

$\Delta z = z_2 - z_1 = $ increase in elevation

The velocities may be calculated by the continuity equation shown previously.

The relationship may be expressed in different words: when velocity and elevation increase, the pressure (pressure head) decreases.

The flow-work term may require some explanation. Consider the section of a stream tube shown above. Work is required to push weight W into the section at area 1. This work is force times length of path, or

$$p_1 A_1 \frac{W}{\gamma_1 A_1} = \frac{W p_1}{\gamma_1}$$

In the same time interval (steady flow), an equivalent weight W leaves the section at area 2 and does the amount of work

$$\frac{W p_2}{\gamma_2}$$

The net amount of work or energy supplied to the section

$$\frac{W p_1}{\gamma_1} - \frac{W p_2}{\gamma_2}$$

compensates for the change in elevation and the change in kinetic energy.

For many engineering problems (those which involve flow of gases), a change in elevation has a negligible effect, resulting in the following simplified equation:

$$\frac{v^2}{2g} + \frac{p}{\gamma} = \text{const}$$

This equation may readily be modified as follows:

$$\frac{\rho v^2}{2} + p = \text{const}$$

where $\rho = $ specific mass or density $= \gamma/g$, the form mostly used in aerodynamics.

The first term is usually called *velocity pressure* or *dynamic pressure*.

The Bernoulli theorem may be expanded to cover cases in which the losses are not negligible and part of the mechanical energy is lost by friction. The equation may then be used as follows:

$$\frac{v_1^2}{2g} + \frac{p_1}{\gamma_1} + z_1 = \frac{v_2^2}{2g} + \frac{p_2}{\gamma_2} + z_2 + h_f$$

where h_f = friction head, ft, gh_f representing the mechanical energy per unit mass of fluid lost by friction.

To calculate pressure p_2 in the preceding example, the friction head must also be subtracted, resulting in a lower pressure than is obtained without losses.

The Continuity Equation

The continuity equation is the mathematical representation of the law of conservation of mass. For a control volume which is fixed in space it can be written in the form of the equation

$$\frac{\partial}{\partial t} \int_{cv} \rho + \oint_{cs} \rho v \cdot d\mathbf{A} = 0$$

where $\partial/\partial t$ represents differentiation with respect to time, ρ denotes the density of the fluid, and v and $d\mathbf{A}$ represent the flow velocity (a vector) and the vectorial element of area of the control surface CS bounding the control volume CV. A closed area is assumed to be oriented by the outward normal field of unit vectors $\mathbf{n}$ ($d\mathbf{A} = \mathbf{n} \, dA$, where dA represents an area element). In a steady flow, the term involving the time derivative (the first term) is equal to zero.

The Momentum Equation

The momentum equation represents the mathematical formulation of the law of conservation of linear momentum. For a control volume which is fixed in space it can be written as

$$\frac{\partial}{\partial t} \int_{cv} \rho v \, dV + \oint_{cs} v(\rho v \cdot dA) = \int_{cv} g\rho \, dV - \oint_{cs} \mathbf{n}p \, dA$$

where g denotes the gravitational force per unit mass, p denotes the pressure, and ρ denotes the density of the fluid.

The Energy Equation

The energy equation for a fixed control volume and a steady flow can be written as

$$\dot{Q} = \oint_{cs} \left(h + \frac{v^2}{2} + gZ \right) \rho v \cdot d\mathbf{A} + p$$

where $\dot{Q}$ denotes the time rate of heat transfer, h represents enthalpy per unit mass, ρ denotes the density of the fluid, and p represents the power produced by the fluid within the control volume.

The Angular-Momentum Equation

The angular-momentum equation, written for a steady flow through a fixed control volume in which a turbomachine rotates about z axis, is

$$M_z = \oint_{cs} v_\theta r \rho \mathbf{v} \cdot d\mathbf{A}$$

where M_z represents the torque exerted by the external forces about the z axis, v_θ denotes the tangential component of the flow velocity $\mathbf{v}$, and r denotes the radius measured from the axis of rotation (z-coordinate axis).

Similarity Laws

For the design and development of fluid-flow machinery and components, test and experimental data are required. It is often economical to gather the desired information from the test of a scaled-down model.

For a model test to be valid, the fluid flow should be similar. The flow is similar when the ratio between the various fluid forces is maintained. For instance, consider a fluid-flow problem in which the inertia and the viscous forces are the only fluid forces of significance. Flow similarity would be obtained when the ratio of these two forces is the same in the model as in the full-scale case. It may be shown that this ratio is expressed by the Reynolds number, which is a dimensionless group of variables entering the problem as defined above. Flow similarity would then be obtained when the Reynolds numbers are equal.

Fluid friction and pressure drop in a pipe are the classic example for such a problem. Fluid flow in two similar pipes (equal shape and equal relative roughness in turbulent flow) is similar when the Reynolds numbers are equal, resulting in equal pipe-friction factors as shown by the Moody diagram. Reynolds number is defined as follows:

$$\mathrm{Re}_d = \frac{\rho v d}{\mu}$$

where ρ = specific mass or density = γ/g
    γ = specific weight, $\mathrm{lb/ft^3}$
    g = 32.174 $\mathrm{ft/s^2}$
    v = velocity, ft/s
    d = characteristic length dimension, i.e., inside diameter for a circular pipe, ft
    μ = absolute viscosity, $\mathrm{lb \cdot s/ft^2}$

The Moody diagram is a plot made from experimental results (model test). It makes it possible to calculate the pressure drop in a pipe by looking up the friction factor for a similar flow case (equal Reynolds number) which has been tested.

Many practical fluid-flow problems require an equal Reynolds number for similarity. This includes machinery-casing tests to study flow distribution, tests for flow resistance of bodies, and losses in passages due to friction. It is important that compressibility effects remain negligible, as will be explained below.

To maintain an equal Reynolds number for the model test, it is required to change the value of some of the variables in order to compensate for the decrease in length dimension. This may be achieved in many different ways. The velocity may be increased to the limit imposed by consideration of compressibility effects;

the test may be made at a higher pressure level or density, at a different temperature level (viscosity); or a different test fluid may be used.

In general, many variables enter a fluid-flow problem. The variables contained in the Reynolds number such as density and viscosity are not the only ones. Dimensional analysis shows how the variables may be arranged in dimensionless groups. The Reynolds number is one of these dimensionless groups. The theory tells that the value of these groups which represent ratios of fluid forces should all be maintained for the model to have strict similarity. This is not possible to achieve in practice, and consideration is concentrated on those groups which represent the forces of significance to a particular flow problem.

For instance, in the model testing of ship hulls, it is important to maintain the value of the Froude number, which expresses the ratio of inertia force and gravitational force. Gravitational effects are an important factor in the motion of free surface waves (water surface waves). Froude number is defined as follows:

$$\text{Fr} = \frac{v^2}{gL}$$

where v = velocity, ft/s
g = acceleration of gravity, ft/s^2
L = characteristic length dimension, ft

Equal Froude number means that a ship model having a smaller length would have to be tested in a water tank at a lower speed.

Another dimensionless group of variables is the Mach number, which represents the (square root of the) ratio of inertia and elastic forces in an ideal gas, such as air. Model tests of aerodynamic shapes at high speed where the compressibility effects become most important should take place at the same Mach number. Mach number is defined as follows:

$$\text{Ma} = \frac{v}{a}$$

where v = velocity, ft/s
a = sound velocity, ft/s

The principle of physical similarity is applied in the performance test of a gas turbine or a centrifugal compressor, in which test data are corrected (reduced) to a reference temperature, which is different from the temperature under which the test was made. Fluid velocities are proportional to the rotor-tip speed, and sound velocity is proportional to the square root of the temperature. The standard correction factors are such that the Mach number is maintained between a test point and a corrected point.

There exists another useful dimensionless group of variables known as the Strouhal number. It is used in analysis of sources of excitation of systems in which a fluid flows past circular cylinders or other blunt objects. Its definition is

$$\text{St} = \frac{fd}{v}$$

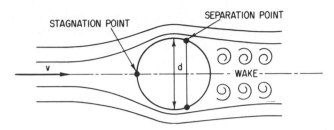

FIG. 4-42

where f = vortex-shedding frequency, s^{-1}
  d = diameter of cylinder, ft
  v = velocity, ft/s

For circular cylinders the Strouhal number is a known function of the Reynolds number Re_d. This experimentally determined relationship can be found in textbooks of fluid mechanics. It allows us to determine system excitation frequency caused by vortex shedding.

When a model test is run at the same Reynolds number as that of the prototype, the Strouhal numbers of both the prototype and the model are the same. This fact allows us to relate the vortex-shedding frequencies of these two systems.

Separation of Flow

In a decelerating flow over a smooth surface the adverse pressure gradient can produce flow reversal if the loss of momentum of the fluid resulting from the viscous friction at the surface is large enough. Laminar boundary layers are more susceptible to separation than turbulent boundary layers.

In a flow over a bluff body such as a circular cylinder (or a sphere) at Reynolds number based on cylinder diameter above 30 or 20, the separated-flow region forms a wake which develops a periodic pattern known as the *Kármán vortex street*. This flow pattern, which represents eddies that break away alternately on either side of the cylinder, is shown in Fig. 4-42. As was mentioned earlier, the dimensionless expression for the vortex-shedding frequency (Strouhal number St $= fd/v$) is a function of the Reynolds number on cylinder diameter.

In a flow over a streamlined body, separation can occur in the region of increasing pressure (adverse pressure gradient) or, in supersonic flows with shock waves, as a result of interaction between a shock wave and the boundary-layer flow.

Drag and Lift

A body immersed in a fluid and moving relative to it with a steady motion experiences forces transmitted from the fluid. The resultant force on the body may have two components. One component is in the direction of fluid motion, called *resistance* or *drag*. There may be a component at a right angle to the fluid motion, called *lift*.

The steady flow of fluid, air or water, past a *sphere* may serve as an example (see Fig. 4-42). The sphere is symmetrical to the direction of motion. The resultant force is the resistance, or drag. Drag is expressed as follows:

$$D = \tfrac{1}{2}\rho v^2 A C_D$$

where ρ = specific mass or density = γ/g
  γ = specific weight, lb/ft^3
  g = 32.174 ft/s^2
  v = velocity, ft/s
  A = projected body area, ft^2
  C_D = drag coefficient

Drag is referred to the dynamic pressure $\tfrac{1}{2}\rho v^2$ (see Bernoulli's theroem). The projected area in this case is the frontal area of the sphere, or $\pi d^2/4$. The drag coefficient is determined experimentally and expresses the drag as a fraction of the total dynamic-pressure force.

As the fluid flows around the sphere, the velocity will change in the vicinity of the surface. The fluid will come to a halt in front center, the so-called stagnation point, and it will reach maximum velocity at the diameter d. According to Bernoulli's theorem, there is a corresponding change in static pressure. The pressure buildup at the stagnation point is equal to the dynamic pressure.

With an ideal nonviscous fluid there would be no drag in this case. The velocity and pressure distribution would be the same on the back of the sphere as in front, and there would also be a stagnation point in back center. The drag is due to the direct and indirect effects of the fluid viscosity. The direct effect is the skin friction or shear in the boundary layer, and the indirect effect is the pressure distribution caused by boundary-layer separation. The pressure will be less on the downstream side of the sphere than in front because of the flow separation.

If the changes in pressure are small compared with the total pressure or if the Mach number is small, the drag coefficient depends only on the Reynolds number Re having the diameter d as a characteristic length dimension. The drag coefficient may then be found from an experimentally determined plot in function of the Reynolds number (Fig. 4-43). The steady flow of fluid past an *airfoil section* may serve as another example (Fig. 4-44). The airfoil section may be the section of an airplane wing or a turbomachinery blade. Profiles of this sort are designed to yield maximum lift and minimum drag. The profile may or may not be symmetrical with respect to its chord line, but a certain angle of attack is required to obtain lift.

It has been explained that the fluid is subject to velocity and pressure changes when flowing around a body. The pressure distribution around an airfoil section is such that the resultant

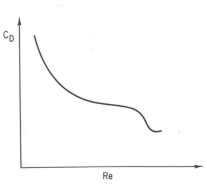

FIG. 4-43

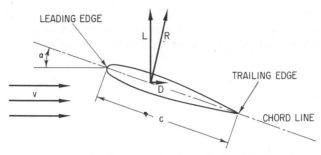

FIG. 4-44 c = chord; α = angle of attack; L = lift; D = drag; R = resultant force.

pressure force is larger for the bottom surface than for the top surface, yielding a net resultant force with a large lift component.

Similarly to the previous example, lift and drag are expressed as follows:

$$L = \tfrac{1}{2}\rho v^2 A C_L$$

$$D = \tfrac{1}{2}\rho v^2 A C_D$$

Again, A is the projected body area but, in this case, the "wing" area or, for the airfoil section, chord times unit length of span.

As explained previously, the drag is due to the direct and indirect effects of the fluid viscosity. It is called *profile drag* to distinguish it from *induced drag*. Induced drag is due to the end effects of a three-dimensional case such as a complete airplane wing. The airfoil section is considered to be part of a very long span, a two-dimensional case. The published airfoil-section data contain profile drag only.

The lift and drag coefficients have been determined experimentally for many different airfoil sections by the National Aeronautics and Space Administration (NASA) and research institutes. The data are plotted in function of the angle of attack in a range of Reynolds numbers. Applications for the data include the design of airplane wings and turbomachinery blading, specifically the blading of axial-flow compressors.

Losses in Flows in Ducts

The energy equation for a one-dimensional, steady, incompressible flow in a duct with losses present was written in the subsection "Bernoulli's Theorem." It is

$$gz_1 + \frac{v_1^2}{2} + \frac{p_1}{\rho_1} = gz_2 \frac{v_2^2}{2} + \frac{p_2}{\rho_2} + h_f$$

where h_f denotes the head loss due to friction.

The head loss h_f can be considered to be made up of two losses: (1) that resulting from the viscous friction along the walls over which the fluid flows and (2) that which occurs as a result of eddies produced by separation of flow or by secondary flow (such as exists in a flow through bent pipes). Separation of flow in a duct can occur whenever the cross section of the duct changes abruptly. The head loss that occurs as a result of eddies is known as the *form head loss*. (Information on the friction loss and the form head loss can be found in Sec. 3.)

MECHANICAL VIBRATIONS

In all rotating or reciprocating machinery, a condition of mechanical vibration will be present because of the force and energy levels inherent in the equipment. The question as to whether these vibrations prove to be harmful or not depends on several factors. These factors can best be expressed in terms of the dynamic motions of the simple harmonic oscillator when subjected to a sinusoidal, time-varying, external force. A typical representation of this system is shown in Fig. 4-45 with the standard nomenclature used in the vibration field.

$$\omega_N = \sqrt{\frac{Kg_0}{W}} = \text{undamped natural frequency}$$

$$\tau = \frac{2\pi}{\omega_N} = \text{period of free vibration}$$

$$\omega_d = \omega_N\sqrt{1 - \zeta^2} = \text{damped natural frequency}$$

$$c_c = 2\sqrt{\frac{KW}{g_0}} = \text{critical damping}$$

$$\zeta = \frac{c}{c_c} = \text{damping ratio}$$

where W = weight of mass in Fig. 4-45, lb
K = spring constant in this system, lb/in
g_0 = standard gravitational constant (386 in/s^2)
ω = frequency, rad/s
c = damping coefficient, lb·s/in

By utilizing the above definitions, the equation of motion for this system is

$$\frac{W}{g}\ddot{x} + c\dot{x} + Kx = P \sin \omega t$$

where P is the maximum amplitude of the exciting force, lb, and t is time, s. The dots represent differentiation with respect to time t. In general, the damping ratio ζ is small, and the steady-state response of the system is given by the usual equation:

$$x = \frac{P}{K} \frac{1}{\sqrt{[1 - (\omega/\omega_N)^2]^2 + [2\zeta(\omega/\omega_N)]^2}} \sin (\omega t - \phi)$$

$$\phi = \tan^{-1} \frac{2\zeta(\omega/\omega_N)}{1 - (\omega/\omega_N)^2} = \text{phase angle}$$

The second fraction in the expression for x above is frequently referred to as the *magnification factor* for the system. A plot of this parameter and of the phase angle as functions of the frequency ratio ω/ω_N is given in Fig. 4-46.

The successive peaks of the curves for the magnification factor (often termed the *response curve*) located at or near the condition of forcing frequency ω equal to the natural system frequency ω_N are termed a condition of *resonance*. This phenomenon of resonance explains the critical nature of vibration in the design of mechanical systems, since the deformation and stress of a member may be multiplied by a considerable factor (5 to 10) above the static reaction to a steady load of intensity P, thus inducing an eventual fatigue failure. The associated concept of tuning the system also stems from consideration of the response curves. By adjusting either the exciting frequencies or the system natural frequency so that the ratio ω/ω_N is significantly removed from the unity value, the system is said to be detuned. In the case $\omega/\omega_N > 1.5$, the response to the forcing function will, in reality, be an attenuation of the input motions.

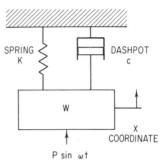

FIG. 4-45

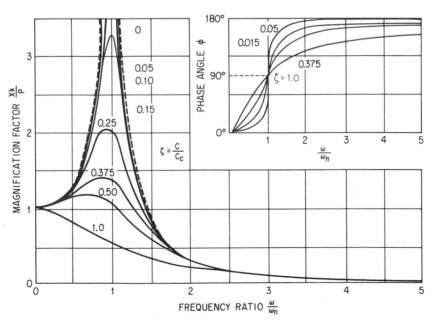

FREQUENCY RATIO $\frac{\omega}{\omega_n}$

FIG. 4-46 (*Courtesy of L. S. Marks,* Standard Handbook for Mechanical Engineers, *7th ed., ed. by Theodore Baumeister et al., McGraw-Hill Book Company, New York, 1967.*)

Formulas for calculating the natural frequency of commonly used beams of uniform cross section due to lateral vibration are given below (W = weight, lb; W_b = weight of beam, lb; W_s = weight of spring, lb; w = weight per unit length, lb/in; l = length of beam, in; I = moment of inertia of cross section, in^4; E = Young's modulus, lb/in^2; k = stiffness of spring, lb/in; ω_N = angular natural frequency, rad/s):

A. Weightless beam with concentrated load (weight)

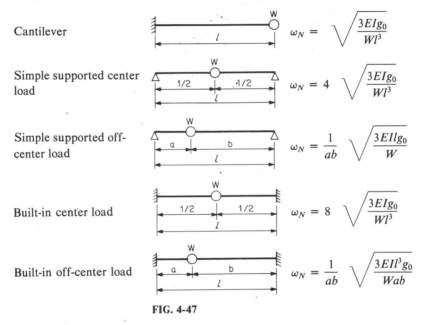

Cantilever $\qquad \omega_N = \sqrt{\dfrac{3EIg_0}{Wl^3}}$

Simple supported center load $\qquad \omega_N = 4\sqrt{\dfrac{3EIg_0}{Wl^3}}$

Simple supported off-center load $\qquad \omega_N = \dfrac{1}{ab}\sqrt{\dfrac{3EIlg_0}{W}}$

Built-in center load $\qquad \omega_N = 8\sqrt{\dfrac{3EIg_0}{Wl^3}}$

Built-in off-center load $\qquad \omega_N = \dfrac{1}{ab}\sqrt{\dfrac{3EIl^3g_0}{Wab}}$

FIG. 4-47

B. Spring or beam with concentrated load (weight)—weight of beam included

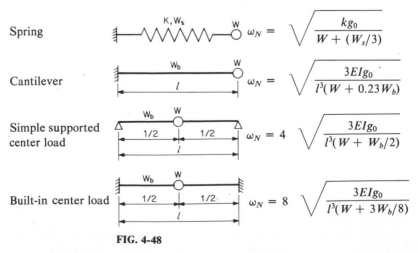

Spring $\qquad \omega_N = \sqrt{\dfrac{kg_0}{W + (W_s/3)}}$

Cantilever $\qquad \omega_N = \sqrt{\dfrac{3EIg_0}{l^3(W + 0.23W_b)}}$

Simple supported center load $\qquad \omega_N = 4\sqrt{\dfrac{3EIg_0}{l^3(W + W_b/2)}}$

Built-in center load $\qquad \omega_N = 8\sqrt{\dfrac{3EIg_0}{l^3(W + 3W_b/8)}}$

FIG. 4-48

A final useful parameter, in discussing the characteristics of the single-degree-of-freedom system, is the logarithmic decrement of damping. It is a measure of the energy-absorption capability of the system. This quantity is defined as the logarithmic ratio of successive peaks of the free-vibration response of the system to a step input. It also may be defined by means of the response curve in terms of the half-power points. Thus the decrement δ is given by either of the following:

$$\delta = \ln \frac{x_i}{x_{i+1}} = \frac{\pi(\omega_2 - \omega_1)}{\omega_N} = \frac{2\pi\zeta}{\sqrt{1 - \zeta^2}} \cong 2\pi\zeta$$

where ω_2 = frequency above ω_N where the magnification factor is 0.707 times the value at ω_N

ω_1 = frequency below ω_N where the magnification factor is also 0.707 times the value at ω_N

The damping in a system can also be determined from the sharpness of the response of that system near the resonant frequency ω_N as indicated in Fig. 4-49. If $\Delta\omega$ is determined at the half-power point,

$$R_i = R_{max}(\sqrt{2}/2) \qquad \text{and} \qquad 2\zeta = \frac{1}{Q} = \frac{\Delta\omega}{\omega_N}$$

where the quantity Q is called the quality factor.
When $R_i = R_{max}/2$,

$$2\zeta \approx \frac{1}{\sqrt{3}} \cdot \frac{\Delta\omega}{\omega_N}$$

The above equations are reasonably accurate for values of $\zeta < 0.1$.

Many of the above concepts carry over into linear multiple-degree-of freedom systems and into continuously elastic systems in the neighborhood of resonance. It is clearly evident that for lightly damped systems the number of resonant peaks of the response curve will coincide with the number of discrete lumped masses in the system. To investigate response near resonance, the use of equivalent viscous damping as an energy-absorption device is frequently justified.

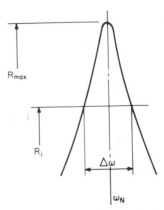

FIG. 4-49 Response curve showing the bandwidth.

The case of torsional vibration (Fig. 4-50a) is analogous to that of lateral vibration, wherein the free vibration is governed by the equation

$$I\ddot{\theta} + c\dot{\theta} + K\theta = 0$$

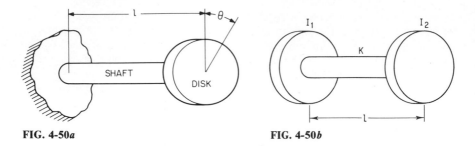

FIG. 4-50a   **FIG. 4-50b**

The natural frequency of this system in hertz (cycles per second) is

$$f_N = \frac{1}{2\pi} \sqrt{\frac{K}{I}} = \frac{1}{2\pi} \sqrt{\frac{GI_p}{Il}}$$

where I = mass moment of inertia of disk
   I_p = polar moment of inertia of shaft
   G = shear modulus of shaft

For the system of two disks and a shaft (Fig. 4-50b) the frequency is

$$f_N = \sqrt{\frac{(I_1 + I_2)GI_p}{I_1 I_2 l}}$$

For more complex torsional systems the tabular method due to Karl Holzer is used to compute the several frequencies of the system.

SHOCK RESPONSE OF MECHANICAL SYSTEMS

Shock, as the term is used in structural analysis, refers to the motion of a structure under investigation due to sudden transient (in time) excitation. This excitation can be defined in terms of displacement, velocity, acceleration, or force. The shock response is transient in nature and usually short-lived.

The equation governing the response of the single-degree-of-freedom system of Fig. 4-45 is

$$\frac{W}{g}\ddot{x} + c\dot{x} + kx = F(t)$$

The preceding subsection on vibration analysis deals with the special cases in which $F(t) = 0$ or $F(t) = P \sin \omega t$. The former case is the one in which the system is undergoing free vibration at its natural frequency due to some initial excitation at time $t = 0$.

The excitation on an item of heavy machinery often results from motion of the ground upon which its support structure sits. A simplified representation of this

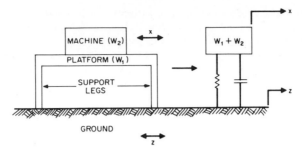

FIG. 4-51   Model of a single-degree-of-freedom system.

system is shown in Fig. 4-51. If the ground motion is $z(t)$, then the governing equation is

$$\frac{W}{g}\ddot{x} + c(\dot{x} - \dot{z}) + k(x - z) = 0$$

If we define $y = x - z$ to be the relative motion between ground and mass, then the governing equation becomes

$$\frac{W}{g}\ddot{y} + c\dot{y} + ky = -\frac{W}{g}\ddot{z}$$

which is the same form as the previous equation.

A simple transient excitation which can produce a shock response is shown in Fig. 4-52. It consists of a time period $t_1 - t_0$ in which the applied force builds up, a period $t_2 - t_1$ in which the force acts at its maximum value, and a period $t_3 - t_2$ in which the force decays back to zero.

If the rise period $t_1 - t_0$ and the decay period $t_3 - t_2$ are large relative to the natural period τ of the system ($\tau = 2\pi/\omega_N$), then the excitation will not cause a dynamic response. The dynamic forces are essentially zero, and the system is governed by the equation

$$kx = F$$

whence the maximum response is

$$x = \frac{F}{k}$$

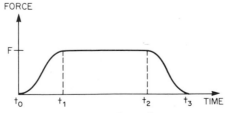

FIG. 4-52   Simple transient pulse.

In practice, rise and decay periods greater than about 3 times the natural period produce no dynamic amplification. Thus, if the lowest natural frequency is above a threshold value such that the longest natural period is a factor of 3 shorter than the rise time of the pulse, the structure can be analyzed statically. This condition arises occasionally in analyzing the response of equipment to seismic loading.

If the rise and decay periods are short relative to the natural period, then the shape of the curve of $F(t)$ can be simplified to the well-known *step pulse* of Fig. 4-53. The magnitude of the system response is related to the duration $t_2 - t_1$ of the pulse. If the duration is long relative to the natural period of the system, response is as shown in Fig. 4-54. The peak amplitude of response is twice the static response; i.e., the *dynamic amplification factor* is 2. In an undamped system, the structure continues to vibrate at its natural frequency and with the peak amplitude of $2 F/k$. In a lightly damped system, the peak response is essentially twice the static response initially, but with time the response amplitude decays to the static value of $x = F/k$. Heavily damped systems exhibit lower peak dynamic amplification factors and more rapid decay to the static-response level.

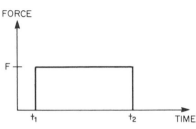

FIG. 4-53   Step pulse.

Removal of the pulse causes another transient response. The undamped system continues to vibrate at its natural frequency, now with response governed by the relationship

$$x = \left[\left(\frac{F}{k} \right) \cdot 2 \cdot \sin \frac{\pi(t_2 - t_1)}{\tau} \right] \sin \omega_N(t - \phi)$$

Thus, depending on the ratio $\dfrac{t_2 - t_1}{\tau}$, the residual vibration can have amplitudes ranging from zero to $\dfrac{2F}{k}$.

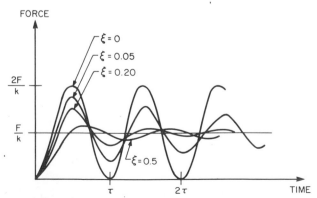

FIG. 4-54   Response to step pulse.

The damped system gradually decays to zero response amplitude.

Another special case of interest is that in which the rise time $t_1 - t_0$, duration of maximum force $t_2 - t_1$, and decay $t_3 - t_2$ all approach zero. The resultant excitation is an *impulse*. An impulse is characterized by a total pulse time length that is much less than the system's natural period. Impulse excitations are typical of impact or collision types of shocks. Application of an impulse to a body at rest is equivalent to a sudden change of velocity; i.e., the system response is that of free vibration subject on an initial condition of zero displacement and nonzero velocity. The magnitude of the initial velocity is

$$v_0 = \frac{Ig}{W}$$

where I is the total impulse, defined as

$$I = \int_0^{t_3} F(t)\, dt$$

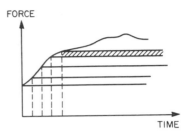

FORCE

TIME

FIG. 4-55 Arbitrary excitation as a series of step functions.

A more general transient excitation is illustrated in Fig. 4-55. The mathematical solution for the undamped response of the single-degree-of-freedom system can be determined by superposing solutions to a series of rectangular step pulses. The result is expressed in terms of *Duhamel's integral*. For a lightly damped system in which $\sqrt{1 - \zeta^2} \approx 1$, Duhamel's integral is expressed as

$$x(t) = x_0 \cos \omega_N t + \left(\frac{v_0}{\omega_N}\right) \sin \omega_N t$$

$$+ \left(\frac{g}{W\omega_N}\right) \int_0^t F(\xi) \exp\left[-\zeta\omega_N(t - \xi)\right] \sin \omega_N(t - \xi)\, d\xi$$

In no case, however, does the dynamic amplification factor exceed twice the maximum applied amplitude. This fact has led to a belief among some designers that the effects of shock loading can be conservatively accounted for by designing the system to resist a load equal to twice the maximum static load. However, this is true only for systems that can be represented by single-degree-of-freedom models. This is not often the case in real systems. For example, in systems that must be modeled as two-degree-of-freedom systems, such as those in Figs. 4-56 and 4-57, dynamic amplification factors greater than 2 routinely occur.

The analysis of the response of elastic structural systems to transient excitation is a complex subject far beyond the scope of this brief introduction. Among many fine sources on the subject the interested reader is referred to C. M. Harris & C. E. Crede, *Shock and Vibration Control Handbook*, McGraw-Hill Book Company, New York, 1976.

Complex structures are usually analyzed in terms of sophisticated spring-mass-damper models that have many moving mass points, i.e., degrees of freedom, to be considered.

It is characteristic of complex structures that they have many natural frequencies of vibration. Associated with each of these natural frequencies is a char-

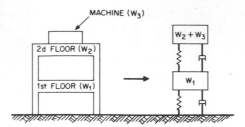

FIG. 4-56 Model of a two-degree-of-freedom system.

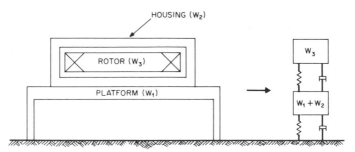

FIG. 4-57 Model of a two-degree-of-freedom system.

acteristic deflection pattern of vibration known as a *mode shape.* Calculation of the linear elastic response of structures to shock excitation is simplified by the fact that once the natural frequencies and mode shapes are known, it is possible to develop a mathematical transformation of coordinates such that the response of the structure can be characterized in terms of a system of single-degree-of-freedom models.

Within this mathematical artifice, each system has a natural frequency corresponding to one of the actual frequencies, structural motion is characterized by a *generalized coordinate,* and excitation is developed by a *generalized excitation.* These latter two characteristics are special linear combinations of the physical motion and excitation on the complex structure. Once the response of these single-degree-of-freedom oscillators has been defined, i.e., once the generalized coordinate motion has been defined in terms of the generalized excitation, the inverse transformation can be imposed to determine the actual response of the structure.

The ability to characterize the response of complex structures in terms of single-degree-of-freedom models has led to a simple technique for conservatively estimating the peak response of complex structures to complex excitations known as *shock spectrum analysis.*

It is a relatively simple matter to estimate the maximum possible response of a single-degree-of-freedom oscillator to a complex shock excitation as a function of the natural frequency of the oscillator. For a multiple-degree-of-freedom structure, one can then estimate the peak response, in terms of generalized coordinate motion, in each vibrational mode. Transformation from generalized coordinates to physical coordinates yields an estimate for the peak motion due to shock of each degree of freedom of the structure. An upper bound to this estimate is obtained through the use of absolute values of both the generalized coordinate deflections and the transformation coefficients.

This practice is especially popular in estimating the response of structures to complex ground motions due to earthquakes.

NOISE AND SOUND MEASUREMENTS

Noise is generally defined in terms of undesirable sound. As such, it tends to interfere with the useful functions associated with operating heavy machinery. Noise may be either structure-borne or airborne. Structure-borne noise is usually closely associated with machine vibrations. Airborne noise principally affects personnel and hinders their activity in the vicinity of the noise source. Airborne noise is also frequently an adequate gauge by which deterioration in performance of a machine may be detected. A distinct increase in noise level often indicates internal distress due to component failure or excessive wear with an operating unit.

The level of noise or sound is commonly expressed in *decibels* (dB). The word *level* usually refers to decibel. Decibel is a ratio quantity of the sound in question and a reference, which is stated or implied. Although the decibel is used to describe the magnitude of sound-pressure level, specifically decibel is associated with power ratio or quantities which are directly proportional to the power.

Power Level

The power level (PWL) is defined as

$$PWL = 10 \log \frac{W_x}{W_{re}} \text{ (dB)}$$

where W_x is the acoustic power in watts, the logarithm is to the base 10, and W_{re} is the reference power, $W_{re} = 10^{-12}$ W. Presently there is no instrument for directly measuring the power level of a sound source. Power level can be computed from sound-pressure measurements.

Sound Intensity

The sound-intensity level (SIL) is defined as

$$SIL = 10 \log \frac{I_x}{I_{re}} \text{ (dB)}$$

where I_x is the acoustic intensity of the sound wave as the average power transmitted per unit area in the direction of wave propagation, W/m^2, the logarithm is to the base 10, and I_{re} is the reference intensity in 10^{-12} W/in^2.

Sound-Pressure Level

Since the acoustic power and the intensity level are proportional to the square of the sound pressure, the sound-pressure level (SPL) is defined as

$$\text{SPL} = 20 \log \frac{P_x}{P_{re}} \text{ (dB)}$$

where P_x is the root-mean-square of the sound pressure in newton per square meter (N/m^2) for the sound in question, the logarithm is to the base 10, and P_{re} is the reference sound pressure; for airborne sound this reference is generally 20 $\mu N/m^2$ (2.9×10^{-9} lb/in^2). A reference pressure of 1 barye (14.5×10^{-6} lb/in^2) has gained widespread acceptance in conjunction with sound measurements in liquids and with transducer calibration.

Sound Level

The apparent loudness that is attributed to a sound varies not only with the sound pressure but also with the frequency content of the sound. A pressure level of a sound at one frequency appears different in loudness from the same pressure level at another frequency. To compensate for this difference, a sound-level meter is equipped with networks that "weigh" the sound-pressure levels at various frequencies (see Fig. 4-58).

It is customary to measure the sound level with either A, B, or C weighing. C weighing has the minimum suppression. A sound-level measurement with no weighing network represents a true sound-pressure reading.

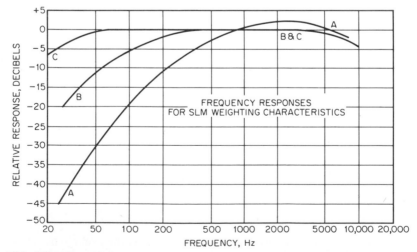

FIG. 4-58 Frequency-response characteristics in the American National Standard specification for sound-level meters, ANSI S1.4-1971.

Measurement

Several sound-level-measuring standards have been established to determine the noise produced by a device, machine, or apparatus. One of these standards is the American National Standard ANSI S1.2-1962 (R 1971) method for the physical measurement of sound. This procedure provides for the use of a sound-level meter in combination with an octave-band filter set to determine the sound level in each octave band of the audible-frequency range.

This standard provides for determination of the sound-power level of a sound source, based upon sound-pressure-level measurements. The standard provides methods for correcting the sound-pressure recordings, based upon the surrounding background noise and acoustic measurement environment.

To guide readers in the field of sound measurement it is recommended that they be familiar with the available equipment. Instrument manufacturers and local representatives offer a wide range of literature to guide users of their instruments.

Combining Decibels

With multiple sound or noise sources, there is a frequent need to combine a noise level of one sound source with another one or with that of several different noise sources. Combining decibels is to be done on an energy basis. The procedure involves converting the decibel readings to power, adding or subtracting them as the situation requires, and then converting back to decibels in question.

Refer to Fig. 4-59 for adding noise levels by the use of a graph. For example, two noise sources are measured separately with A-weighted scale as 80 dBA and 76.5 dBA. What would the total combined noise be? The difference in sound levels is 3.5 dB, and the graph gives a value of 1.6 dB, which should be added to the higher sound level. The combined level is 81.6 dBA.

A periodic inspection and calibration of the sound measurement and analyzer instruments will help users to gain the best results.

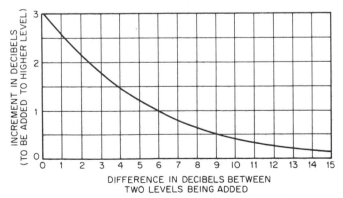

FIG. 4-59 Graph for combining noise levels.

How Much Noise Is Acceptable?

Because of the hearing damage that could be caused by continuous exposure to high noise level, the secretary of labor reissued (May 1971) the following limits on noise promulgated under the Walsh-Healey Act:

1. The maximum acceptable noise level for an 8-h day is 90 dBA (decibels measured on the A scale) for a standard sound-level meter at slow response.
2. Exposure to impulse on impact noise should not exceed 140 dBA.

ROTATING MACHINERY

With the introduction of high-speed, heavily loaded rotating machinery, the problem of the dynamic response of the rotors is becoming increasingly significant. The basic principles may be understood from consideration of the simple single-disk shaft as illustrated in Fig. 4-60.

By equating centrifugal forces acting on the disk with the elastic restoring forces of the shaft, the following equation of motion is obtained (see Fig. 4-60):

$$\frac{W}{g}\omega^2(X + e) = kX$$

If the lateral vibration frequency of the rotor is introduced, $\omega_N^2 = kg_0/W$, the more usual form for the displacement is obtained:

$$X = \frac{(\omega/\omega_N)^2}{1 - (\omega/\omega_N)^2}e$$

For the case in which ω approaches ω_n, the displacement increases drastically and a condition of resonance exists. This value of ω_N is often termed the *critical speed,* but more accurately it induces a condition of synchronous whirl of the shaft. For the simple case in which the mass of the rotor may be concentrated at the center of the span,

$$N_c = \frac{60}{2\pi}\sqrt{\frac{g_0}{\delta_{st}}} = \frac{187.6}{\sqrt{\delta_{st}}}$$

where N_c = critical speed, r/min
δ_{st} = static shaft deflection = W/k, in

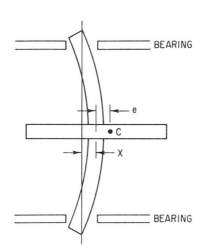

FIG. 4-60 W = weight of rotor; C = location of rotor center of gravity; e = distance between geometric center and C; X = displacement of rotor geometric center from bearing line; ω = rotating frequency of rotor; k = equivalent spring constant for shaft restoring force.

For rotating machinery represented by Fig. 4-61, in which the exciting force is produced by the rotation of an eccentric mass m located at a distance e from the center of rotation, the displacement is given by

$$X = \frac{\frac{m}{M} e \left(\frac{\omega}{\omega_N}\right)^2}{\left[1 - \left(\frac{\omega}{\omega_N}\right)^2\right]^2 + \left[2\zeta\left(\frac{\omega}{\omega_N}\right)\right]^2} \sin(\omega t - \phi)$$

$$\phi = \tan^{-1} \frac{2\zeta\left(\frac{\omega}{\omega_N}\right)}{1 - \left(\frac{\omega}{\omega_N}\right)^2}$$

When $\omega/\omega_N = 1$, $MX/me = 1/2\zeta$ and the response depends only on the amount of damping.

A plot of the response curve and the phase angle as a function of the frequency ratio is given in Fig. 4-62.

From Fig. 4-62 it can be seen that the damped resonant frequency ω_d increases, as the damping ratio ζ increases, with respect to the undamped natural frequency ω_N. This relation can be expressed as follows:

$$\omega_d = \omega_N/\sqrt{1 - 2\zeta^2} > \omega_N$$

The variation between the geometric center and the center of gravity as measured by e gives rise to the concept of balancing rotating machinery. By proper addition or removal of weight on a rotor, the product We can be reduced to minimal limits. It should be noted, however, that each rotor retains some residual unbalance owing to practical considerations and therefore will, at all times, exhibit a response phenomenon. The objective in balancing is to reduce this response to a tolerable minimum.

Balancing may be of the static or the dynamic variety. For the case with all the rotor unbalance in a single plane and on one side of the geometric axis, a static balance in that plane is sufficient for corrective purposes. However, the general case consists of several planes of unbalance, randomly distributed around the geometric axis. This condition is illustrated in Fig. 4-63 for the case of two arbitrarily located unbalance weights.

Generally, the two planes in which balance correction will be accomplished are located toward the ends of the rotor, labeled (1) and (2) in Fig. 4-63. The two corrections at (1) will be $[-W_1(l - a)]/l$ and $[-W_2(l - b)]/l$ and will be coplanar with the respective

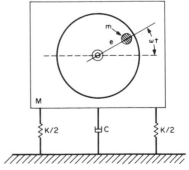

FIG. 4-61 Periodic force due to a rotating unbalance.

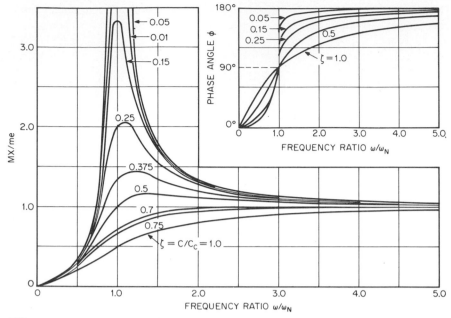

FIG. 4-62   System response with rotating unbalance.

unbalances W_1 and W_2, as indicated by the negative sign. Likewise, the correction at (2) will be $-W_1 a/l$, $-W_2 b/l$. In general, large high-speed rotors can be balanced within fractions of an ounce-inch.

To determine the critical speeds for actual turbomachine rotors which are more complex cases, in which there are several weights W_1, W_2, ..., W_i each concentrated on a single span, a Rayleigh-type energy approach yields, as an estimate of the lowest frequency,

$$N_c = 187.6 \sqrt{\frac{W_1 y_1 + W_2 y_2 + \cdots + W_i y_i}{W_1 y_1^2 + W_2 y_2^2 + \cdots + W_1 y_i^2}} \quad \text{r/min}$$

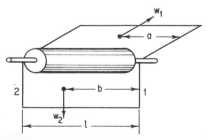

FIG. 4-63 a = distance between W_1 and right balance plane; b = distance between W_2 and right balance plane.

where the y_i are the static deflections of each disk or plane which weighs W_i.

To improve the accuracy of the estimate, the deflection y_i can be computed from dynamic loading, which is initially evaluated on a basis of static deflections. This iterative procedure of improving the accuracy of the y_i by recalculating dynamic loads on the basis of the previous estimate of the shaft-deflection shape is the basis for the Prohl-Myklestad approach to critical-speed calculation. The higher critical speeds can be obtained by substituting the second, third, etc., mode shapes in the basic Rayleigh equation.

A second useful tool in obtaining an estimate of the first critical speed of a shaft with several concentrated weights is that given by Stanley Dunkerley,

$$\frac{1}{N_c^2} \cong \frac{1}{N_0^2} + \frac{1}{N_1^2} + \frac{1}{N_2^2} + \frac{1}{N_3^2} + \cdots$$

where N_0 = critical speed of shaft only

N_1 = critical speed of shaft with W_1 only, located at its proper position on shaft

N_2 = critical speed of shaft with W_2 only, etc.

Utilization of the fact that resonance for synchronous whirl will occur at the same frequency as lateral vibration enables the computation of critical speeds of uniform shafts by means of the formula

$$N_c = a_n \sqrt{\frac{EIg_0}{Wl^4}}$$

where EI = flexural stiffness of the shaft

W = weight per unit length of shaft

l = length of shaft

a_n = constant based on shaft-fixity conditions and the mode number as tabulated below

Shaft fixity	a_1	a_2	a_3	a_4	a_5
Simply supported	94.2	377.	848.	1503.	2356.
Clamped-free	33.6	214.	589.	1156.	1910.
Free-free or clamped-clamped	214.0	589.	1156.	1910.	2848.
Clamped-hinged or hinged-free	147.0	478.	993.	1700.	2597.

Modern computational methods for rotor response evaluations introduce the concepts of bearing-oil-film flexibility, bearing-pedestal stiffness, and damping. Thus the model, the physical idealization of the system involved in dynamic response, is depicted in Fig. 4-64, including the cross-coupling effect of the hydro-dynamic action of the oil film.

The dashpots in Fig. 4-64 imply energy absorption of the resultant dynamic motions and provide a means of evaluation of the rotor's displacement, since the input rotational energy must be equated to the energy removed from the system by means of the dashpots. In general, the bearing-oil-film characteristics are linearized by using an assumption of small oscillations in order to facilitate the computation. The addition of the bearing flexibility usually depresses the lowest frequency of rotor response below that computed for the rigid-bearing critical speed. However, since the additional freedoms introduced greatly affect the normal mode shapes, no general statement can be made for higher modes, and each machine system must be evaluated independently.

These rotor-response computational methods for obtaining dynamic motions

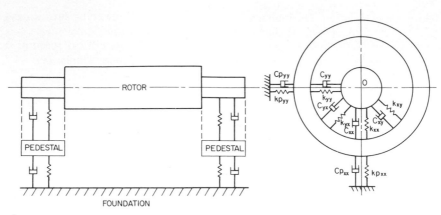

FIG. 4-64   Idealized structure for rotor-response calculation.

also yield dynamic stresses throughout the system since, in general, the deflection curve and/or its derivatives are proportional to the internal-stress patterns of the various components; i.e., for the shaft,

$$\sigma \alpha M = EIy''$$

In practical operation of machinery the dynamics of the equipment is affected not only by the design but also by the physical condition of the hardware and the variables related to the installation. Thus items such as unbalance caused by wear or dirt accumulations, misalignment of couplings, bent shafts, and eccentric jour

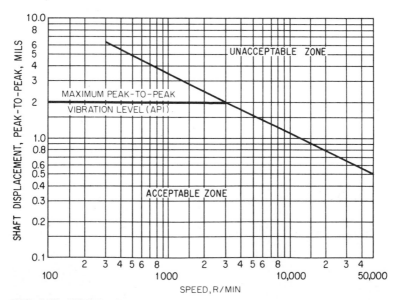

FIG. 4-65   Vibration limits.

nals or bearings often cause unsatisfactory conditions. In the final analysis all factors must be evaluated in terms of machinery operation in field service. A useful criterion for evaluating the dynamic characteristics of a machine is the measurement of the vibration levels on the shaft in the vicinity of the bearing caps.

The American Petroleum Institute (API) specifies maximum shaft-vibration levels (measured at the shaft) as follows:

$$\text{Peak-to-peak mils} = \sqrt{\frac{12{,}000}{\text{r/min}}} \leq 2.0 \text{ mils}$$

Figure 4-65 shows a plot of this relation. (See also subsection "Vibration Limits of Centrifugal Pumps" in Sec. 8.)

BEARINGS AND LUBRICATION

Bearings

Introduction   Bearings permit relative motion to occur between two machine elements. Two types of relative motion are possible, rolling or sliding, each of which depends upon the design of the mechanical bearing element. Thus bearings are classified into two general types: the rolling-contact type (rolling) and the sliding-contact bearing design in which the bearing elements arc separated by a film of oil (sliding). Both can be designed to accommodate axial and/or radial loads. Each has a wide variety of types and designs to fit a wide variation in uses. The selection of a bearing type for application to a particular situation involves a performance evaluation and cost consideration.

There is ample literature available to determine the relative merits of each type. This subsection will provide a general overview of the types of bearings presently encountered in the equipment covered in this edition. Changes take place frequently because many scientists and engineers work constantly to improve the state of the art in bearing design.

Rolling-Contact Bearings   Rolling-contact bearings include ball bearings, roller bearings, and needle bearings. Within each category several variations have been developed for specific applications. Variations in the amounts of radial and thrust load capabilities also exist between specific types. Self-aligning ball or roller bearings, by virtue of their spherically ground outer race, can tolerate misalignment of the shaft or housing.

Rolling-contact bearings consist of four principal components: an outer race, an inner race, rolling elements, and a separator, or spacer, for the rolling elements. The inner ring is mounted on the shaft. The outer ring securely fits in a stationary housing. The facing surfaces of the inner and outer rings are grooved to conform to the rolling-element shape. The rolling elements (with separator) accurately space the inner and outer races and thus enable smooth relative motion to occur (see Fig. 4-66).

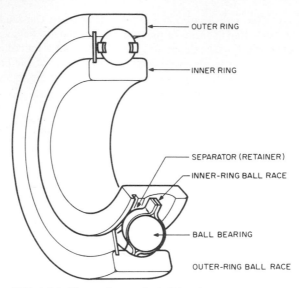

FIG. 4-66   Nomenclature of a ball bearing.

Sliding-Contact Bearings   Sliding-contact bearings are classified into two general types: journal bearings and thrust bearings. Journal bearings support radial loads imparted by the rotating shaft and may also be required to arrest or eliminate hydraulic instabilities which may be encountered in lightly loaded high-speed machinery. The thrust bearings are used for loads parallel to the shaft and may be required to support the full weight of the rotor in cases of vertical machinery.

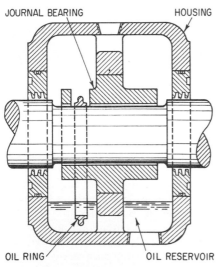

FIG. 4-67   Ring-oil lubrication.

Journal Bearings   The common types of journal bearings are:

Plain journal bearings
Three-lobe journal bearings
Tilting-pad journal bearings

The *plain journal bearing* may be either force-fed from a pressure-lubrication system or ring-oiled by means of a free ring which rests on and rotates with the shaft to serve as a pumping medium, as shown in Fig. 4-67. A section of a typical force-fed journal bearing is shown in Fig. 4-68. This shows a babbitt-lined split-type bearing. Journal-bearing clearance for high-speed machinery, i.e., turbines, centrifugal compressors, pumps, etc., should be not

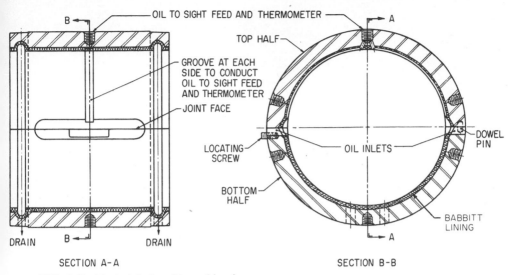

FIG. 4-68   Typical design of journal bearing.

less than 0.0015 in/in of diameter for satisfactory operation. However, a complete analysis should be made to obtain the proper clearance for each case.

The *three-lobe journal bearing* shown in Fig. 4-69 is applied on high-speed machinery, usually when the bearing is end-fed and used in combination with a face-contact seal or when the bearing itself is a shaft seal. The three-lobe design provides, intrinsically, a stable three-point support that is particularly adaptable to very light loads in either horizontal or vertical shafts. Because of the small amount of additional clearance, the end-leakage effect of this bearing can be well controlled and provides good impurity tolerance from the oil system with a means for continuous flushing.

Another design of a three-lobe bearing is shown in Fig. 4-70. It has three equal-diameter bores which are displaced 120° from each other by an amount equal to the bearing clearance less the oil-film thickness. This bearing is center-fed. It is used to support lightly loaded shafts, especially light gear shafts in which the load direction varies with the transmitted torque.

The *tilting-pad journal bearing* consists of three or more separate shoes equally spaced around the circumference of the journal. Its mechanical design and performance follow closely the theory of the tilting-pad thrust bearing. A typical bearing of this kind

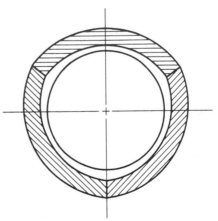

FIG. 4-69   Sketch of a three-lobe journal bearing.

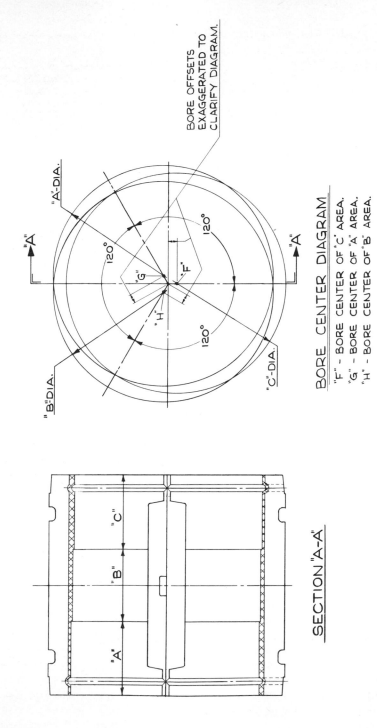

BORE OFFSETS
EXAGGERATED TO
CLARIFY DIAGRAM.

"A"-DIA.

"A"

120°

"G"

"F"

"H"

120°

120°

"B"-DIA.

"C"-DIA.

"A"

BORE CENTER DIAGRAM

"F" - BORE CENTER OF "C" AREA.
"G" - BORE CENTER OF "A" AREA.
"H" - BORE CENTER OF "B" AREA.

SECTION "A-A"

"C"

"B"

"A"

FIG. 4-70 Bore-center diagram. F = bore center of C area. G = bore center of A area. H = bore center of B area.

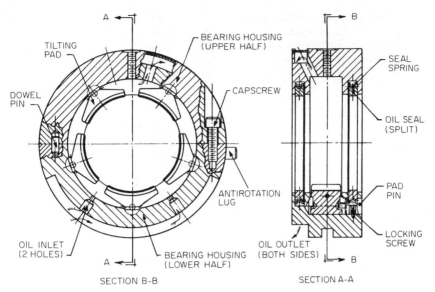

FIG. 4-71 Typical tilting-pad bearing.

utilizes five pads of which one pad is on the bottom, dead center for horizontal shafts, to provide a fixed means of supporting the shaft during alignment. This bearing is considered to be the best design for attaining rotor stability. It also is inherently more tolerant of impurities in the oil system without bearing failure. A typical section of this bearing is shown in Fig. 4-71. This bearing is pressure-fed from a circulation system, and the amount of oil passing through the bearing is controlled by a restriction at the inlet to the bearing shell.

Thrust Bearings   The simplest type of thrust bearing is the sliding-contact type with plain babbitted face, which is used in lightly loaded thrust applications. The second, more complicated design for intermediate thrust loading is found in the tapered-land thrust bearing. This design has a multiple radial series of fixed lands having a tapered inlet edge in the direction of rotation to develop a hydrodynamic film to support the thrust load.

The most common thrust bearing found in high-speed machinery is the tilting-pad thrust bearing. This bearing contains independent-acting shoes with a hardened, pivoted pad backing. Each shoe acts to establish a hydrodynamic film proportional to the speed and loading of the shaft to support the thrust load. A typical bearing section of this type is shown in Fig. 4-72. This bearing may be single-sided in the case of turbines in which unidirectional thrust is encountered or double-sided in the case of compressors or pumps in which the thrust loading may shift, depending on the operating conditions.

Lubrication

Introduction   Lubrication is primarily concerned with reducing resistance between two surfaces moving with relative motion. Any substance introduced on or between the surfaces to change the resistance due to friction is called a *lubri-*

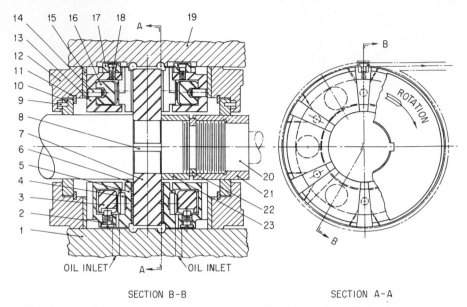

SECTION B-B   SECTION A-A

FIG. 4-72   Section of tilting-pad thrust bearing. (1) Bearing bracket. (2) Leveling-plate set-screw. (3) Upper leveling plate. (4) Shoe support. (5) Shoe. (6) Shoe babbitt (4, 5, and 6 assembled as a unit). (7) Collar. (8) Key. (9) Pin. (10) Oil guard. (11) Snap ring. (12) Thrust-bearing ring. (13) Base ring (in halves). (14) Leveling-plate dowel. (15) Shim. (16) Lower leveling plate. (17) Base-ring key. (18) Base-ring key screw. (19) Bearing-bracket cap. (20) Shaft. (21) Outer check nut. (22) Retaining ring. (23) Inner check nut.

cant. In addition to reducing friction, a lubricant removes excess heat, cleans microscopic wear particles from surfaces, coats surfaces to prevent rust and corrosion, and seals closures to prevent dust and moisture from entering.

The choice of the proper lubricant not only is important to manufacturers in order to enable them to meet their guarantees for performance and reliability but is, of course, of the utmost importance to users of the equipment in keeping their maintenance costs to a minimum and safeguarding machinery against abnormal wear, corrosion, and the effects of contamination. When choosing a lubricant, conditions such as operating speed, load conditions, method of sealing, temperature range, moisture condition, bearing design, and quantity of lubricant all affect the final choice.

It is generally recognized that a specification giving only physical and chemical properties does not guarantee satisfactory performance of any particular lubricant. Manufacturers and users, therefore, must rely on the experience, integrity, and reputation of the lubricant supplier and on the record of satisfactory past performance of the particular type of lubricant offered for a given purpose.

The lubricant should be a first-grade branded product which has previously been used and proved to be satisfactory for the continuous lubrication of similar equipment in the same service. Such experience should have proved the lubricant to be satisfactory, particularly with respect to foaming, rusting, sludging, and separation from water and other impurities.

The brand of lubricant decided upon should be continued in use and should not be changed without compelling reason.

Lubrication Methods   Either splash lubrication or forced-feed oil lubrication is commonly used for rotating machinery such as turbines, pumps, compressors, reduction gears, and worm gears. Splash lubrication is used for relatively slow-speed machinery, while high-speed machinery always requires forced-feed lubrication.

The usual form of splash lubrication employs oil rings. In this arrangement a loose ring rides freely on the journal and dips into a sump in the bearing bracket containing oil. The ring rotates because of its contact with the journal, but at a slower speed. The oil adheres to the ring until it reaches the top of the journal, when it flows onto the shaft (Fig. 4-67).

Ring oiling for small machines is used predominantly when the additional cost of a pumping system cannot be justified. The system enjoys the advantage of self-containment, needing no external motivation for its performance. Cooling coils are sometimes added when the sump temperature may become excessive.

The fully forced, or direct-pressure, system, in which the oil is forced into the bearing under pressure, is used in the majority of large circulation systems. Force feeding increases considerably the flow of lubricant to the bearing, thereby removing the heat generated by the bearing. This system is most reliable in high-speed operations with considerable load.

Grease lubrication is principally used for ball bearings and roller bearings since the housing design and maintenance are simpler than for oil lubrication. As compared with an oil system, there are virtually no leakage problems and no need for a circulation system.

The data in Table 4-7 give desirable viscosities and other specifications for oils. The data in Table 4-8 give grease recommendations for various applications.

Oil Characteristics   A lubricating oil should be a petroleum oil of high quality having guaranteed uniformity, high lubricating qualities, and adequate protection against rust and oxidation. It should be free from acids, alkalies, asphaltum, pitch, soap, resin, and water. The oil must not contain any solid matter or materials that will injure the oil itself or the parts it contacts or impair its lubricating properties. Lubricating oil should not foam, form permanent emulsions, oxidize rapidly, or form sludge. It may contain additives or inhibitors if their use supplements but does not adversely affect the desirable properties and characteristics of an oil.

Horsepower losses, bearing exit temperatures, and oil-film thicknesses decrease with lower viscosity values and increase with higher viscosity values.

When cold starting is important or a product has ring-oiled bearings, a lubricating oil with a high viscosity index should be used. A high viscosity index means that the rate of change of viscosity of an oil with change of temperature is small.

Grease Characteristics   Greases should be high-grade, high-temperature lubricants suitable for application by hand, pressure gun, or hand compression cup.

Greases should remain in the solid state at operating temperatures. Grease components should not separate on standing or when heated below their dropping

TABLE 4-7 Oil Selection

Product	Type of oil	Viscosity, SSU ASTM D 88					Oil temperature °F	
		100°F		130°F		210°F	Minimum operating	Normal to bearings
		Minimum Maximum		Minimum Maximum		Minimum Maximum		
Marine propulsion units: turbine-driven	Turbine*	490 625		220 270		62 minimum	90	110 130
Ship's service turbine-generator sets Marine auxiliaries: direct or gear drive	Turbine	375 525		180 230		54 minimum	90	110 130
Marine propulsion units: diesel-driven	EP, R&O	630 770		270 320		69 minimum	90	110 130
Centrifugal pumps — Direct drive — Ring-oiled bearings — With water cooling	Turbine	250 350		120 155		47 minimum	...	140 160
Centrifugal pumps — Direct drive — Ring-oiled bearings — Without water cooling — Liquids up to 130°F	Turbine	250 350		120 155		47 minimum	...	140 160
Centrifugal pumps — Direct drive — Ring-oiled bearings — Without water cooling — Liquids 131°F and above	Turbine	375 525		180 230		54 minimum	...	140 180
Centrifugal pumps — Direct drive — Forced circulation	Turbine	250 350		120 155		47 minimum	...	110 120
Centrifugal pumps — Direct drive — Forced feed	Turbine	140 180		85 105		42 minimum	90	110 120
Centrifugal pumps — Gear drive — Forced feed	Turbine	250 350		120 155		47 minimum	90	110 120

Equipment			Recommended oil	140 / 180	85 / 105	42 minimum	90	110 / 120
Gears	Helical		Turbine	See Sec. 10, Table 10-4			90	110 / 120
	Worm — Direct drive	Ring-oiled bearings, With water cooling	Turbine	250 / 350	120 / 155	47 minimum	⋮	140 / 160
		Ring-oiled bearings, Without water cooling	Turbine	375 / 525	180 / 230	54 minimum	⋮	140 / 180
		Forced circulation	Turbine	250 / 350	120 / 155	47 minimum	⋮	110 / 120
Turbines	Direct drive	Forced feed	Turbine	140 / 180	85 / 105	42 minimum	90	110 / 120
	Gear drive	Forced feed	Turbine	250 / 350	120 / 155	47 minimum	90	110 / 120
Centrifugal compressors	Direct drive		Turbine†	140 / 180	85 / 105	42 minimum	90	110 / 120
	Gear drive		Turbine†	140 / 180	85 / 105	42 minimum	90	110 / 120
Turbochargers	Lubricated independently		Turbine	375 / 525	180 / 230	54 minimum	⋮	120 / 160
	Lubricated by engine		See engine manufacturer's specifications	SAE 20 or 30 preferred			⋮	120 / 160

*Approximately 300 lb/in Ryder gear machine test.
†Compressors with oil seals, 190 minimum aniline point.

TABLE 4-8   Grease Selection

Component		Type	NLGI No.	Worked penetration Minimum ASTM D 217-68	Drop point Minimum ASTM D566-76	Corrosion test
Ball bearings, roller bearings, oscillating or sliding plain bearings, sliding pedestal supports*		Sodium, lithium, or sodium-calcium soap base	2	265–295	350	Pass federal test method Standard No. 791 Method 5309.2
Governor valve lifting gear	Steam temperature 600°F maximum	Sodium or lithium soap base	2	265–295	350	Pass federal test method Standard No. 791 Method 5309.2
	Steam temperature 600–825°F	Nonsoap base	1 or 2	265–340	500	Pass federal test method Standard No. 791 Method 5309.2
	Steam temperature over 825°F	Silicone	1 or 2	265–340	520	Pass Military G-23827A

*An alternative lubricant for sliding-pedestal supports is a mixture of fine graphite and cylinder or turbine oil mixed to a paste consistency.

point, the temperature at which grease changes from a semisolid to a liquid state. They also should not separate under the action of centrifugal force.

Greases should resist oxidation and must not gum, harden, or decompose. They must not contain dirt, fillers, abrasive matter, excessive moisture, free acid, or free lime.

Oil Maintenance   The lubricating system must be kept clean and free from impurities at all times. The accumulation of impurities will cause lubricant failure and damage to the equipment.

Provision should be made for maximum protection against rust during idle periods. The main lubricating system should be operated at intervals to remove condensation from metal surfaces and coat these surfaces with a protective layer of lubricant. This should be done daily when the variation in day and night temperatures is great and weekly when the variation in day and night temperatures is small. In addition, a unit idle for an extended period of time should, if possible, be operated from time to time at the reduced speeds specified under normal starting procedures.

The use of a suitable oil purifier is recommended. Since some purifiers can alter the properties of lubricating oils, especially inhibited oils, the manufacturer should be consulted before the purifier is selected.

Grease Maintenance   Grease housings should be relubricated routinely when the grease in service is unable to satisfy lubrication demands. The housing should be completely flushed and filled with new grease and any excess worked out before replacing the drain plug. Care should be taken not to overfill the housings, as this will result in a breakdown of the grease to fluid consistency and overheat the bearings.

In some cases, small additions of fresh grease to the housing are sufficient for proper lubrication. When this procedure is followed, the housing should be completely cleaned and new grease added during each major overhaul.

Lubrication Piping   Oil-feed and oil-drain piping is generally of low-carbon steel. Piping used should be pickled (a procedure of cleaning the internal surfaces). If low-carbon-steel piping has not been pickled, the following procedure should be followed:

1. Sandblast pipe along the pipe run.
2. Deburr if necessary.
3. Wash all internal surfaces with a petroleum-base cleaning solvent.
4. Air-blast dry.
5. Visually inspect.
6. If piping is to be stored in house, fog all internal surfaces progressively along the pipe run, through all openings, with an oil-soluble preservative compound.

Turbines

STEAM TURBINES

General

The steam turbine is the most widely used prime mover on the market. In large capacities, it rules without competition; for smaller sizes, the gas turbine and the internal-combustion engine are its only competitors; but for the smallest sizes both the reciprocating steam engine and the internal-combustion engine compete with the steam turbine for the market.

Steam turbines have been designed and built for an output ranging from a few horsepower to more than 1,300,000 kW, with speeds ranging from less than 1000 r/min to more than 30,000 r/min, for inlet pressures from subatmospheric to above the critical pressure of steam, with inlet temperatures from those corresponding to saturated steam up to 1050°F, and for exhaust vacuums up to 29½ inHg.

The turbine requires much less space than an internal-combustion engine or a reciprocating steam engine and much lighter foundations since reciprocating forces on the foundations are eliminated. Another major advantage is the turbine's ability to extract power from the steam and then exhaust all the steam or part of it into a heating system or to a manufacturing process, entirely free from oil.

Simplicity, reliability, low maintenance cost, and ability to supply both power and heat are the main justifications for the industrial turbine. A small factory or a building complex cannot produce electric power as cheaply as a large central-station power plant, but if steam is needed for industrial purposes or for heating, the production of power can be combined with the utilization of extracted or exhaust steam and the power becomes a cheap by-product.

The small noncondensing turbine also occupies a large and important field in power plants and marine installations because it is particularly well adapted to drive variable-speed auxiliaries and because its exhaust steam can be used to supply heat to the feedwater. A further advantage of the auxiliary turbine is its availability and convenience as a standby unit in case of interruptions to the power supply of motor-driven auxiliaries.

Steam Cycles

The Rankine Cycle   Potential energy of steam is transformed into mechanical energy in a turbine. The number of Btu required to perform work at the rate of one horsepower for one hour is 2544; for one kilowatt for one hour it is 3413 Btu.

The enthalpy, or heat content, is expressed as Btu per pound of steam. This is, in effect, the potential energy contained in the steam measured above the conventionally accepted zero point (that of condensed steam at 32°F). Practically it is not possible to release all the energy, so that the end point of heat extraction in a condensing turbine is given by the temperature attainable in the condenser. The considerable amount of energy still contained in the steam at this point cannot be recovered and must be rejected to the cooling water.

The portion of the potential energy that can be used to produce power is called the *available energy* and is represented by the isentropic enthalpy difference between the initial steam condition h_1 and the final condition corresponding to the exhaust pressure h_2. If the condensate enthalpy is h_w, the *ideal Rankine cycle efficiency*, or the *thermal efficiency*, is

$$\eta_R = \frac{h_1 - h_2}{h_1 - h_w}$$

The available energy can be converted into mechanical (kinetic) energy only with certain losses because of steam friction and throttling, which increase the entropy of the steam. The end pressure is therefore attained at a higher steam enthalpy h_2' than with isentropic expansion. The *internal turbine efficiency* then is

$$\eta_i = \frac{h_1 - h_2'}{h_1 - h_2}$$

This efficiency may be reduced to the *external turbine efficiency* by including mechanical and leakage losses not incident to the steam cycle.

Since one horsepower-hour is equivalent to 2544 Btu per hour, the theoretical steam rate of the Rankine cycle in pounds per horsepower-hour is obtained by dividing 2544 by the available energy in Btu. The corresponding value on a pound-per-kilowatthour basis may be found by dividing 3413 by the available energy. To obtain the actual steam rate at the coupling of the turbine, the theoretical steam rate is divided by the external turbine efficiency, which includes the mechanical losses.

To facilitate steam-cycle calculations, standard tables of the thermodynamic properties of steam are reproduced in abstract in Sec. 3. The data contained in these tables are plotted on a *Mollier diagram* (Fig. 5-1), which is employed extensively to solve thermodynamic problems relating to steam turbines.

Example: Determine the performance of a condensing turbine operating on a Rankine cycle based on the following data:

Initial steam pressure	200 psia
Initial steam temperature	600°F
Exhaust steam pressure	2 inHg
Moisture in exhaust steam	5 percent
Exhaust steam temperature	101°F
Measured steam rate	10.5 lb/hp·h

From the Mollier diagram:

Enthalpy at inlet h_1	1322 Btu/lb
Entropy at inlet	1.6767 Btu/°F
Enthalpy at 2 inHg and entropy of 1.6767 (h_2)	936 Btu/lb
Enthalpy at 2 inHg and 5 percent moisture (h_2')	1054 Btu/lb
Enthalpy of saturated liquid at 2 inHg (h_w)	69 Btu/lb

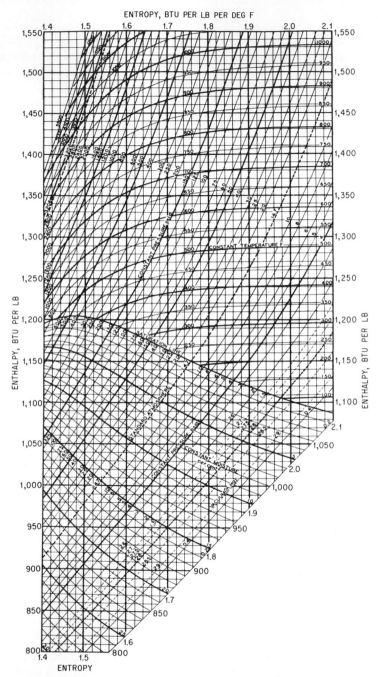

FIG. 5-1 Mollier diagram. *(By permission from 1967 ASME Steam Tables.)*

Calculations:

Isentropic enthalpy drop $= h_1 - h_2 = 1322 - 936 = 386$ Btu/lb

Actual enthalpy drop $= h_1 - h_2' = 1322 - 1054 = 268$ Btu/lb

Ideal Rankine-cycle efficiency $= \eta_R = \dfrac{h_1 - h_2}{h_1 - h_w} = \dfrac{1322 - 936}{1322 - 69} = \dfrac{386}{1253}$

$$= 30.8 \text{ percent}$$

Internal turbine efficiency $= \eta_i = \dfrac{h_1 - h_2'}{h_1 - h_2} = \dfrac{1322 - 1054}{1322 - 936} = \dfrac{268}{386} = 69.5 \text{ percent}$

Rankine-cycle steam rate $= \dfrac{2544}{h_1 - h_2} = \dfrac{2544}{386} = 6.6$ lb/hp·h

External turbine efficiency $= \dfrac{\text{Rankine-cycle steam rate}}{\text{measured steam rate}} = \dfrac{6.6}{10.5} = 63 \text{ percent}$

Improvements in the Rankine cycle may be obtained by raising the initial pressure and temperature. However, to avoid excessive moisture in the low-pressure stages, the increase in pressure must be accompanied by a corresponding increase in temperature. With present alloy steels the upper limit of the cycle is about $1050°$F. The lower limit of the cycle depends on the maximum vacuum obtainable with the available cooling water and rarely exceeds 29¼ to 29½ inHg. The economical limit of the cycle for a particular size of plant may be determined by a study of relative costs and savings.

The Reheat Cycle   The reheat cycle, which is sometimes used for large units, is similar to the Rankine cycle with the exception that the steam is reheated in one or more steps during its expansion.

The reheating may be accomplished by passing the partly expanded steam through a steam superheater, a special reheat boiler, or a heat exchanger using high-pressure live steam. The internal thermal efficiency of the cycle is calculated by totaling the available energy converted in each part of the expansion, as shown on the Mollier diagram, and dividing by the total heat supplied in the boiler, in the superheater, and in the reheat boiler or heat exchanger.

In a plant operating with a steam pressure of 1000 lb/in^2, a steam temperature of $750°$F, and an exhaust pressure of 1 inHg absolute with one stage of reheating to $750°$F at 175 lb/in^2 in a reheat boiler, the increase in thermal efficiency is about 7½ percent. With two reheating stages the improvement over the straight Rankine cycle becomes approximately 10½ percent.

The main advantage of the reheat cycle is that excessive moisture in the low-pressure stages is avoided without employing a high initial steam temperature.

The Regenerative Cycle   In the regenerative, or feed-heating, cycle, steam is withdrawn from the turbine at various points to supply heat to the feedwater. A considerable gain in economy may be obtained by using this cycle because the extracted steam has already given up part of its heat in doing work in the turbine and because the latent heat of the steam condensed in the feedwater heaters is conserved and returned to the boiler, thus reducing the heat loss to the condenser.

The cycle efficiency may be calculated using a method similar to that already mentioned, but in connection with this cycle it is customary to design a flow diagram and to prepare a complete heat balance of the plant. In small and medium-sized plants, one or two extraction heaters may be used in addition to the exhaust heater which serves the steam-driven auxiliaries, and in large plants up to seven heaters may be employed.

Additional plant economies result from reduced size of the condenser. From the viewpoint of steam generation, however, the load on the boiler is slightly increased to compensate for the steam extracted to the feed heaters. Furthermore, the higher temperature of the feedwater, while reducing the size of the economizer, also decreases the boiler efficiency by raising the lower level of the combustion gas cycle. This conflict between turbine and boiler cycle efficiencies may be removed by installing an air heater, which restores this lower level and permits the full benefit of the more economic method of regenerative feed heating.

Regenerative feedwater heating. The basic principles of this cycle have been discussed above. There is an optimum temperature to which the condensate can be heated. When this limit is exceeded, the amount of work delivered by the extracted steam is reduced and the benefit to the cycle gradually diminishes. If we assume, as an example, steam conditions of 400 lb/in^2 and 750°F at the throttle and a 29-inHg vacuum, the most favorable feedwater temperature is about 240°F for one stage of feedwater heating, 290°F for two stages, 320°F for three stages, and 330°F for four stages, as shown in Fig. 5-2.

As the number of heating stages is increased, the savings become proportionately less, as illustrated by the curves. For the steam conditions noted above, the cycle is improved by a maximum of 6 percent with one stage, 7¾ percent with two

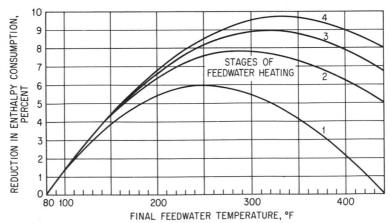

FIG. 5-2 Reduction in enthalpy consumption due to regenerative feedwater heating (steam conditions: 400 lb/in^2, 750°F, 29 inHg).

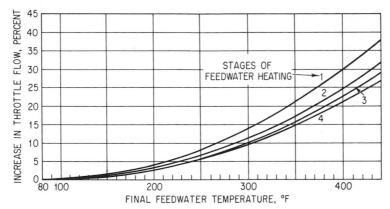

FIG. 5-3   Increase in steam flow to turbine due to regenerative feedwater heating (steam conditions: 400 lb/in^2, 750°F, 29 inHg).

stages, 9 percent with three stages, and 9¾ percent with four stages. For this reason, it is not economically sound to install more than one or two heaters for a small-capacity turbine. Furthermore, the overall plant economy may limit the maximum feedwater temperature. With the condensate heated to a higher temperature because of the increased number of feed-heating stages, the temperature difference available to the economizer, usually provided in the boiler, becomes less; therefore, less heat will be extracted from the flue gases by the economizer. The resulting increase in stack loss and corresponding decrease in boiler efficiency may thus more than outweigh the improvement in the turbine cycle. The use of air preheaters instead of economizers to recover the stack loss makes it possible to obtain the full benefit from the regenerative feed-heating cycle.

Regenerative feedwater heating affects the distribution of steam flow through the turbine. The steam required to heat the feedwater is extracted from the turbine at various points, determined by the temperature in the corresponding feed-heating stage. The extracted steam does not complete its expansion to the vacuum at the turbine exhaust; thus somewhat less power is delivered than with straight condensing operation. To obtain equal output, the steam flow to the turbine must therefore be slightly increased, as shown in Fig. 5-3, which refers to the same steam conditions as in Fig. 5-2. It may be noted from Fig. 5-3 that, for instance, with one stage of feedwater heating to the optimum temperature of 240°F, it is necessary to add about 7½ percent to the throttle flow and that with two stages the increase is about 10½ percent, etc.

On the other hand, a certain percentage of the total steam flow is extracted; thus the flow to the condenser is reduced as shown in Fig. 5-4. For one and two feed-heating stages in the above example the decrease in steam flow to the condenser is about 8 and 10½ percent respectively, as compared with straight condensing operation. The tube surface and size of the condenser can therefore be reduced by similar amounts.

Furthermore, the redistribution of the flow benefits the turbine; the first stages, which usually operate with partial admission, can easily handle more steam effi-

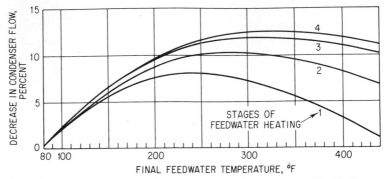

FIG. 5-4   Decrease in steam flow to condenser due to regenerative feedwater heating (steam conditions: 400 lb/in², 750°F, 29 inHg).

ciently, and the last stage in particular will gain in efficiency, mainly because of a decrease in leaving loss resulting from less flow to the condenser.

Fuel savings of 5 to 10 percent, increasing with steam pressure and the number of heating stages and decreasing with superheat, may be obtained by the use of regenerative feedwater heating. The additional equipment is simple and inexpensive; therefore, this cycle is generally employed in preference to straight condensing operation.

Classification of Turbines

To broaden the understanding of turbines and to assist in the preliminary selection of a type suitable for a proposed application, Table 5-1 has been prepared. In this table the general field of application is shown, with corresponding steam and operating conditions that may be provided for in the design of the turbine.

As an example, an industrial plant may use a moderate amount of power which can be obtained at low cost from the steam required for some chemical process; in this case a condensing high-pressure turbine with single or double extraction would be selected, with steam pressure, temperature, and extraction corresponding to the desired conditions. As an alternative, a noncondensing back-pressure turbine might be considered, particularly when power and steam requirements are nearly balanced. The advantage of this type of plant is a less expensive turbine and the elimination of condensing equipment.

In recent years, the superposed, or topping, turbine has found considerable favor in large power stations and industrial plants to provide additional power or process steam and, incidentally, to improve station economy. This turbine is usually of the high-speed multistage type. Because of the small specific volume of the steam at high pressure, it becomes possible to concentrate a large amount of power in a turbine and boiler plant of relatively small physical dimensions; thus in many cases plant capacity may be greatly increased without extensions to existing buildings.

Small turbines for auxiliary drives are usually of the single-stage noncondensing type exhausting at atmospheric or slightly higher pressure into a deaerating

TABLE 5-1 Classification of Steam Turbines with Reference to Application and Operating Conditions

Basic type	Operating condition	Steam condition	Application
Condensing	High-pressure turbine (with or without extraction for feedwater heating)	100–2400 psig; saturated, 1050°F; 1–5 inHg absolute	Drivers for electric generators, blowers, compressors, pumps, marine propulsion, etc.
	Low-pressure turbine (with high-pressure insert)	Main: 100–200 psig; 500–750°F; 1–5 inHg absolute Insert: 1450–3500 psig; 900–1050°F; 1–5 inHg absolute	Electric utility boiler-feed-pump drives
	Low-pressure bottoming turbine	Atmospheric, 100 psig; saturated, 750°F; 1–5 inHg absolute	Drivers for electric generators, blowers, compressors, pumps, etc.
	Reheat turbine	1450–3500 psig; 900–1050°F; 1–5 inHg absolute	Electric utility plants
	Automatic-extraction turbine	100–2400 psig; saturated, 1050°F; 1–5 inHg absolute	Drivers for electric generators, blowers, compressors, pumps, etc.
	Mixed-pressure (induction) turbine	100–2400 psig; saturated, 1050°F; 1–5 inHg absolute	Drivers for electric generators, blowers, compressors, pumps, etc.
	Cross-compound turbine (with or without extraction for feedwater heating, with or without reheat)	400–1450 psig; 750–1050°F; 1–5 inHg absolute	Marine propulsion
Noncondensing	Straight-through turbine	600–3500 psig; 600–1050°F; atmospheric, 1000 psig	Drivers for electric generators, blowers, compressors, pumps, etc.
	Automatic-extraction turbine	600–3500 psig; 600–1050°F; atmospheric, 600 psig	Drivers for electric generators, blowers, compressors

feed heater or to a heating system. In many cases small turbines are used as standby or quick-starting emergency units or for variable-speed applications.

Final selection of a turbine type depends on the turbine's place and functions in the scheme of the complete steam plant. When all pertinent information has been collected and the many factors bearing on the problem have been closely estimated, a tabulation of various alternatives may be prepared.

As a general principle, the investment and operating costs of a less complicated and less expansive turbine and boiler plant may be balanced against increments of savings anticipated with more elaborate equipment. Thus, the selection of the turbine may be based on optimum economy.

Turbine Performance

The problem of determining the tentative performance of steam turbines is frequently encountered in connection with preliminary estimates for proposed industrial steam plants, investigations pertaining to extensions and modernization of existing plants, and comparisons between steam and electrically driven auxiliaries, which are sometimes required in heat-balance calculations for power plants or for ship propulsion.

At this stage available information usually is incomplete or based on assumptions; therefore, a short method yielding results sufficiently accurate to serve as a yardstick in comparing different alternatives may be justified. Following the same reasoning, the actual design of the turbine and its physical dimensions are as yet unknown. As a preliminary approach only the required horsepower, the approximate speed, and the steam conditions are available or assumed; hence average turbine efficiencies may be employed.

In the final analysis the turbine manufacturer may suggest alternative performances depending on type of turbine, available standard frame sizes, and other considerations. Steam rates based on guaranteed performance, applying to the selected operating conditions, should be used for final calculations.

From a study of a number of tests and published reports relating to various sizes and types of steam turbines, it is possible to establish fairly consistent efficiency curves applying to groups of similar turbines operated at selected standard conditions. A simple classification may include, for instance, multistage turbines, condensing and noncondensing, in sizes from 200 to 100,000 rated horsepower, used as main units, as shown in Fig. 5-5, and single-stage turbines, noncondensing, in sizes from 25 to 1000 rated horsepower, for auxiliary or mechanical drive, as illustrated in Fig. 5-6.

The curves refer to efficiencies at full load and are based on the assumption that a series of suitable frame sizes is available, permitting the selection of wheel diameter, steam inlet and exhaust sizes, and other design features appropriate to the horsepower in question. Average correction factors for superheat and for vacuum applying to the efficiency curves of multistage turbines are inserted in Fig. 5-5; ordinary methods of interpolation for values between the curves may also be used, but a consistent procedure should be employed to obtain true comparisons.

Partial-load performance of a group of turbines is more difficult to standardize than performance at rated or full load. However, for preliminary estimates a mul-

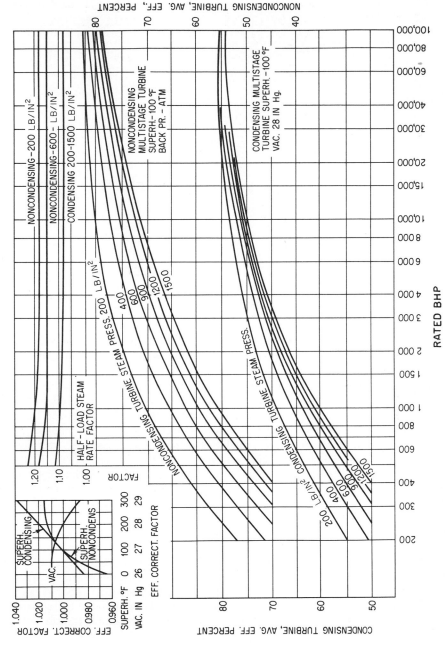

FIG. 5-5 Average efficiency of multistate turbines (gear loss not included).

5-11

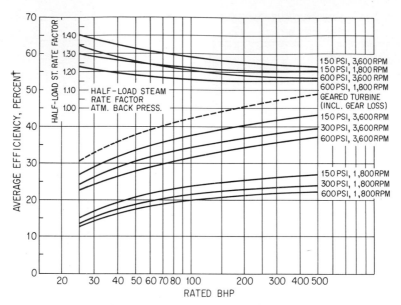

FIG. 5-6 Average efficiency of single-stage turbines (noncondensing dry and saturated steam).

tiplying steam-rate factor for a given partial load (usually one-half load) as shown in the upper part of Fig. 5-5 for multistage and in Fig. 5-6 for single-stage turbines may be employed.

A general method of determining the steam rate at any load is based on the use of the so-called Willans line (see Fig. 5-7). When a turbine is operating under throttle control at constant revolutions per minute, the relation between total steam flow (pounds per hour) and horsepower output may be shown graphically by a nearly straight line which is called the Willans line. Therefore, with the steam rates at full load and one-half load determined with the aid of Fig. 5-5 or 5-6 the total steam flow at these points can be calculated and the Willans line for the turbine can be drawn; sometimes the steam flow at no load is known from tests of similar turbines or can be estimated as a percentage of the full-load steam flow, in which case the no-load instead of the half-load point may be used for the Willans line. Steam rates at any load may then be calculated by dividing the total steam flow obtained from the Willans line by the corresponding horsepower.

Partial-load steam rates are improved by the use of so-called nozzle cutout with multiport governor valves on larger turbines and by hand-operated nozzle valves on small single-stage turbines. Individual Willans lines then apply to each nozzle combination, and the steam-rate curve is stepped in relation to these lines; thus better partial-load steam rates than may be obtained by simple throttle governing result (see Fig. 5-7).

Other types such as high-pressure (topping) turbines and low-pressure condensing (exhaust) turbines are usually built to suit special requirements; efficiencies and performance estimates should therefore be obtained from the turbine manufacturers. Extraction or induction turbines also require special study on the

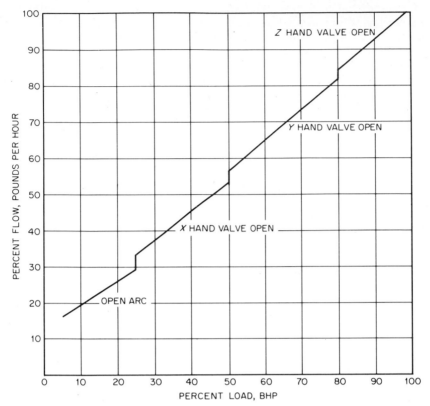

FIG. 5-7   Partial-load Willans lines.

basis of anticipated power load and steam demand as illustrated by a typical case in the subsection "Extraction and Induction Turbines."

Marine turbines may be considered a special class of turbines developed to meet conditions characteristic of ship propulsion. Elaborate heat-balance calculations are always made to determine the overall fuel consumption of the complete plant including its auxiliaries and ship's service requirements (see subsection "Marine Propulsion Turbines").

To illustrate the method of using the curves, a few problems on the selection of turbines have been worked out.

Example 1   Determine the steam rate and total steam per hour of a 500-hp turbine operating at an inlet steam pressure of 300 lb/in², 100°F superheat, exhaust pressure 10 lb/in², and a speed of 3600 r/min.

The steam tables in Sec. 3 and the Mollier diagram are based on absolute pressure; therefore, 15 (14.7) lb is added to the inlet and outlet steam pressures, which are given in pounds gauge in the example. Further, the saturated-steam temperature at 315 psia is 421.8°F; by adding 100°F superheat, a total steam temperature of 522°F is obtained.

From the Mollier diagram the enthalpy at the turbine inlet is 1269 Btu, and by expanding the steam at a constant entropy of 1.577 the enthalpy at the exhaust condition is 1064 Btu. Thus the available energy is 205 Btu. Dividing 2544 by 205 gives a theoretical steam rate of 12.42 lb/hp·h.

From the curves it appears that the required 500 hp would fall in the lower range of a multistage turbine, or if a single-stage turbine is considered, it would be close to the upper limit covered by standard frame sizes, suggesting that both alternatives should be investigated.

Thus, in the case of the multistage noncondensing turbine the average efficiency for the specified condition is about 61.5 percent, obtained from the curve in Fig. 5-5, by interpolation for 300-lb/in^2 steam pressure. Since the efficiency curves in Fig. 5-5 are based on 100°F, the superheat correction factor is 1.00.

The efficiency correction for back pressure is taken as unity for a noncondensing-turbine preliminary calculation. To obtain the actual steam rate, the theoretical steam rate of 12.42 lb is divided by the average efficiency 0.615. A steam rate of 20.20 lb/hp·h may thus be expected, referred to the output at the turbine coupling. The total steam required at the rated load then becomes 10,100 lb/h.

For a single-stage noncondensing turbine of 500 rated horsepower operating at the same steam conditions and speed, the efficiency as indicated in the curve in Fig. 5-6 would be about 39.5 percent, resulting in a steam rate of 31.4 lb/hp·h, or a total steam requirement of about 15,700 lb/h.

This curve also illustrates the advantage of geared turbines as compared with direct drive when the driven machine must be operated at low speed. For instance, a pump used as an auxiliary in a power station may require 500 hp at 1800 r/min, in which case a direct-connected single-stage turbine would have an efficiency of about 24 percent, while a geared turbine may show an efficiency of 49 percent at the gear output coupling. If we assume the same steam conditions as before, the direct-driven unit would require about 26,000 lb of steam per hour as compared with 12,700 lb, or about one-half, for the geared turbine.

When turbines exhaust into feedwater heaters or when the exhaust steam is used for heating or for process steam, it may be necessary to determine the enthalpy and the superheat or moisture in the steam after it has passed through the turbine. The method usually employed is to subtract the heat actually converted to work in the turbine, as expressed by the average (or more correctly the internal) efficiency, from the enthalpy at the inlet condition; this gives the enthalpy available in the exhaust steam. The superheat or moisture is then conveniently obtained from the Mollier diagram at the point of intersection between the remaining enthalpy and the appropriate absolute back pressure.

In the case of the 500-hp multistage turbine operating at 315 psia, 522°F total steam temperature, and 25-psia exhaust pressure with an average efficiency of 61.5 percent, the calculations are made as follows: the available energy is 205 Btu, and with an efficiency of 61.5 percent the work done in the turbine is 126 Btu; deducting this from the enthalpy at the inlet condition, which is 1269 Btu, leaves 1143 Btu in the steam at the turbine exhaust. The intersection on the Mollier diagram at 25 psia and the enthalpy of 1143 Btu is the approximate *end point* of the expansion in the turbine, and the exhaust steam contains about 1¾ percent moisture. In this case, if dry steam is desired at the turbine exhaust, the initial steam temperature must be increased to about 550°F.

As a comparison, the single-stage direct-connected 500-hp turbine would convert to useful work 39.5 percent of the available energy; thus the enthalpy at the turbine exhaust would be 1188 Btu, corresponding to about 55° of superheat, at 25 psia. The interrelations between the various factors entering the problem are easily understood by referring to the Mollier diagram.

Applying this example to an actual case, approximately dry steam may be required at the point where the process steam is used; therefore, after an allowance has been made for radiation loss in the connecting steam pipe, the appropriate initial steam temperature can be selected. If moisture is permissible at the exhaust, a corresponding correction in steam flow is made to ensure that an adequate supply of process heat is available.

Example 2 In connection with a present and future plant, it may be desired to compare the steam rate of a straight condensing turbine driving a 2500-kW generator at two different steam conditions, alternative A, 250 lb/in², 500°F, 28 inHg, and alternative B, 600 lb/in², 750°F, 28.5 inHg. If we assume a generator efficiency of 94.5 percent, the required full-load output of the turbine is about 3550 hp. From the curve in Fig. 5-5, which is based on rated brake horsepower and applies to multistage condensing turbines, average efficiencies of 74.25 and 71.3 percent, respectively, are obtained.

In the first case the vacuum is 28 inHg and the superheat 94°F, which is close to the standard condition used as a basis for this curve; therefore, no correction factors are necessary. The expected steam rate, obtained by dividing the theoretical rate, 6.77 lb/hp·h, by the efficiency, 0.7425, is 9.12 lb/hp·h and the total steam flow 32,350 lb/h.

In the second case, a vacuum correction factor of about 0.993 is approximated from the insert on the curve for 28.5 inHg, and since the superheat is 260°F, the correction factor for superheat is about 1.029. These are multiplying factors; thus the turbine efficiency is increased to 72.8 percent. The steam rate then becomes 7.06 lb/hp·h, and the total steam required is 25,100 lb/h. With turbine-driven generators, the performance is usually expressed in terms of pounds of steam per kilowatthour, which transposition may conveniently be made by dividing the total steam per hour by the kilowatt output at the terminals. Thus the respective steam rates are 12.93 and 10.02 lb/kWh, including the generator loss.

The following tabulation will serve to illustrate the comparison:

Alternative	A	B
Output, kW	2,500	2,500
Output, bhp	3,550	3,550
Steam pressure, lb/in²	250	600
Steam temp, °F	500	750
Vacuum, inHg	28	28.5
Available enthalpy drop, Btu	376	495
Theoretical steam rate, lb/hp·h	6.77	5.14
Turbine efficiency from curve	74.25	71.3
Turbine efficiency corrected, percent	74.25	72.8
Steam rate, lb/hp·h	9.12	7.06
Steam rate, lb/kWh	12.93	10.02
Total steam, lb/h	32,350	25,100

It should be noted, however, that the saving of 7250 lb of steam per hour, or 22.4 precent as shown by this example, does not represent the actual fuel saving corresponding to alternative B. As mentioned in connection with turbine cycles, the heat supplied to the cycle must be considered. For the two examples the initial enthalpies are 1263 and 1380 Btu/lb and the condensate enthalpies 69 and 60 Btu/lb; so the Rankine-cycle efficiencies are 31.5 and 37.5 percent, and with the external turbine efficiencies of 74.25 and 72.8 percent the final external thermal efficiencies are 23.35 and 27.27 percent respectively. So the increase in cycle efficiency is 16.8 percent, and with the same boiler efficiency the saving in fuel 14.4 percent. Further improvement would result if the comparison were based on the regenerative cycle rather than on the straight condensing Rankine cycle.

Turbine Steam-Path Design

The steam turbine is a comparatively simple type of prime mover. It has only one major moving part, the rotor which carries the buckets or blades. These, with the stationary nozzles or blades, form the steam path through the turbine. The rotor is supported on journal bearings and axially positioned by a thrust bearing. A housing with steam inlet and outlet connections surrounds the rotating parts and serves as a frame for the unit.

However, a great number of factors enter into the design of a modern turbine, and its present perfection is the result of many years of research and development. While the design procedure may be studied in books treating this particular subject, a short review of the main principles may serve to compare the various types. This will aid in the selection and evaluation of turbines suitable for specific requirements.

In considering the method of energy conversion, two main types of blading, impulse and reaction, are employed. An impulse stage consists of one or more stationary nozzles in which the steam expands, transforming heat energy into velocity or kinetic energy, and one or more rows of rotating buckets which transform the kinetic energy of the steam into power delivered by the shaft. In a true impulse stage the full expansion of the steam takes place in the nozzle. Hence, no pressure drop occurs while the steam passes through the buckets.

A reaction stage consists of two elements. There is a stationary row of blades in which part of the expansion of the steam takes place and a moving row in which the pressure drop of the stage is completed.

Many turbines employ both impulse and reaction stages to obtain the inherent advantages of each type.

Figure 5-8 illustrates some of the most common types of nozzle and blade combinations used in present turbines. Four of the diagrams, *a*, *b*, *c*, and *d*, apply to the impulse principle, as noted in the legend, and the last one, *e*, shows a type of reaction blading. A constructional difference may also be pointed out: impulse buckets are usually carried on separate disks with nozzles provided in stationary partitions called diaphragms, while the moving reaction blades are generally supported on a rotor drum with the stationary blades mounted in a casing.

The impulse stage has a definite advantage over the reaction stage in handling steam with small specific volume as in the high-pressure end of a turbine or in

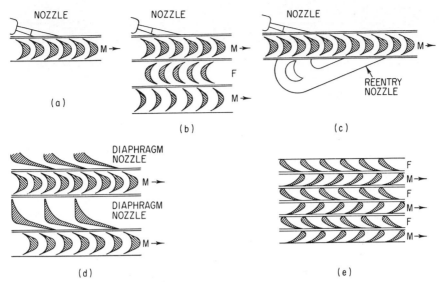

FIG. 5-8 Main types of turbine blading (F = fixed row; M = moving row). (*a*) Impulse turbine: single velocity stage. (*b*) Impulse turbine: two velocity stages. (*c*) Reentry impulse turbine: two velocity stages. (*d*) Impulse turbine: multistage. (*e*) Reaction turbine: multistage.

cases in which the enthalpy drop per stage is great; thus small single-stage turbines are always of the impulse type. The stage may be designed for partial admission with the nozzles covering only a part of the full circumference; therefore, the diameter of the wheel may be chosen independently of the bucket height. Used as a first stage in a multistage turbine, the impulse stage with partial admission permits adjustment of the nozzle area by arranging the nozzles in separate groups under governor control, thus improving partial-load performance.

The dominating principle in turbine design involves expression of the efficiency of the energy conversion in nozzles and buckets or in reaction blades, usually referred to as stage efficiency, as a functon of the ratio u/C. The blade speed u, feet per second, is calculated from the pitch diameter of the nozzle and thus determines the size of the wheel at a given number of revolutions per minute and C, also in feet per second, is the theoretical velocity of the steam corresponding to the isentropic enthalpy drop in the stage, expressed by the formula

$$C = 223.8\sqrt{\text{Btu}}$$

Figure 5-9 illustrates average stage efficiencies which may be attained in various types of turbines operating at design conditions. The losses that are represented in the stage-efficiency curves are due to friction, eddies, and flow interruptions in the steam path, plus the kinetic energy of the steam as it leaves a row of blades.

Part of the latter loss can be recovered in the following stage. Additional losses not accounted for in the stage-efficiency curves are due to windage and friction of the rotating parts and to steam leakage from stage to stage. With the exception

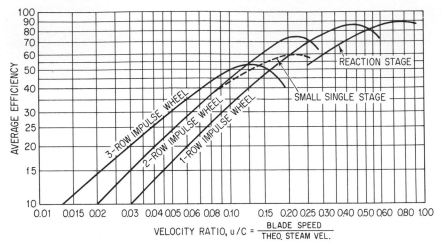

FIG. 5-9 Average efficiency of turbine stages.

of the kinetic energy that may be recovered, all losses are converted to heat with a corresponding increase in the entropy of the steam.

From the group of curves of Fig. 5-9 it follows that the maximum combined efficiency for various types of stages is attained at different velocity ratios. This ratio is highest for reaction stages and lowest for three-row impulse wheels. This implies that for equal pitch-line speeds the theoretical steam velocity or the stage enthalpy drop must be lowest for reaction stages and highest for three-row wheels to maintain the maximum possible efficiency. At this maximum efficiency, the three-row wheel can work with many times the steam velocity and a correspondingly larger enthalpy drop compared with a reaction stage.

The maximum efficiency of reaction stages may exceed 90 percent at a velocity ratio of 0.75, as shown in Fig. 5-9. However, such values can be attained only with a great number of stages. Hence, reaction stages are normally not designed for a higher velocity ratio than 0.65. A section of reaction blading is shown in Fig. 5-8e.

Single-row impulse stages have a maximum efficiency of about 86 percent at a velocity ratio of 0.45. Figure 5-8a shows a combination of impulse buckets with a Delaval expanding nozzle, and Fig. 5-8d shows multistage impulse blading with nonexpanding nozzles.

Let us assume, as an example, a blade speed of 500 ft/s, corresponding to a turbine wheel with 32-in pitch diameter operating at a speed of 3600 r/min; the optimum steam velocity would be $500/0.45 = 1100$ ft/s. The kinetic energy of the steam may be expressed in Btu by the relation Btu $= (C/223.8)^2 = 1100^2/50,000 = 24$; thus the enthalpy drop utilized per stage at the point of maximum efficiency is about 24 Btu for the above condition.

In the case of a turbine operating at high steam pressure and temperature, exhausting at low vacuum, the available energy may be approximately 500 Btu; therefore, about 20 single-row impulse stages would be required for maximum efficiency. Obviously the pitch diameter of the wheels cannot be chosen arbitrarily, but this example illustrates the method of dividing the energy in a number of steps called pressure stages. The turbine would be classified as a multistage impulse turbine.

Figure 5-9 further shows one curve labeled "two-row" with an extension in a broken line referring to small single-stage turbines and one curve marked "three-row impulse wheel." These refer to so-called velocity-compounded stages as illustrated by Fig. 5-8b and c. The purpose of the two- and three-row and also the reentry stage is to utilize a much greater enthalpy drop per stage than that possible in a single-row impulse stage. When the enthalpy drop per stage is increased, the velocity ratio is reduced and the kinetic energy is only partly converted into work in the first row of revolving buckets; thus the steam leaves with high residual velocity. By means of stationary guide buckets the steam is then redirected into a second, and sometimes a third, row of moving buckets, where the energy conversion is completed.

In the so-called helical-flow stage, with semicircular buckets milled into the rim of the wheel, and also in the reentry stage shown in Fig. 5-8c, only one row of revolving buckets is used. This type of velocity compounding is sometimes employed in noncondensing single-stage auxiliary turbines.

The curve marked "two-row impulse wheel" indicates that a maximum stage efficiency of about 75 percent may be attained at a velocity ratio of approximately 0.225. At this condition, the two-row velocity-compounded stage will utilize about 4 times as much energy as a single-row impulse stage. When we compare the efficiencies on the basis of operating conditions as defined by the velocity ratio, it appears from the curves that the two-row wheel has a higher efficiency than a single-row wheel when the velocity ratio is less than 0.27.

Occasionally, in small auxiliary turbines operating at a low speed ratio, a three-row stage may be used. The curve marked "three-row" indicates a maximum efficiency of about 53 percent at a speed ratio of about 0.125. Apparently, at this point the efficiency of a two-row wheel is almost as good; thus the three-row stage would be justified only at still lower speed ratios, that is, for low-speed applications.

The design of a turbine, especially of the multistage type, involves a great many factors which must be evaluated and considered. A detailed study of the steam path must be made, and various frictional and leakage losses which tend to decrease the efficiency, as well as compensating factors such as reheat and carry-over, must be computed and accounted for in the final analysis of the performance of the turbine. Stresses must be calculated to permit correct proportioning of the component parts of the turbine, and materials suitable for the various requirements must be selected.

Single-Stage Turbines

Single-stage turbines, sometimes called *mechanical-drive* or *general-purpose turbines,* are usually designed to operate noncondensing or against a moderate back pressure. The principal use of these turbines is to drive power plant and marine auxiliaries such as centrifugal pumps, fans, blowers, and small generator sets.

They may also be applied as prime movers in industrial plants, and in many cases small turbines are installed as standby units to provide protection in case of interruption of the electric power supply.

They are built in sizes up to 1500 hp and may be obtained in standardized frames up to 1000 hp with wheel diameters from 12 to 36 in. Rotational speeds vary from 600 to 7200 r/min or higher; the lower speeds apply to the larger wheel sizes used with direct-connected turbines, and the higher speeds are favored in geared units. The bucket speed usually falls between 250 and 450 ft/s in direct-connected turbines operating at 3600 r/min and may exceed 600 ft/s in geared turbines.

The efficiency of a turbine generally improves with increasing bucket speed as noted by referring to efficiency versus velocity ratio curves in Fig. 5-9; thus it would seem that both high revolutions and large diameters might be desirable. However, for a constant number of revolutions per minute the rotation loss of the disk and the buckets varies roughly as the fifth power of the wheel diameter and for a constant bucket speed almost as the square of the diameter. Thus, in direct-connected turbines with the speed fixed by the driven unit, the rotation losses may become the dominating factor in selecting the wheel size for maximum efficiency. On the other hand, when reduction gears are adopted, the velocity ratio may be increased by means of higher revolutions, sometimes even with smaller wheel diameter; thus considerably higher efficiencies may be expected, as shown by the dashed curve in Fig. 5-6. Since the rotation losses vary approximately in direct relation to the density of the steam surrounding the wheel, it follows that small wheel diameters should be used particularly for operation at high back pressure.

Turbine manufacturers have complete test data on standard sizes of small turbines on which steam-rate guarantees are based. Knowing the characteristics of different turbines, they are in a position to offer suggestions regarding the most suitable type and size to choose for specific requirements.

The single-stage turbine is simple and rugged and can be depended on to furnish many years of service with a minimum of maintenance expense. The few parts which may require renewal after long periods of operation, for instance, bearings, carbon rings, and possibly valve parts, are inexpensive and easy to install. It is also comparatively simple to exchange the steam nozzles to suit different steam conditions, as sometimes encountered in connection with modernization of old plants, or to adapt the turbine to new conditions due to changes in process-steam requirements.

Steam-Rate Calculations   Approximate steam rates of small single-stage turbines (less than 500 hp) may be computed by the following general method:

1. The available energy, $h_1 - h_2 = H_a$, at the specified steam condition is obtained from the Mollier diagram.

2. Deductions are made for pressure drop through the governor valve (12.5 Btu), loss due to supersaturation C_s (about 0.95), and 2 percent margin (0.98). The remaining enthalpy drop is called net available energy H_n.

3. The theoretical steam velocity C, ft/s, is calculated, based on net available energy H_n. The formula for steam velocity is $C = 223.8 \times \sqrt{H_n}$.

4. The bucket speed u, ft/s, is calculated from the pitch diameter, in (of the nozzles), and the r/min.

5. The velocity ratio u/C is calculated and the "basic" turbine efficiency E is obtained from an actual test curve similar to those given in Fig. 5-9.

6. The "basic" steam rate for the turbine is calculated from the formula

$$\text{Basic steam rate} = \frac{2544}{H_nE} = \text{lb/hp·h}$$

7. The loss horsepower for the specific turbine size is estimated from Fig. 5-10, corrected for back pressure as noted on the diagram.

8. The actual steam rate of the turbine at the specified conditions is

$$\text{Basic steam rate} \times \frac{\text{rated hp + loss hp}}{\text{rated hp}} = \text{lb/hp·h}$$

Example: As a matter of comparison with the short method of estimating turbine performance, the same example of a 500-hp turbine with a steam condition of 300 lb/

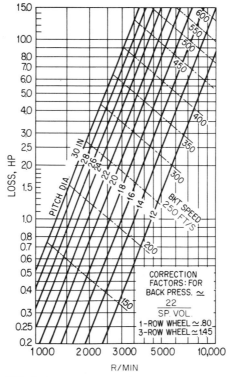

FIG. 5-10 Rotational loss, average for single-stage turbines (two-row wheel; atmosphere exhaust).

in^2, 100°F superheat, and 10 lb/in^2 back pressure at a speed of 3600 r/min may be selected (see "Turbine Performance," Example 1). It is further assumed that a frame size with a 24-in-pitch-diameter two-row wheel is used.

The available energy is 205 Btu; subtracting a 12.5-Btu drop through the governor valve leaves 192.5 net Btu, which corresponds to a theoretical steam velocity $C = 223.8 \times \sqrt{192.5} = 3104$ ft/s.

The bucket speed $u = 3600 \times 24 \times \pi/60 \times 12 = 377$ ft/s. Thus the velocity ratio $u/C = 377/3104 = 0.12$. From Fig. 5-9 the approximate efficiency 0.47 is obtained on the curve marked "two-row impulse wheel" at $u/C = 0.12$.

The supersaturation loss factor C_s (due to the expansion of the steam into the supersaturation state) is a function increasing with the initial superheat and decreasing with the available enthalpy, in this case about 0.96; a margin of 2 percent may also be included, thus the

$$\text{Basic steam rate} = \frac{2544}{192.5 \times 0.47 \times 0.96 \times 0.98} = 30.0 \text{ lb/hp·h}$$

The rotational loss of a 24-in-pitch-diameter wheel at 3600 r/min, determined from Fig. 5-10, is about 6.3 hp. This diagram is based on atmospheric exhaust pressure; therefore, a correction factor must be applied as noted. At 10-lb back pressure the specific volume of the steam is about 16.3 ft^3/lb. Thus

$$\text{Loss hp} = 6.3 \times \frac{22}{16.3} = 8.5$$

$$\text{Steam rate of turbine} = 30.0 \times \frac{500 + 8.5}{500} = 30.5 \text{ lb/hp·h}$$

The use of the short method and Fig. 5-6 results in this case in a steam rate of 31.4 lb/hp·h, which is about 3 percent higher than that obtained by calculations applying Figs. 5-9 and 5-10; both methods are consistent and may serve the purpose for which they are suggested.

Multistage Condensing Turbines

The most important application of the steam turbine is that of serving as prime mover to drive generators, blast-furnace blowers, centrifugal compressors, pumps, etc., and for ship propulsion. Since the economic production of power is the main objective, these turbines are generally of the multistage type, designed for condensing operation; i.e., the exhaust steam from the turbine passes into a condenser, in which a high vacuum is maintained.

The dominating factor affecting the economy, which may be expressed in terms of station heat rate or fuel consumption, is the selection of the steam cycle and its range of operating conditions, as previously discussed in connection with turbine cycles. For smaller units the straight condensing Rankine cycle may be used; for medium and large turbines the feed-heating, regenerative cycle is preferred; and in large base-load stations a combination of a reheating, regenerative cycle may offer important advantages.

If we assume average economic considerations, such as capacity of the plant and size of the individual units, load characteristics, and amount of investment, the initial steam conditions may be found to vary approximately as follows:

Small units	150 to 400 lb/in^2; 500 to 750°F
Medium units	400 to 600 lb/in^2; 750 to 825°F
Large units	600 to 900 lb/in^2; 750 to 900°F
Large units	900 to 3500 lb/in^2; 825 to 1050°F

Similar conditions may prevail with reference to the vacuum; smaller units may operate at 26 to 28 inHg in connection with spray ponds or cooling towers, while larger turbines usually carry 28 to 29 inHg and require a large supply of cooling water.

These general specifications are equivalent to an available enthalpy drop varying from about 350 Btu to a maximum of about 600 Btu. Therefore, the modern condensing turbine must be built to handle a large enthalpy drop; hence a comparatively large number of stages is required to obtain a high velocity ratio consistent with high efficiency, as indicated in Fig. 5-9. Incidentally, the average efficiency curves of condensing multistage turbines in the lower part of Fig. 5-5 cover a range from 363 Btu at 200 lb/in^2 to 480 Btu at 1500 lb/in^2.

As shown in Fig. 5-11, the overall efficiency of multistage turbines is sometimes expressed as a function of the so-called quality factor, which serves as a

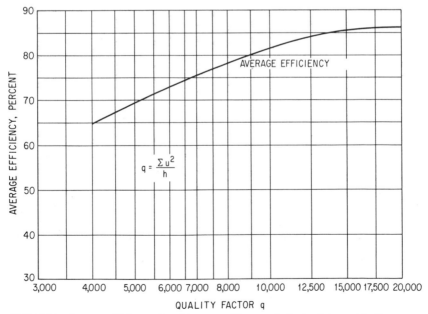

FIG. 5-11   Average efficiency of multistage turbines on the basis of the quality factor.

convenient criterion of the whole turbine in the same manner as the velocity ratio applies to each stage separately. The quality factor is the sum of the squares of the pitch-line velocity of each revolving row divided by the total isentropic enthalpy drop. The pitch-line velocity is expressed in feet per second and the enthalpy drop in Btu.

The curve is empirical, determined from tests of fairly large turbines, and indicates average performance at the turbine coupling. It may be used to evaluate preliminary designs with alternative values of speed, wheel diameters, and number of stages or to compare actual turbines when pertinent information is available. To obtain consistent results the size and type of the turbine must be considered; generally, the internal efficiency improves appreciably with increased volume flow, and the mechanical efficiency also improves slightly with increased capacity; thus a size factor should be applied to the efficiency curve to correlate units of different capacity, or individual efficiency curves based on tests may be used for each standard size.

Example: Determine provisional dimensions of a 3000-hp 3600-r/min condensing turbine operating at 400 lb/in², 750°F, and 28 inHg. A turbine efficiency of 73 percent is desired; thus, for a size factor of, say, 95 percent, the required efficiency is 77 percent, corresponding to a quality factor of about 7500. The available enthalpy is 460 Btu; consequently the sum of velocity squares is 7500 × 460 = 3,450,000. Various combinations of bucket speed and number of moving rows may be selected; for instance, a bucket speed of 500 ft/s corresponding to a pitch diameter of about 32 in would require 14 rows of buckets; 475 ft/s equals 30¼-in diameter with 15 rows, etc.

The pitch diameter usually increases gradually toward the exhaust end; therefore, the so-called root-mean-square diameter is used in these calculations. In this example the diameters would be adjusted in relation to the flow path through the turbine and the number of stages, perhaps 14, resulting in the most satisfactory bucket dimensions and in general compactness of design. This discussion illustrates the general principle of the interdependence of diameters and number of stages for a required turbine efficiency.

In analyzing the design of a condensing turbine as shown in Fig. 5-12, the first stages must be suitable for steam with comparatively high pressure, high temperature, and small specific volume. The last stage, on the other hand, presents the problem of providing sufficient area to accommodate a large-volume flow of low-pressure steam. Taking a large enthalpy drop in the first stage by means of a two-row velocity stage as shown in this particular case results in a moderate first-stage pressure with low windage and gland leakage losses. Furthermore, the remaining enthalpy drop, allotted to the following stages, also decreases; i.e., the velocity ratio improves, and thus a good overall turbine efficiency results from this combination.

Extraction points for feed heating may be located in one or more stages as required, and provision may also be made to return leakage steam from the high-pressure gland to an appropriate stage, thus partly recovering this loss by work done in succeeding stages.

The journal bearings are of the tilting-pad type with babbitt-lined steel pads. They are made in two halves and arranged for forced-feed lubrication. The tur-

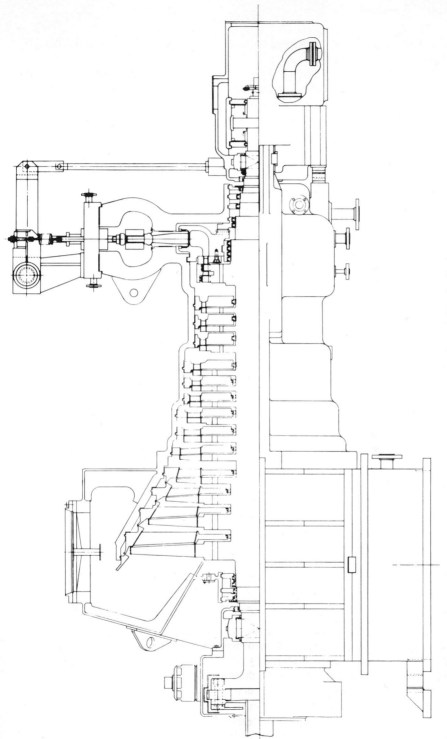

FIG. 5-12 Multistage condensing turbine (56,000 kW, 3600 r/min, 1250 psig, 950°F, 2.5 inHg absolute).

bine-shaft seals are of the stepped-labyrinth type, with the labyrinths flexibly mounted.

The turbine casing is divided horizontally with the diaphragms also made in two halves, the upper ones being dismountable with the top casing. The turbine support is arranged to maintain alignment at all times. The turbine is anchored at the exhaust end, and the casing is permitted to expand freely with changes in temperature.

Group nozzle control, operated from a speed governor by a hydraulic servomotor, results in economic partial-load performance combined with desirable speed-governing characteristics.

This condensing turbine represents a logical application of design principles to obtain maximum efficiency by the proper selection of wheel diameters and number of stages and by proportioning the steam path to accommodate the volume flow of steam through the turbine.

Superposed and Back-Pressure Turbines

Superposed and back-pressure turbines operate at exhaust pressures considerably higher than atmospheric and thus belong to the general classification of noncondensing turbines. Relatively high efficiency is required; therefore, these turbines are of the multistage type. The small single-stage auxiliary turbines previously described are also of the noncondensing type, but of a much simpler design, suitable for less exacting steam conditions.

The main application of superposed turbines, often referred to as *topping turbines,* is to furnish additional power and to improve the economy of existing plants. Since boilers usually fail or become obsolete long before the turbines they serve, it has proved economically sound in many plants to replace old boilers with modern high-pressure high-temperature boilers supplying steam to a new superposed turbine with its generator. The superposed turbine may be an extracting unit supplying such steam to process and its exhaust steam to the existing condensing turbines operating at the same inlet conditions as before. A considerable increase in plant capacity and improvement in station economy is thus obtained with a comparatively small additional investment.

Superposed turbines have been built in sizes of 500 kW and above. The initial steam conditions may vary from 600 to 2000 lb/in^2 with steam temperatures from 600 to 1050°F; the exhaust pressure may range from 200 to 600 lb/in^2 and must correspond to the initial pressure of the existing plant. Topping units are usually arranged to serve a group of turbines but may also be proportioned for individual units.

Investigations in connection with proposed topping units may cover various aspects, for instance, determination of additional capacity obtainable with assumed initial steam conditions or, conversely, selection of initial steam conditions for a desired increase in power. Incidentally, the improvement in station heat rate is also calculated for use in evaluating the return on the proposed investment. However, this evaluation involves heat-balance calculations for the complete plant including the feed-heating cycle adjusted to the new conditions.

To indicate the possibilities of the superposed turbine the following example is suggested. An existing plant of 5000-kW rated capacity is operating at 200 lb/in^2, 500°F, and 1½ inHg absolute condenser pressure. If we assume a full-load steam rate of 13.0 lb/kWh based on two 2500-kW units, the total steam flow is about 65,000 lb/h. Determine the additional power to be expected from a topping unit operating at 850 lb/in^2, 750°F initial steam condition at the turbine throttle, and exhausting into the present steam main.

The available energy of the high-pressure steam is 147 Btu, corresponding to a theoretical steam rate of 23.2 lb/kWh. If we assume a generator efficiency of about 94 percent and a "noncondensing" turbine efficiency of 63 percent, approximated from the curve sheet in Fig. 5-5, the steam rate becomes about 39 lb/kWh. Incidentally, the enthalpy at the turbine exhaust, calculated from the efficiency, is about 1272 Btu; according to the Mollier diagram, this corresponds to about 508°F at 215 psia; thus the initial steam temperature of 750°F selected for the topping unit matches approximately the 500°F assumed at the existing steam header.

Based on a total steam flow of 65,000 lb/h and a steam rate of 39 lb/kWh, the increase in power is about 1665 kW at the full-load condition. Thus the increase in capacity is 33.3 percent; likewise, the combined turbine steam rate is 9.75 lb/kWh, an improvement of 25 percent. To calculate the corresponding fuel saving, additional data for the boiler and plant auxiliaries would be required.

The approximate size of the unit may be arrived at by the quality-factor method referred to in Fig. 5-11. By applying an appropriate-size factor, the topping turbine may in this case be designed for an efficiency of, say, 67 percent, corresponding to a quality factor of about 4500. With an available enthalpy drop of 147 Btu the sum of the velocity squares is 660,000. Because of the comparatively small volume flow and the high density of the steam, small wheel diameters are used; thus the bucket speed is rather low. If we assume, for instance, 350 ft/s, corresponding to about 22½-in pitch diameter at 3600 r/min, the number of stages required would be about 5; and at 300 ft/s with 19-in pitch diameter the number of stages would be 7, etc. (Provisional inlet and outlet connections can be determined from Fig. 5-28, thus indicating the general overall dimensions of the turbine.)

Back-pressure turbines, frequently of fairly large capacity, are often installed in industrial plants where a large amount of process steam may be required. In this case, the electric power required to operate the plant may be obtained from the process steam as a by-product at very low cost. Since good economy is important, these turbines are generally of the multistage type. The usual range of initial pressure is from 200 to 900 lb/in^2 with corresponding steam temperatures from 500 to 900°F. The back pressure, which depends on the requirements of the process steam, may fall between the limits of 5 and 150 lb/in^2.

The approach to the problem is to estimate the amount of power that can be obtained from the process steam with various initial steam conditions. In this manner a balance between available steam and power demand is determined, and as a preliminary step the appropriate initial steam condition is selected. A check on the enthalpy at the turbine exhaust then indicates possible adjustment of the initial steam temperature to obtain approximately dry steam at the point where the process steam is used. Occasionally, heavy demands for steam in excess of the power load may be provided for by supplying the additional steam through a reducing valve directly from the boilers. Supplementary power for peak loads may be obtained from an outside source or from a condensing unit.

Extraction and Induction Turbines

Many industrial plants requiring various quantities of process steam combined with a certain electric power load make use of extraction turbines. It is possible to adapt the extraction turbine to a great variety of plant conditions, and many different types are built, among them noncondensing and condensing extraction turbines with one or more extraction points and automatic and nonautomatic extraction; additionally, in certain urban areas, extraction turbines are used by the utility company to supply steam to buildings in the neighborhood of the plant.

A related type of turbine, the so-called mixed-flow or induction turbine, with provision for the use of high-pressure and low-pressure steam in proportion to the available supply, may also be mentioned in this connection. Generally, the low-pressure steam is expected to carry normal load, and high-pressure steam is admitted only in case of a deficiency of low-pressure steam. Even in case of complete failure of the low-pressure supply the turbine may be designed to carry the load with good economy on high-pressure steam alone.

The most frequently used extraction turbine is the single automatic-extraction condensing turbine as shown in Fig. 5-13. For design purposes it may be considered as a noncondensing and a condensing turbine, operating in series and built into a single casing. Because of the emphasis placed on compactness and comparatively simple construction, the number of stages is usually limited. The performance may therefore not be quite equal to the combined performance of a corresponding back-pressure turbine and a straight condensing turbine built in two separate units. On the other hand, the price of the extraction turbine is also less than the total price of two independent units would be.

Guarantees of steam rate for condensing and noncondensing automatic-extraction turbines are always made on a straight condensing or a straight noncondensing performance, respectively, obtained with no extraction but with the extraction valve wide open, that is, not functioning to maintain the extraction pressure. This nonextraction performance guaranteed for an automatic-extraction turbine will not differ much from that for a straight condensing or a noncondensing unit of the same capacity and designed for the same steam conditions.

The complete performance of an extraction turbine can be represented by a diagram such as Fig. 5-14 in which the output is expressed in percentage of rated capacity and the throttle flow in percentage of that at full load without extraction. The line labeled "0% extraction at const. extr. press." represents the performance of the turbine when no steam is extracted but with the extraction valve acting to hold extraction pressure at the bleed connection.

The guaranteed steam flow for nonextraction, with the pressure at the bleed point varying with the load, that is, with the extraction valve wide open, is also plotted as a broken line on Fig. 5-14. This line intersects the zero-extraction line at full load, while at partial loads the throttle flow for nonextraction is less than for zero extraction. The reason for this is that the low-pressure end of the turbine has been designed for the steam flow which at full load, nonextraction, with the

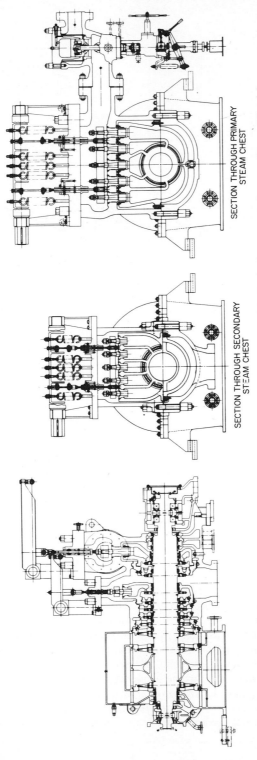

SECTION THROUGH PRIMARY STEAM CHEST

SECTION THROUGH SECONDARY STEAM CHEST

FIG. 5-13 Single automatic-extraction turbine (20,000 bhp, 10,600 r/min, 1500 psig, 800°F, 2 inHg absolute, automatic extraction at 400 psig).

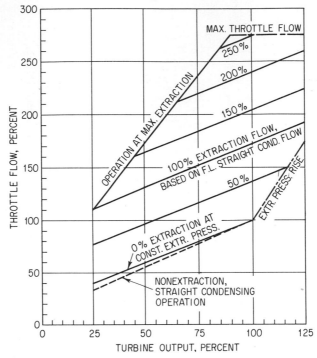

FIG. 5-14   Throttle flow versus output of condensing automatic-extraction turbine.

extraction valve wide open, will give the extraction pressure required. If the steam flow through the low-pressure end of the turbine is decreased, as at partial loads, the absolute pressure at the extraction point would decrease in proportion to the steam flow if it were not for the action of the extraction valve, which throttles the steam to maintain the required extraction pressure. This throttling loss occurs when operating with zero extraction, but not when operating nonextraction.

When steam is extracted from a turbine carrying a given load, the throttle flow must increase, but the increase is not equal to the amount extracted. For a given turbine and set of steam conditions, the increase in throttle steam over that required for zero extraction will bear nearly a constant ratio to the amount extracted. This ratio is called the *extraction factor*. As the extraction pressure is raised from exhaust pressure to inlet pressure by extracting at points of progressively higher pressure, the extraction factor increases from 0 to 1.

The line labeled "operation at max. extraction" represents the performance when all steam entering the throttle, except the cooling steam, is extracted. The line "max. throttle flow" represents the maximum flow which the high-pressure section can pass when the turbine is operated with its normal steam conditions. The corresponding limit for the low-pressure section is the one titled "extr. press. rise." The turbine can operate in the region to the right of this limit but will not then maintain normal extraction pressure. For any given load the flow to exhaust is maximum at zero extraction, so that the maximum flow through the exhaust section for which the turbine must be proportioned is determined by the maximum load to be carried with minimum extraction.

Similar diagrams may be constructed to apply to other combinations such as double automatic and mixed-flow turbines. As an example, lines of "constant induction flow" would be located below and parallel to a line of "zero induction flow" in the case of mixed-pressure or induction turbines.

Low-Pressure Turbines (with High-Pressure Insert)

Electric utility boiler-feed pumps require large blocks of power which can be most economically supplied by a steam-turbine driver. Such a unit is illustrated in Fig. 5-15.

Normal operation is with low-pressure steam extracted from the main turbine driving the generator. The steam chest for this steam is in the upper half of the casing. Operation at low power output, i.e., somewhat less than 50 percent, causes extraction steam pressure from the main turbine to decrease until there is an insufficient supply to drive the pump. At this point, full boiler-pressure steam is admitted through the high-pressure insert located in the lower half of the casing. As the plant load is decreased further, a point is reached when the extraction steam pressure is too low and the nonreturn valves close to prevent a backflow through the low-pressure steam chest into the main turbine.

Calculation methods for sizing a feedwater pump and its turbine driver are readily available for interested persons but are somewhat beyond the scope of this handbook.

Marine Propulsion Turbines

Marine propulsion turbines are multistage turbines of the condensing type specifically designed for shipboard application and are usually direct-connected to a marine-type reduction gear, which reduces the relatively high turbine speed to the ship propeller speed.

For smaller powers up to about 10,000-hp output single-casing turbines may be used, but the majority of modern units are of the cross-compound type.

They are further characterized by having a reversing turbine built into the unit, usually in the low-pressure turbine of the cross-compound unit. The reversing turbine is designed to develop 75 to 100 percent of design torque at full-ahead-power steam flow and at 50 percent speed.

Figure 5-17a shows a cross section of the low-pressure turbine of a 40,000-hp cross-compound unit with a reversing element. Figure 5-17b shows the high-pressure turbine.

Control System   Modern marine turbines are provided with a control system that adapts the necessary features of remote and local control as used for steam power plants. The position of the steam valves is controlled by hydraulic actuators, which may be operated remotely from the bridge by electrical means or from the engine-room control console by either electrical or mechanical means. When

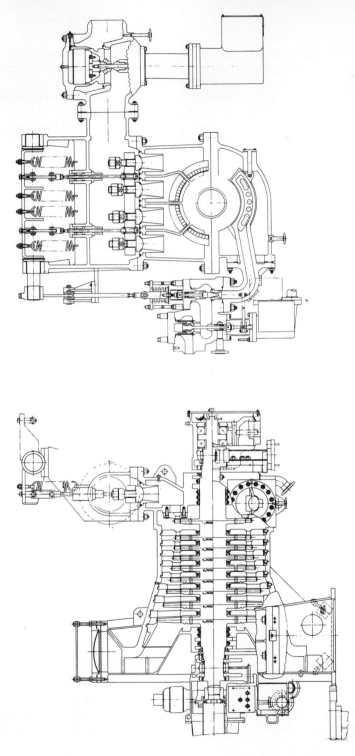

FIG. 5-15 Low-pressure turbine with high-pressure insert (10,000 bhp, 5200 r/min, 105 psig, 623° F, 3 inHg absolute).

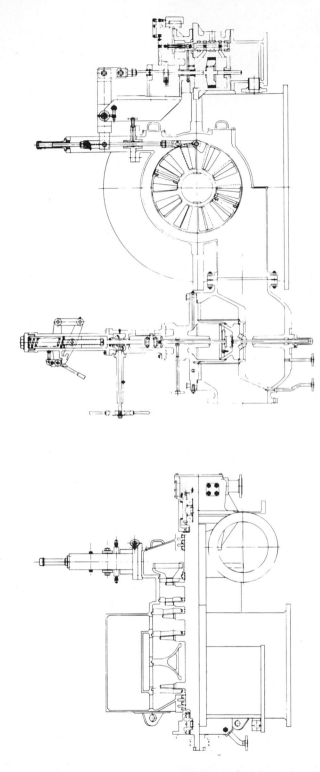

FIG. 5-16 Low-pressure bottoming turbine (9000 bhp, 8700 r/min, 45 psig, 375° F, 3.5 inHg absolute).

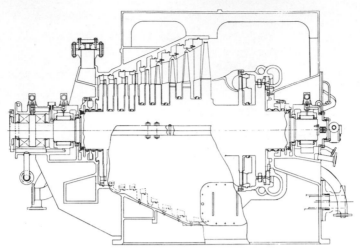

FIG. 5-17a Cross section of low-pressure marine turbine with reversing element.

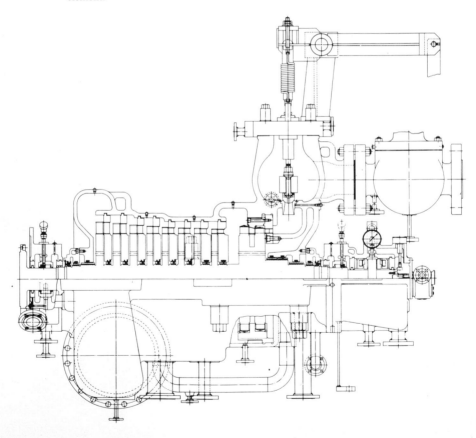

FIG. 5-17b Cross section of high-pressure marine turbine.

operating in either mode, the actuator will respond to an overspeed signal from the speed-control system, limiting the maximum speed.

Turbine Governors

The governor is the "brains" behind the "brawn" of the turbine. The governor may sense or measure a single quantity such as turbine speed, inlet, extraction, induction, or exhaust pressure, or any combination of these quantities and then control the turbine to regulate the quantities measured. Shaft-speed governors are the most common. A simple speed governor will first be considered.

Mechanical Governors   In the direct-acting mechanical governor shown in Fig. 5-18 speed is measured by spring-loaded rotating weights. As the weights are rotated, they generate a force proportional to the product of their mass, the radius of their rotation, and the square of their speed of rotation. Under steady-state conditions the weight force is balanced by the opposing force of the weight spring, and the governor stem remains stationary.

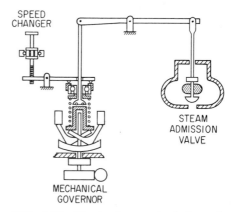

FIG. 5-18   Direct-acting mechanical speed governor.

If some load is removed from the turbine, the turbine would speed up and the governor weights would move outward. As the governor weights move outward, their force is further increased, but the force of the weight spring increases even faster and soon limits the travel of the weights. The movement of the weights is transferred through the governor stem and connecting linkage to the turbine control valve to reduce the flow of steam to the turbine, limiting the turbine-speed increase.

If some load is added to the turbine, the turbine will slow down and the governor weights will move inward. As the governor weights move inward, their force is further decreased, but the force of the weight spring decreases even faster and limits the travel of the weights. The movement of the weights is transferred through the governor stem and connecting linkage to the turbine control valve to increase the flow of steam to the turbine, limiting the turbine-speed decrease.

For any constant setting of the weight spring a certain change in speed is required to provide a full travel of the governor weights. This change in speed between the full-load and no-load speed of the governor is called either the *governor droop* or, when expressed as a percentage of the full-load speed, the *governor regulation.* When units are operated in parallel, any changes in the total load will be shared by the units in inverse proportion to their individual governor regulation. Thus for equal load sharing all units should have equal governor regulation, or in the case of dissimilar units their respective governor regulation can be set to assure proper load sharing.

Frictional forces in the governor, in the connected linkage, and in the control valve must be overcome before the weights can move. This means that the governor will not react to small speed changes. This small range of speed in which no governor action occurs is called the *governor dead band*.

Mechanical-Hydraulic Governors   For most applications the direct-acting mechanical governor does not develop enough force to operate the turbine control valve, so that a force amplifier, or servomotor, is needed. In the governor shown in Fig. 5-19 movement of the governor stem causes the servomotor relay valve to move, directing operating oil to one side of the servomotor power piston and opening the other side of the power piston to drain. The power-piston movement is fed back through the servomotor linkage to the relay valve, using the speed-governor stem as a fulcrum, returning the pilot valve to neutral.

Another type of servomotor is used in the governor shown in Fig. 5-20. In this governor movement of the governor stem causes changes in the speed-governor oil-line pressure. This is possible because the pilot valve has a larger capacity than the orifice which supplies the speed-governor oil line. The speed-governor oil pressure acts on the relay piston against the relay-valve spring to position the relay valve and cause the power piston to move. As in the preceding servomotor the power-piston movement is fed back to the relay valve, returning it to neutral. The pilot valve on the speed-governor stem is subject to loading from the pressure in the speed-governor oil line. This pressure loading makes the pilot valve harder to move and increases the governor regulation.

Another common governor is a pressure governor. Inlet- or exhaust-pressure governors are commonly used on turbines driving generators when the unit speed

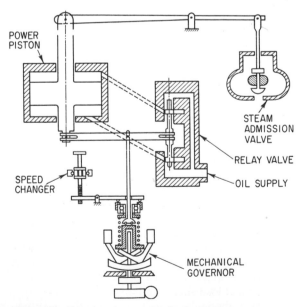

FIG. 5-19   Speed governor with direct-acting servomotor.

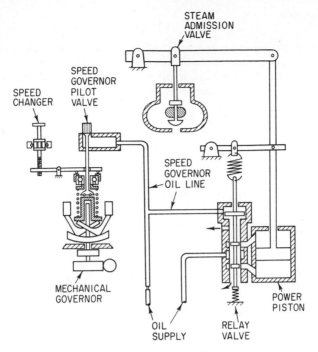

FIG. 5-20 Speed governor with hydraulic servomotor.

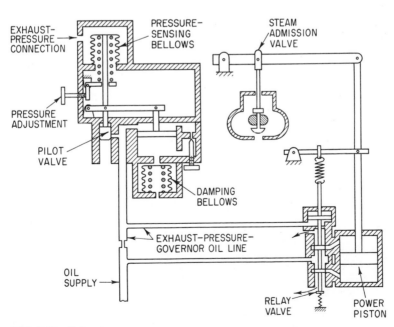

FIG. 5-21 Exhaust-pressure governor.

is held constant by operating in parallel with other generators. In the example of an exhaust-pressure governor shown in Fig. 5-21 the exhaust pressure working on a spring-loaded bellows operates a pilot valve which causes changes in the exhaust-pressure-governor oil-line pressure. The exhaust-pressure-governor oil-line pressure, in turn, controls the inlet-nozzle-valve servomotor. A decrease in the exhaust pressure relaxes the sensing bellows, moves the pilot valve in the opening direction, and causes a drop in the exhaust-pressure-governor oil-line pressure. The decrease in the exhaust-pressure-governor oil-line pressure lowers the servomotor pilot valve. This causes the servomotor power piston to move upward, opening the nozzle valves. As the nozzle valves are opened, more steam passes through the turbine to maintain the desired exhaust pressure within the limit of the governor regulation. The action of the exhaust-pressure governor is damped by a piston with a bypassing needle valve and by a spring-loaded bellows.

Electrohydraulic Governors   These governors employ an electric speed-sensing element whose signal is transmitted to an electrohydraulic actuator which drives the same primary relay valve as that shown in the other governors. Such electric governors have the advantage of a wider speed range and more precise control than have mechanical-hydraulic governors. Electrohydraulic governors may use hydraulic oil systems developed for mechanical governors, in which case the oil pressure would be in the neighborhood of 100 psig. Such a system is shown in Fig. 5-22. It is used for a boiler-feed-pump turbine. It receives a feedwater demand signal (e.g., 4 to 20 mA dc) and uses it to set unit operating speed during normal operation (40 to 100 percent speed) or to set manually the valve position

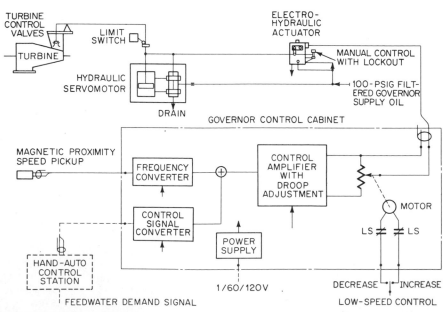

FIG. 5-22 Electrohydraulic governor.

from 0 to 100 percent speed or the minimum governor set point during start-up. The system consists of:

1. A governor control cabinet which contains the electronic circuits necessary for establishing start-up valve position and normal speed control
2. A hand-automatic control station to facilitate manual control of the unit
3. A magnetic proximity speed pickup and a gear mounted on the turbine shaft which produce a frequency signal proportional to the unit speed
4. An electrohydraulic actuator mounted on the hydraulic servomotor which positions the servomotor relay valve in response to the control signals from the electronic circuits of the governor
5. An hydraulic servomotor for positioning the valve gear operating the steam control valves in response to the electrohydraulic actuator

An electrohydraulic governor using control oil pressure in the range of 1000 to 1500 psig is shown in Fig. 5-23. It is used in the control system of a process-industry turbine-driven unit. This system may be used with modular control valves. Steam is admitted to the turbine by any of several valves; the opening of each one is determined by the governor servoamplifier. The system consists of:

1. A governor control cabinet which contains the electronic circuits to establish a speed set point by a signal from the controller which measures process variables
2. A manual control station which may be used to establish a speed set point
3. Magnetic speed pickups and a gear mounted on the turbine shaft which produce a frequency signal proportional to the unit speed
4. An electrohydraulic relay valve mounted on the hydraulic servoactuator (see Fig. 5-39 below)
5. A feedback transducer mounted on the hydraulic servomotor that returns an electric signal of valve position to servoamplifier circuits in the governor control cabinet

The complete servoactuator–steam-control-valve module is adjusted at assembly so that the control valve will be closed in the absence of a control signal. The electric control system is interlocked with the emergency shutdown system so that the steam control valves act as a backup to the trip-and-throttle valve during emergency shutdown, i.e., absence of steam, electricity, or oil.

Extraction Governors   Another form of pressure governor is the extraction governor. This governor could operate, through a servomotor, a set of nozzle valves (secondary valves) which as they are opened pass more steam through the later stages of the turbine and less into the extraction line. Normally when an extraction governor is fitted, it must be coordinated with a speed governor to assure complete control of the turbine. This can be done by using a master regulator to connect the speed-governing and extraction-governing systems, as shown in Fig. 5-24. Pistons in the master regulator receive control-pressure signals from the

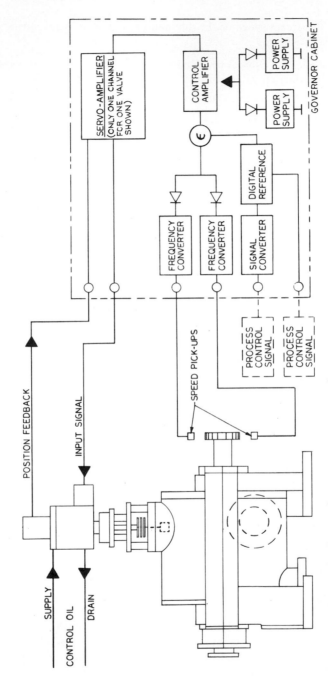

FIG. 5-23 Electrohydraulic control system process reset speed governor.

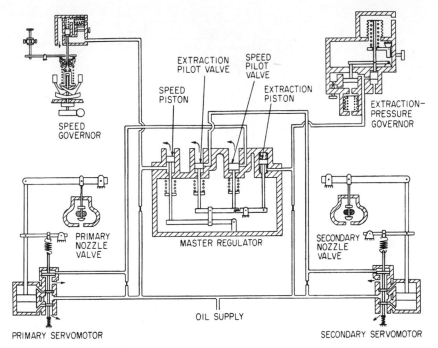

FIG. 5-24   Extraction governor with master regulator.

speed governor and from the extraction-pressure governor. The control-piston movements are transmitted through the regulator linkage to the pilot valves which control the pressure in the servomotor-control oil lines. These controlled pressures cause the servomotors to make the necessary corrections in the nozzle-valve settings.

Overspeed Trips   Mechanical overspeed trip devices have been used for safety against disasters caused by runaway turbines. The unbalanced and spring-loaded plunger is probably the most common (see Fig. 5-25). Mounted in a hole through the rotor and across the axis, it is held in position by a spring until the turbine speed is sufficient for the plunger unbalance to generate centrifugal force greater than the spring force. It then pops out a short distance and strikes a lever system which mechanically or hydraulically actuates the turbine trip valve.

A variation of the plunger type is shown in Fig. 5-26; actually two plungers are used, each attached to a surrounding ring. The rings are attached to each other so that they can move only in opposite directions. A spring-loaded plunger that would be tripped by a shock (e.g., an earthquake) in its tripping direction will be opposed by the other spring-loaded plunger.

Figure 5-27 illustrates an hydraulic trip valve used with a mechanical shaft-mounted trip. When trip speed is reached, the plunger (or ring) mounted in the trip body moves out and strikes the knob mounted on the valve stem. Axial motion to the right unseats the valve to dump oil from the trip circuit. Contemporary

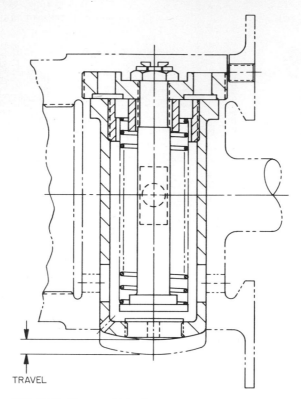

FIG. 5-25 Overspeed trip plunger.

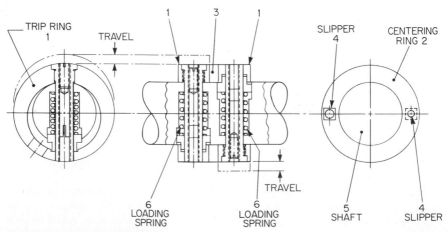

FIG. 5-26 Two-ring (shockproof) overspeed trip.

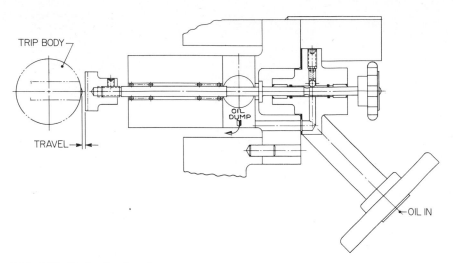

FIG. 5-27   Hydraulic trip valve.

designs of overspeed trips favor electrical non-shaft-contacting trips. These use electrical circuitry that actuates the hydraulic trip valve connected to the steam trip valve (see Fig. 5-23).

Steam and Exhaust Pipes

To determine the size of steam and exhaust pipes for steam turbines, two methods may be employed. Preliminary estimates are usually made by assuming empirical steam velocities, while final calculations are based on permissible pressure drop in the complete pipeline including the loss in valves, fittings, and bends.

Many other factors are involved in the design of steam piping, such as type or class of construction as covered by various specification and codes (ANSI code for pressure piping). As a general rule, cast-iron fittings are permissible up to 250 lb/in^2 and 450°F, and carbon steel fittings are used for 300 to 600 lb/in^2 and a maximum temperature of 750°F. For pressures above 600 lb/in^2 and temperatures over 750°F alloy steels are required. Welded-joint construction is usually employed for high-pressure and high-temperature piping.

The diagram in Fig. 5-28 may be used to determine approximate pipe sizes when detailed information regarding length of pipe and number of valves, bends, and fittings is not available. The diagram is based on a steam velocity of 150 ft/s for steam pipes and noncondensing exhaust pipes and 300 ft/s for condensing exhaust pipes; if other values are selected, the flow will be in direct proportion to the steam velocities. Dry and saturated steam volumes have been used; therefore, corrections should be made for moisture or superheat. These may be made in the case of superheated steam by using an "equivalent" pressure, which can be obtained directly from the steam tables or approximated from the tabulation on the diagram. The pipe size is actual internal diameter; for "nominal" size the appropriate diameter found from tables in Sec. 1 may be used. However, this

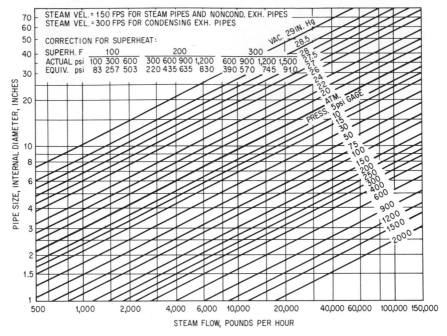

FIG. 5-28   Steam and exhaust pipe sizes (dry and saturated steam; correction to be made for superheat and quality).

would be an unnecessary refinement for preliminary estimates. Equivalent areas of exhaust openings may be obtained from Table 1-2 in Sec. 1.

Example: Find steam and exhaust pipe sizes for the 500-hp multistage turbine referred to in "Turbine Performance": steam pressure 300 lb/in^2, 100°F superheat, exhaust pressure 10 lb/in^2, steam flow 10,000 lb/h, 1¾ percent moisture in the exhaust.

By starting at the bottom of Fig. 5-28 at 10,000 lb/h and using the correction for 100°F superheat, the equivalent steam pressure is 257 lb/in^2, and the steam pipe size is found to be 2.4-in internal diameter. Extra strong steel pipe, 2½ in nominal size, has an internal diameter of 2.32 in and would also meet the requirements for pressure and temperature.

For the exhaust pipe the moisture, which is 1¾ percent, is deducted, making the corrected flow about 9825 lb/h. The required internal diameter of the exhaust pipe at 10 lb/in^2 is 7.4 in; thus a standard 8-in pipe would be selected, providing a margin for possible overload or for operation at a lower back pressure.

After layouts have been made, the pressure drop in the steam and exhaust pipes can be estimated with the aid of the alignment diagram shown in Fig. 5-29. This diagram gives the pressure drop in pounds per square inch per 100-ft length of pipe. The total loss in the pipe is obtained by measuring the lengths from the drawing and adding the loss in valves and fittings expressed in equivalent length of straight pipe.

To estimate the length of straight pipe equivalent to the pressure drop in terms of pipe diameters, the following values may be used. The equivalent length is in feet; therefore, the pipe diameter must be also expressed in feet.

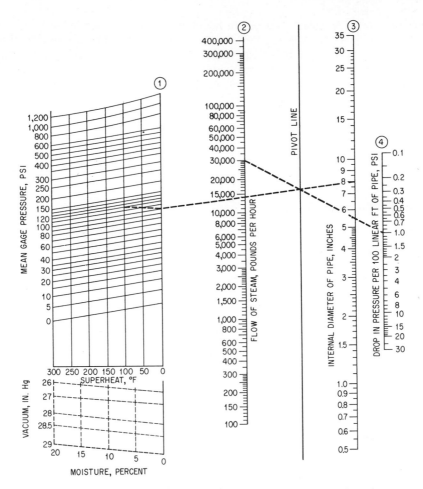

FIG. 5-29 Pressure drop in steam piping, based on the Gutermuth formula $W = 324 \sqrt{Pcd^5}$. W = steam flow, lb/h; P = pressure drop, lb/in²; c = density, lb/ft³; d = actual internal pipe diameter, in.

Example: Steam flow, 30,000 lb/h; initial pressure, 125 lb/in²; superheat, 100°F. Final pressure desired is 115 lb/in². Equivalent length of pipe, 1000 ft; pressure drop per 100 ft, 1 lb; average gauge pressure, 120 lb/in².

Find necessary pipe diameter: Join 30,000 on scale 2 to 1 on scale 4. Locate intersection of 120-lb diagonal and 100° vertical, and proceed horizontally to scale 1. Project line from this point on scale 1 through intersection on pivot line to scale 3. Final answer on scale 3: 8-in-diameter pipe. (*Redrawn from* Power's Data Sheets *by courtesy of* Power.)

Globe valve = 400 × diameter   90° pipe bend = 25 × diameter

Gate valve = 20 × diameter   Standard elbow = 100 × diameter

The examples worked out in Fig. 5-29 serve to illustrate its use. By checking the preliminary pipe sizes previously determined for the 500-hp turbine, the pressure drop per 100-ft length would be about 20 lb/in² for a 2.32-in actual inside diameter or 2½-in nominal-size extra heavy pipe and about 6.5 lb/in² for a 3-in extra heavy pipe. The 8-in standard-weight exhaust pipe will have a pressure drop of about 0.40 lb/in² per 100-ft equivalent length. If we assume that a maximum pressure drop of about 10 percent is permissible in the steam pipe, the 2½-in size would be sufficient up to about 150-ft equivalent length, and if a 5 percent drop is allowed in the exhaust pipe, the 8-in size would be ample for about 300 ft.

Other considerations, such as heat loss from pipes carrying highly superheated steam or from large-sized pipes conveying process steam for long distances, may present special problems in connection with the final selection and design of steam piping for turbines, but the examples given may serve to illustrate the methods commonly used.

Turbine Mechanical Design

Casings and Internals   Turbine casings serve to contain the steam, support the interstage diaphragms, and house the gland labyrinths, the steam admission valves (except in large electric utility units), and the journal and thrust bearings. Contemporary designs are divided horizontally into an upper cover and a lower case, bolted together. The so-called steam chest is integral with the cover and houses the steam admission valves. It is closed with a bolted cover that supports the valve-operating gear.

Thermal expansion of the casing must be provided for not only longitudinally but also crosswise. This may be accomplished with sliding pedestals or flexible supports, or both. The casing is anchored axially at one end and laterally at both ends in the vertical plane of the rotor centerline. All pipe attached to the case must be arranged with sufficient flexibility or sliding freedom to prevent excessive strain in the pipe or case (see *Reference Codes and Standards,* NEMA SM-23, for allowable forces and moments of piping).

Provision must be made for drainage of water formed during the warming up of the case and piping and the water formed in the steam as it expands to the exhaust pressure (see subsection "Steam Sealing and Drainage Systems").

Uncontrolled extraction openings must be protected by nonreturn valves primarily to prevent entry of water into the turbine casing. The water cools the lower portion of the casing, causing it to deflect upward sufficiently to rub the interstage labyrinths or, in extreme cases, to cause severe damage to the blading. Controlled extraction openings must be protected in the same way except that their size usually favors a swing check valve in a horizontal run of pipe below the turbine.

A turbine with an upward-facing exhaust must provide water drainage to the condenser hot well, for a condensing unit, or a nonreturn-valved line to a subatmospheric line for drainage during start-up with atmospheric exhaust, for a backpressure unit.

The efficiency of the turbine can be improved and the erosion of the rotor blading reduced by properly proportioned and located water-collection rings between stages in the portion of the steam path operating in the moist-steam region, as shown in Fig. 5-30. Erosion is reduced by the centrifugal slinging off and trapping of the water that wets the rotating blades in the preceding stage. Water that the rings drain to the bottom of the case may be passed through orifice holes in suc-

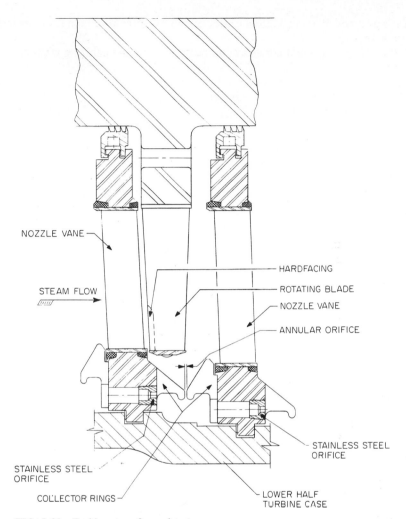

NOZZLE VANE

STEAM FLOW

HARDFACING

ROTATING BLADE

NOZZLE VANE

ANNULAR ORIFICE

STAINLESS STEEL
ORIFICE

STAINLESS STEEL
ORIFICE

COLLECTOR RINGS

LOWER HALF
TURBINE CASE

FIG. 5-30 Turbine stage for moist steam.

ceeding diaphragms or through the casing wall to a piping manifold below the turbine.

Safety often dictates that the exhaust hood have a sentinel valve (Fig. 5-13) to warn of a dangerously high exhaust pressure or, as shown in Fig. 5-12, an explosion disk to vent the steam flow in the event of a condenser failure.

The diaphragms are split horizontally with an antileakage key extending from the interstage labyrinths to the nozzle vanes and, in larger diaphragms, from the vanes to the outside diameter. Usually the key is extended slightly into the labyrinth ring to serve as an antirotation device. To obtain true centerline support of the diaphragms, the lower halves are supported on keys at the horizontal split. A key is also added at the bottom center of the diaphragm to hold it on centerline crosswise.

Each diaphragm half consists of an outer ring, a vane half ring with outer and inner shrouds, and an inner half ring that holds the interstage labyrinth. These several parts are welded together to form an integral whole. Regardless of steam conditions, the vanes and shrouds are usually stainless steel (12 percent chromium). The inner and outer half rings may be stainless in high-moisture portions of the steam path or carbon steel in superheated-steam portions.

Blading and Rotors Turbine blading is designed to match conditions in the section of the turbine where it is located. In the first (high-pressure) stage, in which the blade is subjected to shocks from varying steam pressures as it passes the several inlet nozzle groups, it is short and sturdy. Blading of intermediate stages becomes increasingly longer to accommodate the increased specific volume of the steam as it approaches the exhaust end. Low-pressure end stages are those long enough to require tapering from base to tip to meet centrifugal-force requirements; they are twisted to accommodate the increasing peripheral velocities from base to tip.

Vibration considerations. For so-called constant-speed turbines, the blades may be tuned so that their natural frequencies and harmonics are not in resonance with the running speed, the passing frequency of nozzle partitions or struts, any other weak or strong variations of steam pressure around the circumference of the stage, or stimuli from wheel disk and torque variations.

For variable-speed turbines, all the considerations used for constant speed must be expanded to cover the speed range in which the unit may operate. Figure 5-31 is a Campbell diagram for an intermediate stage, indicating the difficulty of avoiding resonance. Excitations (fundamental or multiple) can produce excessive stress levels. The diagram shows the natural frequencies of the blades and the excitation that could result in a resonant condition. The natural frequencies are based on groups of blades shrouded together. These frequencies can vary slightly from blade to blade because of machining tolerances. The frequency bands T_I, T_{II}, and R_0 take these differences into account.

For short-nozzle blades the nozzle-passing frequency (NPF), and for long blades the low per-revolution (number/rev) excitations, are possible. The NPF excitation is the steam impulse on a blade as it passes each nozzle in the diaphragm in one full revolution. The low per-revolution excitations are possible (1) as multiples of running speed or (2) as multiples of equally spaced discontinuities in the flow path. The horizontal split forms 2/rev.

Intersection of the exciting frequency with blade natural frequencies in the speed range indicates the existence of a resonance. Blading must be designed and manufactured to resist the resulting alternating stresses when operating in resonance. When resonances occur, vibration amplitude and stress increase until they are limited by damping in the structure and material.

Rotor disks are subject to disk vibration, which is actually a wave traveling in a direction opposite to the rotation of the disk and so stands still in space. The cause then is a constant axial force that bends the disk in resonance with a natural frequency.

Each disk is subjected to a differential pressure from one side to the other; this

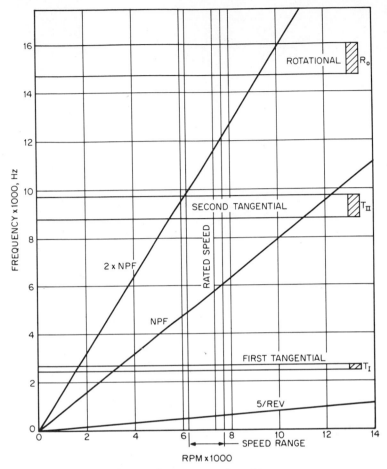

FIG. 5-31 Campbell diagram for one stage of a turbine.

pressure must be counterbalanced by the thrust bearing. Impulse turbines are built with holes (placed in a lower-stressed area) to reduce this unbalance force by bleeding upstream steam to downstream.

Variable-speed turbines with flexible rotors (e.g., in marine propulsion), which must operate at any speed, are grooved at each end of a labyrinth, as shown in Fig. 5-32a and c, to reduce the tendency to bow in the event of a rub.

The rotor must be given a heat-indication test to make sure that the rotor forging is uniform throughout and that the rotor will not bow because of temperature changes within the operating range of the turbine.

Blading operating in the moist-steam region may be protected to a large extent by adding hardfacing to the inlet edge of the blade near its tip. The extent of the hardfacing is determined by examination of the velocity triangles of steam and water flow and blade motion.

<u>Protection from internal shaft currents</u>. Rotors sometimes suffer severe dam-

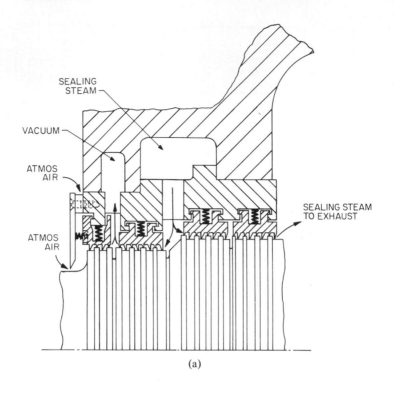

SEALING
STEAM

VACUUM

ATMOS
AIR

ATMOS
AIR

SEALING STEAM
TO EXHAUST

(a)

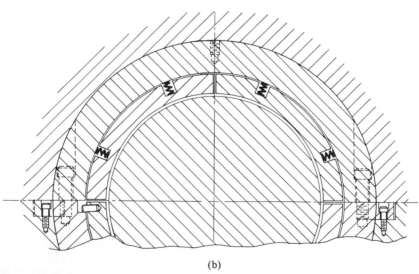

(b)

FIG. 5-32 Shaft end glands. (*a*) Exhaust end. (*b*) Sectional view normal to axis. (*c*) High-pressure end.

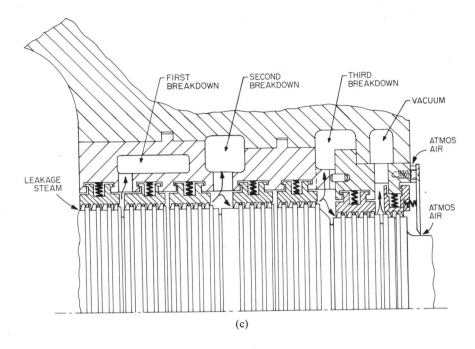

FIRST BREAKDOWN SECOND BREAKDOWN THIRD BREAKDOWN VACUUM ATMOS AIR LEAKAGE STEAM ATMOS AIR ATMOS AIR VACUUM

(c)

age to bearings and other parts because of the destructive effects of shaft currents produced internally by electromagnetic and electrostatic effects. Although many cases may be cited, this phenomenon is not inherent in all machinery, and it can be controlled by relatively simple precautions and procedures.

Two basic types of shaft currents cause destructive deterioration of rotating equipment: electrostatic and electromagnetic induced currents.

Electrostatic currents are produced by the periodic discharge of electrostatic potentials built up between the rotor and stationary parts. This usually occurs on condensing-type turbines when moisture particles impinge upon the rotating parts. When the electrostatic potential builds up to levels high enough to break down the resistance path between rotating and stationary parts, an electrical discharge occurs, causing the removal of a small amount of material. Repeated discharges of this type cause progressive surface deterioration, resulting in a frosty appearance and eventually in a more serious destruction of the surface until failure occurs. The path of least resistance is usually through the smallest oil film, as at journal or thrust bearings or at governor gears, coupling teeth, or driven gears.

The voltage involved is in the range of 30 to several hundred volts. This destructive process is usually relatively slow, and its progress sometimes can be observed on shaft-position probes. The machinery can be protected against electrostatic discharge by the installation of shaft brushes that provide a low-resistance path between rotating and stationary elements.

Electromagnetically generated, internally self-induced shaft currents may be much more destructive than electrostatic currents. Rotating machinery may produce very high induced circulation currents from the effect of a magnetic rotor rotating in a stator or of a rotor rotating in a magnetic housing. The currents normally manifest themselves as heat, but if a low-resistance path is produced to facilitate high-circulation currents, very destructive heat concentration may be produced at the areas of highest resistance. For example, if a rotor contacts a stator, the resistance of the current path may be so decreased that the current magnitude is high enough to weld contacting parts together. This type of destruction does not always occur when rotors rub, indicating that electromagnetic fields and orientation necessary to produce high currents are required.

Precautions to prevent destructive shaft currents

1. *Electrostatic shaft currents*. Shaft brushes can be installed to prevent the buildup of a high voltage between the shaft and the case. Brushes should preferably contact the shaft end where sliding velocities are low. Brushes should be routinely maintained to sustain their protective capabilities.

2. *Electromagnetic shaft currents*. Rotors should be demagnetized prior to installation. Stationary parts including casings, diaphragms, and bearing housings should be demagnetized to the lowest possible levels. Extreme care should be taken when welding close to rotating machinery to make sure that high welding currents do not pass through the casing or rotating parts.

Materials. Rotors of mechanical-drive and smaller generator-drive turbines are usually machined from a single forging of chromium-nickel-molybdenum-vanadium steel, for those operating below 750°F, or from chromium-molybdenum-vanadium steel, for those operating above 750°F. Turbines using high-moisture steam (in nuclear service) often are made of 12 percent chromium steel, with bearing journals and thrust collars either inlaid or sleeved with low-carbon steel to guard against "steel wooling" types of failure.

Blading and shrouding are generally made of 12 percent chromium stainless steel (Types 403, 410, and 422) because of its inherent strength against vibratory stress. The history of stress-corrosion cracking of these steels indicates that Types 403 and 410 should not be hardened in excess of R_C 22 and that Type 422 should not be hardened above R_C 30. Shot peening of critical areas of blading, stress relief of shrouds, and shot peening of tenons help protect blading against stress-corrosion cracking; the same processes are helpful in rotors.

Areas in a turbine that are alternately wet and dry and are subject to relatively high tensile stresses are most highly susceptible to stress-corrosion cracking.

Glands and Labyrinth Packing Examination of the cross-section drawing of any turbine will bring out the methods used to prevent leakage of steam where the shaft extends from the casing, where control valve stems pass through covers, where diaphragms encircle the rotor between stages, and at the outside diameter of rotor blading.

Shaft end glands are as shown in Fig. 5-32c for the high-pressure end (both ends of a back-pressure turbine) and Fig. 5-32a for the exhaust end of a condensing turbine. (The source of sealing steam and vacuum to remove the air in-leakage at the outermost labyrinth is shown in Figs. 5-41 and 5-42.) Each labyrinth is in quarter segments with a slight gap between them to provide differences in thermal expansion of the different materials of the housing and the labyrinth. Coil springs keep each segment as close to the shaft as the T support in the housing permits. Steam or air pressure presses each labyrinth axially against the housing toward its low-pressure side, which effects an axial seal. Since a portion of the outside diameter sees a higher pressure than the average pressure through the labyrinth, the coil spring is assisted in holding the segment to the shaft, provided, of course, that the weight of the lower segments does not overcome the spring and differential steam pressure. If the axial-seal face is too close to the inlet side of the labyrinth, the differential radial steam pressure can suffice to hold the labyrinth segments open, thus greatly increasing leakage.

Journal and Thrust Bearings   Journal bearings may be plain, lobed, grooved, of the three-lobe type, or of the tilting-pad type. These types are illustrated in Sec. 4.

The high rotative speeds of turbines favor tilting-pad journal bearings as a means to prevent oil-film-induced vibrations up to and including tripping speed. It should be remembered that heat flows through the turbine shaft and is removed by the oil flow in the bearings; thus, different units may require different bearing clearances.

Thrust bearings are of the tilting-pad type, with a preference for integral-thrust collars. Oil flow is controlled by an orifice at the inlet to the bearing, and the oil should be removed at the outer diameter of the collar by a collection groove known as an oil-control ring (see Sec. 4).

The thrust to be taken by the thrust bearing consists of the mean pressure difference across the individual rotor disks, the steam-flow thrust across the blading, and the pressure thrust from differences in rotor diameter in the several pressure regions through which the rotor passes. For units coupled together by a flexible coupling (each unit having its own thrust bearing), the force necessary to slip the teeth of a dental coupling or the deflection force of a diaphragm coupling must be added to the algebraic sum of axial-thrust forces.

Steam Control Valves   Steam flow to the turbine is controlled by valves in the inlet steam chest. These valves are arranged to open sequentially so that the increase in flow area produces a directly proportional increase in steam flow, as shown in Fig. 5-33. The upper marked point is less than 100 percent throttle flow to allow for governor control of speed as well as for variations in steam conditions. This point is usually considered the maximum rated turbine output. The lower marked point is the normal rating of the turbine. This point is kept as close to the valve-opening point as practical to reduce throttling losses (it must be remembered that the valve-opening point will change slightly with changing steam conditions).

The bar-lift type of valve chest is illustrated in Fig. 5-13. A cam-lift design in which each individual valve is raised by its own cam is illustrated in Fig. 5-34. The grid-valve type of control valve is illustrated in Fig. 5-16.

The bar-lift valve is the simplest type. There are only two lifting stems with their consequent seal leakages; however, steam-flow forces are generated in the steam chest. These forces may buffet the valves or cause them to spin, with consequent wear of the bar and of the valve supports. The wear may change the

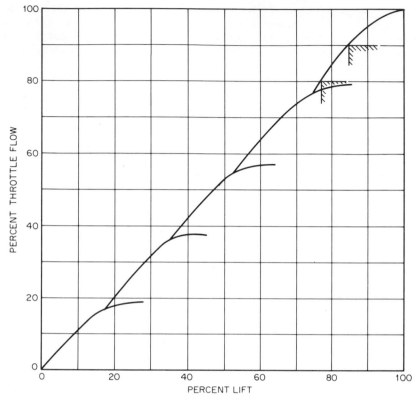

FIG. 5-33   Valve-lift curve.

steam flow characteristic of the turbine and cause breakage of the valve stems. A splitter vane is located at the inlet to the steam chest to prevent direct impingement of the entering steam on the end of the bar and the first valve.

Cam-lift valves are used when a larger mechanical advantage is needed because of larger flows and a small number of valves. These valves can be balanced and thus reduce the force required to open them.

A grid valve is desirable when large volumes of steam must be controlled at low pressure. It consists of two disks, one stationary and one rotating. The stationary disk, which is quite thick axially, has uniformly spaced openings covering somewhat less than half of its inlet face; at its outlet face the openings are expanded to cover the full circumference with steam to the succeeding stage. The rotating disk is relatively thin and has varying-size openings that progressively match openings in the stationary disk. As the disk is rotated, it produces steam flow versus rotation corresponding to Fig. 5-33.

Trip-and-Throttle Valves   Any turbine must be connected to its steam supply line on the downstream side of a stop valve. There is also a trip-and-throttle valve between the stop valve and the turbine steam inlet. This valve serves two purposes. First, it serves as a throttle valve to admit steam gradually to the turbine after

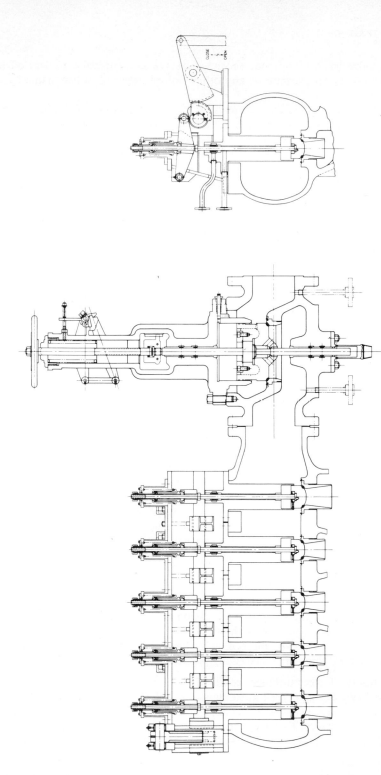

FIG. 5-34 Cam-lift steam control valves.

the stop valve has been opened. Second, it serves as a trip valve to shut off all steam on loss of oil pressure or any other failure actuated either manually or electrically.

Admission of steam to the turbine must be controlled to allow gradual warming of the inlet piping and turbine casing. Warming drains in the piping and casing must be opened to a sump until water ceases to discharge. Governor valves should be open to allow drainage until the turbine reaches the set speed.

Fig. 5-35 depicts an inverted trip-and-throttle valve (shown closed) that trips automatically upon failure of oil pressure. The manual trip lever and its linkage are shown engaged. When the trip lever is pulled, the spring will pull down the central stem bushing, which, in turn, pulls the main and pilot valves onto their seats. Loss of oil pressure allows the spring-loaded oil-cylinder piston to trip the valve just as the manual lever does. The trip action must be completed in a small fraction of a second.

Figure 5-36 depicts an inverted trip-and-throttle valve (shown closed) that is tripped on loss of oil pressure and automatically reset as oil pressure is reestablished. Construction of the valve is similar to the one shown in Fig. 5-35 except that when the valve is open, the portion of the stem in the packing-box area has a step that seals against the lower end of the uppermost bushing, thus preventing gland leak-off during operation of the turbine.

Figure 5-37 depicts an inverted stop volve which is used ahead of a trip-and-throttle valve to completely shut off all steam to the unit. It is similar in construction to a trip-and-throttle valve except that the manual handwheel mechanism is omitted.

Servomotors   Servomotors are required to lift the steam admission valves against the differential steam pressure and the valve-closing springs. These forces may amount to several thousand pounds under ideal conditions, and a factor of 2 or more (to allow for friction) is usually used to size a servomotor.

Figure 5-38 is an hydraulic servomotor used with 120-lb/in^2 control oil. The main servopiston is on the left; it is moved in the opening direction (down) as its pilot valve is raised to admit control oil above the piston; it is moved in the closing direction (up) as its pilot valve is lowered to drain control oil from above the piston so that the valve-closing springs can pull upward on the piston.

The illustrated servomotor is known as a single-acting servo because it is opened by hydraulic force and closed by spring force. Note that the main pilot valve is moved by a small servomotor with its own pilot valve by action of the speed governor.

Figure 5-39 illustrates a steam-admission-valve module which contains an hydraulic servomotor operating with 1000- to 1500-psig control oil, a servomotor relay valve, and a valve-position feedback transducer. The module as shown may be one of several which are individually positioned as required (see Fig. 5-23).

Turning Gears   Larger turbines should be equipped with a turning gear to spin the rotor slowly during (1) the warming-up period, (2) the shutting-down period, and (3) any standby period short enough that the rotor does not cool to the final shutdown temperature. Turning gears are electric-motor-driven and may have an attached lubricating-oil pump or a separate motor-driven pump mounted on the

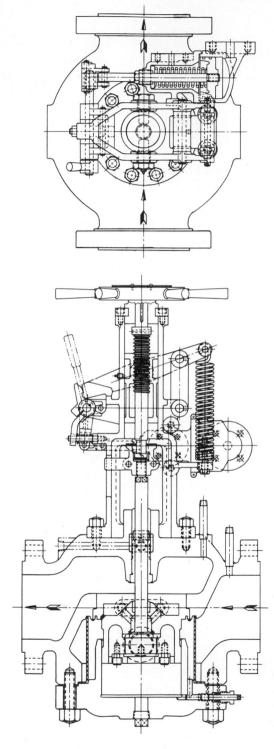

FIG. 5-35 Trip-and-throttle valve with oil-cylinder trip. *(Courtesy Gimpel Corporation.)*

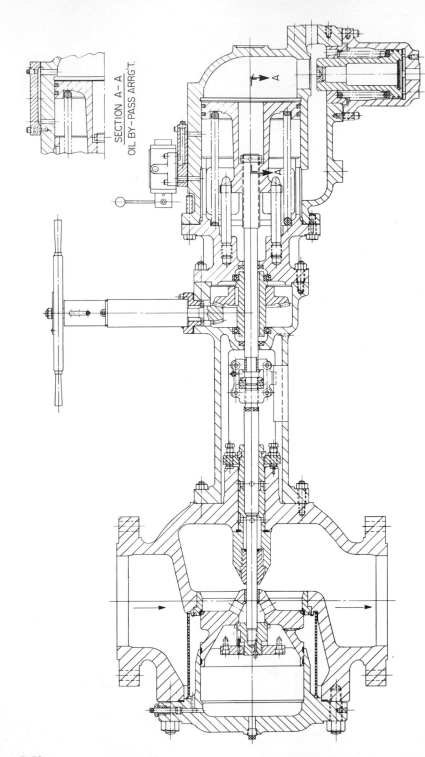

SECTION A-A
OIL BY-PASS ARRG'T.

FIG. 5-36 Trip-and-throttle valve with large oil cylinder and hydraulic exerciser. (*Courtesy Gimpel Corporation.*)

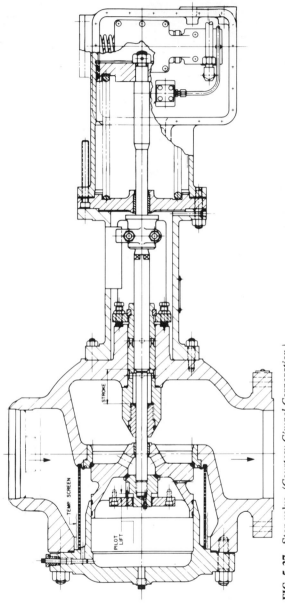

FIG. 5-37 Stop valve. *(Courtesy Gimpel Corporation.)*

STROKE

TEMP. SCREEN

PILOT LIFT

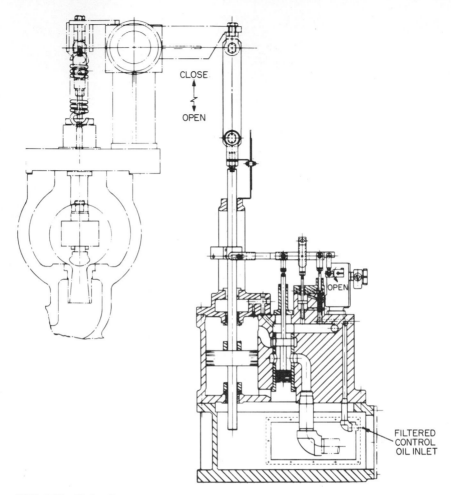

FIG. 5-38   Hydraulic servomotor.

oil reservoir. They usually have an automatic disconnect arrangement to protect against an unintended steam admission to the turbine that would otherwise back-drive the gear and motor to destruction.

If the rotor is not kept turning during the mentioned periods, it will absorb heat from hotter sections of the casing and reject heat to cooler locations such as the condenser tubes, thus causing the rotor to bow somewhat and be out of balance. A bowed rotor may take many hours of slow running with steam to eliminate the bow.

Figure 5-40 shows a typical turning gear used on a large mechanical-drive turbine. The motor is mounted on top and drives through a spur-gear reduction, a worm-gear reduction, and a second spur-gear reduction containing an idler gear to the rotor-mounted gear.

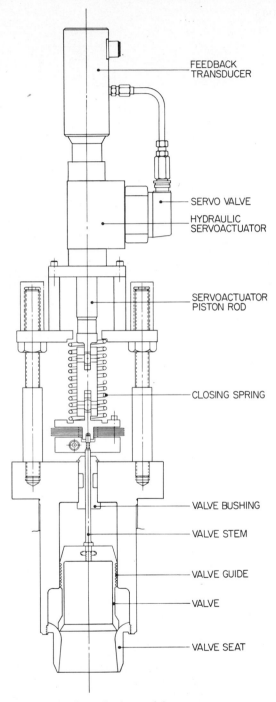

FEEDBACK
TRANSDUCER

SERVO VALVE

HYDRAULIC
SERVOACTUATOR

SERVOACTUATOR
PISTON ROD

CLOSING SPRING

VALVE BUSHING

VALVE STEM

VALVE GUIDE

VALVE

VALVE SEAT

FIG. 5-39 Control-valve module.

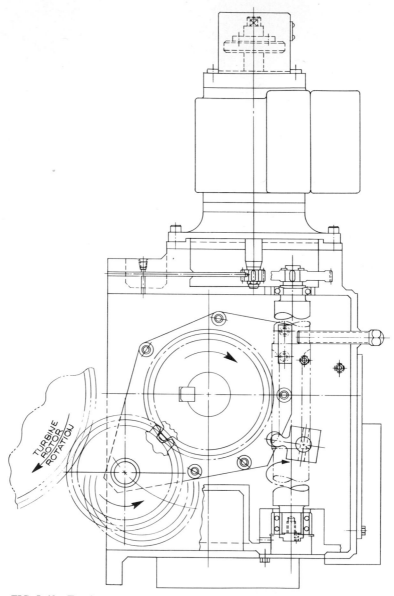

FIG. 5-40   Turning gear.

The idler gear is mounted in a swinging cradle to move it into or out of engagement. With the turning gear driving, the idler is held in engagement by tooth reactions; with the turbine rotor overspeeding, the idler is thrown out of engagement.

See Table 5-2 for materials of construction for turbine components.

Steam Sealing and Drainage Systems

All steam turbines must be provided with sealing and drainage systems to prevent leakage of steam from within and to remove the water formed during the warm-up of the casing and connected steam piping. Condensing steam turbines also must include means to remove the water formed as the steam expands below saturation in the blading. Typical sealing and drainage systems are shown in Figs. 5-41 and 5-42.

The shaft seals are the major ones. Seals must also be provided to control steam leakage from the stems of trip-and-throttle valves and control valves. Stepped-type labyrinths are normally used in contemporary multistage-turbine shaft seals, whereas carbon rings are normally used for single-stage turbine shaft seals. Labyrinth seals are used to control steam leakage at the shaft between stages.

Valve stems are usually sealed with close-clearance bushings arranged along the stem with leak-off spaces between bushings. The outermost leak-off space is held below atmospheric pressure by an eductor in the sealing system.

Steam-Seal Regulator A gland-sealing steam-pressure regulator is necessary for an isolated unit, whereas units in a station may all be controlled by a single regulator properly sized. Figure 5-43 illustrates an hydraulically operated regulator that throttles turbine inlet steam to gland-seal pressure during start-up or during light-load periods. During high-load periods leakage through the glands increases and must be dumped by the regulator into the condenser.

The regulator senses the steam-seal pressure in the bellows chamber (filled with water to protect the bellows from live steam) and raises or lowers the pilot valve of the hydraulic servomotor. The steam valve on raising admits high-pressure steam into the seal line; on lowering, it dumps sealing steam into the line to the condenser.

Lubrication and Control Oil Systems

Most present-day turbines have a combined oil system to provide lubrication to bearings and high-pressure oil to actuate the controls. A typical system such as this is shown in Fig. 5-44 for a turbine-driven boiler-feed pump and in Fig. 5-45 for a petrochemical plant.

An hydraulic control system contains servomotors, relays, and spring-loaded actuators, all requiring varying quantities of relatively high-pressure oil for their operation. This is usually provided for by a self-contained oil system as an integral part of the turbine. Much lower oil pressures are required for lubrication; these are obtained by a pressure-reducing valve that bleeds down from the high-pressure portion of the system or from a separate low-pressure pump that feeds the low-pressure portion of the system.

Special oils have been developed for use in turbine lubrication systems. Such oils contain rust and oxidation inhibitors.

Turbines to be used in contaminated atmospheres, especially those containing ammonia, will require special construction materials for those parts in contact with the lubricating or control oil.

TABLE 5-2 Materials of Construction

Part	Fossil service			Nuclear service		
	Material	ASTM specification*	Temperature limit (°F)	Material	ASTM specification*	Remarks†
Steam chests	Cast carbon steel	A216, Grade WCB	To 750	Cast carbon steel	A216, Grade WCB	Inlaid with Type 309 stainless steel as noted below
	Carbon steel plate	A283, Grade D	To 750			
	Cast carbon-molybdenum	A217, Grade WC1	751–825			
	Cast 1¼ chromium, ½ molybdenum	A217, Grade WC6	826–950	Cast 12 percent chromium	A351, ACI-CA6NM	
	Cast 2¼ chromium, 1 molybdenum	A217, Grade WC9	951–1000			
Intermediate case	Cast carbon steel	A216, Grade WCB	To 750	Cast carbon steel	A216, Grade WCB	Inlaid with Type 309 stainless steel as noted below
	Carbon steel plate	A283, Grade D	To 750			
	Cast carbon-molybdenum	A217, Grade WC1	751–825	12 percent chromium steel	A351, ACI-CA6NM	
Exhaust case	Cast carbon steel	A216, Grade WCB	To 750	Carbon steel plate	A283, Grade D	Inlaid with Type 309 stainless steel as noted below
	Carbon steel plate	A283, Grade D	To 750			
Turbine rotor	Forged alloy steel	A470, Class 7	To 750	12 percent chromium steel forging	A470	MIL-S-860B, Class G; journals inlaid with carbon steel; temperature to 750°F
	Forged alloy steel	A470, Class 8	To 1000			
Turbine blades (stellited as required)	Stainless steel	A276, Type 403	To 900	Stainless steel	A276, Type 403	
	Stainless steel	AISI, Type 422	To 1000	Stainless steel	AISI, Type 422	
Blade shroud	Stainless steel	AISI, Type 422	To 1000	Stainless steel	AISI, Type 422	
Inlet-nozzle ring and diaphragms	Carbon steel plate	A283, Grade D	To 750	Stainless steel	A176, Type 405	
	Carbon steel plate	A515, Grade 70	To 750	Stainless steel	A276, Type 405	

	Carbon-molybdenum steel plate	751–875	A204, Grade B		
	1¼ chromium, ½ molybdenum steel forging	876–950	A182, Grade 11		
	2¼ chromium, 1 molybdenum steel forging	951–1000	A182, Grade F22		
Inlet-nozzle ring and diaphragm vanes	Stainless steel	To 900	A276, Type 405	Stainless steel	A276, Type 405
Labyrinth packing	Leaded nickel bronze	To 850	B271 or B584, Alloy no. 976	Leaded nickel bronze	B271 to B584, Alloy no. 976
Labyrinth packing springs	Inconel X-750	To 1050	AMS-5698		
Journal bearing	Steel Tin-base babbitt		B23, Grade 2	Steel Tin-base babbitt	B32, Grade 2
Governor valves	Stainless steel with a stellited surface	To 1000	AISI, Type 422	Stainless steel with a stellited surface	AISI, Type 422
Governor-valve seats	2¼ chromium, 1 molybdenum with a stellited surface	To 1000	A182, Grade F22	2¼ chromium, 1 molybdenum with a stellited surface	A182, Grade F22

*These ASTM specifications are closest to Transamerica Delaval specifications. In all cases the materials supplied by Transamerica Delaval are of equal or superior quality in meeting the requirements of the application.

†The following contact areas of the carbon steel parts are inlaid with Type 309 stainless steel:

1. Nozzle ring to steam chest
2. Diaphragms to case
3. Labyrinth seal rings to case packing-box areas
4. Case to case at horizontal split

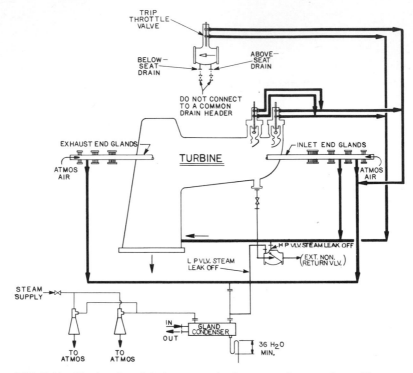

FIG. 5-41   Gland seals and drains: noncondensing automatic-extraction turbine.

Filtration of the control oil to remove particles as small as 10 μm is usually provided for as a maintenance preventive.

Monitoring Systems

Continuous monitoring of pressures, temperatures, speed, vibration, alignment, position, and levels throughout the entire machinery train is desirable in the interest of reliability and availability. The signals from the parameters being monitored may be recorded, may actuate visual or audible alarms, and may actuate shutdown trips to prevent serious failures. Pressures, including differential pressures, and temperatures to be monitored should be selected with care, keeping in mind the effect of location, elevation, and flow conditions of the fluid within the pipe or vessel involved.

Electrical and/or electronic sensors for sensing and measuring any of the quantities mentioned above are available in the marketplace. Table 5-3 indicates the types of instruments available for monitoring.

Figure 5-46 shows schematically a vibration- and shaft-speed-monitoring system as is used for turbine-driven boiler-feedwater pumps. The mechanical integ-

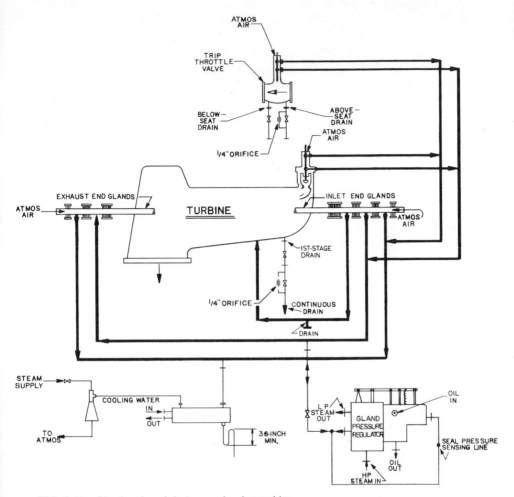

FIG. 5-42   Gland seals and drains: condensing turbine.

rity of pump and turbine is continuously monitored for vibration, shaft axial position, shaft eccentricity, and speed. Signals from 12 noncontact eddy-current-type probes located at critical points can be transmitted to a control room. It is important that all sensors be insensitive to thermal movement, twisting, or other position changes during operation of the unit.

Figure 5-47 illustrates a system that may be applied to either a dental type or a diaphragm type of flexible coupling to monitor alignment variations between two rotors. This "hot alignment" indicating system measures coupling misalignment under actual operating conditions. The basic components of the system are shown in Figs. 5-48 and 5-49:

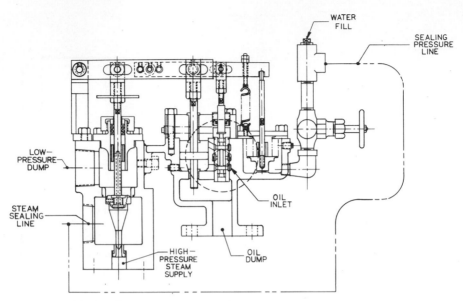

FIG. 5-43   Gland-sealing steam-pressure regulator.

1. Inductive proximity probes, rotating with the coupling, measure the amplitudes of the once-per-revolution variations in gap due to misalignment at each end

2. Power rotary transformer that couples power to the probe and to the shaft electronics without mechanical contact

3. Signal rotary transformer that feeds out from the shaft signals proportional to the misalignment at each end

4. Shaft electronic circuits

5. Marker probes at "twelve o'clock" and "three o'clock" provide X and Y reference-square waves

6. Probe target

The system has the following important advantages. Since the probes rotate with the coupling, each probe "looks" at the same area all the time, so that there is no dependence upon surface flatness or machining accuracy. Misalignment is measured at running speed, under "hot" conditions, so that there is no need to calculate shaft radial position, thermal expansions, or machine-casing distortion. For diaphragm couplings, axial position is measured to show the umbrella effect. Alignment can be accurately and conveniently checked at any time without necessitating a shutdown.

Foundations, Soleplates, and Bedplates

Turbines and their driven machines, such as compressors, pumps, and electric generators, must be fastened to a stable support in order to maintain the alignment of the rotors within allowable limits. This is accomplished by mounting on a bedplate or soleplates, which must be supported, in turn, by a suitable foundation. Foundations are usually of reinforced concrete and are usually designed by a specialist specifically for a given installation, taking into account local physical conditions and codes. Occasionally, structural steel or a combination of concrete and steel is used for a foundation; it is subject to the same requirements of deflection and strength.

Bedplates must accept the static forces of the weight of the machinery, the forces and moments of the attached piping, and the dynamic forces due to torque, thermal growth, fluid pressures, and seismic or other shocks. Their design must consider the location of the supports provided by the foundation in addition to the static and dynamic forces from the rotating machinery. The design criterion is usually more one of deflection than one of strength. All beam spans, plate surfaces, and cantilevers should be designed so as not to resonate with any fundamental or harmonic frequency generated by the mounted equipment or transmitted from without. The designer must also be sure that the stresses and deflections to which the bedplate will be subjected during lifting, shipping, and installation with all equipment mounted will be within safe and acceptable limits. For convenience during installation, a jacking bolt should be provided adjacent to each foundation hold-down bolt.

Soleplates are individual steel plates set in the foundation under each of the machinery support points. They are set in place on the rough concrete of the foundation and leveled by jacking screws. Hold-down bolts set in the concrete foundation pass upward through holes in the soleplates. When the soleplates are set in final position, grout is poured over the rough foundation to fill the space below and around them. A nonshrink grout should be used. Shims between the machinery feet and the soleplates may be required for final alignment. After the grout has set, the foundation bolts are tightened.

Handling

All parts weighing 50 lb or more must be provided with a means for handling from raw material through all stages of manufacturing including the finished unit. Failure to plan handling may cause serious accidents to the workers or the material, or both.

Each step of the manufacturing, testing, shipping, field-erection, and future-maintenance process must be considered. If special tools are required, they are provided.

Useful Documents

Documents that are usually furnished by a turbine manufacturer for the owner's use are:

1. Technical manual
2. Test agenda: shop
3. Outline drawing

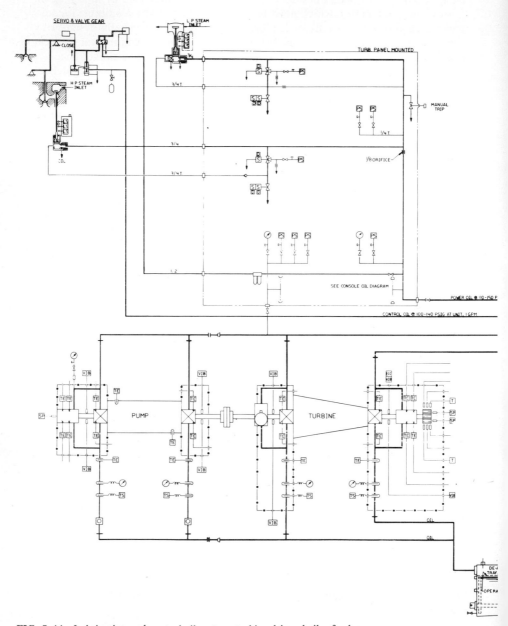

FIG. 5-44 Lubrication and control oil system: turbine-driven boiler-feed pump.

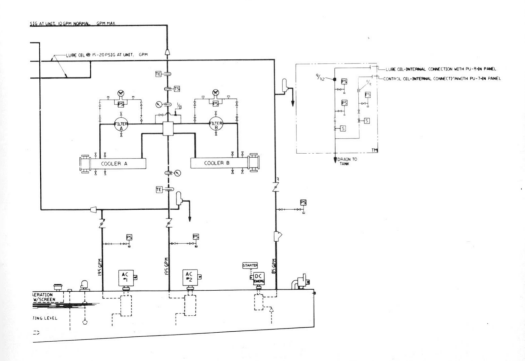

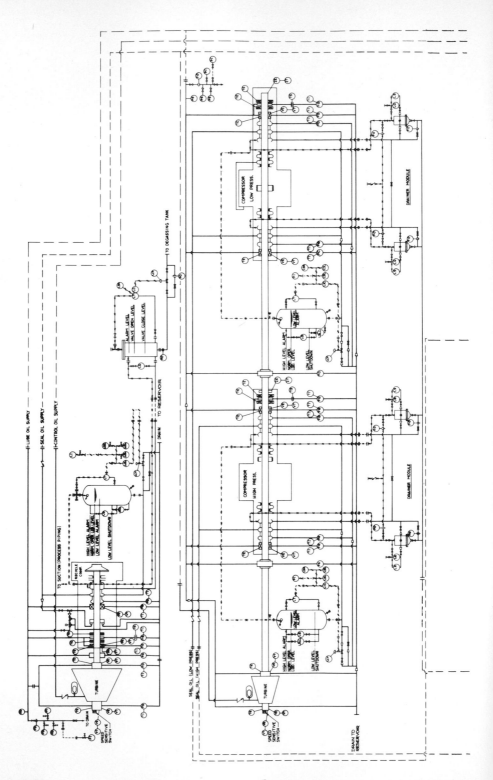

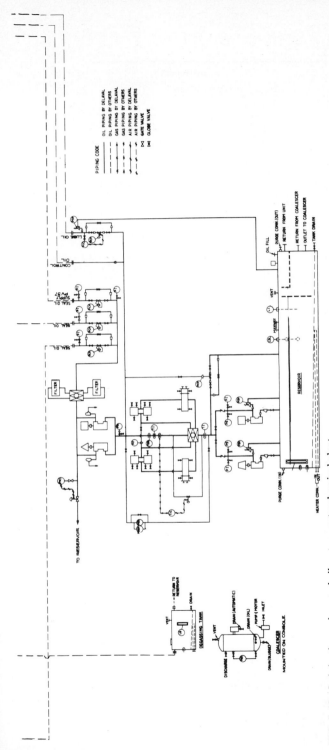

FIG. 5-45 Lubrication and control oil system: petrochemical plant.

TABLE 5-3

Item	Instrument
Pressure	Strain gauge; load cell
Temperature	Thermocouple
	Resistance temperature detector (RTD)
Speed	Magnetic; proximity; speed pickup
Vibration*	Proximity; eddy-current probes
Alignment	Inductive proximity probes
Position	Inductive proximity probes

*Under some vibration conditions, the wave shapes generated may not be sinusoidal to an extent that would require Fourier analysis to allow determination of and effect of harmonics.

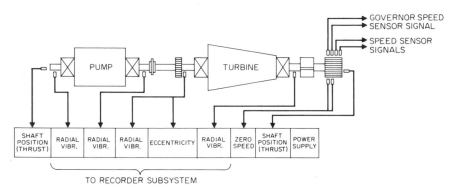

FIG. 5-46 Vibration-monitoring system.

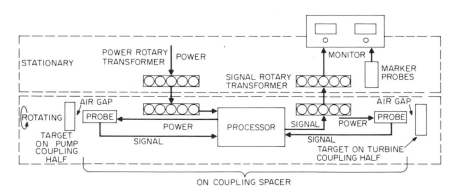

FIG. 5-47 Hot alignment system.

4. Assembly drawing

5. Allowable forces and moments: piping

6. Thermal movements

7. Loading diagram: foundation

8. Erecting procedure: shop

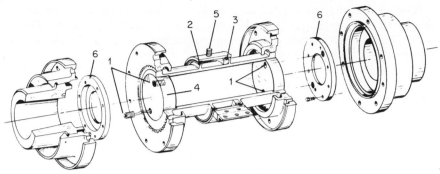

FIG. 5-48 Alignment-monitoring system: dental coupling. *(Courtesy of the Indikon Company, Inc.)*

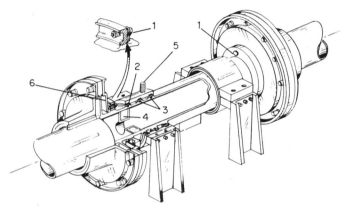

FIG. 5-49 Alignment-monitoring system: diaphragm coupling. *(Courtesy of the Indikon Company, Inc.)*

9. Erecting procedure: field
10. Preservation for shipment
11. Preservation for storage
12. Alignment to driven equipment
13. Initial operation: field
14. Maintenance

Preservation for Shipment or Storage

In the interest of economy, machinery is shipped assembled if it is not too large or heavy. It may be mounted together with several components on a common support and shipped as a unit. On arrival at the jobsite, it may be stored for considerable periods of time and under varying ambient conditions. During both shipping and storage, protective measures must be taken to prevent damage that can occur in differing degrees depending on length of storage and serverity of conditions.

Turbines are normally shipped with the rotor blocked to prevent axial movement of the rotor against the thrust bearing. Shims of paper or plastic are placed at the journals to take up the clearance in the bearings. Valve gear is blocked to set valves. Internal parts are coated with a rust preventive. All openings are closed to prevent entry of foreign material. External machined surfaces are protected against rust. Instruments, gauges, and other small parts are usually removed, individually wrapped, and packed for off-unit shipment.

Long-term storage should be in a heated building. Sensitive parts should be disassembled from the turbine, placed within moisture barriers with desiccant bags, and inspected periodically.

Waterborne shipment requires boxing and desiccant protection.

System Considerations

Marine Propulsion Systems   See Sec. 7.

Turbine-Driven Electric Generators   Direct-connected units with rigid couplings must be connected together so that the elastic curve of their centerlines is not distorted by improper machining of the coupling faces and rabbets or by improper tightening of the coupling bolts. If either or both of the rabbets are not concentric with the journal, in a three-bearing rotor system the shaft with only the outboard journal bearing will swing in a circle at its outboard end (see Fig. 5-50). This is also true if either coupling face is not square with its axis. The bearings of a three-bearing set must be placed at different elevations to accommodate the elastic curve, as can be seen in Fig. 5-51.

Short circuits and out-of-phase synchronizing of a generator with the grid to which it is being connected can cause transient torques several times the full-load torque of the unit. The magnitude of these torques is subject to calculation and should be provided for in the design of the unit.

Maintenance

Preventive Maintenance   This is a matter of establishing an inspection and service routine to detect and correct conditions which could cause malfunction or breakdown of any part of the unit.

Before a unit is started initially or while it is shut down, every precaution should be taken against corrosion. For this purpose steam valves should be kept tight and checked regularly for leakage. If leakage is detected, measures such as inserting a blind flange downstream of the valve, with a leak-off between the flange and the valve, should be taken. The atmosphere in an idle turbine will be saturated with moisture from the condenser hot well. This moisture will condense on internal surfaces as their temperature changes because of external ambient-temperature changes. To prevent this will require blanketing the turbine internals with a positive flow of dry gas. The lubrication system should be operated for a few minutes each day and the rotor rotated to lubricate the journals.

After the unit has been placed in service, a periodic service and inspection schedule should be established for lubrication of the valve-lifting gear and all

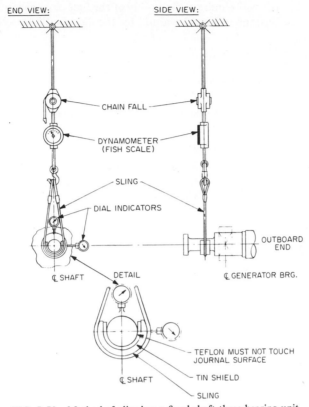

END VIEW: SIDE VIEW:

CHAIN FALL

DYNAMOMETER
(FISH SCALE)

SLING

DIAL INDICATORS

OUTBOARD
END

₵ SHAFT   DETAIL   ₵ GENERATOR BRG.

TEFLON MUST NOT TOUCH
JOURNAL SURFACE

₵ SHAFT   TIN SHIELD

SLING

FIG. 5-50 Method of aligning a fixed-shaft three-bearing unit.

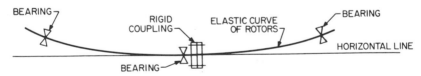

BEARING   RIGID   ELASTIC CURVE   BEARING
COUPLING   OF ROTORS

HORIZONTAL LINE

BEARING

FIG. 5-51 Three-bearing turbine-generator rotors.

exposed control linkages and for testing and checking the accuracy of tripping and governing mechanisms. The slide joints of a pedestal need lubrication, and all oil, control, and steam valves must be kept free. The condition of the oil in the sump should be checked on a regular schedule to determine whether or not any sludge from oxidation of the lubricating oil has collected on the bottom of the oil sump.

Corrective Maintenance This involves making repairs and necessary inspections to return the unit to normal operation after a failure or malfunction. Usually inspection and repair will require disassembly of the unit or of one or more components; in doing this, the technical manual for the unit should be followed.

Reference Codes and Standards

1. American Petroleum Institute

    a. API Standard 612   *Special Purpose Steam Turbines*
    b. API Standard 614   *Lube, Sealing & Control Systems for API 612*
    c. API Standard 611   *General Purpose Steam Turbines*

2. National Electrical Manufacturers Association

    a. Publication No. SM 12   *Direct-Connected Steam Turbines, Synchronous Generator Units, Air Cooled*
    b. Publication No. SM 23   *Steam Turbines for Mechanical Drive Service*

3. U.S. Navy

    a. MIL-T-17523   *Turbine, Steam, General Auxiliary (Naval Shipboard Use)*
    b. MIL-T-17600   *Turbines, Steam, Propulsion, Naval Shipboard*
    c. MIL-T-24398   *Turbine, Steam and Reduction Gear, Auxiliary Generator-Drive (Naval Shipboard Use)*

4. American Society of Mechanical Engineers

    a. STD TWDPS-1   *Turbine Water Damage Prevention: Fossil Fueled Units and Nuclear Units*

5. American Bureau of Shipping

    a.   *Rules for Building and Classifying Steel Vessels*

6. American Society for Testing and Materials

    a.   *Properties of Materials*

7. American Iron and Steel Institute

    a.   *Chemistry of Materials*

GAS TURBINES, TURBOCHARGERS, AND EXPANDERS

Gas Turbines

Gas turbines have achieved ever-increasing importance. Their applications are many, and their design varies from great simplicity to a certain sophisticated complexity. They are found in the following *areas of application:*

Propulsion of military, commercial, and civil aircraft
Peaking-power facilities
Compressor and pump drive
Total-energy systems
Vehicle drive for locomotives, ships, trucks, off-road vehicles, and automobiles

Process air and gas generation
Airborne and ground-support equipment for aircraft

Gas turbines have found their application because of some or all of the following potential *advantages* over competitive equipment:

Small size and weight per horsepower
Self-contained units of moderate first cost
Reliability through turbomachinery components
Easy maintenance
Instant power
Ability to burn a variety of fuels
No cooling water required

Some of the simpler gas-turbine arrangements are described below.

The basic cycle may be demonstrated by the simple open-cycle *single-shaft gas turbine* sketched in Fig. 5-52. Ambient air is compressed in the compressor *C*. Heat is added to the compressed air at essentially constant pressure by burning fuel in the combustion chamber *CC*. The hot combustion gas is expanded in the turbine *T*. The power produced by the turbine is larger than the power required to drive the compressor. The difference in power is available to drive the load *L*.

The turbine *T* may be split into two turbines in series. Figure 5-53 is a sketch of the simple open-cycle *two-shaft gas turbine*. The turbine T_1 is sized to drive the compressor *C*. Turbine T_1, compressor *C*, and the combustion chamber *CC* constitute a gas generator. The turbine T_2 drives the load *L*. Turbine T_2 is mechanically independent of the gas generator, may run at different speeds, and is often called a *power turbine*.

The gas generator produces a combustion gas (essentially air) at moderately elevated pressure and temperature. The energy contained in this gas may be put to uses other than driving a power turbine. The gas may be expanded to a high velocity in a nozzle for the propulsion of aircraft. Figure 5-54 is a sketch of a simple *aircraft gas turbine* (turbojet).

On the other hand, the aircraft gas generator, highly developed for aircraft use, has found interesting ground applications. In combination with an industrial power turbine, a gas-turbine arrangement as shown in Fig. 5-53 results.

The *efficiency* of the gas turbine poses a certain problem. As the output is the

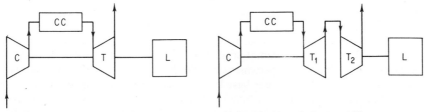

FIG. 5-52 Open-cycle single-shaft gas turbine.   **FIG. 5-53** Two-shaft gas turbine.

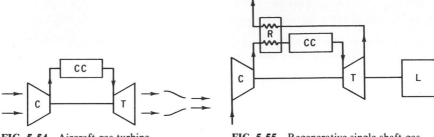

FIG. 5-54   Aircraft gas turbine.

FIG. 5-55   Regenerative single-shaft gas turbine.

difference of comparatively large powers, the component efficiencies must be very high and the cycle conditions are liable to be stringent.

The efficiency of the gas turbine is also improved by the adoption of a more complex cycle. As the larger part of the fuel energy leaves with the exhaust gas, waste-heat recovery in one form or another will improve the system efficiency greatly.

Waste heat may be recovered by means of a regenerator which preheats the compressed air with the exhaust gas before it enters the combustion chamber. The efficiency is improved because less fuel has to be burned to bring the gas up to turbine inlet temperature. Figure 5-55 is a sketch of a *regenerative* open-cycle single-shaft *gas turbine*.

Waste heat may be recovered by means of a waste-heat boiler in the exhaust stack. The steam generated by the waste-heat boiler may drive a steam turbine for additional power, it may be used in a process, or it may be used for heating and air-conditioning a building.

Utilization of the exhaust heat has led to the adaptation of the total-energy concept to gas turbines. A *total-energy system* is a system of high overall thermal efficiency designed around a gas turbine which meets (the total) different and varying energy requirements.

The combustion of fuel in the airstream is the simplest and most common way to add heat to the cycle. As an alternative, heat may be added entirely by heat exchange. The heat source may be a fired heater burning a cheap fuel such as coal. In this case the cycle gas does not become contaminated and may be used over and over again in a closed cycle. In a closed cycle, the pressure level may be raised and varied, leading to a smaller unit with good part-load efficiency. Gases other than air may be used as a cycle gas, which facilitates the use of an atomic reactor as a heat source for the gas turbine. Figure 5-56 is a sketch of a regenerative *closed-cycle gas turbine*.

The compressor may be a centrifugal or an axial type or a combination of both, and the turbine may be a radial or an axial type or a combination of both. Turbomachinery is high-flow machinery on which the small specific size and weight advantage of the gas turbine is based. The use of radial-flow machinery, centrifugal compressor, and radial turbine has been restricted more and more to smaller units as the gas turbine has grown in size. The field of high power belongs to the highly efficient axial compressor and axial reaction turbine.

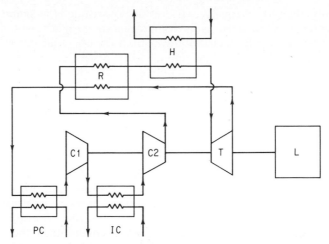

FIG. 5-56 Regenerative closed-cycle gas turbine. $C1$, $C2$ = compressors, T = turbine, L = load, H = heater, R = regenerator, PC = precooler, IC = intercooler.

Gas-turbine development is a continuing process. Component efficiencies, pressure ratios, and turbine inlet temperatures have increased, and there is an intensive search for ever better high-temperature materials. Much progress has been made in cooling turbine blades and rotors.

Industrial gas turbines range in size from a few hundred horsepower to about 30,000 kW, although smaller and larger units are being built.

BIBLIOGRAPHY

Diesel and Gas Turbine Catalog, vol. 32, Diesel and Gas Turbine Progress, Milwaukee, Wis., 1967.
Gas Turbine International Magazine, bimonthly publication by Gas Turbine Publications, Inc., Stamford, Conn.
Shepherd, D. G.: *Introduction to the Gas Turbine,* 2d ed., D. Van Nostrand Company, Inc., Princeton, N.J., 1960.

Turbochargers

Turbochargers are used to increase the operating pressure level of internal-combustion engines, thereby increasing the power output of the engine. The turbocharger serves to uprate the engine or to restore sea-level performance at high altitudes. At the same time, a saving in specific fuel consumption is achieved.

Main industrial applications are on two- and four-cycle diesel engines, gas engines, and dual-fuel engines.

Performance parameters vary, and close cooperation between turbocharger and engine manufacturers is required in order to adjust the turbocharger to an individual application.

Basically, the turbocharger is a gas turbine consisting of a compressor and a turbine with the engine replacing the combustion chamber as shown in Fig. 5-57. The air consumed by the engine is drawn from the atmosphere, compressed by compressor *C*, and discharged through a cooler in some designs into the intake manifold of the engine. The exhaust gas from the engine is expanded in turbine *T* and is exhausted to the atmosphere.

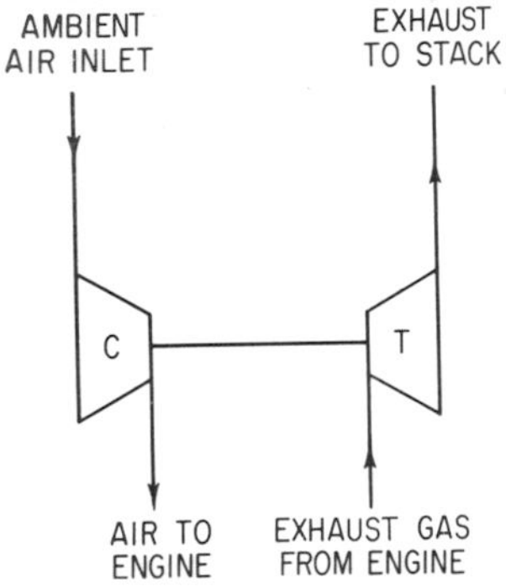

Typically, there is no mechanical connection between the shaft of the turbocharger and the engine. The power produced by the turbine matches the power absorbed by the compressor. This balance adjusts itself by speed variation.

Typical pressure ratios used were 1.5 to 3.0; lately modern turbochargers have used pressure ratios of 3.2 to 3.5 and higher. The turbine pressure ratio is somewhat smaller than the compressor pressure ratio because of the pressure drop in the engine.

The compressor consists of a single centrifugal stage and the turbine of a single radial or axial stage which may be arranged between or outboard of the turbocharger bearings. If the pressure ratio exceeds the capability of the single stages, two turbochargers of different standardized sizes may be used in series.

A cross section of a Delaval turbocharger is shown in Fig. 5-58. A mixed-flow centrifugal-compressor stage and a mixed-flow radial-turbine stage are arranged back to back on one side of the bearing case.

FIG. 5-57   Turbocharger.

AMBIENT AIR INLET

EXHAUST TO STACK

C   T

AIR TO ENGINE   EXHAUST GAS FROM ENGINE

Expanders

Expansion of gas in a turbine produces work and lowers the temperature of the gas stream as energy is removed. Turbines which produce work from the expansion of process gases and which serve the recovery of process waste energy are often called *expanders*. Some of these expanders are of considerable horsepower size. Representative gas conditions are inlet temperature = 1000°F, inlet pressure = 300 psia, and exhaust pressure = atmospheric or above.

Turboexpanders are part of low-temperature process equipment and refrigerators and are widely used in the cryogenic industry. Typical applications are air-separation plants for the production of gaseous and liquid oxygen and nitrogen when the turboexpander operates on an air or nitrogen stream down to the vicinity of −300°F. Applications involving the lowest temperature are helium liquefiers in which the turboexpander may operate at a temperature as low as −450°F.

The single-stage radial turbine has almost become a standard, although some axial turboexpanders have been built. The turbine is arranged outboard of the bearing case and separated from it by a seal. Most designs have oil-lubricated sleeve bearings. The use of bearings lubricated by the cycle gas is very attractive, and several experimental units with gas-lubricated bearings have been built. The load horsepower is of secondary importance and is usually absorbed by a single-

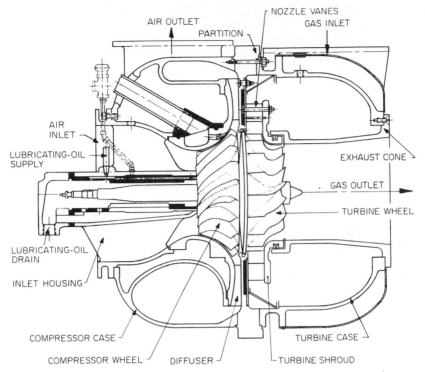

FIG. 5-58 Delaval turbocharger.

stage centrifugal compressor arranged outboard of the bearing case at the opposite end. The compressor may compress atmospheric air and dissipate the load by
throttling, or some use may be made of the horsepower by compressing seal or
process gas. Some larger units have been built with a load-absorbing generator
driven through a reduction gear. Small units may dissipate the load by an oil
brake.

Flow and horsepower sizes vary over a wide range. Turboexpanders with 4- to
6-in-diameter turbine wheels are typical. Turboexpanders have been built with
wheel diameters over 17 in and as small as $\frac{5}{16}$ in. Miniature turboexpanders with
wheel diameters below 1 in have a rotative speed above 100,000 r/min when only
gas-lubricated bearings make a successful design possible.

BIBLIOGRAPHY

Gas Turbine Engineering Handbook, Gas Turbine Publications, Inc., Stamford, Conn.,
     1966.

Gas and Oil Engines

Introduction

Gas and oil engines are prime movers of the reciprocating internal-combustion type. They cover a broad range of output horsepower and speeds and are in widespread use in applications for which their combination of good fuel economy at full and partial load and their compactness, durability, reliability, and reasonable first cost make them the most economical and effective choices.

Oil engines, more commonly referred to as *diesel engines,* have a dominant position as propulsion plants for heavy-duty transportation and mobile equipment. Diesel, dual-fuel, and gas engines, however, are widely used as primary power sources for generators, pumps, compressors, and similar equipment. In the industrial field gas and oil engines compete with gasoline engines, electric motors, gas turbines, steam turbines, and hydraulic turbines. The choice depends upon many factors and circumstances, the selected unit normally being the one that will give the best economy over its anticipated lifetime while being compatible with the requirements of the purchaser.

This section will concern itself mainly with oil and gas engines of the medium-speed range for industrial and marine use. The medium-speed range is generally accepted to be the 300- to 750-r/min range, with horsepower output ranging from a few hundred to more than 20,000 bhp per unit. The slow-speed range (90 to 300 r/min) covers most direct-drive marine main propulsion units which may have unit outputs higher than 40,000 bhp at approximately 100 r/min. High-speed units (750 to more than 3000 r/min) are used in the automotive and industrial fields and range from a few horsepower to approximately 4000 bhp. Engine manufacturers usually concentrate their efforts in one or two of these speed ranges because of the wide differences in production technology and equipment needed for economical production of the various engine types.

Classification of Engines

Reciprocating internal-combustion engines are classified according to their fuel type and ignition method. There are also many subclassifications which group engines according to cycle arrangements and modifications, mechanical arrangements, and other design features. The more important main classifications of medium-speed engines are the following:

Diesel Engines   Diesel engines are compression-ignition engines which operate on liquid fuel. The compression ratio and the resulting temperature of the compressed-air charge are high enough to ignite suitable fuels when they are injected into the cylinders near the end of the compression stroke. Diesel engines do not really operate on the diesel cycle, which stipulates combustion at constant pressure. High-speed diesel engines approach constant-volume combustion, or the Otto cycle, and engines running at very low speed can approach combustion at constant pressure. In most diesel engines the combustion phase of the actual cycle is a mixture of the Otto and the diesel cycles and is often referred to as a combination, or limited-pressure, cycle.

Gas Engines   Gas engines use gaseous fuel which is spark-ignited. Combustion takes place at essentially constant volume (Otto cycle). Compression pressures may be as high as those in diesel engines, but more commonly they are lower. Autoignition of the air-gas mixture must not take place, as the resulting detonation would rapidly damage or destroy the engine. The fuel gas is mixed into the combustion air during the induction part of the engine cycle or injected into the cylinder during the compression stroke. The first arrangement is common on four-stroke-cycle engines; the latter, on two-stroke-cycle engines. High-voltage energy, usually from magnetos or solid-state electronic devices and transformers, fires one or more spark plugs per cylinder for ignition of the combustible mixture.

Dual-Fuel Engines   Dual-fuel engines have two modes of operation. One is operation as a diesel engine; in the other mode a "pilot" injection of liquid diesel fuel ignites as in a diesel engine and subsequently ignites the main charge of fuel gas and air. The pilot fuel usually constitutes 5 to 7 percent of the total fuel energy input at full load. Dual-fuel engines generally resemble diesel engines but have additional equipment for the control of combustion air and fuel gas. Controls are provided for switching from one mode to the other, and in some dual-fuel engines the pilot-fuel–fuel-gas ratio may be varied to select any fuel proportion from minimum pilot fuel to full diesel operation.

Engine Subclassifications

Some of the more common subclassifications of gas and oil engines are:

Four-Stroke-Cycle Engines   The four-stroke engine cycle is completed in two crankshaft revolutions; it consists of the following piston strokes:

1. Induction (suction-intake) stroke, normally aspirated engines; (inlet) stroke, supercharged engines
2. Compression stroke
3. Expansion (power) stroke
4. Exhaust stroke

Ignition and combustion take place near the end of the compression stroke and the beginning of the expansion stroke.

Two-Stroke-Cycle Engines   The two-stroke cycle is completed in one revolution of the crankshaft; it consists of two piston strokes:

1. Compression stroke
2. Expansion (power) stroke

Scavenging and combustion-air recharging take place near the end of the expansion and the beginning of the compression stroke. Ignition and combustion take place near the end of the compression and the beginning of the expansion stroke.

Supercharged Engines   A supercharged engine is one in which the density of the combustion-air charge is increased over that of the surrounding atmosphere by precompression of the air before it is inducted into the cylinders. The purpose of supercharging is to increase the power output from a given cylinder size and/or to maintain rated output at higher altitudes. Two general methods of supercharging are used:

1. Mechanically driven blower which is normally driven by an engine auxiliary output shaft. It may also be driven by an electric motor or other separate prime mover.
2. Exhaust-turbine-driven blower in which the blower is driven by a turbine which derives its power from the engine exhaust gases. The turbine and the blower are mounted on the same shaft; this combination is called a turbocharger.

Naturally Aspirated Engines   A naturally aspirated engine is one in which the combustion air inducted into its power cylinders is of a density not greater than that of the surrounding atmosphere. Two-stroke-cycle engines usually have a blower which forces air through the cylinders for scavenging and recharging, but they are termed *naturally aspirated* if this does not result in a substantially higher than atmospheric combustion-air density (pressure) at the beginning of the compression stroke.

Aftercooled Engines   The term *aftercooled* is used when referring to an engine equipped with equipment for cooling the combustion air after it has been supercharged and before it is inducted into the power cylinders. The purpose of aftercooling is to improve internal cooling, increase combustion-air density, and reduce detonation tendencies in gas-burning engines.

Number and Arrangement of Cylinders   Engines may be classified according to the number of power cylinders and the arrangement of those cylinders. Two common arrangements are:

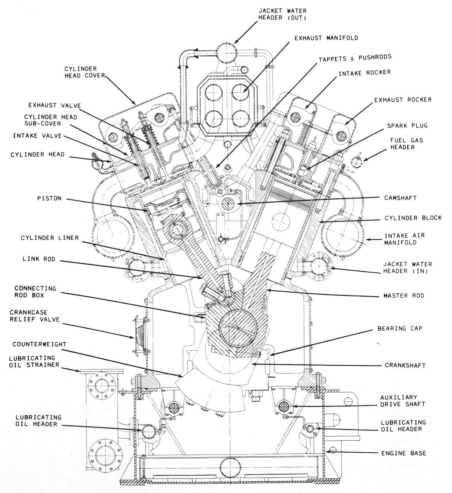

FIG. 6-1   Enterprise spark-ignited gas engine.

1. *V-type* engines, in which the cylinders are arranged in two banks parallel to the crankshaft, the banks forming a V when viewed from the end of the engine. (See Figs. 6-1 and 6-2.)

2. *In-line* engines, in which the cylinders are in line in a single bank, generally in a vertical position parallel to the crankshaft. (See Fig. 6-3.)

There are other less common arrangements such as horizontal opposed and radial, but these are of declining importance in the medium-speed field.

Valve Type and Location   Four-stroke-cycle engines have intake and exhaust valves of a type known as *poppet valves*. These valves and their associated parts are normally part of the cylinder head, an arrangement known as *overhead valves*. A simpler arrangement having the valves in the block *(side valves)* does not permit a sufficiently compact combustion chamber for high-compression engines and gives poorer performance than overhead valves, in which the compression ratio is

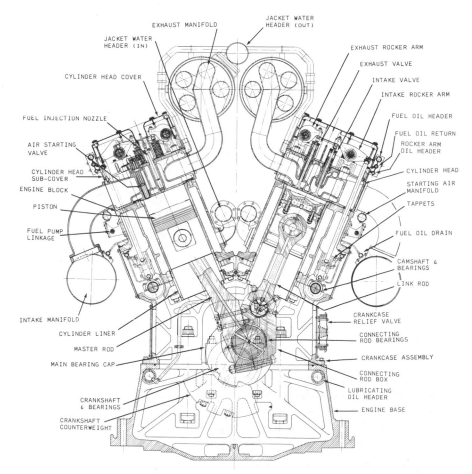

FIG. 6-2 Enterprise V-type four-stroke-cycle turbocharged and aftercooled diesel engine.

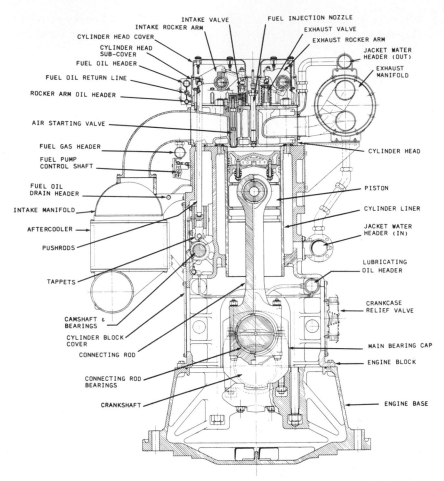

FIG. 6-3 Enterprise in-line four-stroke-cycle turbocharged and aftercooled diesel engine.

more suitable. The cylinder heads may have two, three, or four valves. Generally current high-output engines have four valves (two intake and two exhaust).

Engine Description

No matter what the engine classification, certain parts of any engine are basically the same, and their method of operating is essentially alike. Fuel, in gaseous or liquid form, is introduced into a power cylinder and ignited. The resulting expansion of the burning fuel mixture transforms the chemical energy of the fuel into mechanical energy by forcing a piston downward, which causes a crankshaft to rotate. This rotation is then utilized to drive some mechanical device, performing work. The basic parts of a gas or oil engine are:

1. Cylinder block
2. Cylinder heads
3. Pistons and connecting rods
4. Crankshaft
5. Valve mechanisms
6. Intake and exhaust manifolds
7. Fuel pumps or metering valves

Crankshafts may be bedded in the base of the engine and retained by shell-type bearings which are held in place with bearing caps. In some applications, crankshafts are underslung. The cylinder block provides support for the cylinder liners, pistons, and cylinder heads. The cylinder head usually contains the intake and exhaust valves and rocker assemblies and forms the upper part of the combustion chamber. It is normally water-cooled by jacket water circulating through passages in the head. The cylinder head may also contain fuel injectors (diesel and dual-fuel engines), gas admission valves (dual-fuel and spark-ignited engines), air-starting valves, and spark plugs (gas engines).

In V-type engines, connecting rods may be in either a side-by-side or a master-and-articulated-rod arrangement. In the side-by-side design, each piston has a connecting rod, and each cylinder pair is attached to the same crankshaft journal. The master-and-articulated-rod design uses a master rod which is connected to the crankshaft and a link rod which is attached to a link pin in the link box on the master rod.

Terms and Definitions

A great number of terms are used in connection with engines and their performance. Some of the more important ones are:

Horsepower (hp)   A unit of work rate. One horsepower in the U.S. customary system of units equals 550 ft·lb/s (33,000 ft·lb/min). The metric horsepower equals 75 kg·m/s, which is slightly less than the U.S. customary unit.

Brake Horsepower (bhp)   The brake horsepower of an engine is the horsepower delivered to the driven equipment by the engine output shaft. The relationship between brake horsepower, torque, and output shaft speed is

$$bhp = \frac{2\pi \times T \times r/min}{33,000}$$

where T = output shaft torque, ft·lb

Indicated Horsepower (ihp)   Indicated horsepower is a measurement of the gross rate of work done on the piston by the gases in the cylinders. Its determination requires the use of an engine-cylinder-pressure indicator capable of producing a plot of cylinder pressure versus volume for a complete engine cycle. The net working pressure is called mean indicated pressure (mip) or indicated mean effective pressure (imep).

$$\text{ihp/cyl} = \frac{\text{imep} \times LAN}{33{,}000}$$

where L = piston stroke, ft
A = piston area, in^2
N = number of power strokes per minute
imep is given in lb/in^2

Friction Horsepower (fhp) Friction horsepower is normally defined as the difference between indicated horsepower and brake horsepower. It is a measure of power lost between engine cylinders and the output shaft. Part of this loss is caused by friction between engine parts; the rest is usually power required by the engine-driven auxiliaries such as oil and water pumps. A rough measure of friction horsepower can be determined by plotting the total fuel flow (lb/h) versus bhp. The curve should be extended to zero fuel flow, which will indicate a negative horsepower. This negative horsepower is the fhp.

Mechanical Efficiency The ratio of brake horsepower to indicated horsepower.

Brake Mean Effective Pressure (bmep) Brake mean effective pressure is calculated from the following formula:

$$\text{bmep} = \frac{\text{bhp} \times 33{,}000 \times C}{LANn} \qquad \text{lb/in}^2$$

where C = 1 for 2-cycle and 2 for 4-cycle
L = piston stroke, ft.
A = piston area, in^2
N = number of revolutions per minute
n = number of cylinders

Thermal Efficiency The thermal efficiency of an internal-combustion engine equals the ratio of energy (work) output rate over energy (heat) input rate. Input and output are usually expressed in Btu per time unit. Fuels used in engines have both a low and a high heating value; it is therefore proper to designate which value is being used. Normal custom for oil and gas engines is to use the low heating value for both liquid fuels and gaseous fuels. Brake thermal efficiency is based on the brake-horsepower output; indicated thermal efficiency, on the indicated horsepower.

Brake Thermal Efficiency Brake thermal efficiency divided by indicated thermal efficiency equals the mechanical efficiency of the unit.

The theoretical cycle-efficiency relationship to the compression ratio for the Otto cycle (spark-ignition engine) is

$$\text{Cycle efficiency} = 1 - \left(\frac{1}{r}\right)^{k-1}$$

where r = compression ratio
k = isentropic constant for working medium (1.4 for air)

The actual relationship of compression pressures and brake mean effective pressure to thermal efficiency cannot be expressed by a formula as it varies from engine to engine and from condition to condition.

Brake-Horsepower Guarantees   Brake-horsepower capacity and fuel-consumption guarantees are usually made by the manufacturer, either on the basis of tests conducted on development and/or original models of an engine or as the result of factory tests conducted on an engine prior to shipment. Guarantees for naturally aspirated engines operating below 1500-ft altitude are generally made contingent on an atmospheric pressure of not less than 28.25 inHg and an air intake temperature not exceeding 90°F. The bhp capacity will be reduced on naturally aspirated engines operating above this altitude or temperature. Guarantees for American turbocharged and aftercooled engines are based on site conditions regardless of elevation and ambient temperature or pressure. Most European engines are rated in accordance with the ISO 3046/1-1975 standard.

Limited-Speed-Range Characteristics of Stationary Units   Most stationary engines are used to drive equipment with a limited range of speed requirements, such as generators, pumps, and blowers. Therefore, the need for a wide speed range is not significant, and the engine can be designed to operate near the optimum point for best engine efficiency. An exception is the fixed-pitch-propeller application for main-engine marine service.

Speed-Power Relationships   In general, the higher the required power output, the larger the engine and the slower the engine speed. High-speed engines are usually smaller and have lower power outputs.

Engine Performance and Rating Limitations

Engine power ratings are determined by certain practices and usages which provide a yardstick for the measurement of engine performance, forming a common ground of reference for the manufacturer and the purchaser of engines. The type of engine must be considered when determining how it will be rated. Important factors are bmep, engine speed, and piston speed. A naturally aspirated engine will present certain problems which are not significant when a turbocharged engine is considered. The naturally aspirated engine is usually limited by its air supply and may be rated at the power output at which it first shows evidence of exhaust smoke. On the other hand, the turbocharged engine may be limited by the air supply that can be retained in the power cylinders (degree of turbocharging if excessively high). Engine-strength limitations, or the point at which an engine will produce the greatest power output without any harmful mechanical or thermal stresses, may determine the power rating. The speed rating is based on mechanical stresses and the piston's and the piston ring's ability to receive adequate lubrication and properly seal combustion gases. Excessively high piston speeds will prevent lubricants from reaching vital areas such as the ring-belt area. Another factor to be considered is the bearing loading limits.

If a gas or a dual-fuel engine can be adequately cooled, detonation will become the primary factor in determining engine rating. The gas used, ignition timing, the air-fuel mixture, the combustion-chamber design, engine speed, and inlet-air-manifold temperatures are some of the factors that will affect detonation in a gas-fueled engine.

Effects of Supercharging and Inlet-Air Cooling and Ratings   Supercharging and/or cooling of the inlet air will improve the engine rating by placing a denser air charge in the power cylinders, which with additional fuel will increase the power output of a given size of cylinder. In rating an engine, performance is calculated at an elevation of not more than 1500 ft and a temperature not exceeding 90°F. Supercharging will improve performance at sea level and will compensate for loss of atmospheric pressure above this altitude. Additionally, it will provide more air during the scavenging period, after the power stroke, thereby providing improved cooling to the cylinder surfaces and the exhaust valves.

Effects of Altitude on Power Ratings   Altitude has little effect on a turbocharged engine if the turbocharger is properly applied; however, for naturally aspirated engines an increase in altitude will result in decreased power output. As the altitude increases, air density decreases, resulting in a reduction of intake-air charge density. The same effect is true for increased temperatures.

Load and Fuel Relationship to Supercharging   Supercharging allows for a greatly increased bmep rating and usually will result in a slightly improved specific fuel consumption. If the engine is placed in an overloaded condition, however, the specific fuel consumption may increase, and economy of operation will be degraded.

Relationship of Continuous Rating and Overload Operation to Thermal Efficiency   When a diesel engine is operating in an overload condition, the thermal efficiency may be reduced, and this may also be the case with dual-fuel and spark-ignited engines. Engines are usually designed to operate at their best thermal efficiency at 75 percent or more of their continuous rating.

Effects of Operation above Continuous Rating   If an engine is operated for any great length of time above the continuous-rating power level, the higher mechanical and thermal stresses will cause abnormal wear and possible damage to the engine, as well as a loss in operational economy.

Economic Relationship of Unit Price to Fuel Consumption   As in other situations, the application of the engine will be the controlling element of the unit-price–fuel-consumption ratio. If the engine is going to be in continuous or nearly continuous service, the rate of fuel consumption is a large factor in the selection of an engine. On the other hand, if the engine is to be used infrequently, as the prime mover for standby equipment, emergency auxiliary use, or such, then the rate of fuel consumption is of little practical importance as related to the unit price of the engine, and the engine which proves least expensive while still providing the capability to fulfill its intended use would be the logical selection.

Torsional Vibrations and Critical Speeds

When a torsional pendulum consisting of a length of shafting fixed at one end and carrying a heavy disk at the free end is given an angular displacement, a restoring moment is induced in the shaft and tends to return the disk to the original equilibrium position. The frequency at which the disk oscillates when released is known as the *torsional natural frequency*. For every shaft there is a unique stiffness, and for every disk a unique inertia; when they are combined, they possess a unique torsional natural frequency.

If the disk is displaced once and then released, it will oscillate at that amplitude repeatedly provided there is no damping. In real life, however, there are in existence friction and other damping effects which will cause the amplitude to decrease progressively until the disk comes to its original equilibrium position. This is known as *damped vibration*.

Since we are interested mainly in reciprocating engines (which include diesel, gas, dual-fuel, steam, and gasoline engines), we will describe only the torsional mass-elastic system of this machinery. The system may be considered to consist of disks of inertias due to the front-end drives, the individual crankthrow with its rotating and reciprocating masses, the flywheel, and the driven equipment. Between each of these inertias are the connecting shafts, each with its own distinct torsional stiffness. As with any mass and spring system, there exist numerous modes of natural frequencies. One group of inertias may oscillate against the remainder at one frequency, and two groups of inertias at each end may oscillate against the group in the middle at some other frequency. The highest natural frequency will be that when each inertia would oscillate against the others. In practical terms, only the lower modes of natural frequency are of interest.

Owing to the crank configuration and firing order of reciprocating engines, there exist various harmonics which, depending on the operating speed, cause torsional excitations of a certain frequency. Take, for example, a six-cylinder four-cycle engine. Depending on the firing order, some harmonics are more prominent and others less prominent. Because of the complexity of this phenomenon, we will not explore it here, but it is easy to visualize its major orders of the third, sixth, and ninth harmonics, since it has three firings per revolution. When the engine is operating at, say, 600 r/min, the three firing pulses will cause an excitation of 3 × 600 = 1800 vibrations per minute (v/min). If the natural frequency of the system is 1800 v/min or near 1800 v/min, operation at 600 r/min will cause a resonant condition with the third harmonic, or third order, critical. The effect of resonance is large amplitudes of torsional oscillation. Since the excitation will persist, the large amplitudes experienced will be continuous, resulting in torsional-fatigue failure of the crankshaft.

To avoid exciting the natural frequency, two things can be done. One is avoid running at 600 r/min as in the above example. If the operating speed is changed to 500 r/min, the third harmonic will excite only at a frequency of 1500 r/min, which is nowhere near the natural frequency of 1800 v/min. As mentioned before, we are faced with numerous other harmonics, so when the operating speed is changed, it is important to avoid getting into resonance with another harmonic. The alternative is to modify the mass-elastic system so that we do not have a resonance at 1800 v/min as in the above example. This can be accomplished by changing some of the shafting stiffness or the inertias. Since the engine usually is standardized, these stiffness and inertia changes can be realized only with components external to the engine. In the real world, as in a generating plant, the operating speed is not usually changeable; therefore modification of the mass-elastic system is the usual approach to avoid resonance.

Torsional vibration is different from other forms of vibration in that there may be no visible or audible indication that dangerous vibrations are present. Furthermore, an engine-generator set with dangerous torsional vibrations may not show any defects during assembly inspection, yet shaft failure may occur after many hours of operation. It is apparent that when applying a reciprocating engine to drive any equipment, the torsional natural frequencies for the system proposed must be thoroughly and rigorously analyzed. The engine builder is normally charged with this responsibility, although the mass-elastic information of the driven equipment must be furnished to the engine builder by the manufacturer of the driven equipment. The user of the equipment must not alter the arrangement of the total system by modifying shafting, changing couplings, or altering either the driving or the driven equipment. Likewise, the operating-speed range must not be changed. If such modifications must be made, the details of the changes should be sent to the engine builder so that a thorough analytical investigation can be made to ensure feasibility and subsequent satisfactory operation.

Engine Design

Combustion Chambers   The engine builder will design combustion chambers to meet the requirements of the particular engine. In general, diesel and dual-fuel engines have combustion chambers to make maximum use of the air charge, while spark-ignited engines have chambers designed to minimize or suppress detonation tendencies of the fuel.

The shape of the space forming the combustion chamber when the piston is at its closest approach to the cylinder head and the volume contained therein in relation to the piston displacement volume are very important in their effects on performance. The volume of the combustion chamber relative to the volume displaced by the piston establishes the compression ratio of the engine, which is an important factor in the starting characteristics of a diesel engine. High compression ratios also tend to increase the thermal efficiency of engines, but this advantage must be weighed against the detrimental effects of high maximum combustion pressures and the possibility of encountering conditions which result in detonation of gaseous fuels or excessive mechanical stress in rods and bearings on diesels.

The compression ratio

$$R_c = \frac{sv + cv}{cv}$$

where R_c = nominal compression ratio
cv = clearance volume
sv = swept volume

Valve Arrangements   Two configurations are in general use: a two-valve head and a four-valve head, the latter becoming more widely used on high-output engines to overcome the breathing restrictions present with two-valve arrangements. The prevailing design is the overhead valve, actuated by rocker arms and pushrods. In a four-stroke-cycle engine the exhaust valve is held open during the exhaust stroke, following the power stroke, and the inlet valve is opened on the next downward, or intake, stroke. The valves remain open for longer periods than the corresponding piston strokes to provide time for opening and closing and to utilize the inertia effects of the high velocity at which the gases flow. The opening and closing points of both valves are therefore displaced somewhat from dead center. The exhaust valve is timed to open before the piston reaches bottom dead center of the power stroke. This provides a blowdown period to release the gases from the cylinder and reduce the pressure approximately to exhaust manifold pressure by the end of the power stroke. Some useful work near the end of the power stroke is lost by this early valve opening, but additional negative work during the exhaust stroke is avoided by not requiring the piston to ascend against back pressure. The exhaust valve remains open past top dead center, allowing the kinetic energy of the outflowing gases to reduce the pressure within the combustion chamber below atmospheric pressure and decrease the exhaust gases remaining that would dilute the next charge. The inlet valve is opened before top dead center to take advantage of the pulling effects of the exhaust gases at the end of the exhaust period. An overlap results between the opening of the intake valve and the closing of the exhaust valve which allows scavenging of the combustion chamber of gases with fresh air from the intake system and also provides cooling of the piston crown and cylinder-head fire deck.

Piston speed. Piston speed is determined by

$$\text{Piston speed} = \text{stroke in feet} \times \text{r/min} \times 2$$

At piston speeds for which the frictional losses are the least, the engine is operating below its point of maximum specific output; therefore, a compromise is made in design to achieve a combination of high output with an acceptable frictional loss. Constant-speed heavy-duty engines are now designed with piston speeds up to approximately 1800 ft/min, and variable-speed high-output engines will operate with even higher piston speeds. Engines running at very high piston speeds will normally have shorter lifetimes than more conservatively rated engines because of higher wear rates.

Supercharging Supercharged intake-air systems are the predominant systems for medium- and large-sized engines and exist to a great degree in small engines. Supercharging provides the engine with combustion air at a greater density than that of the surrounding air and, with increased fuel input, provides an increased power output for a given size of cylinder.

The turbocharger is a self-contained unit composed of a gas turbine and a centrifugal blower attached to a common shaft. The exhaust gas from the power cylinders of the engine is discharged through manifolds to the turbine, which makes use of some of the energy in the exhaust gas that would otherwise be wasted. This salvaged energy is used to drive the blower, which furnishes all the air required by the engine at a pressure above atmospheric.

There are two general classifications of supercharging systems: the exhaust-gas turbine-driven blower and the mechanically driven blower. In the former, the supercharger is driven by the exhaust gases from the engine through a turbine which is shafted to a compressor. The mechanically driven blower may be driven from the engine or separately by an electric motor or other prime mover. Turbine-driven blowers may be of a constant-pressure or a pulse type, or of a combination thereof. The constant-pressure type receives the entire exhaust output through its turbine at a constant pressure, whereas the pulse type receives its driving force through multiple pipe manifolds, thereby receiving a series of pulses to drive it. The pulse converter, or multipulse system, makes use of multiple pipe manifolds from the cylinder to a point near the turbine inlet, where by means of nozzles and a diffuser the pulses are smoothed out and fed to a constant-pressure turbine.

Pressure-charging and scavenging a four-cycle diesel or gas engine accomplishes two purposes. First, it scavenges out the hot residual gases otherwise left in the cylinder at the end of the exhaust stroke and replaces these with cool, fresh air. Second, it fills the cylinder with an air charge of higher density at the end of the inlet stroke. The provision of a greater amount of fresh air permits the combustion of a greater amount of fuel and, consequently, a higher output from a turbocharged engine than from one not so equipped.

The valve timing of an engine arranged for pressure charging differs primarily from that of the same engine normally aspirated in that the exhaust valves of the pressure-charged engine close later and the inlet valves open earlier. Thus the valve overlap, or period when both valves are open, is considerably greater, permitting nearly perfect scavenging when the piston is near top dead center. Timing of the valves and dimensions of the exhaust manifold in the Buchi system are so proportioned that timed pressure fluctuations are induced in the manifold. Both valves are open when the pressure in the exhaust manifold is at a minimum, thus permitting scavenging with a lower blower pressure than would otherwise be possible. The constant-pressure turbocharging system requires uniform flow through the turbine at steady pressure. In its simplest form, exhaust from all the cylinders is collected in a large chamber and connected to a radial in-flow or axial-flow turbine. Features of the three different systems are as follows:

The pulse, or Buchi, system is very rapid in response, allowing very fast load or speed changes to be applied to the engine.

The constant-pressure system at designed conditions at higher bmep may be more efficient than the Buchi system, resulting in better specific fuel consumption and lower exhaust temperatures. However, as the operating point of power output or speed is reduced from the design condition, efficiency may fall well below that of the Buchi system. Further, transient response is poor even at design conditions.

The pulse converter, or multipulse system, attempts to achieve the best features of the other two systems. Efficiency is improved over a wide range, and transient response is only slightly less than that of the Buchi system.

Scavenging the combustion space with cool air effects a degree of cooling of the cylinder head, cylinder walls, valves, and pistons. Since more air is trapped in the cylinder, a greater amount of fuel can be burned and greater power developed without harmful effects on engine parts due to excessive heat.

No control over the turbocharger on diesel engines is necessary, as the correlated action of the turbine and blower is entirely automatic. On spark-ignition or dual-fuel engines, control of the air-fuel ratio is required, usually by throttling turbocharger air delivery or by slowing down the turbocharger by bypassing exhaust gas around the turbine. This latter method of reducing air delivery is called *wastegating*. The speed and output of the turbocharger vary automatically with variations in load or speed, or both, of the engine. No consideration need be given to the direction of rotation of the turbocharger when applied to a direct-reversing engine. The turbocharger rotates in one direction only regardless of the direction of rotation of the engine.

Firing Order   Engine firing order is influenced by the requirements for crankshaft balance. The position of each throw along the crankshaft will determine the firing order, depending on the direction of engine rotation. Firing sequence is chosen to distribute the power impulses along the length of the engine in order to minimize vibration. Consideration is also given, when selecting firing order, to the fluid-flow pattern in the intake and exhaust manifolds. Balancing of the crankshaft may be further improved by adding counterweights to the shaft to offset the eccentric masses of metal in the crankthrows.

The arrangement of the throws of the crankshaft for a multicylinder four-stroke-cycle engine is chosen with regard to the firing order of the cylinders and the angular interval between firing impulses. The average number of degrees through which the crankshaft must turn between successive firings is found by dividing 720° (the two revolutions completed by the crankshaft during a complete cycle) by the number of cylinders for an in-line engine or by the number of cylinders in a single bank for a V-type engine. Regardless of the number of pistons in a four-cycle engine (even numbers only being considered), two pistons must arrive at top dead center in unison so that one cylinder is ready to fire 360° after the other cylinder fires. Half of the cylinders will fire on each revolution of the crankshaft.

Air-Fuel Mixture Requirements   Dual-fuel and spark-ignited gas engines usually operate with air-fuel mixtures on the lean side of the ideal (stoichiometric) ratio. This is done to reduce the cylinder exhaust temperatures and to reduce the tendency toward detonation.

Diesel engines operate with air-fuel ratios much leaner than the stoichiometric ratio. This is necessary because the nonhomogeneous distribution of fuel in the combustion air makes a large surplus of air necessary in order to have complete combustion of the injected fuel.

Ignition Systems   Most ignition systems are of either the magneto or the solid-state electronic type, the latter coming into more general use as designs are improved. Reliability, good spark-plug life, firing accuracy, proper voltage output, and cost are the primary considerations for an ignition system, and the requirements for a particular installation will help determine the best type to use. A conventional magneto system poses problems owing to erosion and wear on the breaker points and to difficulties involved in providing ignition wiring that will withstand the stress and wear placed on it with attendant tendencies for high-tension circuits to create safety hazards. The breakerless magneto eliminates most of the point problems by using breaker points only for starting and then using pulse generators to provide the necessary voltage to the spark plugs.

The solid-state ignition system eliminates many of the problems found in magneto systems. A typical solid-state system uses an alternator to provide current, which is then rectified to charge a tank, or firing, capacitor and a smaller trigger capacitor. A trigger vane rotates with the drive shaft, and as it passes a trigger coil (one for each cylinder), it induces a small current which gates a silicon control switch. This allows the trigger capacitor to discharge, triggering the gate on a silicon control rectifier. The voltage from the tank capacitor then discharges through the primary winding of the transformer, which is mounted on or near the spark plug. This induces the necessary high voltage in the secondary windings of the transformer to fire the spark plug. This system eliminates long high-tension leads by mounting the transformer on or near the spark plug.

Starting Systems   Many methods are used to bring engines up to the speed necessary for starting. Compressed air is usually employed, and in such cases opening the air-starting valves automatically admits compressed air to the power cylinders in a predetermined "firing order" sequence.

Medium- and large-sized high-speed engines may be brought up to starting speed by various methods such as an attached electric motor, air motor, or gasoline engine. Engines driving direct-current generators are sometimes cranked by motorizing the generator from a separate source of direct current.

Air starting. This system is the one most often used in low- and medium-speed engines. Most engine builders design their engines for starting air pressures of 250 to 520 lb/in², although some older engines used pressures ranging up to 1000 lb/in². In this system, compressed air from a suitable source is piped to the engine and flows to the cylinders through valves controlled by an air distributor. Air is admitted to the cylinder through an air admission valve and forces the piston down, turning the crankshaft.

Motor starting. Electric motors or air motors driven by either compressed air or fuel gas are geared to the engine to rotate the crankshaft until the engine is brought up to starting speed. On large engines, a number of motors may be used.

Automatic starting. In some installations, particularly when the engine is driving standby or emergency pumps, generators, etc., it may be necessary to have automatic-starting provisions incorporated in the engine. In systems employing a direct air admission system, only compressed air may be used. The introduction of flammable gases or oxygen may well result in an explosion because of gas in the power cylinders and in the exhaust system.

Cooling-Water Systems   Cooling-water systems are required to maintain the engine at a desired operating temperature by cooling the cylinders and cylinder heads and to provide a cooling method to remove heat from lubricating oil and intake air. The water system used for cooling the engine proper is normally referred to as the *jacket-water system,* and the one used for cooling lubricating oil and combustion air is called the *cooling-water system.* Water supplied to the cold side of jacket water, lubricating oil, and aftercooler heat exchangers is called *raw water.* For diesel engines, a single-loop cooling system, in which the jacket water is also used to cool the aftercooler and the lubricating oil, can be used.

Many design considerations must be observed when specifying the particular system to be used. The system should be designed so that all the water flows through the engine at all times. The water flow should never be throttled to raise the outlet temperature. To maintain a uniform heat transfer and a high efficiency of cooling through the engine jackets, only soft water or treated water should be circulated through the engine. The character of the water should be such that there will be no deposit in the water spaces. The water should be free of corrosive properties, and a pH value between 8.25 and 9.75 should be maintained for minimum corrosion.

In designing a cooling-water system and in selecting equipment for it, many factors must be considered, such as the following:

1. *Water supply.* What are the hardness and corrosive characteristics? Is the supply ample or limited? Is it seawater, river water, or lake water? Is it brackish?

2. *Atmospheric conditions.* What are maximum and minimum wet- and dry-bulb temperatures? What are the average wind velocity and its direction? Is the locality subject to dust storms?

3. *Space available.* Can all the equipment be installed indoors? In either case, how much space is available to accommodate the cooling equipment?

4. *Heat recovery.* Is it desirable to recover heat from the jacket water or exhaust muffler for space heating or for other uses around or near the plant?

Equipment to dissipate heat from jacket water has become highly specialized. The amount of heat to be removed from engine jackets and from the lubricating oil and the limiting water temperatures should always be obtained from the engine manufacturer. It is then the cooling-equipment manufacturer's entire responsibility to supply adequate heat-exchange equipment to dissipate the specified amount of heat for actual service conditions during the life of the equipment.

There are two general types of cooling systems, the open system and the closed system.

Open system. In this system water is drawn from an atmospheric source such as a cooling-tower basin, river, lake, or sea, pumped through the heat source, and then returned to the atmospheric source. It is an open-ended piping system.

Closed cooling system. In this system water is pumped through a heat source and on to a heat exchanger such as a radiator or shell-and-tube cooler and then back to the pump suction. This is truly a closed-loop piping system.

Jacket-water systems. There are two basic jacket-water systems, a single-loop and a two-loop system. The open cooling system is not recommended for either of these jacket-water systems. The recooling of water by cooling tower or spray pond is a process that depends upon evaporation. Continued evaporation and addition of makeup water increases the concentration of either hardness or impurities and causes the water to become increasingly objectionable for use in the engine jackets.

Scale-forming materials will not remain in solution but will be deposited on the jackets of the engine, retarding the transfer of heat from the metal walls to the cooling water. This is detrimental to the engine; therefore, only the closed-loop type of system is recommended for the jacket-water system.

Single-loop system. This type of system as shown in Fig. 6-4 is a typical single-loop cooling system, in which the jacket water is subcooled in the heat exchanger and then passed through the lubricating-oil cooler and the aftercooler, which are connected in parallel; the flow is then combined before it enters the engine. This system can be used only on a diesel engine application in which higher intake manifold temperatures can be used. In designing this system care must be taken to ensure that the correct temperature differentials and flows exist, and these parameters must be specified by the engine builder. Advantages of this system are that it minimizes the amount of piping and pumps and eliminates the need

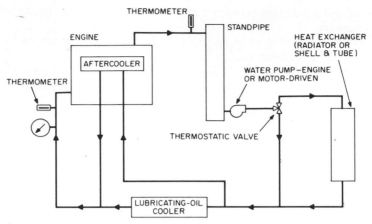

FIG. 6-4   Single-loop cooling-water system.

for a lubricating-oil thermostat. A disadvantage is the lack of flexibility on changing operating temperatures in portions of the system. The heat exchanger for this system may be either a radiator, using air to cool the water, or a shell-and-tube heat exchanger using cooling-tower water or lake, river, or sea water as the raw-water source.

Two-loop system. The two systems are shown in Fig. 6-5. The jacket-water system is a closed loop in itself. In this system the engine is the sole heat source. The water is pumped through a thermostatic valve to control the engine outlet temperature, specified by the engine builder, then through the heat exchanger, and back to the engine. The heat exchangers that can be used for this system are the same as those described for the single-loop system.

Both the jacket-water systems described above have a standpipe, which serves

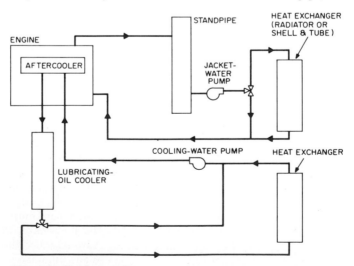

FIG. 6-5   Two-loop cooling-water system.

as (1) an expansion tank and (2) a deaeration tank and (3) provides positive suction to the jacket-water pump.

The second loop is a separate independent system of the closed-loop design that may be used to furnish a medium for aftercooler and lubricating-oil cooling. This system is not directly an engine-cooling system but indirectly provides engine cooling by removing heat from the lubricating oil after the oil has removed heat from pistons, bearings, etc. The heat exchanger used for this system depends on the application, such as diesel, dual-fuel, or gas, regard being given to manifold temperature requirements. This circuit can use either an open system or a closed system since both the aftercooler and lubricating-oil cooler are cleanable.

Heat recovery. Heat may be removed from the cooling water for use in space heating or for other purposes. The outlet water from the engine is passed through radiators or other heat-exchanging devices before being cooled further for return to the engine.

Exhaust-heat recovery is also becoming more and more attractive as a means of increasing overall system efficiency, particularly in view of the current cost of fuels. Such heat-recovery systems, called *total energy systems*, can achieve overall thermal efficiencies of 75 percent or higher.

Lubrication Systems

In oil and gas engines the primary function of the lubricating oil is to prevent contact between moving parts, to cool critical areas, and to remove contaminants from within the engine. Oil is drawn from a sump, either the engine base itself or an external tank, by a motor- or engine-driven pump, then forced through a cooler-filter system and introduced into the engine system at a controlled pressure (see Fig. 6-6). It is led through drilled passages or tubing into the crankshaft to lubricate main and connecting-rod bearings. On large units the connecting rods contain an axial passage which conducts oil to the piston pin and the underside

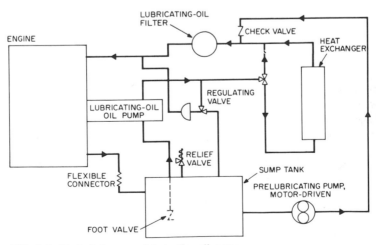

FIG. 6-6 Typical dry-sump lubricating-oil system.

of the piston crown, where combustion heat in the piston is removed. Lubrication of cylinder liners and piston rings is of prime importance. If oil is allowed to pass the piston rings in excessive quantities, harmful deposits may result from combustion of some oils; this also increases oil costs.

Auxiliary tubing leads oil from the main inlet header to the camshaft bearings, cams, tappets, rocker arms, gears, and various accessory drives. Oil normally returns to the engine base and/or sump tank by gravity flow; however, some marine engines may require a return or scavenging pump to accomplish this.

The main lubricating-oil pump may be engine-driven or motor-driven. On a constant-speed engine, an engine-driven pump is generally used, whereas on variable-speed engines motor-driven pumps are most often called for. Direct-reversing marine engines almost always use electric-motor-driven pumps. The lubricating-oil filters can be either simplex or duplex, depending on the application. There should also be a suction strainer between the sump tank and the lubricating-oil pump; it may either be built into the sump tank or be a separate basket strainer installed in the line between the sump and the pump.

Controls

Engine-governing systems for oil and gas engines may be broadly divided into two categories: those which regulate engine speed only, regardless of external conditions, and those which control not only speed but load as well to achieve a preplanned program of power output in response to one or more external conditions or demands.

Governors for medium-speed engines, because of the force level required to regulate fuel input, usually employ a highly responsive speed sensor. Normally, this is a flyball configuration which has very little force output but which operates a servosystem, normally hydraulic, that can deliver sufficient force to operate the fuel control. In a diesel engine equipped with individual fuel-injection pumps, the governor servo acts through linkage to rotate the injection-pump plungers. By means of helical grooves on the plungers, the length of the pumping portion of the plunger stroke, and hence the quantity of fuel delivered, is controlled. In addition to the regular helical groove, a second (timing) helical groove can be located at the top of the plunger directly above the regular helical groove. This helical groove has various configurations to meet varying engine requirements. Generally, it is designed to provide a retard in the fuel-injection timing on dual-fuel engines and at low revolutions per minute on marine diesel engines. On a gas engine, whether spark- or pilot-oil-ignited (dual fuel), the servo acts to control the position of a fuel-gas metering valve or valves, to control fuel-gas flow.

A speed-only governing system normally acts to maintain a desired steady-state speed regardless of external loading. When the unit's load capacity is reached, the fuel input will remain at maximum and the speed will drop. When transient or sudden loads are applied or removed, there are small temporary speed increases or decreases, the magnitude of which depends on the engine and the rotating mass as well as on the governor configuration.

Alternating-current power generation presents a special situation when a standard hydraulic governor is used. When two or more alternators are operated in parallel, that is, connected to the load, only one unit can control the frequency of the system if stability is to be maintained. Therefore, a function termed *speed droop* is built into governors for such applications. The effect of the speed-droop system is to cause the governor-speed set point to decrease as load is applied; that is, as the servo acts to increase fuel, it also, through linkage, acts to reduce speed setting. Since the unit is tied to a constant-frequency load, it cannot change speed; therefore, it tends to operate at a fixed load, with the frequency-controlling unit, which has no speed droop, accepting all system-load changes. If automatic load-change sharing is to be obtained on all plant units using this system, line-frequency deviation must be accepted since all units must operate with speed droop.

Speed-load governing systems are generally used in applications in which the engine speed as well as the load is a variable. Controllable-pitch propellers, compressors, pumps, direct-current generators, and mechanical drives are examples of this type of application. The governor is arranged to provide an output signal, either electric or hydraulic, which bears a predesigned relationship with engine speed and/or load and which operates to control this load. By careful design of the system, a power output which will take best advantage of engine performance characteristics and load requirements can be automatically obtained. Modifications to the power program can be achieved automatically in response to atmospheric or subsystem performance changes.

A special case is that of the electrohydraulic governing system used for alternator service. It consists basically of an electrical-load- and/or frequency-sensing device, a computer which translates these inputs into a single electrical output, and an electrical device which operates the governor servo. With this system, more than one alternator may be operated in parallel, with all units sharing system-load changes equally and with no basic speed change. Automatic start and synchronization may be accomplished, with the oncoming unit automatically assuming its programmed share of the total load. This is an extremely fast and stable method and may be used when precise control of frequency is mandatory, as for computer or radar service.

Governor-System Performance   The performance of a governing system depends not only on the type of governor used but equally upon the configuration of the engine and the governor match to that engine. The basic system parameters are control of basic steady-state speed, the amount of speed change during a load change, and the rapidity of return to the set speed after such a load change has occurred. The response of the governor servosystem itself varies with type, the mechanical-hydraulic type generally being the slowest and the electric governor the fastest. The engine type, the time required to convert a fuel-input change to a torque-output change, the inertia of the rotating assembly, the basic speed, and the number of cylinders all influence the total response time of the system. For example, a new fuel setting will result in a new torque output in a diesel engine much sooner than in a gas engine because fuel is admitted on inlet stroke in a gas engine, whereas the diesel engine produces torque from the fuel almost as soon as it is injected. Furthermore, the fuel-gas-admission system in a gas engine requires time to fill and drain in response to the governor action. The more cylinders an engine has and the higher its basic speed setting, the less time is required to produce a new total torque output from a fuel-input change and the faster its response. The inertia of rotating parts is quite important. Larger flywheels will reduce the speed excursion but increase recovery time. When turbocharged engines accept very large load changes, the inertia of the turbocharger becomes important since it requires time to produce enough air to allow the new fuel input to burn.

Since a diesel engine with an electric governor can control speed more precisely than any other prime mover, it is by far the best choice when such control is required. The unit can be designed to achieve control such as absolutely no reduction of basic speed through the load range, ± 0.3 percent control of steady-state speed, 1-cycle speed deviation for an instantaneous 50 percent load change, and recovery time in the 1-s range. It may do this in parallel or in isolated operation.

The gas engine, equipped with mechanical-hydraulic governing such as is normally used in municipal or industrial power generation, is somewhat wider in control range but still much faster than water, steam, or gas-turbine units.

Although the controls installed on any particular engine will vary with the type, model, manufacturer, and requirements of the installation, certain controls, discussed below, are common to nearly all engines.

Overspeed Governor   Engines may be equipped with an overspeed trip which automatically shuts off fuel or otherwise stops the engine if it operates at a speed in excess of some predetermined value. This overspeed governor should be separate from the regular speed governor.

Fuel-Selector Control (Dual-Fuel Engines)   This control mechanism permits selection of either diesel or dual-fuel mode of operation. Some controls permit selection of any proportion of fuel, ranging from dual-fuel with minimum pilot oil to full diesel operation (see Fig. 6-7).

Fuel-Air-Ratio Control   This control is used on dual-fuel and spark-ignited engines to control the volume of intake combustion air to achieve the necessary ratio of air to fuel (see Fig. 6-8).

Ignition Control (Spark Advance)   This control regulates ignition advance relative to engine speed. On engines so equipped, automatic control is possible. Ignition advance may be manually set, automatically set, or preset, depending on the particular requirements of the installation.

Alarm and Safety Shutdown Controls   These controls provide for visual and/or audio alarm and shutdown of the engine when some malfunction, such as excessive temperatures or insufficient pressures, occurs. A wide variety of control features are available, and the extent of an installation will depend on the needs of the plant and the desires of the operator. Most systems utilize pneumatic relays coupled to sensing devices such as pressure switches, vibration switches, and temperature-sensing elements. The sensing elements are set for predetermined values, and if these values are met, they will initiate an automatic shutdown of the engine by cutting off the fuel or by other means. The plant may utilize an audio alarm system which will sound at a point before shutdown occurs, thus giving the operator an opportunity to correct the condition before the engine is shut down and so prevent loss of engine-operating time. A system of pneumatic relays, connected in a series circuit and pressurized with air, gas, or oil, is installed. Each relay is controlled by a sensing device for a particular function. If a sensing device detects

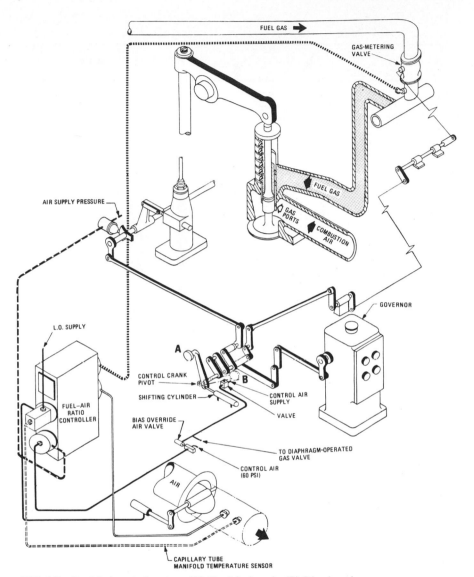

FIG. 6-7 Dual-fuel control system. **(A)** Dual-fuel mode. **(B)** Diesel mode.

an unsatisfactory condition, it will trip at the preset value, causing its relay to vent. This releases pressure on the system; fuel will be cut off, and the engine will stop. A feature of this system is that the relay involved will indicate itself, informing the operator as to the cause of the stoppage and shutdown. The overspeed governor normally is connected to this system, and if the engine overspeeds, the governor will trip a sensing device, which in turn vents its relay, and the engine is shut down.

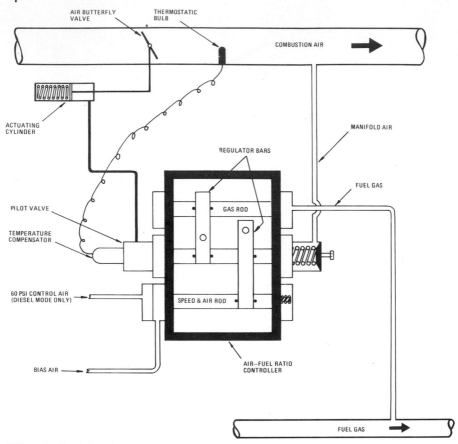

FIG. 6-8   Fuel-air-ratio control.

Starting Controls   Engine-starting controls are provided, the exact installation depending on the requirements and specifications of the engine. Because of the different starting methods available, there is no standard arrangement.

Temperature Controls (Thermostatic Valves)   Various arrangements may be designed to achieve temperature control of cooling water and lubricating oil. Thermostatically controlled valves can be used to modulate flow, depending on the temperature of the fluid, by directing the flow through heat exchangers, radiators, etc., when the temperature rises above a preset value.

Load Controls   There may be incorporated in an engine controls which will signal the driven equipment (particularly a compressor) that the engine is approaching an overload. This signal will activate control circuits or mechanisms in the equipment to cause it to unload before the engine is overloaded.

Lubricants

The high unit and thermal loading of various parts of the modern medium-speed engine imposes stringent requirements on lubricating oil. To fulfill these requirements, oil contains a rather large percentage of special additives. These are compounds which resist oxidation at high temperatures, resist foaming under severe agitation, impart great wear resistance to highly loaded areas such as cam lobes and tappet rollers, neutralize corrosive-acid compounds formed by combustion, and dissolve and/or carry in suspension to the filter the dirt and foreign matter that invariably find their way into the system. In addition to these requirements, the oil must maintain a film between moving parts and itself be noncorrosive to the wide array of materials used in the engine; it is also a primary cooling medium in many interior areas such as the piston crown, main bearings, connecting-rod bearings, and piston pins.

An additional requirement for oil used in gas engines is that the oil, when exposed to the heat of combustion, must not form deposits which build up on the combustion chamber. Such deposits form hot spots which may cause preignition in a gas engine and thereby lower the load-carrying ability of the engine.

The viscosity of an oil, or its resistance to shearing, primarily affects its ability to maintain a film (load-carrying ability) and its ability to flow into small clearances. Furthermore, the oil must not change its viscosity excessively with temperature changes. It must be free-flowing enough at low temperatures to allow the unit to be started with adequate lubrication, but when the unit is warmed up, it must not thin so much as to reduce its load-carrying ability. This characteristic of oils is called the *viscosity index* and normally is a property of the base oil stock. Additives that can improve the viscosity index are available. These are called *VI improvers*. Multiviscosity oils fall into this category.

Oils are classified by various agencies, generally in categories broadly describing overall oil performance. The American Petroleum Institute (API) has defined four types of service conditions normally encountered and has developed oil specifications meeting the requirements of each. United States military specifications (MIL specs) are frequently more important to large-engine owners. These are also rather wide descriptions of oils which have been found by test and experience to function acceptably under various service conditions. Table 6-1 indicates the various designations of lubricating oil.

It may be seen from the above that selection of a lubricating oil for a given unit is a complex problem, involving engine design, service conditions, fuels, and the properties of the oil. Therefore, it is normal for the owner, the lubricating-oil manufacturer, and the fuel-oil manufacturer to work closely together in selecting an oil for a particular job on the basis of the engine manufacturer's lubricating-oil requirements.

Fuels

Oil and gas engines, as their names imply, utilize two categories of fuel, liquid and gaseous. Within these categories, however, is a wide variety of fuel classifications. Since fuel, probably more than any other single factor, affects engine

TABLE 6-1

Letter designation	API engine service description	ASTM engine oil description
CA for diesel engine service	Light-duty diesel engine service: Service typical of diesel engines operated in mild to moderate duty with high-quality fuels. Occasionally gasoline engines in mild service have been included. Oils designated for this service were widely used in the late 1940s and the 1950s. These oils provide protection from bearing corrosion and from high-temperature deposits in normally aspirated diesel engines when using fuels of such quality that they impose no unusual requirements for wear and deposit protection.	Oil meeting the requirements of MIL-L-2104A, for use in gasoline and naturally aspirated diesel engines operated on low-sulfur fuel. The MIL-L-2104A specification was issued in 1954.
CB for diesel engine service	Moderate-duty diesel engine service: Service typical of diesel engines operated in mild to moderate duty but with lower-quality fuels which necessitate more protection from wear and deposits. Occasionally gasoline engines in mild service have been included. Oils designated for this service were introduced in 1949. Such oils provide necessary protection from bearing corrosion and from high-temperature deposits in normally aspirated diesel engines with higher-sulfur fuels.	Oil for use in gasoline and naturally aspirated diesel engines. It includes MIL-L-2104A oils when the diesel engine test was run with high-sulfur fuel.
CC for diesel engine service	Moderate-duty diesel and gasoline engine service: Service typical of lightly supercharged diesel engines operated in moderate to severe duty. Certain heavy-duty gasoline engines have been included. Oils designed for this service were introduced in 1961 and used in many trucks and in industrial and construction equipment and farm tractors. These oils provide protection from high-temperature deposits in lightly supercharged diesels and also from rust, corrosion, and low-temperature deposits in gasoline engines.	Oil meeting requirements of MIL-L-2104B. It provides low-temperature antisludge, antirust, and lightly supercharged diesel engine performance. The MIL-L-2104B specification was issued in 1964.
CD for diesel engine service	Severe-duty diesel engine service: Service typical of supercharged diesel engines in high-speed, high-output duty requiring highly effective control of wear and deposits. Oils designed for this service were introduced in 1955 and provide protection from bearing corrosion and from high-temperature deposits in supercharged diesel engines when using fuels of a wide quality range.	Oil meeting Caterpillar Tractor Co. certification requirements for superior lubricants (Series 3) for Caterpillar diesel engines. It provides moderately supercharged diesel engine performance. The certification of Series 3 oil was established by the Caterpillar Tractor Co. in 1955. The related MIL-L-45199 specification was issued in 1958.

performance, economy, and life, the selection of a fuel is of the utmost importance.

Gaseous Fuels   Natural gas. Of the gaseous fuels, natural gas is by far the most widely used. It is composed mainly of methane (CH_4) and has an octane rating of approximately 110 and a methane number (MN) of about 90. The MN for 100 percent methane is, of course, 100. The MN is a measure of gaseous-fuel knock sensitivity. One hundred percent hydrogen has an MN of 0 and cannot be run undiluted in an engine because of detonation intensity. Natural gas burns extremely clean, thus reducing contamination, corrosion, and abrasion of engine parts. Lubricating-oil life is significantly extended, with a resultant economy of operation. Since the combustion of natural gas produces less infrared radiation than that of most liquid fuels, thermal loading of combustion-chamber surfaces is less.

The source of supply is frequently a transmission pipeline or a commercial distribution system; therefore, installation costs are reduced by the elimination of storage requirements. However, natural gas for engines other than gas-compressor engines is becoming more difficult and costly to obtain.

Manufactured gas. The more commonly used fuels of this classification are propane, butane, and sewage gas. Two main factors affect the use of these fuels: octane rating and the inclusion of unstable components in the fuel. As propane has an octane rating of about 100 and butane about 92, the detonation limit and/or load-carrying ability for these fuels is somewhat lowered. This disadvantage can be offest to a degree by decreasing the compression ratio of the engine, but this is nearly always done at the expense of efficiency. Both fuels burn clean and have the same general advantages as natural gas.

Sewage, or digester, gas is a by-product of the sewage treatment process and makes an acceptable engine fuel with many of the advantages of natural gas. It is usually produced on the jobsite. Sewage gas contains little more than half the Btu content of natural gas. It requires higher admission pressures, and it displaces more combustion oxygen than does natural gas, and therefore an adjustment to the apparent air-fuel ratio is necessary.

In some instances manufactured gas contains unstable compounds which produce sporadic detonation at high loads. Sewage gas contains compounds of sulfur which may be corrosive to engine materials, particularly copper alloys.

Liquid Fuels   Diesel fuel is a term which encompasses a range of fuel oils suitable for use in diesel and dual-fuel engines. This type of fuel must be injected into the combustion chamber in an atomized state, vaporized, and then caused to ignite by the heat of compression. The overall ability to ignite promptly is measured by a value called the *cetane number,* the lower cetane numbers indicating fuels which are more prone to produce undesirable knock in an engine. Distillate diesel fuels are normally in the 40- to 60-cetane range.

The heat values of diesel fuels vary widely, and this variation may have a marked effect on load-carrying ability and/or thermal efficiency for any given engine. Fuels having less volumetric heat content produce less load-carrying ability for a given governor position and a higher fuel consumption on a pounds-per-bhp basis for a given load. The volumetric heat content is proportional to the API

TABLE 6-2

Gravity, API	Specific gravity at 60°F	Weight of fuel, lb/gal	Low heating value	
			Btu/lb	Btu/gal
44	0.8063	6.713	18,600	125,000
42	0.8155	6.790	18,560	126,200
40	0.8251	6.870	18,510	127,300
38	0.8348	6.951	18,460	128,500
36	0.8448	7.034	18,410	129,700
34	0.8550	7.119	18,360	130,900
32	0.8654	7.206	18,310	132,100
30	0.8762	7.296	18,250	133,300
28	0.8871	7.387	18,190	134,600
26	0.8984	7.481	18,130	135,800
24	0.9100	7.578	18,070	137,100
22	0.9218	7.676	18,000	138,300
20	0.9340	7.778	17,930	139,600
18	0.9465	7.882	17,860	140,900
16	0.9593	7.989	17,790	142,300
14	0.9725	8.099	17,710	143,600
12	0.9861	8.212	17,620	144,900
10	1.000	8.328	17,540	146,200

NOTE: It should be understood that heating values for a given gravity fuel may vary somewhat from those shown in these listings.

gravity index of the fuel, the higher-gravity fuels having less volumetric heat, as shown in Table 6-2.

Diesel fuels are classified by the American Society for Testing and Materials (ASTM) into groups bearing numerical designations: Number 1 fuel oil, Number 2 fuel oil, etc. In nearly all cases fuels of lower viscosity, higher cetane ratings, and less volumetric heat fall into the lower numerical classes. Number 1 fuel oil is used primarily for high-speed cold-starting engines; Number 2, for slower medium-speed engines, etc. Engine builders design and tune their product for one given fuel oil, and unless an engine is specifically modified for a different fuel, less than optimum performance will be the normal result if other than the recommended fuel is used.

With an ASTM Number 4, 5, or 6 fuel, we enter the range of residual, crude, heavy, or bunker fuels. *Residual fuels* are normally highly viscous, and when cold they will require heating for pumping, filtering, and injection. Since the cetane number is low, warmer combustion air is needed to reduce knock. However, the volumetric heat content is higher than that of distillate fuels. Residual fuels may contain undesirable contaminants, even salt water, and usually must be centrifuged and filtered to remove these contaminants before they are used. There may be present nonremovable compounds, notably sulfur, vanadium, and sodium, whose oxides are corrosive under certain conditions. Care in oil selection can sometimes reduce these undesirable effects.

The compatibility of different residual-fuel stocks is of particular concern to operators of marine installations. Certain fuels, particularly those which have been through a thermal-cracking process, may become unstable in storage.

Because marine operators may receive their fuel supply from many different sources, it is very important that they select a fuel that will not react unfavorably with the fuel already on board in bunkers. Crude, or unrefined, fuels, generally are very good fuels, requiring only heating and filtering.

Applications

The process of selecting an engine which will be suitable for an application must take into consideration a number of factors in addition to the engine's ability to produce the required horsepower. Other major factors are output shaft rotational speed, fuel requirements, physical size, weight, and service record.

Estimates of first costs must be weighed against fuel and maintenance costs. In some cases it will be found most economical to use a compact high-speed engine with a reduction gear for driving a certain piece of equipment. In other cases, a medium-speed direct-drive engine may be found the most economical choice. The first cost of the latter choice will usually be higher, but the better fuel consumption and greater durability, normal for the medium-speed engine, may more than compensate for the higher first cost, especially if the number of running hours per year is high.

Consideration must be given to auxiliary systems, switchgear, and other equipment required and to the cost of this installation. In addition, the compatibility of the engine selected with existing facilities may require special design features and must be considered.

For any power application, the following conditions should be satisfied when selecting size and number of engines:

1. The choice must produce sufficient horsepower to satisfy the maximum demand for the power.
2. It should provide the necessary power at the lowest total cost for both operating expenses and fixed charges.
3. It should provide for future growth.
4. It should provide enough units to permit one unit on line and at least one unit down for maintenance (preferably one unit on standby and one unit down for maintenance) without reducing the plant demand.
5. It should consider the fuel source and availability.

Installation

Building Stationary engines are generally installed in buildings which have been specifically constructed for this purpose. While the architectural design of the building will depend upon a great many factors, certain features which in general are common to all sites should be incorporated:

1. Sufficient clearance and an adequately sized overhead crane should be provided for maintenance, taking into account the distances necessary for piston removal and similar disassembly of the engine.

2. Provisions should be made for future installation of additional units (if such addition is provided for in expansion plans). Also, any one unit should be able to be removed without disturbing the other units.

3. There should be adequate ventilation.

4. There should be protective guards around flywheels and other exposed moving machinery.

Foundations   It is accepted practice for engine builders to provide bolt-location drawings for engine foundations. Before final details of the foundation design are established, the bearing capacity and suitability of the footings on which the foundations will rest should be determined. In all cases, the purchaser should consult an expert in soil mechanics. It may be found advisable to modify the manufacturer's drawings to meet special requirements set by local conditions. The engine builder on request will furnish information regarding the values of horizontal and vertical unbalanced inertia forces and the deadweight of the machinery to be supported.

Piping Systems   The engine builder normally furnishes suitable piping schematics for each engine in which minimum pipe sizes for all service lines are recommended. In addition, the following should be observed:

1. Piping must not cause strain at the mounting points on reciprocating or rotating auxiliary equipment, nor should a heavy piece of auxiliary equipment be supported by engine service piping.

2. Whenever there is a possibility of strain, flexibility must be designed into the piping.

3. Chill rings must not be used in welded pipe joints for lubricating-oil systems, as they tend to retain scale, welding slag, and beads, which can come loose as the pipe vibrates and becomes hot during engine operation.

All lubricating-oil and fuel-oil system piping should be pickled after fabrication to remove varnish, mill scale, welding debris, dirt, and grease. Serious engine damage can result from failure to clean this piping properly. Other piping systems should be thoroughly wire-brushed. Pickling should be done under controlled conditions, using the phosphoric acid process or equal; then the interior should be neutralized. Immediately after pickling, the pickled surfaces should be coated with a rust-preventive compound which is soluble in the lubricating oil that will be used in the engine. Just prior to start-up the entire lubricating-oil system must be flushed with lubricating oil and double-checked for cleanliness before permanent connection to the engine.

Engine Alignment with Driven Equipment   The engine builder will furnish instructions for alignment of the engine on its subbase or foundation and, if the builder is also furnishing the driven equipment, alignment procedures for aligning the engine with this equipment. In other cases, the producer of the driven equipment will provide alignment procedures for units of its supply. Engine-alignment procedures will include instructions for adjusting crankshaft-web deflections in the engine. When flexible couplings are used, the coupling manufacturer will normally furnish information concerning alignment procedures for the couplings.

Operation and Maintenance

A broad knowledge of the engine and its auxiliary systems is necessary to permit the operator to operate the engine properly and safely. All valves and cutoffs, strainers, filters, etc., must be properly identified and their operating position in the systems understood.

Before they are started for the first time, all systems should be checked, and after it has been determined that each system is ready for operation, the engine should be barred over manually for a minimum of two complete revolutions to ensure that there are no restrictions. After other necessary preparations preliminary to starting have been made and the engine has been started, it should be run at a speed as low as practical while all gauges are observed. The engine speed should be clear of torsional vibration frequencies and increased slowly while pressures and temperatures are observed to make sure that all parts and systems are working properly.

Manufacturers issue specific instructions for each of their engines, and such instructions are based on experience. To secure the utmost in reliability and efficiency, these instructions should be read, understood, and followed.

Cleanliness is an absolute necessity, particularly for fuel and lubricating-oil systems. A large number of delays and repairs result from neglecting this important area.

Exhaust Emissions

In 1963 the U.S. Congress legislated the Clean Air Act, which empowered the federal government with the responsibility of improving ambient-air quality through the reduction of air pollution. An amendment of 1970 expanded the act to provide the Environmental Protection Agency (EPA) with power to establish national air-quality standards and to set maximum limits on various air contaminants, including exhaust emissions from new, modified, and reconstructed stationary sources.

In 1979 the EPA published in the *Federal Register* (40 CFR Part 60), under the title "Stationary Internal Combustion Engines: Standards of Performance for New Stationary Sources," proposed regulations to limit NO_x emissions from internal-combustion engines. While these regulations are not as yet law, the Clean Air Act, through the EPA, provides local air-quality districts with the power to enact rules and regulations regarding local air quality, including exhaust-emissions regulations.

The complete combustion of any hydrocarbon fuel in an internal-combustion engine would result in exhaust gases composed of harmless carbon dioxide (CO_2), water vapor (H_2O), and nitrogen (N_2). Except in special cases, however, combustion is not complete and other combustion products, many of which if given in large concentrations could constitute a health hazard in some cases, are produced.

Among the combustion products classed as air pollutants are oxides of nitrogen (NO_x), carbon monoxide (CO), oxides of sulfur (SO_x), and nonmethane hydrocarbons (NMHc). Oxides of nitrogen and sulfur are of significant concern in certian areas since, in combination with water, nitric acid and sulfuric acid are produced. Also, the reduction of nitrogen dioxide in combination with hydrocarbons and sunlight produces photochemical smog. The proposed federal regulations for internal-combustion engines would limit the emissions of NO_x to 600 ppm for diesel and dual-fuel engines. These limits would be compared with exhaust-gas measurements which are corrected to 15 percent oxygen in the exhaust, 75 gr of water vapor to 1 lb of dry air (humidity correction), and 85°F ambient-air intake temperature. The limit would apply to engines with a thermal efficiency of 35 percent or lower. At thermal efficiencies greater than 35 percent the NO_x limit is raised linearly.

Most four-stroke-cycle spark-ignition engines, tuned for fuel efficiency, produce from 14 to 18 gr/bhp·h of NO_x. Many dual-fuel engines, because of their greater insensitivity to air-fuel-ratio variations, can produce from 4 to 14 gr/bhp·h of NO_x.

Much work has been and continues to be done in an effort to reduce engine-exhaust emissions without incurring a major fuel penalty, and significant progress is being made by many engine manufacturers, research labratories, government agencies, and users. Much of this progress is due to the impetus created by local air-quality districts and the proposed federal regulations through the *New Source Performance Standards*.

Because NO_x is the primary ingredient in the production of photochemical smog and because it can only be created through the reaction of oxygen and nitrogen under high-temperature conditions (2000°F), most urban NO_x comes from heat engines, including internal-combustion engines.

Successful NO_x-reduction techniques to date include modifications of engine parameters such as retarded ignition, increased combustion-inlet-air cooling, exhaust-gas recirculation, water injection, and derating. Among other exhaust-gas aftertreatment methods are the use of catalytic converters, the use of exhaust-gas scrubbers, both wet and dry, and ammonia injection into the exhaust, both with and without a catalytic converter to convert $NO_x + NH_3$ to N_2 and H_2O. Many of these methods are proven technology, while others are as yet developmental.

The prospective engine purchaser should make certain that proper inquiries are made with the local air-quality management district for complete information on requirements for securing the necessary permits for construction and operation.

Marine Propulsion Systems

General

Other sections have dealt with details of the design, manufacture, testing, and operation of machinery classified as prime movers, power transmitters, and energy converters. For productive purpose this machinery is always applied in combinations of at least two, forming machinery systems. The intent of this section is to provide a general introduction to the utilization of the components in marine propulsion systems and the system considerations involved in their use. The total subject is much too large for treatment here, but numerous publications on various aspects of marine propulsion systems are available for further study. The American Bureau of Shipping (ABS), the regulatory body for commercial ships built in the United States, publishes annually a book entitled *The Rules for Building and Classing Steel Vessels*. The *Transactions* of the Society of Naval Architects and Marine Engineers is suggested as another of the many sources available.

The Marine Propulsion System

The marine propulsion system is the combination of machinery used to move vessels through water, both on the surface and below it. As defined here, the system is limited to large-powered ships used for commercial and military purposes. Commercial ships are surface vessels, basically classified as bulk-cargo ships, container ships, and tankers. The military uses similar types of auxiliary ships to service combat ships ranging from relatively small destroyers to huge aircraft carriers and submarines.

Modern-day propulsion systems utilize gas and steam turbines and diesel engines as prime movers, reduction gears as power transmitters, and propellers in conjunction with thrust bearings as energy converters. Multiple drivers are commonly employed, generally in like pairs. Combination drives such as steam-and-gas turbines and diesel-and-gas turbines have been or are in use.

The speeds of modern commercial ships range from about 15 to 20 kn. Bulk carriers and tankers, because of their hull design, are the slower ships. The latest-design container vessels, in contrast, can sustain speeds of 20 kn. Propulsive power ranges from about 10,000 to 50,000 shp. Military ships are capable of considerably faster speeds and greater power.

The energy source for the prime movers is oil and coal, the latter of which has received increased consideration since the oil energy crisis. Nuclear energy is presently limited to military use.

Steam Turbines   At about the turn of the twentieth century the steam turbine began to replace the steam engine as the prime mover. An immediate advantage was the elimination of the unbalanced forces associated with the steam engine, while subsequent developments provided the technology to utilize high steam pressures and temperatures. These developments, in addition to an ability to utilize low exhaust pressures, provided significant thermodynamic improvements. System and operating efficiencies were greatly improved by the resulting lighter weight and reduced space requirements. The steam turbine was also much easier to maintain.

The first such turbines were mammoth direct-drive turbines. Although they were able to develop more than 15,000 hp with steam pressures of 235 lb/in^2, three-quarters of the ship's hull length was required for the engine room. Turbine speeds were initially below 200 r/min, which was a compromise with propeller requirements. Both components operated at much less than optimum efficiency. The development of the reduction gear soon phased out direct-drive turbines.

Figure 5-17a and b depicted cross sections of a modern multistage condensing turbine. Marine turbines are most commonly used as cross-compound units consisting of a high-pressure and a low-pressure turbine, and Fig. 7-1 shows the cross-compound arrangement. As the designations imply, the turbines are designed for different steam conditions and operate at different speeds. The high-pressure-turbine exhaust is the inlet steam for the low-pressure turbine. Marine turbines are further distinguished by having reversing turbines for propelling backward (astern) and for ship maneuvering. These reversing turbines are built into the low-pressure turbines of cross-compound units.

The steam conditions of present-day central-electric-generating-station turbines vary considerably with their application. Pressures vary from about 150 to 3500 lb/in^2, and temperatures are limited to about 1000°F maximum. Marine propulsion turbines, whose development generally lags behind that of central-station turbines, currently use steam conditions of 850 lb/in^2 and 950°F for the larger-powered ships. Pressures up to 1450 lb/in^2 are required for reheat-cycle systems. The marine industry, which is noted for its decade-step development, is

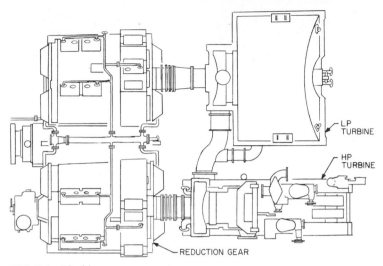

LP
TURBINE

HP
TURBINE

REDUCTION GEAR

FIG. 7-1 Machinery arrangement.

currently contemplating systems with steam-pressure conditions of 2050 lb/in^2. Efficiency is again the prime motivating factor.

Turbine speeds for commercial ships are about 6000 r/min for the high-pressure turbine and 3600 r/min or lower for the low-pressure turbine. The astern turbine is designed to provide 75 to 100 percent of the ahead torque at 50 percent of ahead speed. Turbines for military ships are of higher power and speed to reduce their weight.

Single-turbine drives and the low-pressure turbines of cross-compound drives exhaust to water-cooled condensers. Water is supplied to the condensers by circulating pumps on the slower ships and by direct scoops on the faster ships.

Purchase specifications of marine turbines emphasize efficiency, reliability, and ruggedness. Such specifications are warranted, for the turbines are required, in addition to maintaining efficient and continuous operation at design powers, to withstand the rigors of heavy seaways, ship maneuvering, and crash-ahead or crash-astern emergencies.

Turbine shipboard foundations require the same ruggedness and consideration. Their design, which is complicated by elevation requirements of the reduction gear and proximity of the condensers, must not only support the turbines but also maintain precise alignment with the reduction gear. The high-pressure turbine is usually mounted on a cradle for structural reasons. The foundation of the low-pressure turbine has the condenser either supporting the turbine or hung from the turbine, the choice usually being dictated by the size and weight of the condenser. A third method has the foundation supporting both turbine and condenser. In this case, a suitable expansion joint is required between the turbine and the condenser. Chocks are fitted between the turbine feet and the foundation to locate precisely

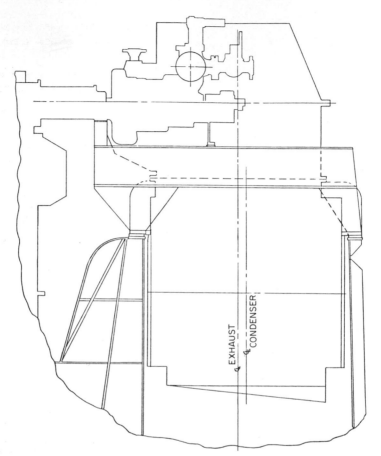

FIG. 7-2   High-pressure and low-pressure foundation cross sections.

the turbine relative to the reduction gear. Figure 7-2 shows high-pressure and low-pressure foundation cross sections.

Turbine Controls   Marine turbines are provided with a system capable of either remote or local control. Operation from the bridge is by electrical signal, while engine-room operation may be either electrical or mechanical.

Electric control is automated. Control logic is transmitted electrically to hydraulic actuators, which position the control valves for either ahead or astern operation. Valve position and speed signals are fed back to control logic. The system includes pumps coupled to the turbines which sense and limit speed. The limited-speed range is from about 60 to 112 percent of rated speed. Electric overspeed trips set at 125 percent speed are incorporated as backup protection.

Later versions use higher hydraulic pressures and electric speed sensing and limiting in lieu of pumps. This design reduces actuator size and response time.

Reduction Gears   The need for a reduction gear was evident shortly after the

steam turbine had been adopted for propulsion service. Both the size of the turbine and efficiency prompted the search for a device to utilize optimum designs
of turbine and propeller. The development of the reduction gear progressed steadily. One of the earliest applications was a gear which reduced a 1450-r/min turbine speed to a propeller speed of 73 r/min and transmitted 1095 hp.

Various gear arrangements and types were tried during this development. Figure 7-3a depicts the earliest and simplest arrangement. Its present-day applications are with diesel engines and marine turbine-driven generators. Figure 7-3b
is a single reduction with two inputs. Figure 7-3c shows an epicyclic type of gear
used in combination with the fixed parallel axis in either the first or the second
reduction. Figure 7-3d is an articulated gear, Fig. 7-3e a locked-train gear, and
Fig. 7-3f the single-input version of the locked-train gear.

A typical double-reduction articulated gear is shown in cross sections in Fig.
7-4. It consists of two first-reduction input pinions meshing with two first-reduction gears. Each gear is connected to a second-reduction pinion with a dental-type
flexible coupling at each end of a quill shaft which passes through the second-
reduction pinion. This arrangement enables independent movement of both
reductions—thus the term *articulated*. The first-reduction pinions are connected
to the turbines by flexible couplings of the dental, diaphragm, or flexible-disk
type. The low-speed gear, commonly called the *bull gear,* is connected to the propeller by bearing-supported shafting.

The gear case is designed to provide the support for the bearings necessary to
keep the meshing elements in precise alignment under all conditions of operation.
It also provides an oiltight enclosure. The articulated gear shown has its case split
into three planes. The bull gear is on the lowest plane, the first-reduction gears
and second-reduction pinion on the second plane, and the first-reduction pinion
on the highest plane. The first-reduction pinion can be rolled around the first-
reduction gears to accommodate turbine spacing and optimum location. The
extreme case results in a two-plane arrangement.

The double-input, double-reduction locked-train arrangement is shown in cross
section in Figs. 10–3 and 10–4. It differs from the articulated type in that the
power input of each first-reduction pinion is split betweeen two first-reduction
gears, which are coupled in similar manner to two second-reduction pinions. Four
second-reduction pinions mesh with the bull gear.

This arrangement, although increasing the number of rotating parts and bearings, is lighter, more compact, and more efficient than an articulated gear for the
same power conditions. Its application was initially limited to use on naval combat
ships. As industry power requirements increased, the weight difference of larger
gears became a significant cost factor. This, coupled with the arrangement's suitability for ships with low engine-room overhead, made it competitive for commercial use, especially in the latest designs of container vessels.

The locked-train gear is characterized by an A-frame construction which provides structural strength and accessibility. This design does require limits in the
ratio relationship between the first and second reductions, but for most applications total ratio requirements are attainable in two reductions.

The term *locked train* is a misnomer and frequently is interpreted to mean
"locked-in torque," which in reality is not possible. It probably stems from timing:

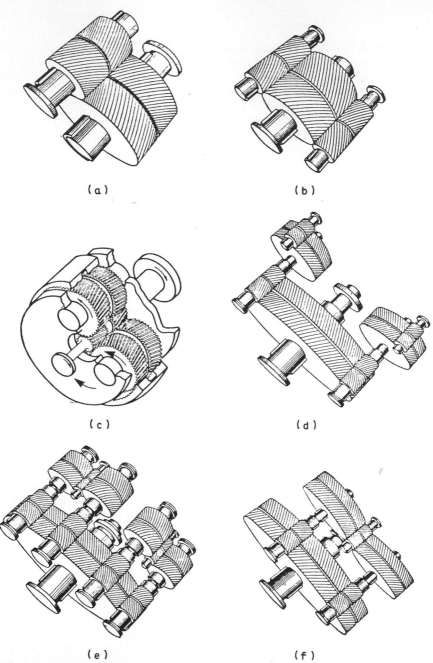

FIG. 7-3 Arrangements of reduction gears. (*a*) Single reduction, single input. (*b*) Single reduction, double input. (*c*) Single reduction, planetary. (*d*) Double reduction, double input, articulated. (*e*) Double reduction, double input, locked train. (*f*) Double reduction, single input, locked train.

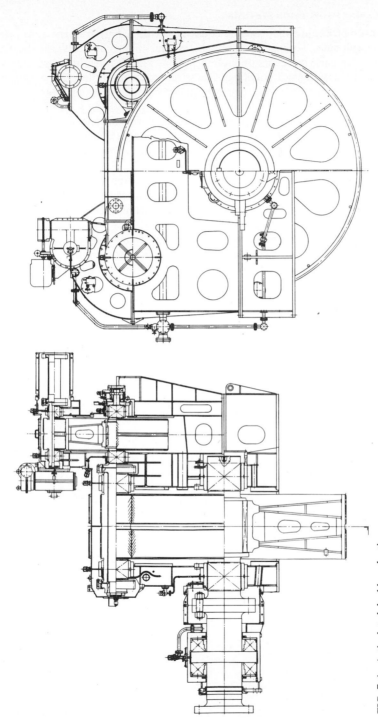

FIG. 7-4 Articulated double-reduction gear.

an assembly procedure that assures equal power distribution in the split power paths on each side. Timing is simply the vernier movement between the first-reduction gear and the second-reduction pinion. This feature is incorporated in the design of the quill-shaft couplings. Relative movements of less than 0.001 inch are possible by selection of the numbers and difference in numbers of coupling teeth.

Reduction gears are rigidly bolted to ship foundation structures, thus enhancing the stiffness designed into their cases. Naval architects design foundations to isolate them from the strains of the ship's hull under seaway conditions. Figure 7-5 shows a typical reduction-gear foundation.

Internal alignment between meshing elements, which is crucial for reliable operation, is achieved by carefully fitted chocks. Factory-established tight wire and pin reference measurements are reproduced during installation aboard ship. A typical chock is shown in Fig. 7-6.

The installation of assembled reduction gears is now routine practice. An advantage is the ability to assemble a gear in a clean environment without disrupting engine-room construction. Figure 7-7 depicts a large gear being installed aboard ship.

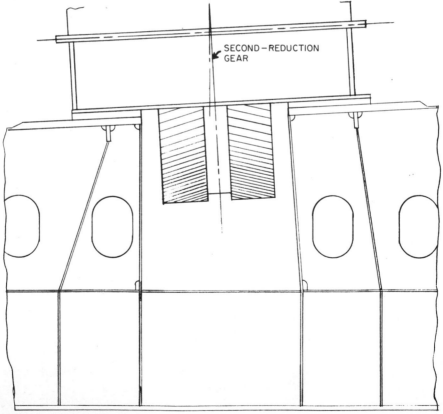

SECOND – REDUCTION GEAR

FIG. 7-5 Typical reduction-gear foundation.

Shafting   The propeller is connected to the reduction gear by shafting, the length of which is determined by the location of the propeller and the engine room. Fig-

ure 7-8 shows shafting arrangements with two different propeller locations. Figure 7-8*a* is the arrangement of shafting of multiscrew ships and single-screw ships with transom sterns. The shafting is extended for a considerable distance outboard to provide adequate clearance between the propeller and the hull. Figure 7-8*b* shows the more typical arrangement of bulk-cargo ships and tankers.

Engine-room locations also vary. Tankers usually have their engine rooms well aft and have relatively short shafting lengths. Bulk-cargo engine rooms, which are closer to midships, require longer shafting lengths.

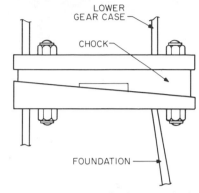

FIG. 7-6   Typical reduction-gear chock.

The shafting located inside the ship is termed *line shafting*. The thrust shaft, in cases in which the thrust bearing is aft of the gear, is also part of the shafting system. The section on which the propeller is mounted is termed the *propeller,* or

FIG. 7-7   Large gear being installed aboard ship.

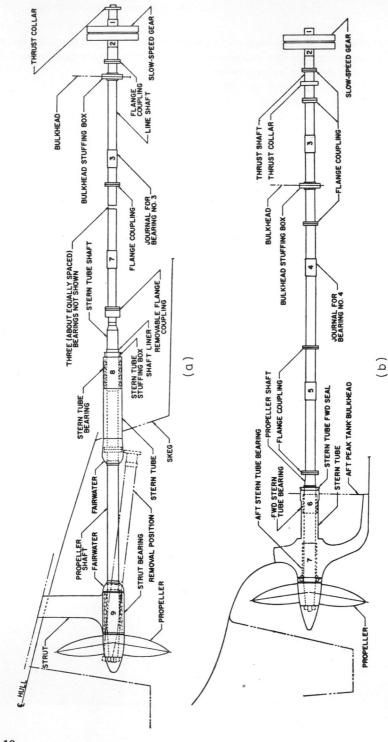

FIG. 7-8 Propeller shafting. (*a*) Existing area with strut bearings. (*b*) Existing area without strut bearings.

tail, shaft. The shaft passing through the stern tube on a strut-supported propeller is termed the *stern-tube shaft.*

Shaft sections have integral flanges on each end and are solidly bolted to one another, forming in essence a single continuous shaft.

Bearings are used to support the shafting and maintain proper alignment. They are divided into two groups, those within the ship and those outside. They bear names related to the shaft sections that they support, such as strut, stern-tube, and line bearings. Figure 7-9 shows a typical line shafting bearing. Strut bearings are either water- or oil-lubricated; in the latter case they require sophisticated seals. Inboard bearings utilize oil lubrication. Seals or stuffing boxes are incorporated in stern tubes to prevent the entrance of seawater.

The fundamentals of shafting design consist of sizing for torque transmission and support. This requires determination of the number of bearings needed and their relative locations. The advent of the computer simplified and shortened this design effort.

Propeller   The propeller is that component of the system which converts the torque developed by the prime mover to propulsive thrust. The design of a propeller is complex. Generally, the most efficient propeller is the largest and the slowest. Hull lines and propulsion machinery are included in trade-off considerations. The final design is usually influenced by systematic model testing to achieve the optimum propulsive system.

The number of blades of propellers varies from four to seven. The number selected is frequently influenced by torsional vibration considerations.

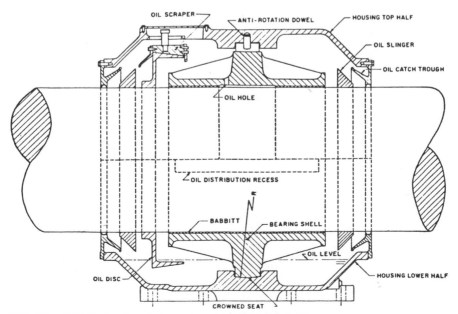

FIG. 7-9   Self-aligning line shaft bearing with oil-disk lubrication.

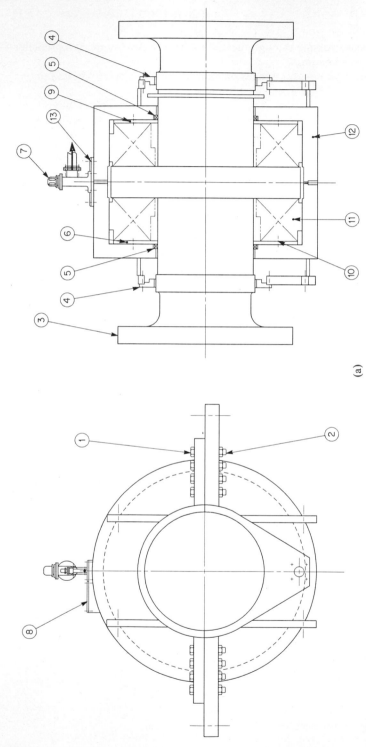

FIG. 7-10 Thrust bearing. (*a*) Thrust-bearing elements: (1) Bolt (body-bound). (2) Nut. (3) Shaft. (4) Oil deflector. (5) Oil-seal ring. (6) Filler piece. (7) Sight-flow indicator. (8) Inspection cover. (9) Filler piece. (10) Cap screw. (11) Thrust bearing. (12) Thrust case. (13) Gasket.

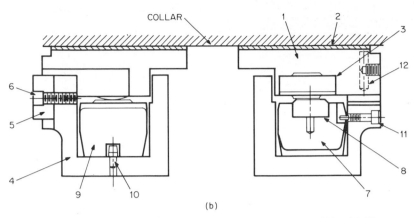

FIG. 7-10 (*continued*)   (*b*) Thrust elements: (1) Shoe. (2) Shoe babbitt. (3) Shoe support. (4) Base ring. (5) Base-ring key. (6) Base-ring key screw. (7) Upper leveling plate. (8) Upper-leveling-plate plug. (9) Lower leveling plate. (10) Lower-leveling-plate plug. (11) Leveling-plate setscrew. (12) Base-ring bumper dowel.

The majority of modern ships have fixed-blade propellers. Detachable-blade types are selected in some instances to facilitate repair work. Controllable-pitch and reversible-pitch propellers are used where reversing capabilities are not part of prime movers such as nonreversing diesels and gas turbines. Contrarotating propellers are efficient, but their use has been limited to military research.

Main Thrust Bearing   The thrust developed by the propeller is transmitted to the ship's hull by means of a thrust bearing generally referred to as the *main thrust bearing*. The bearing is the self-leveling type. Figure 7-10*a* and *b* depicts a cross-sectional view and details of the bearing parts. A secondary purpose of the thrust bearing is to position the bull gear in its case.

The location of the thrust bearing can vary. In the larger-powered propulsion systems the thrust bearing is most commonly located just aft of the bull gear, but it can be located further aft in the shafting. Sometimes it is just forward of the bull gear; the thrust housing is then fabricated as part of the lower gear case.

Thrust bearings are designed for unit loads up to 400 lb/in^2, requiring sizes in excess of 60-in diameter for high-powered vessels. Thrust-bearing foundation design requires both strength and stiffness for propulsion-thrust transmission.

System Consideration

High-Speed Alignment   Thermal-growth analyses are made of the gear, turbines, and their foundations. These results, in addition to rotor reactions, provide data for the cold alignment of the components. The turbine-gear arrangement usually requires substantial coupling-spool piece lengths. This allows practical alignment tolerances well within the misalignment capability. Figure 7-11 shows typical high-speed-alignment data of a large-powered ship.

Low-Speed Alignment   Since the bull gear becomes part of the shafting as a result of the solid-coupling method, the low-speed-alignment analysis treats more

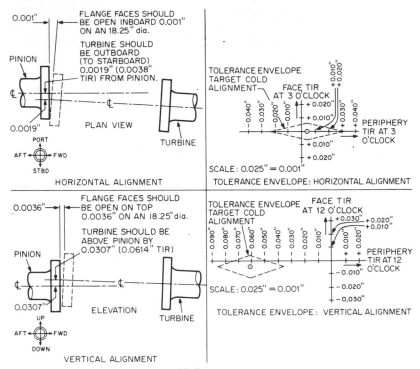

FIG. 7-11 High-pressure-turbine cold alignment.

than thermal growth. To maintain internal alignment of the reduction-gear elements, the static loads on the bull-gear bearings must be kept nearly equal. This is accomplished by controlling the shafting-bearing reactions by relative vertical positioning and relative location of the bearings. Generally, the location of the first line shaft bearing aft of the gear provides the necessary flexibility to minimize shafting influence on the bull-gear bearing.

The use of the computer greatly simplifies this analysis. It can be done in conjunction with design of the shafting. The resulting data provide data for cold-bearing reactions.

Calibrated hydraulic jacks or load cells are used for determining bearing reactions. The gap-and-sag method is still a good way to check athwartship alignment.

Torsional Analysis Severe vibration problems experienced in the past, especially with reciprocating steam and diesel engine drives, has established torsional analysis as an important design consideration of the marine propulsion system. The large rotating masses of the turbines, gear, and propeller connected by significant lengths of shaft constitute a system capable of vibration. A significant source of excitation by the propeller and diesel engines necessitates, by design, the location of criticals sufficiently outside of or in acceptable ranges within the operating speed. Detailed treatment is beyond the intent of this section. Generally, a torsional analysis consists of the following:

1. Calculation of the polar moments of inertia of all rotating parts.
2. Calculation of spring rates of all parts, primarily shafts, which have significant torsional flexibility.
3. Establishing a system of masses connected by torsional springs.
4. Reducing the system to an equivalent system of a single speed for analysis purposes. This is accomplished on the basis of equal energy with the propeller speed as the equivalent speed.
5. Determination of the significant resonant frequencies. This is done by using Holzer's method of analysis. Its use is considerably simplified by the employment of computers.
6. Determination of the magnitudes of vibratory torques at the significant resonant frequencies.

Figure 7-12 depicts a schematic of a cross-compound-turbine drive and its equivalent model for analysis purposes.

Steam-turbine drives usually encounter only one significant source of torsional excitation, the propeller. Modern-day precision gear cutting eliminates gear-

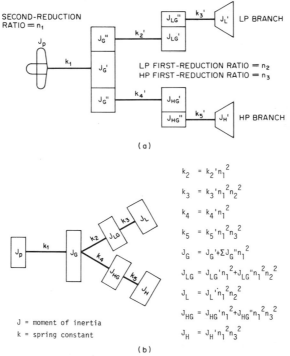

$$k_2 = k_2' n_1^2$$

$$k_3 = k_3' n_1^2 n_2^2$$

$$k_4 = k_4' n_1^2$$

$$k_5 = k_5' n_1^2 n_3^2$$

$$J_G = J_G' + \Sigma J_G'' n_1^2$$

$$J_{LG} = J_{LG}' n_1^2 + J_{LG}'' n_1^2 n_2^2$$

$$J_L = J_L' n_1^2 n_2^2$$

$$J_{HG} = J_{HG}' n_1^2 + J_{HG}'' n_1^2 n_3^2$$

$$J_H = J_H' n_1^2 n_3^2$$

J = moment of inertia
k = spring constant

(b)

FIG. 7-12 Model illustrations. (*a*) Schematic illustration of a geared turbine-driven propulsion system. (*b*) Equivalent six-mass system with all branches referred to the propeller's revolutions per minute.

tooth-spacing errors as significant sources of tortional excitation. Investigation of two of the modes of vibration is usually sufficient for turbine drives. The first mode almost always has a node in the shafting between the bull gear and the propeller. This can be located at a very low propeller speed that is not in the operating range and can be passed through rather quickly. The second mode may be treated by tuning the low-pressure and high-pressure branches. This produces a node at the bull gear, precluding excitation of these branches by the propeller.

The diesel drive requires a much more complicated analysis. The excitation forces of the diesel are significant and more complex. In addition to the fundamental cylinder-firing frequencies, the problem of the loss of cylinder firing must be examined. Mass-elastic modeling is also different, in that the cylinders themselves are significant masses separated by shafting.

Longitudinal Analysis   Vibration problems diagnosed as longitudinal were encountered with the advent of larger ships. Longitudinal analysis then became a standard design consideration of the marine propulsion system.

Similarly to that of the torsional system, the mass of the propeller is connected to the mass of the machinery by elastic shafting which constitutes a system capable of vibration. The propeller for this mode of vibration is the only excitation source.

Longitudinal resonances can cause such deleterious effects as (1) accelerated wear of gears, couplings, and thrust bearings, (2) damage to turbines when relative movements exceed design clearances, and (3) fatigue cracking of piping and machinery supports including foundations and hull structures.

Generally, the longitudinal analysis follows the procedure of the torsional analysis involving the calculation of masses and spring rates, the latter including those of the thrust-bearing foundations.

Either the Holzer method of analysis or the mechanical-impedance method can be used to calculate the system frequencies. Figure 7-13 is a schematic of a longitudinal system.

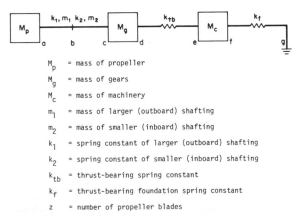

M_p = mass of propeller

M_g = mass of gears

M_c = mass of machinery

m_1 = mass of larger (outboard) shafting

m_2 = mass of smaller (inboard) shafting

k_1 = spring constant of larger (outboard) shafting

k_2 = spring constant of smaller (inboard) shafting

k_{tb} = thrust-bearing spring constant

k_f = thrust-bearing foundation spring constant

z = number of propeller blades

FIG. 7-13   Representation of a geared-turbine-propulsion shafting system for longitudinal-vibration analysis.

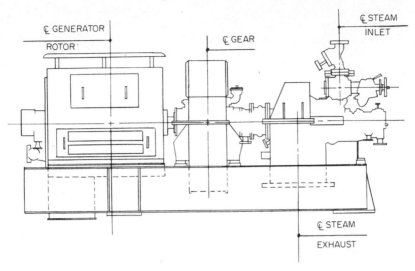

FIG. 7-14 Typical steam-driven generator.

Satisfactory shafting systems free of deleterious longitudinal vibration are usually attainable with shafting and foundation design. Thrust-bearing resonance changers can be incorporated to vary the system frequency if necessary.

Auxiliary Systems

Lubrication The components of the marine propulsion system are lubricated by pressure-fed oil. Pumps are usually motor-driven with backups. In some applications pumps are driven by drives taken off the reduction gear; these have motor-driven backup pumps. System oil pressure is regulated by gravity-feed tanks which provide roll-down lubrication in cases of complete power failure.

Power-Generating Sets Ship power is generally provided by steam-turbine-geared generator sets. Diesel-driven generators provide backup power and also power when the ship's boilers are not in operation. A typical steam-driven generator drive is shown in Fig. 7-14.

Pumps

CENTRIFUGAL PUMPS

General

The centrifugal pump is one of the most versatile types of machinery for industry. Every plant has operating a multitude of pumps of this type, and modern civilization could not be visualized without this equipment.

Compared with other types of pumps, i.e., reciprocating and rotary pumps, centrifugal pumps operate at relatively high speeds and consequently are smaller and lighter when designed for comparable capacity and head. Required floor space, weight, initial cost, and building costs are therefore reduced.

Owing to their relatively high speed, centrifugal pumps are usually direct-connected to the driver, the majority being electric-motor-driven. Having no reciprocating parts, centrifugal pumps are inherently balanced. There are no internal rubbing parts, and because running clearances are relatively large, wear is minimized. The liquid is delivered in a steady stream so that no receiver is needed to even out pulsations.

In contrast to positive-type displacement pumps, centrifugal pumps develop a limited head at constant speed over the operating range from zero to rated capacity, and excessively high pressures cannot occur. They can therefore be started against a closed discharge valve but should be operated at this condition for a minimum period (see subsection "Minimum Flow-Through Pump"). Generally, the bearings are located outside the casing, so that the liquid does not come in contact with the lubricating oil and is not contaminated by it.

Classification

There are three general classes of pumps, depending on the configuration of the pump impellers:

Centrifugal or radial pump
Mixed-flow pump
Axial-flow pump

Those classes can be subclassified according to

Number of stages: single-stage pump, multistage pump
Arrangement of liquid inlet: single-suction pump, double-suction pump
Position of shaft: horizontal pump, vertical pump (dry-pit type), vertical pump
  (submerged type)

Specific Speed

Specific speed is a correlation of pump capacity, head, and speed at *optimum* efficiency which classifies pump *impellers* with respect to their geometric similarity corresponding to the classification mentioned above (see also Fig. 8-1).

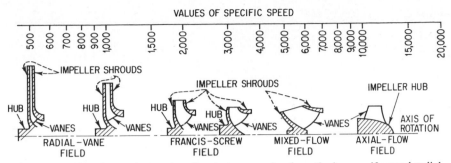

FIG. 8-1 Profile of several pump-impeller designs, ranging from the low-specific-speed radial flow on the left to the high-specific-speed impeller design on the right, placed according to where each design fits on the specific-speed scale. *(Courtesy of the Hydraulic Institute.)*

Specific speed is a number usually expressed as*

$$\text{Specific speed } N_s = \frac{N\sqrt{Q}}{H^{3/4}} \quad \text{or} \quad N_s = \frac{N\sqrt{Q}H^{1/4}}{H}$$

where N_s = specific speed

N = rotative speed, r/min

Q = flow, gal/min, at or near optimum efficiency

H = head, ft per stage

The specific speed of an impeller is defined as the revolutions per minute at which a geometrically similar impeller would run if it were of such a size as to discharge one gallon per minute against one foot head.

Specific speed is indicative of the shape and characteristics of an impeller, and it has been found that the ratios of major dimensions vary uniformly with specific speed. Specific speed is useful to the designer in predicting required proportions and to the application engineer in checking the suction limitation of pumps.

Impeller form and proportions vary with specific speed, as shown in Fig. 8-1.

Pumps are traditionally divided into three classes: the centrifugal or radial-flow, the mixed-flow, and the axial-flow, but it can be seen from Fig. 8-1 that there is a continuous change from the radial-flow impeller, which develops pressure principally by the action of centrifugal force, to the axial-flow impeller, which develops most of its head by the propelling or lifting action of the vanes on the liquid.

In the specific-speed range of approximately 1000 to 4000, double-suction impellers are used as frequently as single-suction impellers.

Table 8-1 gives values of $H^{3/4}$ for the accurate determination of the specific speed N_s.

Figure 8-2 may be used to find specific speed with sufficient accuracy for practical purposes without calculating the head to the three-fourths power. The point

*The value of $H^{3/4}$ may be found in Table 8-1.

TABLE 8-1 Values of $H^{3/4}$

Head	$H^{3/4}$	Head	$H^{3/4}$	Head	$H^{3/4}$	Head	$H^{3/4}$
0	0	50	18.8	225	58.1	1,050	184.4
1	1.00	52	19.4	230	59.0	1,100	191.0
2	1.68	54	19.9	235	60.0	1,150	197.4
3	2.28	56	20.5	240	61.0	1,200	203.8
4	2.83	58	21.0	245	62.0	1,250	210.2
5	3.34	60	21.6	250	62.9	1,300	216.4
6	3.83	62	22.1	260	64.8	1,350	222.7
7	4.30	64	22.6	270	66.6	1,400	228.8
8	4.75	66	23.2	280	68.4	1,450	234.9
9	5.20	68	23.7	290	70.2	1,500	241.0
10	5.62	70	24.2	300	72.0	1,550	247.0
11	6.03	72	24.7	310	73.9	1,600	252.9
12	6.45	74	25.2	320	75.7	1,650	258.8
13	6.85	76	25.7	330	77.4	1,700	264.7
14	7.24	78	26.2	340	79.2	1,750	270.5
15	7.62	80	26.8	350	80.9	1,800	276.3
16	8.00	82	27.3	360	82.6	1,850	282.0
17	8.38	84	27.7	370	84.4	1,900	287.7
18	8.73	86	28.2	380	86.1	1,950	293.4
19	9.09	88	28.7	390	87.8	2,000	299.0
20	9.45	90	29.2	400	89.4	2,100	310.2
21	9.80	92	29.7	410	91.1	2,200	321.2
22	10.2	94	30.2	420	92.8	2,300	332.1
23	10.5	96	30.7	430	94.4	2,400	342.8
24	10.8	98	31.1	440	96.1	2,500	353.5
25	11.2	100	31.6	450	97.7	2,600	364.1
26	11.5	105	32.8	460	99.3	2,700	374.5
27	11.8	110	33.9	470	101	2,800	384.9
28	12.2	115	35.0	480	103	2,900	395.1
29	12.5	120	36.2	490	104	3,000	405.3
30	12.8	125	37.4	500	106	3,100	415.4
31	13.1	130	38.5	520	109	3,200	425.4
32	13.5	135	39.6	540	112	3,300	435.3
33	13.8	140	40.6	560	115	3,400	445.2
34	14.1	145	41.8	580	118	3,500	455.0
35	14.4	150	42.8	600	121	3,600	464.7
36	14.7	155	43.9	620	124	3,700	474.4
37	15.0	160	45.0	640	127	3,800	483.9
38	15.3	165	46.0	660	130	3,900	493.5
39	15.6	170	47.1	680	133	4,000	502.9
40	15.9	175	48.1	700	136	4,100	512.3
41	16.2	180	49.2	720	139	4,200	521.7
42	16.5	185	50.2	740	141	4,300	531.0
43	16.8	190	51.2	760	145	4,400	540.2
44	17.1	195	52.2	780	148	4,500	549.4
45	17.4	200	53.2	800	150	4,600	558.5
46	17.7	205	54.2	850	157	4,700	567.6
47	18.0	210	55.1	900	164	4,800	576.6
48	18.2	215	56.2	950	171	4,900	585.6
49	18.5	220	57.1	1,000	178	5,000	594.6

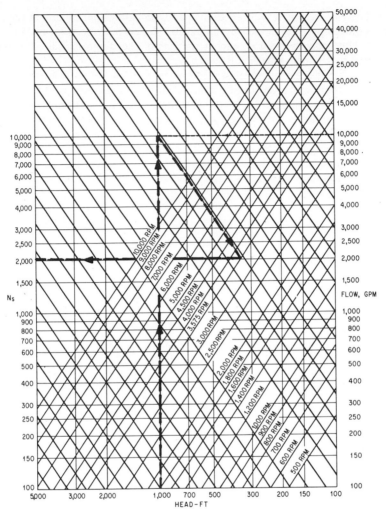

FIG. 8-2 Diagram for determination of specific speed. Using the diagram, plot the head-capacity point; move from this point parallel to heavy lines to correct speed; from there move horizontally to the left and read specific speed. Example (dashed lines): $H = 1000$ ft; $Q = 10,000$ gal/min; $N = 3575$ r/min; $N_s = 2015$.

located by plotting the total head and capacity in gallons per minute at the design point is moved parallel to the sloping lines to the correct speed in revolutions per minute. The specific speed is read at the left of the diagram. The procedure is illustrated by the heavy dashed lines.

For double-suction impellers, total flow is used in the calculation, although historically there has been considerable use of half of the flow in the equation.

For multistage pumps, the head per stage is used in the specific-speed equation. Generally, this is the total head of the pump divided by the number of stages.

Impeller performance curves are intimately related to their types or specific speeds. Higher-specific-speed impellers operating at partial loads have higher heads, require more horsepower, and have lower efficiency. This is illustrated in Fig. 8-3 on a percentage basis for the three general types mentioned above.

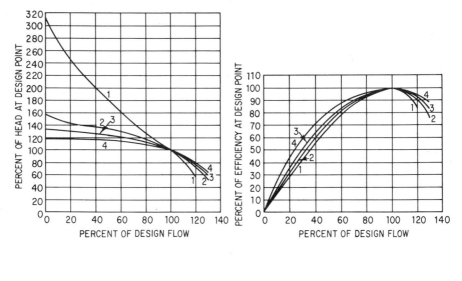

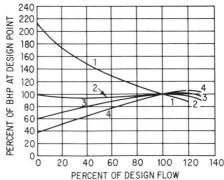

FIG. 8-3 Comparison of performance curves for various types of impellers: (1) propeller; (2) mixed-flow; (3) Francis; (4) radial.

Hydraulics

Definition of Static, Pressure, and Velocity Heads   One of the most useful relationships of hydraulics is the *continuity equation,* which is based upon the principle that after steady conditions in any system have been established, the weight flow of fluid per unit of time passing any point is constant. Since most liquids are practically incompressible, this may be put in equation form as $Q = AV$, where Q = flow, ft^3/s, A = cross-sectional area, ft^2, and V = velocity, ft/s. This equation may be rewritten in the form $V = 0.321Q/a$,* where V = velocity, ft/s, Q = volume flow, gal/min, and a = area of pipe, in^2. This equation is of importance in determining the velocity of the fluid at various points in either the piping or the pump itself.

Another important term is *head,* which is the energy contained in a pound of fluid. It is the height to which a column of the same fluid must rise to contain the same amount of energy that a unit weight of the fluid has under the conditions considered. It may appear in any of three interchangeable forms.

The *potential head* is based upon the elevation of the fluid above some arbitrarily chosen datum plane. Thus, a column of fluid z ft high contains an amount of energy due to its position and is said to have a head of z ft of fluid.

The *pressure head* is the energy contained in a unit weight of the fluid due to its pressure and equals P/γ, where P is the pressure, lb/ft^2, and γ is its specific weight, lb/ft^3. If an open manometer tube is set perpendicular to the flow, the fluid in it will rise to a height equal to P/γ. This is referred to as *static-pressure head.*

The *kinetic* or *velocity head* is the energy contained in a unit weight of the fluid due to its motion and is given by the familiar expression for kinetic energy $V^2/2g$, where V = velocity, ft/s, and g = acceleration due to gravity (32.17 ft/s^2). Values of this velocity head for various velocities are given in Table 3-4. The head may be determined by taking the difference between a reading obtained by a pitot tube facing the flow and a tube at right angles to the flow at the same location.

The total energy of the fluid is equal to the sum of the three heads, or

$$\frac{P}{\gamma} + \frac{V^2}{2g} + z = H$$

Since energy cannot be created or destoyed, H is constant at any point of a closed hydraulic system (losses are neglected). This equation is known as *Bernoulli's theorem.* The various forms of head may vary in magnitude at different sections, but if losses are neglected, their sum is always the same.

When liquid flows through a pipe, there will be a drop in pressure or head due to friction losses. The approximate drop in head for water flowing through various sizes of pipe may be found from Tables 3-10 and 3-11. For viscous fluids the pressure drop in pounds per square inch may be estimated from Fig. 3-12 for various pipe sizes.

*For velocities in pipes from 1 to 72 in in size, see Tables 3-10 and 3-11.

The *total head H* developed by a centrifugal pump is the measure of energy increase of the liquid imparted to it by the pump and is the difference between the total discharge head and the total suction head. Expressed in equation form,

$$H = h_d - h_s$$

where H = total pump head, ft

h_d = total discharge head, ft, above atmospheric pressure at datum elevation

h_s = total suction head, ft, above atmospheric pressure at datum elevation

NOTE: h_d and h_s are negative if the corresponding pressures at the datum elevation are below the atmospheric pressure.

The common datum is taken through the pump centerline for horizontal pumps and at the entrance eye of the suction impeller for vertical-shaft pumps.

Suction lift h_s exists when the total suction head is below atmospheric pressure. The total suction lift as determined on test is the static pressure (vacuum) as measured by a mercury column expressed in feet of the liquid being pumped less the velocity head at the point of gauge connection. This is equivalent to the static lift plus entrance and friction losses in the piping if the water-supply level is below the centerline of the pump. In the case of water-supply level above the pump centerline and at atmospheric presure, suction lift will exist if the entrance and friction losses in the suction piping are greater than the static head.

Suction head h_s exists when the total suction head is above atmospheric pressure. Total suction head as determined on test is the gauge reading expressed in feet of the liquid being pumped plus the velocity head at the point of gauge connection.

The total suction head is equivalent to the static head less entrance and friction losses in the suction piping.

The *total discharge head h_d* as determined on test is the pressure-gauge reading expressed in feet of liquid being pumped plus the velocity head at the point of gauge connection. It is thus equivalent to the static discharge head plus all losses in the discharge piping if the discharge water level is at atmospheric pressure.

In the case of a discharge into a closed vessel under pressure, the total discharge head is equivalent to the static head corresponding to the water level in the vessel, plus the gauge pressure expressed in feet of liquid corrected to this water level, plus the friction losses in the discharge piping.

If the pump discharges to a level below the pump centerline or into a vessel under vacuum, the static head and pressure are taken as negative.

Determination of Head Figure 8-4 shows the usual arrangement for determining the total head developed by a pump working with suction pressure

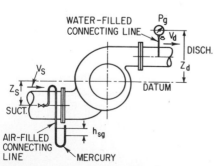

FIG. 8-4 Determination of head.

below atmospheric. Suction head is measured by means of a mercury U tube which is connected to the suction through an *air*-filled tube. Discharge head is measured by means of a Bourdon gauge connected to the pump discharge through a *water*-filled tube.

The total head for the arrangement shown in Fig. 8-4 is derived from the following expressions:

$$h_s = -\frac{\gamma m}{\gamma} h_{sg} - z_s + \frac{V_s^2}{2g}$$

$$h_d = \frac{144P_g}{\gamma} + z_d + \frac{V_d^2}{2g}$$

$$H = h_d - h_s$$

$$= \frac{144P_g}{\gamma} + z_d + \frac{V_d^2}{2g} + \frac{\gamma m}{\gamma} h_{sg} + z_s - \frac{V_s^2}{2g}$$

$$= \frac{144P_g}{\gamma} + \frac{\gamma m}{\gamma} h_{sg} + z_d + z_s + \left(\frac{V_d^2}{2g} - \frac{V_s^2}{2g}\right)$$

where H = total pump head, ft
P_g = discharge gauge reading, lb/in²
h_{sg} = suction gauge reading, ftHg
γ = specific weight of liquid being pumped, lb/ft³
γm = specific weight of mercury, lb/ft³
V_d = average velocity in discharge pipe, ft/s
V_s = average velocity in suction pipe, ft/s
z_d and z_s = elevation, ft

NOTE: For water at 68°F, 1 lb/in² = 2.3107 ft, and $(144/\gamma)P_g = 2.3107P_g$. The specific gravity of mercury at 68°F is $\gamma m/\gamma = 13.57$. Therefore, for water and mercury at 68°F, the formula simplifies to

$$H = 2.3107P_g + 13.57h_{sg} + z_d + z_s + \left(\frac{V_d^2}{2g} - \frac{V_s^2}{2g}\right)$$

Example: Determine the head developed, given a pump with a 10-in suction and 8-in discharge with a capacity of 2000 gal/min. The suction U tube reads 11 inHg and the discharge-pressure gauge 50 lb/in². The distance from the pump centerline to the U-tube connection is 7 in and to the center of the discharge gauge + 18 in. Therefore,

$P_g = 50$ $V_d = 12.8$ ft/s*

$h_{sg} = \frac{11}{12}$ or 0.916 ft $V_s = 8.17$ ft/s*

$z_d = 1.5$ ft $\frac{V_d^2}{2g} = 2.55†$

$z_s = \frac{7}{12}$ or 0.583 ft $\frac{V_s^2}{2g} = 1.03†$

*From Table 3-11.
†From Table 3-4.

Substituting in the formula,

$$H = 2.3107 \times 50 + 13.57 \times 0.916 + 1.5 + 0.583 + (2.55 - 1.03)$$
$$= 115.5 + 12.4 + 1.5 + 0.6 + 1.5 = 131.5 \text{ ft}$$

Determination of Power Required The work required for pumping depends on the total head and the weight or volume of the liquid to be pumped in a given time. This will give a theoretical liquid horsepower (Lhp) as expressed in the following formula:

$$\text{Lhp} = \frac{wH}{33,000}$$

where w = lb liquid pumped per minute
H = total head, ft of liquid

When the liquid is water at $68°$F weighing 62.318 lb/ft^3, the formula becomes

$$\text{Lhp} = \frac{\text{gal/min} \times H}{3960}$$

If a liquid other than water is pumped or water at a temperature other than $68°$F, the formula must be corrected for the specific gravity of the liquid so that

$$\text{Lhp} = \frac{\text{gal/min} \times H \times \text{sg}}{3960}$$

where sg = specific gravity of the liquid referred to water at $68°$F.

If the total head is expressed as pounds per square inch, the specific gravity does not enter into the final equation, which is as follows:

$$\text{Lhp} = \frac{\text{gal/min} \times (\text{total head, lb/in}^2)}{1714}$$

The theoretical horsepower is less than the actual or brake horsepower because of losses in the pump such as friction and leakage. The efficiency of the pump is therefore the ratio of the liquid-horsepower output to the brake-horsepower input, or

$$\text{Efficiency, percent} = \frac{\text{Lhp}}{\text{bhp}} \times 100$$

and the bhp required at the coupling will be

$$\text{bhp} = \frac{\text{Lhp}}{\eta}$$

where η = pump efficiency, percent/100.

If the entire unit is considered, it is necessary to include the efficiency of the driver and any other associated equipment to arrive at the overall efficiency.

Pump Performance

When studying the flow of liquid through a pump impeller, three types of velocities must be considered. The first is the velocity of a point on the impeller and is designated by the symbol U. The second is the velocity of the fluid relative to the casing, known as the absolute velocity and designated by the symbol C. The third is the velocity of the liquid relative to the impeller, known as the relative velocity and designated by the symbol W. The relative velocity W is found by taking the vector difference of the absolute velocities U and C. Subscripts are generally used with these symbols, the subscript 1 designating the velocities at the inlet to the vanes and the subscript 2 at the outlet of the vanes. The angle between the vectors U and W is designated as β and is the angle which the vane makes with a tangent to the impeller. The angle α between vectors C and U represents the angle at which the fluid enters or leaves the wheel. This is illustrated in Fig. 8-5 for a typical impeller.

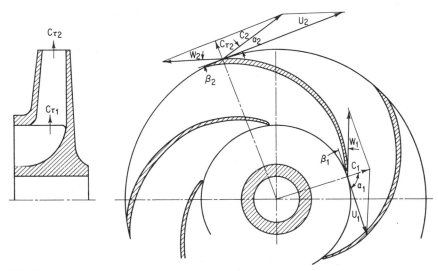

FIG. 8-5

The ideal head developed by the wheel is given by the equation

$$H_i = \frac{U_2^2 - U_1^2}{2g} + \frac{W_1^2 - W_2^2}{2g} + \frac{C_2^2 - C_1^2}{2g}$$

where the first term represents the head due to the centrifugal action, the second that due to the change in the relative velocity, and the third that due to the change in the absolute velocity. The first two terms represent the pressure head which is developed in the impeller, while the last is the velocity head developed in the impeller and converted into pressure in the volute or diffuser. This expression may also be written

$$H_i = \frac{1}{g}(U_2 C_2 \cos \alpha_2 - U_1 C_1 \cos \alpha_1)$$

The actual head developed by the pump will be less than the ideal owing to fluid friction and shock losses and to a circulatory flow which takes place between the vanes.

The basic formula can be expressed in a simplified form for the actual head developed by a pump impeller at rated design conditions as follows:

$$H = \mu \frac{U_2^2}{g}$$

where H = actual head, ft
  U_2 = impeller-tip velocity, ft/s
  g = acceleration of gravity, ft/s² (32.17 ft/s²)
  μ = pressure coefficient (varying from 0.45 to 0.52 for radial-type impellers)

From this it follows that the impeller diameter will be

$$D_2 = \frac{1300}{n} \sqrt{\frac{H}{\mu}}$$

where D_2 = impeller diameter, in
  n = pump speed, r/min

The above equations give the basis for estimating the effect of speed and impeller changes on the head. The weight or volume flow is directly proportional to the revolutions per minute of the pump. It may be observed from the above equations that the head is proportional to the square of the r/min. The horsepower required to drive the pump is the product of the head and weight flow divided by a constant. The efficiency of a pump is little affected by reasonable changes in either the speed or the impeller diameter.

The net effect of changes in the outside diameter of the impeller is similar to that of varying the speed of the unit.

The effect of changes in operating conditions may be summarized by the following equations, where the subscript a refers to the original condition while b refers to the new.

$$\text{Volume or weight flow } Q_b = Q_a \frac{n_b}{n_a} \frac{D_b}{D_a}$$

$$\text{Head } H_b = H_a \left(\frac{n_b}{n_a}\right)^2 \left(\frac{D_b}{D_a}\right)^2$$

$$\text{Horsepower hp}_b = \text{hp}_a \left(\frac{n_b}{n_a}\right)^3 \left(\frac{D_b}{D_a}\right)^3$$

$$\text{Efficiency } \eta_b = \eta_a$$

where n = speed, r/min
  D = impeller outside diameter, in

These relations may be applied over the entire range of a pump characteristic curve but should be used only for relatively small changes in speed or impeller diameter.

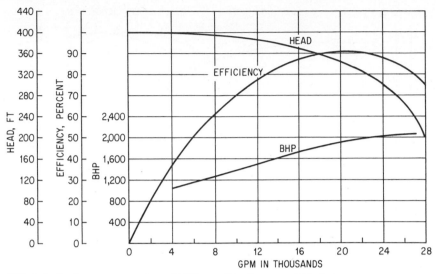

FIG. 8-6   Performance curve for a 24/20 centrifugal pump at 885 r/min.

Example: Assume that a pump is delivering 2500 gal/min of water against a head of 150 ft when running 1760 r/min with an efficiency of 81 percent. The brake horsepower is then 117. The outside diameter of the impeller is 13½ in. What would the performance be if the impeller diameter were reduced to 13 in and the pump speeded up to 1800 r/min?

$$Q_b = Q_a \frac{n_b D_b}{n_a D_a} = 2500 \frac{1800}{1760} \frac{13}{13.5} = 2460 \text{ gal/min}$$

$$H_b = H_a \left(\frac{n_b}{n_a}\right)^2 \left(\frac{D_b}{D_a}\right)^2 = 150 \left(\frac{1800}{1760}\right)^2 \left(\frac{13}{13.5}\right)^2 = 146.7 \text{ ft}$$

$$\text{hp}_b = \text{hp}_a \left(\frac{n_b}{n_a}\right)^3 \left(\frac{D_b}{D_a}\right)^3 = 117 \left(\frac{1800}{1760}\right)^3 \left(\frac{13}{13.5}\right)^3 = 112 \text{ hp}$$

$$\eta_b = \eta_a = 81 \text{ percent}$$

Figure 8-6 shows a typical performance curve for a centrifugal pump. It illustrates the variations of head, power, and efficiency as a function of capacity at constant speed. These curves are called *characteristic curves* of pump performance, and they are determined experimentally for each pump type.

Figure 8-7 shows diagrammatically the characteristic curves of a pump at two speeds N_1 and N_2, where points of similar flow conditions are related according to the formulas given above.

Parallel and Series Operation

When pumping requirements are variable, it may be desirable to install several small pumps in parallel rather than use a single large pump. When the demand

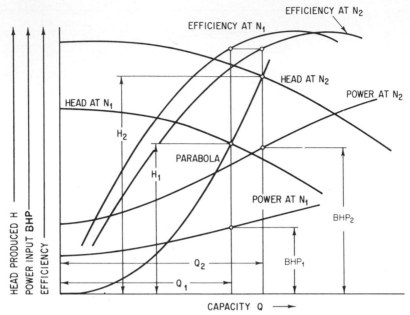

FIG. 8-7 Characteristic curves at two speeds N_1 and N_2.

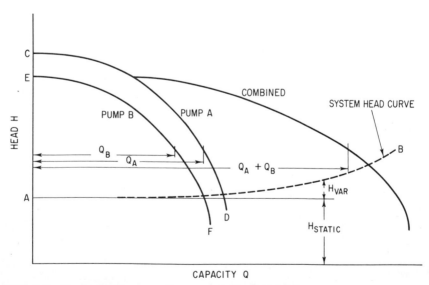

FIG. 8-8 Head-capacity curves of pumps operating in parallel.

drops, one or more smaller pumps may be shut down, thus allowing the remainder to operate at or near peak efficiency. If a single pump is used with lowered demand, the discharge must be throttled (for constant speed), and it will operate at reduced efficiency. Moreover, when smaller units are used, opportunity is provided during slack-demand periods for repairing and maintaining each pump in turn, thus avoiding the plant shutdowns which would be necessary with single units. Similarly, multiple pumps in series may be used when liquid must be delivered at high heads.

In planning such installations, a head-capacity curve for the system must first be drawn. The head required by the system is the sum of the static head (difference in elevation and/or its pressure equivalent) plus the variable head (friction and shock losses in the pipes, heaters, etc.). The former is usually constant for a given system, whereas the latter increases approximately with the square of the flow. The resulting curve is represented as line *AB* in Figs. 8-8 and 8-9.

Connecting two pumps in parallel to be driven by one motor is not a very common practice, and offhand such an arrangement may appear more expensive than a single pump. However, it should be remembered that in most cases it is possible to operate such a unit at about 40 percent higher speed, which may reduce the cost of the motor materially. Thus, the cost of two high-speed pumps may not be much greater than that of a single slow-speed pump.

For units to operate satisfactorily in parallel, they must be working on the portion of the characteristic curve which drops off with increased capacity in order to secure an even flow distribution. Consider the action of two pumps operating in parallel. The system head-capacity curve *AB* shown in Fig. 8-8 starts at *H* static when the flow is zero and rises parabolically with increased flow. Curve *CD* represents the characteristic curve of pump *A* operating alone; the similar curve for pump *B* is represented by *EF*. Pump *B* will not start delivery until the discharge pressure of pump *A* falls below that of the shutoff head of *B* (point *E*). The combined delivery for a given head is equal to the sum of the individual capacities of the two pumps at that head. For a given combined delivery head, the capacity is divided between the pumps and designated as Q_A and Q_B. The combined characteristic curve shown on the figure is found by plotting

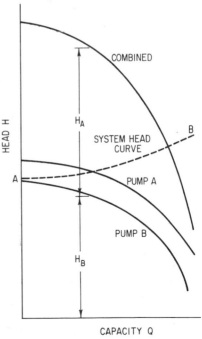

FIG. 8-9 Head-capacity curves of pumps operating in series.

these summations. The combined brake-horsepower curve can be found by adding the brake horsepower of pump A corresponding to Q_A to that of pump B corresponding to Q_B and by plotting this at the combined flow. The efficiency curve of the combination may be determined by dividing the combined power $\gamma(Q_A + Q_B)H/550$ by the corresponding combined brake horsepower (Q taken as ft^3/s).

If two pumps are operated in *series*, the combined head for any flow is equal to the sum of the individual heads as shown in Fig. 8-9. The combined brake-horsepower curve may be found by adding the horsepowers given by the curves for the individual pumps. Points on the combined efficiency curve are found by dividing the combined fluid horsepower $(H_A + H_B)Q\gamma/550$ by the combined brake horsepower; Q is again in ft^3/s.

Pump Performance for Viscous Liquids

As the viscosity of the liquid being handled by a centrifugal pump is increased, the effect on the performance is a marked increase in the brake horsepower, a reduction in the head, and some reduction in the capacity.

Figure 8-10 is taken from the *Hydraulic Institute Standards* and may be used to estimate the magnitude of these effects for a particular liquid. The diagram should be used only for the conventional radial-type impeller and should not be extrapolated or used for nonuniform liquids such as gels, slurries, or paper stock or when the net positive suction head is inadequate.

The procedure for selecting a pump for a given head-capacity–viscosity requirement when the desired capacity and head of the viscous liquid and the viscosity and specific gravity at the pumping temperature are known is as follows. Enter the diagram at the base with the desired viscous capacity Q_{vis} and proceed upward to the desired viscous head H_{vis} in feet of liquid. For multistage pumps use the head per stage. Proceed horizontally (either left or right) to the fluid viscosity, and then go upward to the correction curves. Divide the viscous capacity Q_{vis} by the capacity correction factor C_Q to get the approximate equivalent water capacity Q_w. Divide the viscous head H_{vis} by the head correction factor C_H from the curve labeled "$1.0 \times Q_N$" to get the approximate equivalent water head H_w. Having this new equivalent water head-capacity point, select a pump in the usual manner. The viscous efficiency and the viscous brake horsepower may then be calculated.

To illustrate, assume that it is desired to select a pump to deliver 750 gal/min at 100 ft total head of a liquid having a viscosity of 1000 SSU and a specific gravity of 0.90 at the pumping temperature. Enter the diagram at 750 gal/min, go up to 100 ft head, over to 1000 SSU, and then up to the correction factors:

$$C_Q = 0.95 \qquad Q_W = \frac{Q_{vis}}{C_Q} = \frac{750}{0.95} = 790 \text{ gal/min}$$

$$C_H = 0.92 \text{ for } Q_N = 1.0 \qquad H_W = \frac{H_{vis}}{C_H} = \frac{100}{0.92} = 108.8 \text{ ft head, or roughly 109 ft}$$

Select a pump for a water capacity of 790 gal/min at 109 ft head. The selection should be at or close to the maximum efficiency point for water performance.

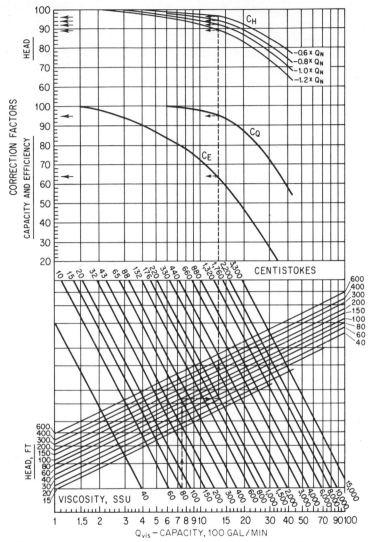

FIG. 8-10 Performance correction diagram. *(Courtesy of the Hydraulic Institute.)*

If the pump selected has an efficiency with water of 81 percent at 790 gal/min, then the efficiency for the viscous liquid will be

$$E_{vis} = C_E \times E_W = 0.635 \times 81 \text{ percent} = 51.5 \text{ percent}$$

The brake horsepower for pumping the viscous liquid will be

$$\text{bhp}_{vis} = \frac{Q_{vis} \times H_{vis} \times sg}{3960 \times E_{vis}} = \frac{750 \times 100 \times 0.90}{3960 \times 0.515} = 33.1 \text{ hp}$$

The procedure to determine the pump performance on a viscous liquid when its performance with water is known is as follows. From the efficiency curve locate the water capacity $(1.0 \times Q_N)$ at which maximum efficiency is obtained. From this capacity determine the capacities $0.6 \times Q_N$, $0.8 \times Q_N$, and $1.2 \times Q_N$. Enter the diagram at the bottom with the capacity at best efficiency $(1.0 \times Q_N)$, go upward to the head developed (in one stage) H_w at this capacity, then horizontally (either left or right) to the desired viscosity, and then proceed upward to the various correction curves. Read the values of C_E and C_Q and of C_H for all four capacities. Multiply each capacity by C_Q to obtain the corrected capacities. Multiply each head by its corresponding head correction factor to obtain the corrected heads. Multiply each efficiency value by C_E to obtain the corrected efficiency values, which apply at the corresponding corrected capacities. The head at shutoff can be taken as approximately the same as that for water. Calculate the viscous brake horsepower from the equation

$$\text{bhp}_{\text{vis}} = \frac{Q_{\text{vis}} \times H_{\text{vis}} \times \text{sg}}{3960 \times E_{\text{vis}}}$$

Net Positive Suction Head, or NPSH

If the pressure at any point inside a pump drops below the vapor pressure corresponding to the temperature of the liquid, the liquid will vaporize. These bubbles of vapor will be carried along to a point of higher pressure where they will suddenly collapse. This phenomenon is known as *cavitation*. It is accompanied by removal of metal in the pump, reduced flow, loss in efficiency, and noise and hence should be avoided. It occurs around the pump suction and inlet edge of the vanes when the absolute suction pressure is low.

The net positive suction head (NPSH) of a pump is the equivalent total head at the pump centerline corrected for vapor pressure. It is found from the equation

$$\text{NPSH} = H_p + H_z - H_{vp} - H_f$$

H_p is the head corresponding to the absolute pressure on the surface of the liquid from which the pump draws. This will be the barometric pressure if the tank is open to the atmosphere or the absolute pressure in the closed tank or condenser from which the pump takes liquid. In a deaerating heater, it is normally the saturated pressure at the existing water temperature. H_z is the height in feet of the fluid surface above or below the pump centerline. If above, it is considered to be plus since the suction head is then increased; if below, it is minus. H_{vp} is the head corresponding to the vapor pressure at the existing temperature of the liquid. H_f is the head lost because of friction and turbulence between the surface of the liquid and the pump suction flange.

In designing a pump installation and purchasing a pump, there are two types of NPSH to be considered. One is the *available* NPSH of the system, and the other is the *required* NPSH of the pump to be placed in the system. The former is determined by the plant designer and is based upon the pump location, fluid temperature, etc., while the latter is based upon suppression pump tests of the manufacturer. To secure satisfactory operating conditions, the available NPSH must be greater than the required NPSH. In higher-energy pumps such as boiler-feed pumps, values of NPSHA should be 1.5 to 2.0 times larger than NPSHR (as normally measured with a 3 percent drop in head).

The calculation of available NPSH will be illustrated by two examples. The head corresponding to a given pressure is given by the equation $H_p = 2.31p/\text{sg}$, where p is the pressure in pounds per square inch and sg the specific gravity of the liquid. The atmospheric pressures corresponding to various altitudes are given in Table 3-16; the vapor pressure and specific gravity of water are given in Table 3-1.

Assume that water at 80°F is to be pumped from a sump. The unit is located at an altitude of 800 ft above sea level, and the suction lift (from water surface to pump centerline) is 7 ft. The pipe losses amount to 1 ft head. What is the available NPSH?

From Table 3-16 the atmospheric pressure at an altitude of 800 ft is 14.27 psia. From Table 3-1 the specific gravity of the water at 80°F is 0.9984, and the vapor pressure is 0.5069 psia.

$$H_p = \frac{2.31p}{\text{sg}} = \frac{2.31 \times 14.27}{0.9984} = 32.97 \text{ ft}$$

$$H_z = -7 \text{ ft (negative since it is a lift)}$$

$$H_{vp} = \frac{2.31p}{\text{sg}} = \frac{2.31 \times 0.5069}{0.9984} = 1.17 \text{ ft}$$

$$H_f = 1 \text{ ft}$$

$$\text{NPSH} = H_p + H_z - H_{vp} - H_f = 32.97 - 7 - 1.17 - 1 = 23.80 \text{ ft}$$

Determine the available NPSH of a condensate pump drawing water from a condenser in which a 28-in vacuum, referred to a 30-in barometer, is maintained. The friction and turbulence head loss in the piping is estimated to be 2 ft. The minimum height of water in the condenser above the pump centerline is 5 ft. The absolute pressure in the condenser is $30 - 28 = 2$ inHg, or 0.982 lb/in². The corresponding specific gravity from Table 3-1 is 0.9945.

$$H_p = H_{vp} = \frac{2.31p}{\text{sg}} = \frac{2.31 \times 0.982}{0.9945} = 2.28 \text{ ft}$$

$$H_f = 2 \text{ ft} \qquad H_z = 5 \text{ ft}$$

$$\text{NPSH} = H_p + H_z - H_{vp} - H_f = 2.28 + 5 - 2.28 - 2 = 3 \text{ ft}$$

A third example is that of a deaerating heater having a water level 180 ft above the pump centerline. The water temperature is 350°F (from Table 3-1 sg = 0.892, and from Table 3-1 vapor pressure = 134.6 psia). The pipe friction loss is 12 ft. Since the water in the deareator is in a saturated condition, this absolute pressure on the surface liquid equals the vapor pressure at 350°F. Then

$$\text{NPSH} = H_p + H_z - H_{vp} - H_f$$
$$= \frac{2.31 \times 134.6}{0.892} + 180 - \frac{2.31 \times 134.6}{0.892} - 12 = 168 \text{ ft}$$

The required NPSH must be determined in most cases by means of suppression tests. The Hydraulic Institute has prepared a series of diagrams to estimate this required head (see Figs. 8-11 to 8-14 inclusive). These diagrams are not to be considered as the highest values which can be obtained by careful design, but they may be used for estimating as they represent average results of good present-day practice.

The use of the diagrams is simple and may be illustrated by an example. A double-suction pump operating at 3600 r/min delivers 1000 gal/min against a total head of 200 ft. What should the minimum NPSH be for satisfactory operation? The specific speed as found from Fig. 8-2 is 2200. By referring to Fig. 8-

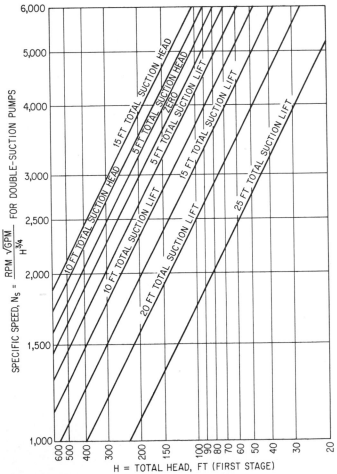

FIG. 8-11 Upper limits of specific speeds: double-suction pumps handling clear water at 85°F at sea level. *(Courtesy of the Hydraulic Institute.)*

11, the point corresponding to this specific speed for double-suction pumps and a total dynamic head of 200 ft gives a 12-ft suction lift as the safe maximum. If the same conditions were applied to a single-suction pump with the shaft through the eye of the impeller, the safe minimum suction condition would require at least a 1 ft positive head (i.e., the suction head would have to be at least + 1 ft rather than − 12 ft; hence the required NPSH would be 35 ft instead of 22 ft).

These curves are based upon handling clear water at 85°F and sea-level barometric pressure. If the water temperature is higher, the difference in head corresponding to the difference in vapor pressures between 85°F and the temperature

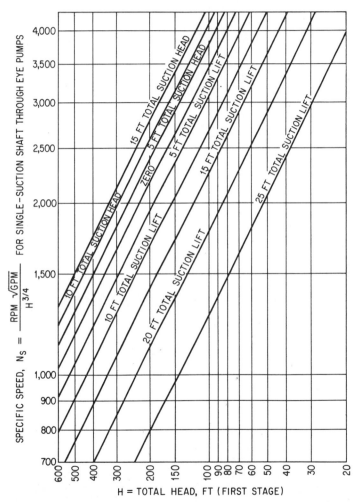

FIG. 8-12   Upper limits of specific speeds: single-suction shaft through eye pumps handling clear water at 85°F at sea level. *(Courtesy of the Hydraulic Institute.)*

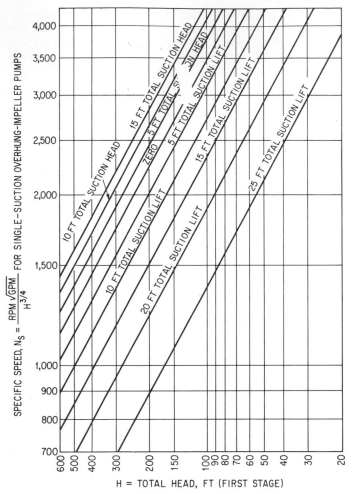

FIG. 8-13 Upper limits of specific speeds: single-suction overhung impeller pumps handling clear water at 85°F at sea level. *(Courtesy of the Hydraulic Institute.)*

of the water pumped should be subtracted from the suction lift or added to the suction head. Also, if the unit is to be located above sea level, the difference in head corresponding to the difference in atmospheric pressures should be subtracted from the suction lift or added to the suction head.

Thus, in the above example, if the water temperature is 140°F and the plant is located at an altitude of 2000 ft, the correction for vapor pressure (see Table 3-1) will be $2.889 - 0.596 = 2.293$ lb/in², and for altitude (see Table 3-16) will be $14.69 - 13.66 = 1.03$ lb/in². The corresponding head change will be 2.31 $(2.293 + 1.03)/0.9850 = 7.8$ ft. For the double-suction pump the maximum suction lift would be $12.0 - 7.8 = 4.2$ ft, and for the single-suction pump the positive suction head would have to be $1.0 + 7.8 = 8.8$ ft.

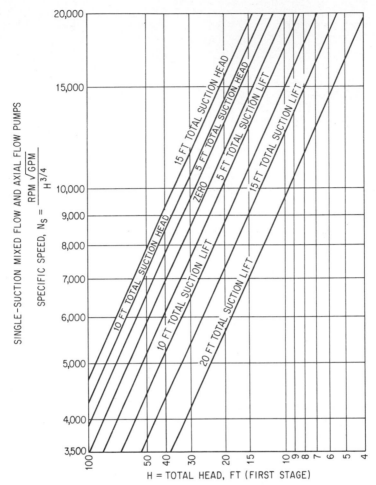

FIG. 8-14 Upper limits of specific speeds: single-suction mixed- and axial-flow pumps handling clear water at 85°F at sea level. *(Courtesy of the Hydraulic Institute.)*

A series of diagrams (Figs. 8-15, 8-16, and 8-17) have been prepared by the Hydraulic Institute to determine NPSH on the basis of the flow, operating speed, and discharge pressure for hot-water and condensate pumps. They may also be used to find the maximum permissible flow for a given available NPSH.

Hot Water Two curves, Figs. 8-15 and 8-16, have been prepared for pumps handling hot water at temperatures of 212°F and above. These curves show the recommended minimum NPSH in feet for different design capacities and speeds.

Figure 8-15 applies to single-suction pumps and Fig. 8-16 to double-suction pumps. These curves serve as guides in determining the NPSH for hot-water pumps and do not necessarily represent absolute minimum values.

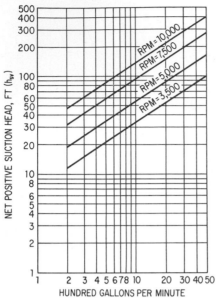

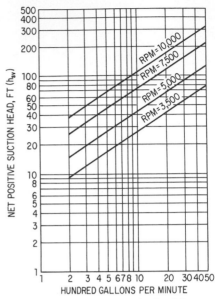

FIG. 8-15   Net positive suction head for centrifugal hot-water pumps, single-suction. *(Courtesy of the Hydraulic Institute.)*

FIG. 8-16   Net positive suction head for centrifugal hot-water pumps, double-suction, first-stage. *(Courtesy of the Hydraulic Institute.)*

Net Positive Suction Head for Condensate Pumps   Figure 8-17 indicates NPSH for condensate pumps with the shaft passing through the eye of the impeller. It applies to pumps having a maximum of three stages, the lower scale representing single-suction pumps and the upper scale double-suction pumps or pumps with a double-suction first-stage impeller.

For single-suction overhung impellers the curve may be used by dividing the specified capacity, if 400 gal/min or less, by 1.2, and if greater than 400 gal/min, by 1.15.

The curve may be used for capacities and speeds other than shown by the relation that, for a definite NPSH, the product of r/min $\times \sqrt{\text{gal/min}}$ remains constant.

Minimum Flow-Through Pump

The difference between the power put into a centrifugal pump and the useful power performed by the pump, or the difference between the brake and water horsepowers, is converted into heat, most of which appears as increased temperature of the water. The brake-horsepower curve of a pump rated at 500 gal/min against a 2600-ft head and the corresponding water-horsepower curve are shown in Fig. 8-18. If we neglect bearing losses, which are minor, the difference between these curves at any capacity represents the horsepower absorbed by the water in

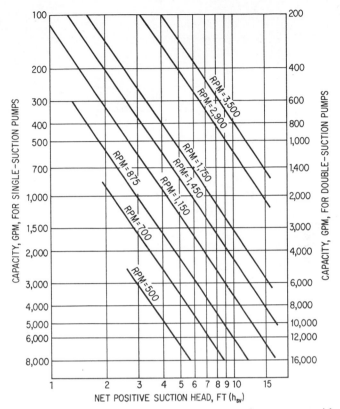

FIG. 8-17 Capacity and speed limitations for condensate pumps with the shaft through the eye of the impeller. *(Courtesy of the Hydraulic Institute.)*

the form of heat. Multiplying these differences by 42.4 gives the Btu generated in the pump per minute. Dividing these values by the flow in pounds per minute gives the temperature rise at each capacity. This curve is plotted in Fig. 8-18.

This temperature rise is generally not important in single-stage pumps, particularly if they are handling cold water, but for pumps handling hot liquids, such as boiler-feed pumps, it may become a serious matter. Then the resulting rapid temperature rise may cause the internal rotating parts to expand more rapidly than the heavier encircling parts so that severe rubbing may occur, or the impeller may even become loose on the shaft. Also the temperature of the water may rise to a point at which the water flashes into steam, causing the pump to become vapor-bound.

The temperature rise in a pump may be calculated from the formula

$$\Delta t = \frac{(1 - \eta)H}{778c\eta}$$

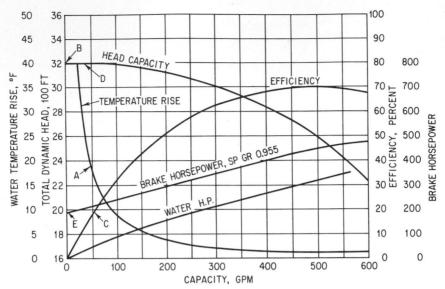

FIG. 8-18

where Δt = temperature rise, °F

η = overall pump efficiency, expressed as a decimal

c = specific heat of fluid being pumped (equals 1.0 for water)

H = total head of pump, ft

The allowable temperature rise of the water before it flashes into steam depends upon the suction conditions, or NPSH, of the pump. NPSH is the net head above the vapor pressure corresponding to the temperature of the liquid handled. As outlined in the subsection "Net Positive Suction Head, or NPSH," every installation has two types of NPSH: that available in the installation and that required by the pump. The maximum allowable vapor pressure at the pump inlet is found by converting the available NPSH into pounds per square inch and adding this to the vapor pressure corresponding to the temperature of the liquid being handled. This pressure is the vapor pressure corresponding to the temperature to which the liquid may be raised before it will flash into vapor.

After the allowable temperature rise has been established, an approximate minimum safe continuous-flow efficiency can be obtained by rewriting the previous equation in the form

$$\eta = \frac{H_{so}}{778(\Delta t)c + H_{so}}$$

where H_{so} = heat at no flow, or shutoff head

c = specific heat of liquid

The flow corresponding to this efficiency is found on the pump-performance curves.

In boiler-feed pumps having single-suction impellers, all facing in the same direction, a leak-off balancing arrangement is used to compensate hydraulic thrust. If the balancing leak-off flow is returned to the suction of the pump, flashing can occur at extremely low rates of delivery. Therefore, the balancing flow is frequently piped to an open heater in the feed system of the pump, where, by flashing, the temperature of the water will be reduced to that corresponding to the pressure and there should be no valve of any kind between the junction of the balancing connection and the heater. Many installations do pipe the balance flow to the pump suction line.

With the advent of the larger, higher-energy boiler-feed pumps of the 1960s and 1970s, it became apparent that the higher vibration and pressure pulsations occurring at partial capacities would affect the minimum-flow setting. Consequently, minimum-flow values of 25 percent of the flow at best efficiency became commonplace, with higher levels in special cases.

Recirculation Connection

The minimum flow through a pump is directed through a recirculation line from a discharge-line connection to the suction source. This line has a recirculation orifice designed to pass the required minimum flow or has a modulating-pressure breakdown valve.

The recirculation connection is in the pump discharge line between the discharge nozzle and the check valve. All valves in the recirculation line must be open whenever the pump is operating under any of the following conditions:

1. Low flows
2. Starting pump
3. Stopping pump

The valves should be opened or closed either manually or by automatic controls. When automatic controls are used, they should be checked at initial starting and occasionally thereafter during starting procedures.

Figure 8-19 shows a diagram of the piping arrangement.

Materials for Pumping Various Liquids

The materials used for pumps must be suitable for the liquids handled to prevent excessive corrosion. The *Hydraulic Institute Standards* give the materials to be used for the more common liquids and should be consulted for selecting the applicable material combination.

From the standpoint of materials pumps may be divided into three basic types: standard fitted (combination of iron and bronze), all iron, and all bronze. Other materials, including corrosion-resisting steels, are listed in the subject standards. Table 8-2 gives a summary of the various materials used for centrifugal pumps.

If the liquid to be handled is an electrolyte, the use of dissimilar metals in close combination, especially those that are widely separated in the galvanic series, should be avoided insofar as possible. The use of bronze and iron in the same pump handling seawater will greatly accelerate the corrosion of the cast-iron parts.

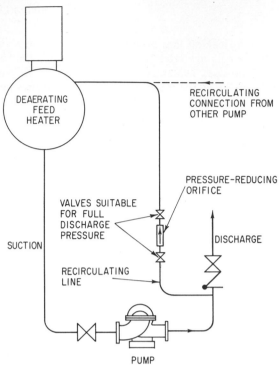

FIG. 8-19 Recirculating connections for boiler-feed service.

TABLE 8-2 Summary of Material Selections and National Society Standards Designations*

Corresponding national society standards designation†			Remarks
ASTM	ACI	AISI	
A48, Classes 20, 25, 30, 35, 40, and 50	. . . . .	. . .	Gray iron—six grades
A339, A395, and A396	. . . . .	. . .	Nodular cast iron—five grades
B143, 1B and 2A; B144, 3A; B145, 4A	. . . . .	. . .	Tin bronze—six grades (includes two grades not covered by ASTM Specifications)
A216—WCB	. . . . .	1030	Carbon steel
A217—C5	. . . . .	501	5% chromium steel
A296—CA15	CA15	410	13% chromium steel
A296—CB30	CB30	. . .	20% chromium steel
A296—CC50	CC50	446	28% chromium steel
A296—CF-8	CF-8	304	18-8 austenitic steel
A296—CF-8M	CF-8M	316	18-8 molybdenum austenitic steel
A296—60T	CN-7M	. . .	A series of highly alloyed steels normally used where the corrosive conditions are severe
. . . .	. . . . .	. . .	A series of nickel-base alloys
. . . .	. . . . .	. . .	High-silicon cast iron
. . . .	. . . . .	. . .	Austenitic cast iron
A439	. . . . .	. . .	Nodular austenitic cast iron
. . . .	. . . . .	. . .	Nickel-copper alloy
. . . .	. . . . .	. . .	Nickel

*Reprinted from *Hydraulic Institute Standards,* 11th ed., Hydraulic Institute, New York, 1965.
†ASTM = American Society for Testing and Materials; ACI = Alloy Casting Institute; AISI = American Iron and Steel Institute.

A table of the galvanic series is given below.

Galvanic Series of Metals and Alloys*

Corroded End (Anodic, or Least Noble)

Magnesium
Magnesium alloys

Zinc

Aluminum 2S

Cadmium

Aluminum 17ST

Steel or iron
Cast iron

Chromium stainless steel, 400 series (active)

Austenitic nickel or nickel-copper cast-iron alloy

18-8 chromium-nickel stainless steel, type 304 (active)
18-8-3 chromium-nickel-molybdenum stainless steel, type 316 (active)

Lead-tin solders
Lead
Tin

Nickel (active)
Nickel-base alloy (active)
Nickel-molybdenum-chromium-iron alloy (active)

Brasses
Copper
Bronzes
Copper-nickel alloy
Nickel-copper alloy

Silver solder

Nickel (passive)
Nickel-base alloy (passive)

Chromium stainless steel, 400 series (passive)
18-8 chromium-nickel stainless steel, type 304 (passive)
18-8-3 chromium-nickel-molybdenum stainless steel, type 316 (passive)
Nickel-molybdenum-chromium-iron alloy (passive)

Silver

Graphite
Gold
Platinum

Protected End (Cathodic, or Most Noble)

*Reprinted from the *Hydraulic Institute Standards,* 11th ed., Hydraulic Institute, New York, 1965

Pump Application

General   As outlined in the subsection "Classification," there are on the market a multitude of centrifugal-pump types. Some of the basic types will be mentioned here.

Single-Stage Double-Suction Pump See Fig. 8-20. This type of centrifugal pump is the most common one and is used for general service in industrial and municipal plants. In the larger sizes, the use of these pumps is almost universal for municipal-water distribution, and pumps of this type are in service in practically every major city of the United States. For heads up to 300 ft or higher, single-stage pumps are used, while for higher heads two or more units are

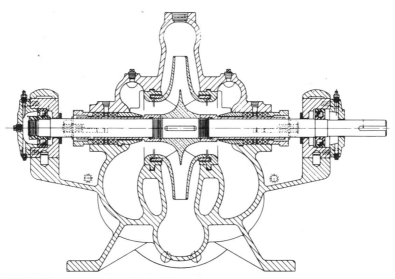

FIG. 8-20 Cross section of a double-suction pump.

FIG. 8-21 Installation of two double-suction pumps arranged in series, motor-driven.

arranged in series. Figure 8-21 shows an installation of two motor-driven units arranged in series.

Boiler-Feed Pumps   For industrial use, multistage split-case pumps having two to six or more stages are employed. See Figs. 8-22 and 8-23.

For utility service the pressures are now generally in the range of 2000 to 5000 lb/in^2, and barrel-type multistage units are used. See Fig. 8-24.

Hydraulic-Pressure Pumps   Pumps similar to those used in boiler-feed service are employed for this application.

Condensate Pumps   Special pumps, which may be either single-suction or double-suction, are used for this service; they are designed to operate at low submergence. Figure 8-25 shows a three-stage single-suction vertical unit of this type.

Nonclogging Pumps   Pumps for this service are designed to assure maximum freedom from clogging. They are usually of the single-suction type. For the larger sizes, mixed-flow pumps are used, as shown in Fig. 8-27.

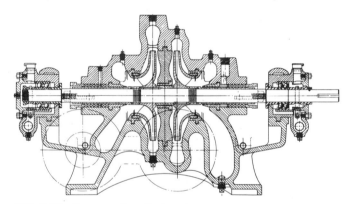

FIG. 8-22   Cutaway view of a two-stage pump.

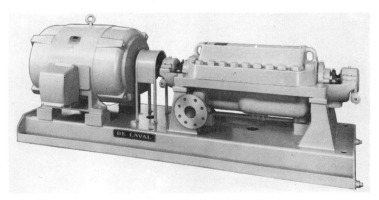

FIG. 8-23   Motor-driven multistage pump.

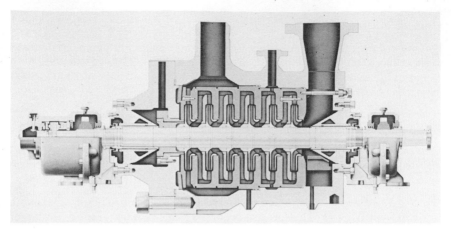

FIG. 8-24 Double-case boiler-feedwater pump.

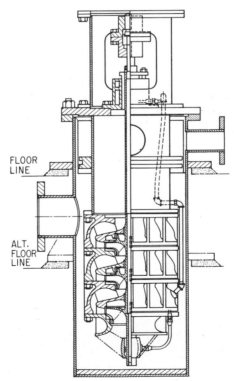

FLOOR LINE

ALT. FLOOR LINE

FIG. 8-25 Typical assembly of a three-stage condensate pump.

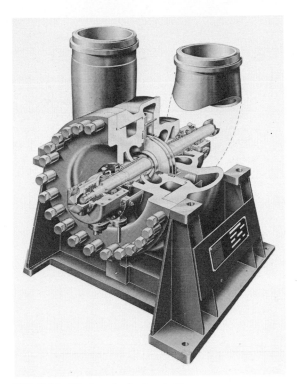

FIG. 8-26   Nuclear feedwater pump.

Design Details   The general arrangement of several basic centrifugal-pump types has been covered under "Pump Application." See Figs. 8-20 to 8-27. Some special design features are discussed below.

Axial Balance

The impeller of a single-suction centrifugal pump has an unbalanced hydraulic thrust directed axially toward the suction. This is due to the difference in pressure of the fluid which has passed through the impeller and of the fluid on the suction side. Several methods have been devised to counteract this force and avoid the use of large thrust bearings.

If a double-suction impeller (Fig. 8-20) is used, the pressures are symmetrical about the centerline and no unbalanced thrust should exist.

On multistage units single-suction impellers placed back to back, as illustrated in Fig. 8-22, may be employed. The axial thrust created in one impeller is thus balanced by the corresponding thrust in the other, any remaining thrust being taken by a small thrust bearing.

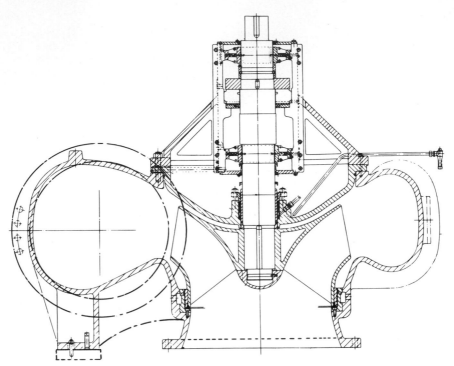

FIG. 8-27 Cutaway view of a mixed-flow pump.

An alternative method of automatically providing a balanced thrust arrangement in boiler-feed pumps is shown in Figs. 8-24 and 8-28. As shown in Fig. 8-28, water at essentially discharge pressure enters the clearance between the rotating and stationary drums at *A*. The water follows a path through the first fixed orifice (*A* to *B*), through the variable orifice (*B* to *C*), and then through the second fixed orifice (*C* to *D*). Chamber *D* is connected by a pipe to the suction source.

Should a condition of increased impeller thrust toward suction occur, the rotor tends to move toward suction, closing the variable orifice between the disk faces (*B* to *C*). By thus reducing the balance flow, the pressure drop between *A* and *B* decreases. The resulting greater pressure at *B* creates an increased thrust in the outboard direction, thereby providing self-compensation for the increased impeller thrust.

A similar type of self-compensating balance occurs if the impeller thrust toward suction should decrease. The rotor is free to move axially and hence permits the variable gap at the disk faces to match the requirements for hydraulic balance.

Still another method of balancing in-line multistage boiler-feed pumps is the use of a fixed-diameter balance drum similar to the *A*-to-*B* portion of Fig. 8-28. This method also is reliable, but it generally requires a larger thrust bearing.

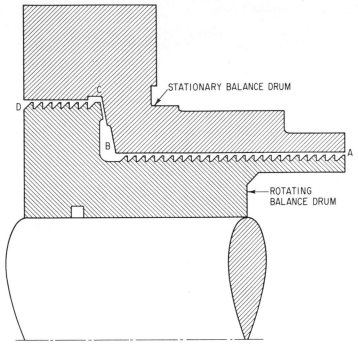

STATIONARY BALANCE DRUM

ROTATING
BALANCE DRUM

FIG. 8-28   Axial balance-drum and disk combination.

Double-Volute Pumps

The pressure within the casing of any pump develops radial forces when operating at capacities other than normal. The result of these radial forces in single-volute casings is shaft deflection. This is especially true if the pump operates for extended periods at other than design capacity, since imbalance from the radial forces then becomes greater. For such applications, double-volute pumps are used (Fig. 8-29).

In the double-volute casing, the water leaving the impeller is collected in two similar volutes, the tongues of which are set 180° apart. The two volutes merge into a common outlet to form the discharge of the pump. Hydraulic forces (indicated by arrows) produced by the pressure in one volute are balanced by equal forces produced by the pressure in the other volute. Thus, radial thrust is counterbalanced and for all practical purposes is eliminated.

The double-volute pumps provide insurance against shaft deflection and savings in repairs and shutdown time.

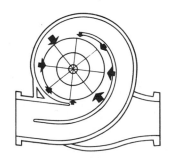

FIG. 8-29   Double-volute pump.

Centrifugal-Pump Stuffing Boxes

Stuffing boxes are located on pumps where the rotating shaft enters the pump case. They contain packing or mechanical seals which control the leakage of fluid from within or of air from without.

Stuffing Boxes   A typical stuffing box using packing, as shown in Fig. 8-30, has a plain throat bushing, seal ring, and packing gland. Figure 8-31 shows a water-cooled stuffing box with quench glands and breakdown bushing. Stuffing boxes may have various combinations of the features shown in Figs. 8-30 and 8-31, depending upon operating conditions.

The innermost packing ring is usually placed against a solid removable throat bushing. The plain bushing of Fig. 8-30 is used for suction lifts and moderate pressures. Higher pressures require the breakdown bushing of Fig. 8-31. It relieves the pressure on the packing, reduces mechanical losses, and lessens wear on the shaft sleeve.

Seal rings are placed between rows of packing (see Fig. 8-30). They provide a space surrounding the shaft for the sealing liquid. This forms a seal which, when a vacuum exists in the suction chamber of the pump, prevents air from entering. In addition, it assures lubrication for the packing. The liquid comes from either an outside source or a high-pressure portion of the pump.

Packing is held in the stuffing box by packing glands. These are usually split, making it possible to remove them without taking the pump apart. Quench glands are used when the liquid being pumped exceeds a safe margin on its vapor pres-

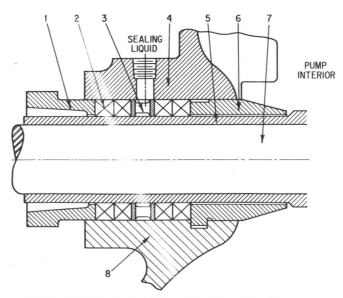

FIG. 8-30   (1) Packing gland; (2) packing; (3) seal ring; (4) pump cover; (5) shaft sleeve; (6) throat bushing; (7) shaft; (8) pump case.

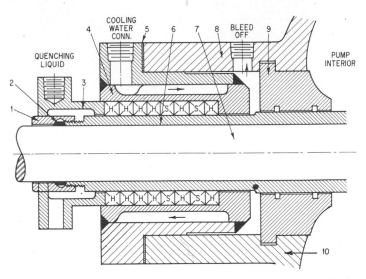

FIG. 8-31 (1) Packing nut; (2) packing ring; (3) quench gland; (4) packing box; (5) gasket; (6) shaft sleeve; (7) shaft; (8) pump cover; (9) breakdown bushing; (10) pump case; H = metallic packing; S = plastic packing.

sure. Glands are pulled into place by gland bolts. Frequently these are swing bolts, making disassembly easier.

Renewable shaft sleeves usually protect the shaft where it passes through the stuffing boxes. Bronze is most commonly used for cold-water applications. High suction pressures, elevated temperatures, dirty water, and many liquids call for special materials.

The packing-ring cross section is square. A good grade of braided asbestos impregnated with graphite is most commonly used, especially for water service. This often comes in a continuous coil which is cut into proper lengths for making rings. The cut should preferably be on a diagonal, with a slight gap to allow for expansion when put in place. Gaps for adjacent rings should be staggered. Metallic packing rings are required for some service, and one of the recommended combinations for high-pressure boiler-feed service is shown in Fig. 8-31. These packings are usually purchased in molded sets from the packing manufacturer. Also note the special water-jacketed stuffing box in Fig. 8-31.

Packing does its sealing along the shaft. It is pliable in order to form itself around the shaft. Cutting down the leakage by tightening up on the packing increases friction, resulting in more power required and increased wear on the shaft or shaft sleeve.

Packing requires a certain amount of leakage to keep it lubricated. Certain applications, such as corrosive acids and inflammable, gritty, or contaminated liquids, cannot allow leakage. Mechanical seals are now being used for these applications as well as for high-pressure water seals.

Mechanical Seals Mechanical seals consist of a stationary and a rotating member with some manner of auxiliary seal. The liquid in the stuffing box is prevented from passing along the shaft by means of the auxiliary seal (an O ring, bellows, wedge ring, or other device). The rotating member is forced against the stationary member by a spring or springs and by pressure in the stuffing box. The liquid tries to leak through the contacting surfaces and forms a liquid-pressure wedge which prevents the seal faces from actually making contact during operation.

Three general types of mechanical seals are used: single seal, double seal, and balanced seal.

A typical single-seal mechanical seal used for most applications is shown in Fig. 8-32. Hydraulic forces on the sealing washer are unbalanced, as shown in Fig. 8-33, limiting the use of single-seal mechanical seals to moderate pressures.

A typical *balanced* mechanical seal used for high pressures (see Figs. 8-34 and 8-35) shows how areas on each side of the sealing washer are adjusted to reduce unbalanced forces by stepping down the shaft. This keeps the pressure between sealing surfaces within allowable limits.

A typical double mechanical seal is shown in Fig. 8-36. Simply stated, this is two single seals mounted back to back with space for an isolating liquid between them. Such seals are used for liquids which have high temperatures or are gritty, corrosive, volatile, contaminating, etc.

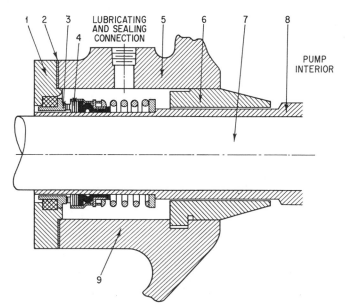

FIG. 8-32 Mechanical seal: (1) packing gland; (2) gasket; (3) stationary seal; (4) rotating seal; (5) pump cover; (6) throat bushing; (7) shaft; (8) shaft sleeve; (9) pump case.

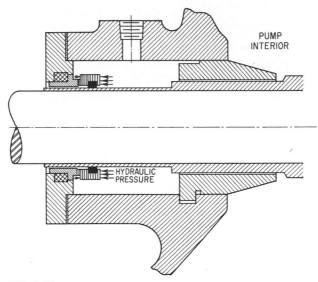

PUMP
INTERIOR

HYDRAULIC
PRESSURE

FIG. 8-33

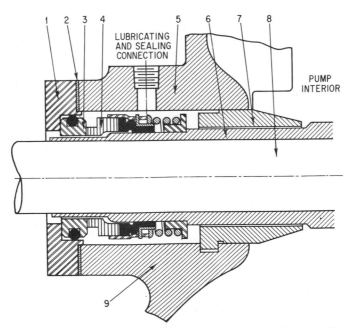

1 2 3 4

LUBRICATING
AND SEALING
CONNECTION

5 6 7 8

PUMP
INTERIOR

9

FIG. 8-34 Balanced mechanical seal: (1) packing gland; (2) gasket; (3) stationary seal; (4) rotating seal; (5) pump cover; (6) shaft sleeve; (7) throat bushing; (8) shaft; (9) pump case.

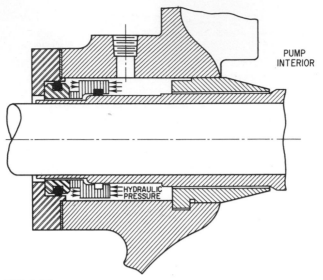

FIG. 8-35

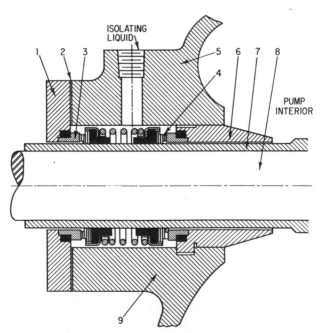

FIG. 8-36 Double mechanical seal: (1) packing gland; (2) gasket;
(3) stationary seal; (4) rotating seal; (5) pump cover; (6) throat
bushing; (7) shaft sleeve; (8) shaft; (9) pump case.

Seals for Boiler-Feed Pumps   Although in the past, the rubbing velocity of the contact surfaces has been a limitation, these velocities increased in the 1970s to 230 ft/s or more, and mechanical-seal applications have increased in boiler-feed pumps, particularly in Europe.

In boiler-feed-pump applications that exceed the limitations of packing or mechanical seals, serrated bushing seals and multifloating ring seals are generally used. Examples of the latter two types are shown in Figs. 8-37 and 8-38, respectively. In both cases, a cool external water supply—generally condensate—is injected into the seals. The cold injection water prevents flashing in the seals that would otherwise occur if the high-temperature water in the pump were permitted to flow through the seals to atmospheric pressure.

In differential-pressure control systems, a constant differential between seal injection-water pressure and seal chamber pressure permits some water to enter the seal chamber and the larger portion of the injection water to pass outboard to an atmospheric drain. This has the advantages of having no out-leakage from the pump and having only filtered water flowing from the seals.

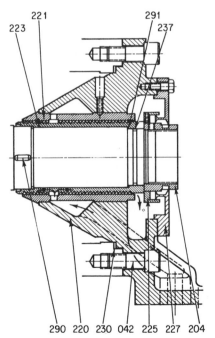

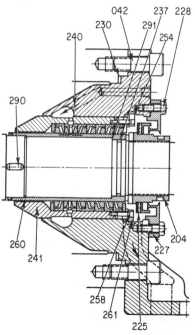

FIG. 8-37 Injection-type serrated bushing seal: (042) capscrew; (204) check nut; (220) housing; (221) bushing; (223) sleeve; (225) guard; (227) cover; (230) gasket; (237) O ring; (290) key; (291) O ring (metallic).

FIG. 8-38 Injection-type multifloating ring seal: (042) capscrew; (204) check nut; (225) guard; (227) cover; (228) bolt; (230) gasket; (237) O ring; (240) housing; (241) seal subassembly; (254) key; (258) capscrews; (260) sleeve; (261) ring; (290) key; (291) O ring (metallic).

In temperature-controlled systems, the amount of seal injection water is regulated by the temperature of the drain water from the seals. One advantage of this system is that normally no cold water enters the pump since all the seal injection water flows outward.

Although the examples in Figs. 8-37 and 8-38 show single cooling-water injection, other variations having double injection and/or additional bleed-off connections are used.

Warm-Up Procedure

In starting boiler-feed pumps, the warm-up procedure plays a vital role. It is important to eliminate temperature gradients within the pump to reduce rotor distortion prior to start-up. These temperature gradients are usually measured by monitoring case temperatures at locations as determined by the manufacturer. The highest temperature differential should not exceed 50°F, and the difference between the inlet temperature and the warmest part of the case should normally be not more than 50°F.

To obtain minimum temperature gradients, the warm-up water should enter the case through a warm-up connection at the bottom of the case and discharge through the suction nozzle. A recommended schematic is shown in Fig. 8-39, with water supplied from the bleed-off connection of the operating pump.

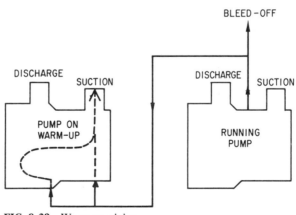

FIG. 8-39   Warm-up piping.

Alternative warm-up paths may be used, but case temperatures should be the controlling factors.

Effect of Operating Temperature on Pump Efficiency at Constant Speed*

Variations in temperature of the fluid pumped cause changes in the specific weight and viscosity, with resultant changes in the performance of the pump.

*Courtesy of the Hydraulic Institute.

Any reduction in specific weight, as caused by an increase in temperature, results in a directly proportional reduction in Lhp (as covered in the subsection "Determination of Power Required") and in input power; so the efficiency is not changed.

Reduced viscosity will have an influence on efficiency, and for pumps in the lower range of specific speed, such as high-pressure multistage boiler-feed pumps and large single-stage hot-water-circulating pumps, reduced viscosity will

1. Increase the internal leakage losses
2. Reduce disk friction losses
3. Reduce hydraulic skin friction or flow losses

The net effect of a reduction in viscosity due to higher temperature will depend on specific speed and on the design details of the pump. When substantiating data are available and when a high degree of understanding exists between the manufacturer and the user, consideration may be given to adjusting the performance data from a cold-water test to hot-water operating conditions on the basis of the following formula:

$$\eta_0 = 1 - (1 - \eta_t)\left(\frac{v_0}{v_t}\right)^n$$

where η_t = efficiency at test temperature, decimal value
η_0 = efficiency at operating temperature, decimal value
v_0 = kinematic viscosity at operating temperature
v_t = kinematic viscosity at test temperature
n = exponent to be established by manufacturers' data (probably in the range of 0.05 to 0.1)

Typical Example of Adjustment of Efficiency for Increased Temperature   A test on water at $100\,°F$ resulted in an efficiency of 80 percent. What will be the probable efficiency at $350\,°F$?

$$\eta_0 = 1 - (1 - \eta_t)\left(\frac{v_0}{v_t}\right)^n$$

$$\eta_0 = 1 - (1 - 0.80)\left(\frac{0.00000185}{0.0000076}\right)^{0.1}$$

$$\eta_0 = 1 - (0.2)0.868$$

$$\eta_0 = 0.826 = 82.6 \text{ percent}$$

Vibration Limits of Centrifugal Pumps

General   Recommendations for upper limits of vibration of centrifugal-pump units under field operating conditions are shown in the curves in Figs. 8-40 and 8-41. Figure 8-40 is for centrifugal pumps handling clean liquids, and Fig. 8-41 is for vertical or horizontal centrifugal nonclog dry-pit pumps.

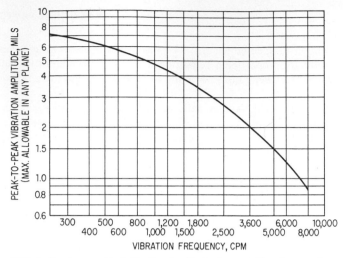

FIG. 8-40 Acceptable field vibration limits for centrifugal pumps handling clean liquids. Frequency corresponds to revolutions per minute when dynamic unbalance is the cause of vibration. Vertical pumps: measure vibration at top motor bearing. Horizontal pumps: measure vibration on bearing housing. *(Courtesy of the Hydraulic Institute.)*

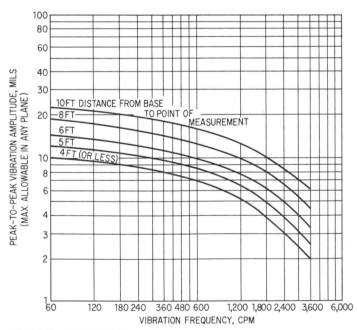

FIG. 8-41 Acceptable field vibration limits for vertical or horizontal centrifugal nonclog pumps. Frequency corresponds to revolutions per minute when dynamic unbalance is the cause of vibration. Vertical pumps: measure vibration at top motor bearing. Horizontal pumps: measure vibration on bearing housing. *(Courtesy of the Hydraulic Institute.)*

In using these curves, the following conditions apply:

1. Curves are applicable when the pump is operating at rated speed within plus or minus 10 percent of rated capacity.
2. Measurements should be made as indicated on the appropriate curve sheet.
3. Piping should be connected so as to avoid strains on the pump.

The curves should be used as a general guide, with recommendations that vibrations in excess of the curve may require investigation and correction. Often more important than the actual vibration itself is the change of vibration over a period of time. Vibrations in excess of the curves may often be tolerated if they show no increase over considerable periods of time and if there is no other indication of damage, such as an increase in bearing clearance.

Factors Affecting Vibration   A number of factors besides physical unbalance of the rotating parts may cause vibration. Among these are:

1. Resonance between the unit and its foundation or piping or resonance within the unit due to natural frequency of the pump, the motor frame, the motor-supporting pedestal, or the foundation. Resonant vibrations may also be caused by other equipment in operation in the area.
2. Operation at or near a critical speed. The amount of vibration observed will depend on the degree of unbalance and damping present. Normal design practice is to avoid a critical speed by approximately 25 percent.
3. Vibrations due to hydraulic disturbances caused by improper design of the suction piping or sump. Disturbances may also be caused by improperly designed valves, piping supports, piping, and other components exterior to the pump. Such vibrations are usually at random frequencies.
4. For the nonclog pumps, sudden increases in the vibration levels may be due to the passage of large solids through the pump. If the vibration condition persists, solids may be lodged in the impeller, and remedial measures should be taken to clear it.

Pump Installation and Operation

In addition to consideration of the correct suction head on the pump (NPSH), other precautions must be followed to ensure satisfactory operation. A few of these factors are:

1. *Accessibility.* The unit should be located where it may be easily inspected and repaired.

2. *Foundation.* The foundation should be heavy and rigid to avoid misalignment and vibration.

3. *Alignment.* The pump and driver should be correctly aligned to avoid excessive wear of the coupling, packing, and bearings. Pumps handling hot liquids should be aligned at their operating temperature.

4. *Piping.* Both suction and discharge lines should be independently supported near the pump to avoid strains on the casing. Piping should have as few bends as possible and be larger than the pump nozzles to reduce head losses.

If an expansion joint is installed in the piping between the pump and the nearest point of anchor in the piping, it should be noted that a force equal to the area of the expansion joint (which may be considerably larger than the normal pipe size) times the pressure in the pipe will be transmitted to the pump proper. Some slip-type couplings have the same effect. This force may exceed the allowable pump loading. If an expansion joint or slip-type coupling must be used, it is recommended that either an anchor be installed between it and the pump or that the joint be restrained or otherwise designed so as to prevent this force from being transmitted to the pump. If properly installed, this will eliminate the objectionable forces mentioned above.

The suction pipe should slope upward to the pump nozzle to avoid pockets in which dissolved air may be liberated (see Fig. 8-42). The reducer at the pump suction nozzle should be eccentric rather than a straight taper for the same reason. Any bends should have a long radius, and on the suction side they should be as far from the pump as possible.

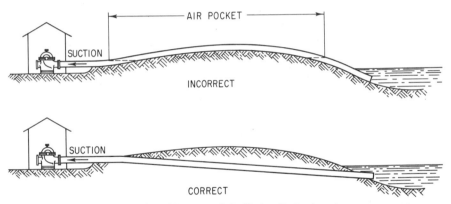

FIG. 8-42 Slope of suction pipe. *(Courtesy of the Hydraulic Institute.)*

5. *Valves.* A check valve and a gate valve should be placed in the discharge line. The former is placed next to the pump to prevent water from running back through the pump if the driver should fail and to protect the pump from excessive line pressure. The latter is used to regulate the flow and in priming.

Priming Before a centrifugal pump can deliver fluid, it is necessary that the suction side of the impeller be submerged and the suction line filled. The pump should never be run unless the impeller is filled with liquid since the wearing rings may rub and seize; also the packing must be lubricated by the liquid leaking past it. If air should leak into the suction line or casing, the unit may become air-bound and lose its prime, i.e., cease delivery. It is then necessary to stop the pump and prime it before starting it up again.

If the impeller is submerged below the water level, it is only necessary to open a petcock leading from the volute to release any entrapped air and ensure that the pump is primed. If the pump is above the water level, it may be primed by using either a water or a steam-jet exhauster or a small priming pump to remove the air from the pump casing. In such cases there must be a shutoff valve in the discharge line. Water from an outside source, such as a reservoir, or from a filled discharge line may be used to fill the suction line and pump. A foot valve or a check valve must then be placed below the water level in the suction line.

The foot valve is installed in the suction line for priming with low- or medium-suction lifts. Foot valves should not be used for high lifts since failure of the driver would cause the water to rush back suddenly and cause water hammer. A screen is placed before the foot valve to prevent foreign matter from lodging in it.

For large units or those located in remote localities, automatic priming devices which maintain the water level in the pump at a safe level continuously are employed.

Operating Difficulties

The following table, taken from the *Hydraulic Institute Standards* gives the causes of common operating difficulties:

1. No water delivered
   a. Pump not primed
   b. Speed too low*
   c. Discharge head too high
   d. Suction lift higher than for which pump is designed
   e. Impeller completely plugged up
   f. Wrong direction of rotation
2. Not enough water delivered
   a. Air leaks in suction or stuffing boxes
   b. Speed too low*
   c. Discharge head higher than anticipated
   d. Suction lift too high (Check with gauges. Check for clogged suction line or screen.)
   e. Impeller partially plugged up
   f. Not enough suction head for hot water
   g. Mechanical defects:
      Wearing rings worn
      Impeller damaged
      Casing packing defective
   h. Foot valve too small
   i. Foot valve or suction opening not submerged deep enough
3. Not enough pressure
   a. Speed too low*

---

*When the pump is direct-connected to an electric motor, check up whether the motor is across the line and receives full voltage. When the pump is direct-connected to a steam turbine, make sure that the turbine receives full steam pressure.

   b. Air in water

   c. Mechanical defects:
   Wearing rings worn
   Impeller damaged
   Casing packing defective

   d. Impeller diameter too small

4. Pump working for a while and then losing suction
   a. Leaky suction line

   b. Water seal plugged

   c. Suction lift too high

   d. Air or gases in liquid

5. Pump taking too much power
   a. Speed too high

   b. Head lower than rating; too much water pumped

   c. Specific gravity or viscosity too high

   d. Mechanical defects:
   Shaft bent
   Rotating element binding
   Stuffing boxes too tight
   Wearing rings worn
   Casing packing defective

ROTARY PUMPS

General

The rotary pump is one of the most versatile and widely used types of pump serving industry today. It is the vital heart of the fluid-power systems providing the muscle for most of the equipment involved in the aerospace age. It is the workhorse of the rapidly expanding fluid-power industry, which is providing much of the energy-transfer systems for today's highly sophisticated machines and tools. It is finding ever-widening use in diversified fields of application such as Navy and marine fuel-oil service, marine cargo, oil burners, crude oil, chemical processing, and lubricating service. Its broadest field of application is in the handling of fluids having some lubricating value and sufficient viscosity to prevent excessive leakage at required pressure.

The rotary pump is built in capacities from a fraction of a gallon to more than 5000 gal/min, with pressures ranging up through 10,000 lb/in^2 and handling viscosities from less than 1 cSt to more than 1,000,000 SSU.

The rotary pump is quite often defined as a positive-displacement type by most authoritative engineering references because of the general employment of characteristic close-running clearances which substantially limit internal leakage. It might be more logical and technically correct to drop the *positive* term and refer to this type simply as a *displacement pump*. In the rotary pump, mechanical displacement of the fluid from inlet to outlet is produced by trapping a slug of fluid between one or more moving elements such as gears, cams, screws, vanes, lobes, plungers, or similar devices within a stationary housing or casing. The rotary motion of the centrifugal pump is combined with the positive-pressure characteristic of the reciprocating pump, resulting in a displacement device which delivers a given quantity of fluid with each revolution of the input shaft.

Unlike the centrifugal pump, it is generally self-priming and produces a delivery not severely affected by pressure variations. Speeds of operation are much higher than normally found in a reciprocating pump, with the result that in many instances direct-connected drivers can be used.

Rotaries are available for pumping practically any fluid that will flow, although their greatest specialty is the handling of very viscous fluids. Rotaries are generally simple, compact, and light in weight.

Classification

There are many different types of pumps which fall in the general category of *rotaries*. It is recommended that the reader refer to the *Hydraulic Institute Standards,* "Rotary Pumps," for a detailed description of the various types. A number of the major types are listed and described briefly here.

1. *Vane (sliding)* (Fig. 8-43). Vanes, blades, or rollers are located in the periphery of a rotor surrounded by a stator to form cavities between two successive vanes which carry fluid from inlet to outlet.

2. *Piston (axial)* (Fig. 8-44). A number of pistons reciprocate within cylinders arranged axially around the periphery of a rotor moving past inlet and outlet ports.

3. *Gear (external)* (Fig. 8-45). Fluid is carried between teeth of two external gears and displaced as they mesh.

4. *Gear (internal)* (Fig. 8-46). Fluid is carried between teeth of one internal and one external gear and displaced as they mesh.

5. *Lobe* (Fig. 8-47). Fluid is carried between one or more lobes on each of two rotors which are timed by separate means.

6. *Screw.* Fluid is carried between screw threads on two or more engaged rotors and is displaced axially as they mesh.

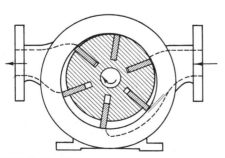

FIG. 8-43   Sliding-vane pump.

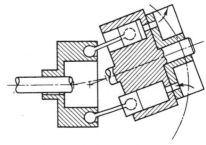

FIG. 8-44   Axial piston pump.

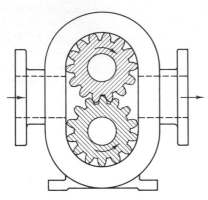

FIG. 8-45   External-gear pump.

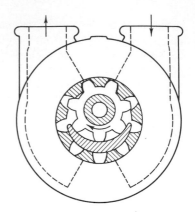

FIG. 8-46   Internal-gear pump.

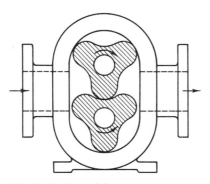

FIG. 8-47   Three-lobe pump.

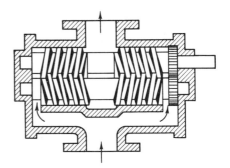

FIG. 8-48   Timed screw pump.

a. Timed (Fig. 8-48). Separate timing gears, located either internally or externally, are required to maintain the proper meshed relationship of the screw threads, and rotors are also generally supported by separate sets of bearings.

b. Untimed (Fig. 8-49). Rotors incorporate the use of generated thread forms which provide a synchronized gearing action, making separate timing unnecessary.

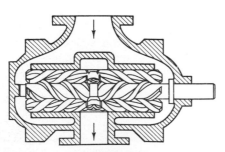

FIG. 8-49   Untimed screw pump.

The Transamerica Delaval CIG pump (see Fig. 8-50) is of the internal-gear type (see Fig. 8-46). In this type of pump, fluid is carried from the inlet to the discharge by a pair of gears consisting of one internal and one external gear. The gears are placed eccentrically to each other and are separated by a crescent-shaped divider which provides a sealing path for the internal and external flow paths.

FIG. 8-50 Cutaway view of two-stage pump.

The internal-gear design is generally known for its quiet operation. Modification of the gear profile also provides for reduction of the trapped oil, eliminating any pressure pulsations and thus reducing the noise level. The design is extremely simple and allows gear sets to be stacked into a multistage arrangement for increased pressure rating. With this arrangement, the pressure loads are distributed to reduce stress on the pump components, thus lengthening pump life. The design also has the inherent feature of providing for a hydrodynamic film buildup on the bearings and external gear ring that eliminates metal contact between the working parts, also adding to pump life. The design also provides for double pump configurations consisting of two independent pumps arranged on a common shaft, each pump having a separate discharge and sharing a common suction.

The Transamerica Delaval GTS pump is of the externally timed screw type (see Fig. 8-51). The construction of this type of pump is conducive to operation on nonviscous liquids such as water that exclude the use of the IMO design.

This design relies upon timing gears for phasing the mesh of the threads and support bearings at each end of the rotors to absorb the reaction forces. With this arrangement, the threads do not come into contact with each other or with the housing bores in which they rotate. This feature, combined with the external location of the timing gears and bearings, which are oil-bath- or grease-lubricated, makes the pump suitable for handling nonviscous, corrosive, or abrasive fluids.

To provide for operation with corrosive or abrasive fluids, the pump housing can be supplied in a variety of materials including cast iron, ductile iron, cast steel, stainless steel, and bronze. Moreover, the rotor bores can be lined with industrial hard chromium for additional abrasive resistance. The rotors also may be supplied in a variety of materials including cast iron, heat-treated alloy steel, stainless steel, Monel, and Nitralloy. The outside diameter of the rotors can be furnished with hard coatings including tungsten carbide, chromium oxide, and ceramics.

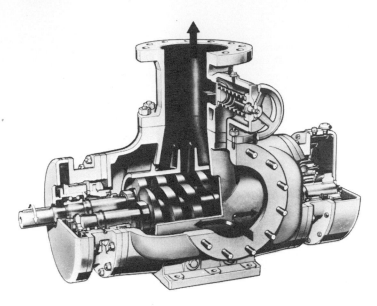

FIG. 8-51 Cutaway view of externally timed screw pump.

The IMO pump (see Fig. 8-52) falls into the untimed-screw category, and it will serve as a base for all further discussion of rotary pumps in general. Because the fundamental characteristics of all rotaries are similar, many IMO pump features can be related to other types of rotaries without comment; however, when characteristics unique to the IMO pump are mentioned, they will be so identified.

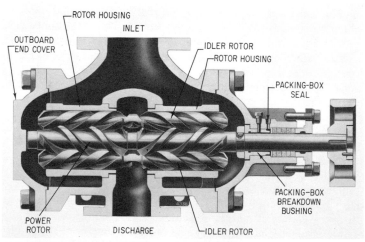

FIG. 8-52 Cutaway view of double-end IMO pump.

Characteristics

The IMO pump normally is offered as a three-screw type having no need for timing gears or conventional support bearings. It is simple and rugged and has no valves or reciprocating parts to foul. It can run at high speeds, is quiet-operating, and produces a steady pulsation-free flow of fluid.

Properly applied, the IMO pump can handle a wide range of fluids from molasses to gasoline, including modern fire-resistant types, even to 5 percent soluble oil in water. It can be made with hardened wear-resistant rotors to handle some types of contamination and abrasives. Wide ranges of flow and pressure are available.

In the IMO pump, as in most screw pumps, it is the intermeshing of the threads on the rotors and the close fit of the surrounding housing which create one or more sets of moving seals between pump inlet and outlet. These sets of seals act as a labyrinth and provide the screw pump with its positive-pressure capability. Between successive sets of moving seals or threads are voids which move continuously from inlet to outlet. These moving voids, when filled with fluid, carry the fluid along and provide a smooth flow to the outlet, which is essentially pulsationless. Increasing the number of threads or seals between inlet and outlet increases the pressure capability of the pump, the seals again acting similarly to classic labyrinth seals.

The flow of fluid through the screw pump is parallel to the axis of the screws as opposed to the travel around the periphery of centrifugal, vane, and gear-type pumps. This axial flow gives the screw pump ability to handle fluids at low relative velocities for a given input speed, and it is therefore suitable for running at higher speeds, with 1750 and 3500 r/min common for IMO pumps.

The fundamental difference between the IMO pump and other types of screw pumps lies in the method of engaging or meshing the rotors and maintaining the running clearances between them. Timed screw pumps require separate timing gears between the rotors to provide proper phasing or meshing of the threads. Some sort of support bearing also is required at the ends of each rotor to maintain proper clearances and proper positioning of the timing gears themselves.

The IMO pump rotors are precision-made gearing in themselves, having mating generated thread forms such that any necessary driving force can be transmitted smoothly and continuously between the rotors without need for additional timing gears. The center or driven rotor, called the *power rotor,* is in mesh with two or three close-fitting unsupported *sealing,* or *idler, rotors* symmetrically positioned about the central axis by the close-fitting rotor housing. This close-fitting housing and the idlers provide the only transverse bearing support for the power rotor. Conversely, the idlers are transversely supported only by the housing and the power rotor.

The real key to all IMO pump operation is the means employed for absorbing the transverse idler-rotor-bearing loads which are developed as a result of the hydraulic forces built up within the pump to move the fluid against pressure. These rotors and the related housing bores are, in effect, partial journal bearings with a hydrodynamic fluid film being generated to prevent metal-to-metal contact. This phenomenon is most often referred to as the *journal-bearing theory,* and IMO pump behavior is closely related to the applied principles of this theory.

The three key parameters of speed, fluid viscosity, and bearing pressure are related exactly as in a journal bearing. If viscosity is reduced, speed must be increased or bearing pressure reduced in order not to exceed acceptable operating limits. For a constant viscosity, however, the bearing-pressure capability can be increased by increasing the speed. It is this phenomenon that gives the IMO its high-speed capability; in fact, with proper inlet conditions, the higher the IMO pump speed the better the performance and the better the life. This is directly opposite to most rotary-pump behavior.

Since the IMO pump is a displacement device, like all rotaries, it will deliver a definite quantity of fluid with every revolution of the power rotor. If no internal clearances exist, this quantity, called *theoretical capacity Q_t,* would depend only upon the physical dimensions of the rotor set and the speed. Clearances, however, do exist, with the result that whenever a pressure differential occurs, there always will be internal leakage from outlet to inlet. This leakage, commonly called slip S, varies with the pump type or model, amount of clearance, fluid viscosity at pumping conditions, and differential pressure. For any given set of conditions, it is usually unaffected by speed. The delivered capacity, or net capacity Q, therefore, is the theoretical capacity less slip.

The theoretical capacity of any pump can readily be calculated with all essential dimensions known. Basically, IMO pump theoretical capacity varies directly as the cube of the power rotor's outside diameter, which is generally used as the pump-size designator. Thus a relatively small increase in pump size can give a large increase in capacity. Slip can also be calculated but usually is based upon empirical values developed by extensive testing.

Performance

Inlet Conditions   The key to obtaining good performance from an IMO pump, as with all other rotaries, lies in a complete understanding and control of inlet conditions and the closely related parameters of speed and viscosity. To ensure quiet, efficient operation, it is necessary to completely fill with fluid the moving voids between the rotor threads as they open to the inlet, and this becomes more difficult as viscosity, speed, or suction lift increases. Basically, if the fluid can properly enter into the rotor elements, the pump will perform satisfactorily.

Remember that a pump does not pull or lift fluid into itself. Some external force must be present to push the fluid into the voids. Normally, atmospheric pressure is the only force present, but in some applications a positive inlet pressure is available.

Naturally the more viscous the fluid, the greater the resistance to flow and, therefore, the lower the rate of filling the moving voids of the threads in the inlet. Conversely, light-viscosity fluids will flow quite rapidly and will quickly fill the moving voids. It is obvious that if the rotor elements are moving too fast, the fill will be incomplete and a reduction in output will result. The rate of fluid flow must always be greater than the rate of void travel or closing to obtain complete filling. Safe rates of flow through the pump for complete filling have been found from experience when atmospheric pressure is relied upon to force the fluid into the rotors. The following table gives these safe axial-velocity limits for various fluids and pumping viscosities.

Fluid*	Viscosity, SSU	Velocity, ft/s
Diesel oil	32	30
Lubricating oil	1,000	12
No. C fuel oil	7,000	7
Castor oil	20,000	2
Cellulose	60,000	½

*For characteristics of fuel oils see Table 8-3.

It is thus quite apparent that pump speed must be selected to satisfy the viscosity of the fluid to be pumped. The pump manufacturer generally must supply the determination of the axial velocity through a screw pump, although the calculation is quite simple when the driving-rotor speed and screw-thread lead are known. The lead is the advancement made along the thread during a complete revolution of the rotor as measured along the axis. In other words, it is the travel of the fluid slug in one complete revolution.

In this handbook, the more general term *fluid* is used to describe the fluids handled by rotaries which may contain or be mixed with matter in other than the liquid phase. The word *liquid* is used only to describe true liquids that are free of vapors and solids. Most of the fluids handled by rotary pumps, especially petroleum oils, because of their complex nature contain certain amounts of entrained and dissolved air or gas which is released as vapor when the fluid is subjected to pressures below atmospheric. If the pressure drop required to overcome entrance losses to push such a fluid into the rotor voids is sufficient to reduce the pressure so that vapors are released in the rotor voids, cavitation results. This leads to noisy operation and an attendant reduction in output. It is therefore very important to be aware of the characteristics of the entrained air and gas of the fluids to be handled. In fact, it is so important that a more detailed study of this relatively complex subject is included below in the subsection "Effect of Entrained or Dissolved Gas on Performance."

Speed   The speed N of a rotary pump is the number of revolutions per minute of the driving rotor. In most instances this is the input shaft speed; however, in some geared-head units the driving-rotor speed can differ from the input shaft speed.

Capacity   The actual delivered capacity of any rotary pump, as stated earlier, is theoretical capacity less internal leakage or slip when handling vapor-free fluids. For a particular speed, this may be written $Q = Q_t - S$, where the standard unit of Q and S is the United States gallon per minute. Again, if the differential pressure is assumed to be zero, the slip may be neglected and $Q = Q_t$.

The term *displacement D* is of some general interest, although it is no longer used in rotary-pump calculations. It is the theoretical volume displaced per revolution of the driving rotor and is related to theoretical capacity by speed. The standard unit of displacement is cubic inches per revolution; thus $Q_t = DN \div 231$. The terms *actual displacement* and *liquid displacement* are also less frequently used for rotary-pump calculations but continue to be used for some theoretical studies. Actual displacement is related to delivered capacity by speed.

TABLE 8-3 Detailed Requirements for Fuel Oils[a]

No.	Grade of fuel oil[b] Description	Flash point, °F min	Pour point, °F max	Water and sediment, % max	Carbon residue on 10% residuum, % max	Ash, % max	Distillation temp, °F 10% point max	90% point max	End point max	Viscosity Saybolt Universal at 100°F max	Universal at 100°F min	Furol at 122°F max	Furol at 122°F min	Kinematic centistokes at 100°F max	100°F min	122°F max	122°F min	Gravity, °API min
1	Distillate oil intended for vaporizing pot-type burners and other burners requiring this grade[c]	100 or legal	0	Trace	0.15	...	420	...	625	...	...	...	...	2.2	1.4	...	...	35
2	Distillate oil for general-purpose domestic heating for use in burners not requiring No. 1	100 or legal	20[d]	0.10	0.35	...	[e]	675	...	...	...	...	...	(4.3)	...	...	...	26
4	Oil for burner installations not equipped with preheating facilities	130 or legal	20	0.50	...	0.10	...	...	...	125	45	...	...	(26.4)	(5.8)	...	...	...
5	Residual-type oil for burner installations equipped with preheating facilities	130 or legal	...	1.00	...	0.10	...	...	...	...	150	40	...	...	(32.1)	(81)	...	...
6	Oil for use in burners equipped with preheaters permitting a high-viscosity fuel	150 or legal	...	2.00[f]	...	...	...	...	...	...	...	300	45	...	...	(638)	(92)	...

Reprinted by permission from Commercial Standard CS 12-48 on Fuel Oils of U.S. Department of Commerce.

[a]Recognizing the necessity for low-sulfur fuel oils used in connection with heat treatment, nonferrous metal, glass, and ceramic furnaces, and other special uses, a sulfur requirement may be specified in accordance with the following table:

Grade of fuel oil	Sulfur, maximum %
No. 1	0.5
No. 2	1.0
Nos. 4, 5, and 6	No limit

Other sulfur limits may be specified only by mutual agreement between the buyer and seller.

[b]It is the intent of these classifications that failure to meet any requirement of a given grade does not automatically place an oil in the next lower grade unless in fact it meets all requirements of the lower grade.

[c]No. 1 oil shall be tested for corrosion for 3 h at 122°F. The exposed copper strip shall show no gray or black deposit.

[d]Lower or higher pour points may be specified whenever required by conditions of storage or use. However, these specifications shall not require a pour point lower than 0°F under any conditions.

[e]The 10 percent point may be specified at 440°F; maximum for use in other than atomizing burners.

[f]The amount of water may be specified at 440°F; maximum for use in other than atomizing burners.

[f]The amount of water by distillation plus the sediment by extraction shall not exceed 2.00 percent. The amount of sediment by extraction shall not exceed 0.50 percent. A deduction in quantity shall be made for all water and sediment in excess of 1.0 percent.

The actual delivered capacity of any specific rotary pump is reduced by

1. Decreasing speed
2. Decreased viscosities
3. Increased differential pressure

The actual speed must always be known and most often differs somewhat from the rated or nameplate specification. This is the first item to be checked and verified in analyzing any pump's operating performance. It is surprising how often the speed is incorrectly assumed and later found to be in error.

Because of the internal clearances between rotors and the housing of a rotary pump, lower viscosities and higher pressure increase slip, which results in a reduced delivered capacity for a given speed. The impact of these characteristics can vary widely for the various types of rotary pumps encountered. The slip, however, is not measurably affected by changes in speed and thus becomes a smaller percentage of the total flow with the use of higher speeds. This is a very significant factor in dealing with the handling of light viscosities at higher pressures, particularly in the case of devices such as the IMO pump which favor high speed. Always run at the highest speed possible for best results and best volumetric efficiency with the IMO pump. This will not generally be the case with rotaries having support-bearing speed limits.

Pump volumetric efficiency V_y is calculated as $V_y = Q/Q_t = (Q_t - S)/Q_t$, with Q_t varying directly with speed.

As stated previously, theoretical capacity of an IMO pump is a function which varies directly as the cube of the power rotor's outside diameter for a standard three-rotor pump configuration. For a constant speed, a 2-in rotor will have a theoretical capacity 8 times that of a 1-in rotor size. However, for a given model, slip varies directly as the square of the rotor size; therefore, the slip of the 2-in rotor is 4 times that of a 1-in rotor with all fluid variables held constant.

On the other hand, viscosity change affects the slip inversely to some power which has been determined empirically. An acceptable approximation for 100 to 10,000 SSU is obtained by using the one-half power. Slip varies directly with approximately the square root of differential pressure, and a change from 400 to 100 SSU will double the slip just as will a differential-pressure change from 100 to 400 lb/in^2.

Pressure The pressure capability of different types of rotary pumps varies widely. Some of the gear and lobe types are fairly well limited to 100 lb/in^2, normally considered low-pressure. Other gear and vane types perform very well in the moderate-pressure range (100 to 500 lb/in^2) and beyond. Some types can operate well in the high-pressure range, while others such as axial piston pumps can work at 5000 lb/in^2 and above. The slip characteristic of a particular pump is one of the key factors that determine the acceptable operating range, which generally is well defined by the pump manufacturer; however, all applications for high pressure should be approached with some caution, and the manufacturer or the manufacturer's representative should be consulted.

The IMO pump is suitable for a wide range of pressures from 50 to 5000 lb/in^2, dependent upon the selection of the right model. Internal leakage can be restricted for high-pressure applications by introducing increased numbers of moving seals or threads between inlet and outlet (see Figs. 8-53 to 8-55). The number of seals between inlet and outlet normally is specified for a particular model in terms of number of closures. The number of closures is increased to obtain higher-pressure capability, which also results in increased pump length for a given rotor size.

IMO pumps generally are available with predetermined numbers of closures versus maximum pressure rating when rated at 150 SSU and 3500 r/min in the 10- to 100-gal/min range:

Number of closures	Maximum pressure, lb/in²
1	100
2	500
3	1500
5	3000
11	5000

Horsepower   The brake horsepower (bhp) required to drive a rotary pump is the sum of the theoretical liquid horsepower and the internal power losses. The theoretical liquid horsepower is the actual work done in moving the fluid from its inlet-pressure condition to the outlet at discharge pressure.

NOTE: This work is done on all the fluid of theoretical capacity, not just delivered capacity, because slip does not exist until a pressure differential occurs. Rotary-pump power ratings are expressed in terms of horsepower (550 ft·lb/s), and theoretical liquid horsepower can be calculated: $tLhp = Q_t \Delta P \div 1714$. It should be noted that the theoretical liquid horsepower is independent of viscosity and is concerned only with the physical dimensions of the pumping elements, the rotative speed, and the differential pressure.

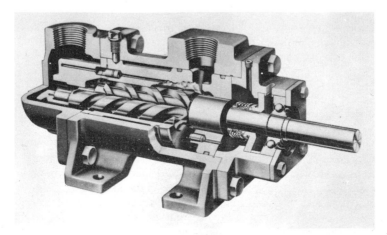

FIG. 8-53 Cutaway view of single-end IMO pump; two closures.

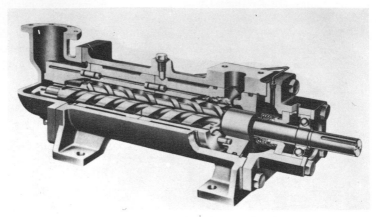

FIG. 8-54   Cutaway view of single-end IMO pump; five closures.

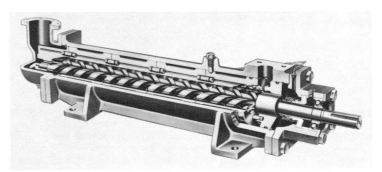

FIG. 8-55   Cutaway view of single-end IMO pump; 11 closures.

Internal power losses are of two types: mechanical and viscous. Mechanical losses include all the power necessary to overcome the mechanical friction drag of all the moving parts within the pump, including rotors, bearings, gears, mechanical seals, etc. Viscous losses include all the power lost from the fluid viscous-drag effects against all the parts within the pump as well as from the shearing action of the fluid itself. It is probable that the mechanical loss is the major component when operating at low viscosities and high speeds while the viscous loss is the larger at high viscosity and slow-speed conditions.

No direct comparison can easily be made between various types of rotary pumps for internal power loss, as this falls into the category of closely guarded trade secrets. Most manufacturers have established their own data on a basis of tests made under closely controlled operating conditions, and they are very reluctant to divulge their findings. In general, the losses for a given type and size of pump vary with viscosity and rotative speed and may or may not be affected by pressure, depending upon the type and model of pump under consideration. These losses, however, must always be based upon the maximum viscosity to be handled since they will be highest at this point.

The actual pump power output (whp), or delivered liquid horsepower, is the power imparted to the fluid by the pump at the outlet. It is computed in the same way as theoretical liquid horsepower, using Q in place of Q_t; hence the value will always be less.

Pump efficiency is the ratio of the pump power output (whp) to the brake horsepower (bhp).

Application and Selection

In the application of rotary pumps certain basic factors must be considered to ensure a successful installation. These factors are fundamentally the same regardless of the fluids to be handled or the pumping conditions.

The pump selection for a specific application is not difficult if all the operating conditions are known. It is often quite difficult, however, to obtain accurate information as to these conditions. This is particularly true of inlet conditions and viscosity, since it is a common feeling that inasmuch as the rotary pump is a positive-displacement device, these items are unimportant.

In any rotary-pump application regardless of the design, suction lift, viscosity, and speed are inseparable. Speed of operation, therefore, is dependent upon viscosity and suction lift. If a true picture of these two items can be obtained, the problem of making a proper pump selection becomes simpler, and it is probable that the selection will result in a more efficient unit.

Viscosity   It is not very often that a rotary pump is called upon to handle fluids having a constant viscosity. Normally, because of temperature variations, it is expected that a range of viscosity will be encountered, and this range can be quite wide; for instance, it is not unusual that a pump is required to handle a viscosity range of 150 to 20,000 SSU, the higher viscosity usually being due to cold-starting conditions. This is a perfectly satisfactory range insofar as a rotary pump is concerned; but if information can be obtained concerning such things as the amount of time during which the pump is required to operate at the higher viscosity and whether or not the motor can be overloaded temporarily, a multispeed motor can be used, or the discharge pressure can be reduced during this period, a better selection can often be made.

Quite often no viscosity but only the type of fluid is given. In such cases assumptions can sometimes be made if sufficient information is available concerning the fluid in question. For instance, Bunker C, or Number 6, fuel oil is known to have a wide latitude as to viscosity and usually must be handled over a considerable temperature range. The normal procedure in a case of this type is to assume an operating viscosity range of 20 to 700 SSF. The maximum viscosity, however, may very easily exceed the higher value if extra heavy oil is used or exceptionally low temperatures are encountered. If either should occur, the result may be improper filling of the pumping elements, noisy operation, vibration, and overloading of the motor.

Although it is the maximum viscosity and the expected suction lift that determine the size of the pump and set the speed, it is the minimum viscosity that determines the capacity. Rotary pumps must always be selected to give the specified capacity when handling the expected minimum viscosity, since this is the point at which maximum slip, hence minimum capacity, occurs. If this rule is not followed, the pump will not meet the requirements of the system unless a considerable margin has been allowed initially in specifying capacity or there is over-capacity available in the pump. The latter is often the case, since practically all rotary pumps are made in certain stock sizes and it is standard practice to apply the next larger pump when a capacity that falls between sizes is specified.

It should also be noted that the minimum viscosity often sets the model of the pump selected since it is more or less standard policy of most manufacturers to downrate their pumps, insofar as allowable pressure is concerned, when handling liquids having a viscosity of less than 100 SSU. This is done for two reasons: first, to avoid the poorer volumetric efficiency as a result of increased slip under these conditions; and second, because a film of the liquid must be maintained between the closely fitted parts which is likely to break down if a combination of low viscosity and high pressure should occur. Although viscosity is not necessarily a definite criterion of film strength, it is generally so used by pump manufacturers.

Entrained Air   As mentioned previously, a factor which must also be given careful consideration is the possibility of entrained air or gas in the fluid to be pumped. This is particularly true of installations in which recirculation occurs and the fluid is exposed to air through either mechanical agitation, leaks, or improperly located drain lines.

Likewise, most fluids will also dissolve air or gas, retaining it in solution, the amount depending upon the liquid itself and the pressure to which it is subjected. It is known, for instance, that lubricating oils under conditions of atmospheric temperature and pressure will dissolve up to 10 percent air by volume and gasoline up to 20 percent.

When pressures below atmospheric exist at the pump inlet, dissolved air will come out of solution, and both this and entrained air will expand in proportion to the existent absolute pressure. This expanded air will accordingly take up a proportionate part of the available volume of the moving voids with a consequent reduction in delivered capacity. (See subsection "Effect of Entrained or Dissolved Gas on Performance.")

One of the apparent effects of handling fluids containing entrained or dissolved air or gas is noisy pump operation. When such a condition occurs, it is usually dismissed as *cavitation;* then, too, many operators never expect anything but noisy operation from rotary pumps. This should not be the case. With properly designed systems of pumps, quiet, vibration-free operation can be produced and should be expected. Noisy operation is inefficient, and steps should be taken to make corrections until the objectionable conditions are overcome. It is true, of course, that some types of pumps are more critical to the handling of air than others; this is usually due to the high inlet losses inherent in these types, but proper design and speed selection can go a great way toward overcoming the problem.

It should be pointed out that if a pump will be called on to handle fluids containing entrained air, this fact should be included in any specifications which may be written and the percentage specified.

Nonnewtonian Fluids The viscosity of most liquids, as, for example, water and mineral oil, is unaffected by any agitation to which they may be subjected as long as the temperature remains constant; these liquids are accordingly known as *true,* or *newtonian,* fluids. There is another class of liquids, however, such as cellulose compounds, glues, greases, paints, starches, slurries, and candy compounds, which changes in viscosity as agitation is varied at constant temperature. The viscosity of these fluids will depend upon the shear rate at which it is measured, and these fluids are termed *nonnewtonian.*

If a fluid is known to be nonnewtonian, the expected viscosity under actual pumping conditions should be determined, since it can vary quite widely from the viscosity under static conditions. One instance concerned the handling of a cellulose product for which the viscosity was given as 20,000 SSU, which was its actual static, or apparent, viscosity. It developed that under actual pumping conditions the viscosity was approximately 500 SSU. No serious harm was done, but a large low-speed pump was installed when a smaller, cheaper, higher-speed unit could have been used.

Since a nonnewtonian fluid can have an unlimited number of viscosity values (as the shear rate is varied), the term *apparent viscosity* is used to describe its viscous properties. Apparent viscosity is expressed in absolute units and is a measure of the resistance to flow at a given rate of shear. It has meaning only if the shear rate used in the measurement is also given.

The grease-manufacturing industry is very familiar with the nonnewtonian properties of its products, as evidenced by the numerous curves wherein *apparent viscosity* is plotted against *rate of shear* that have been published. The occasion is rare, however, when one is able to obtain accurate information as to viscosity if it is necessary to select a pump for handling this fluid.

It is understood that it is practically impossible, in most instances, to give the viscosity of grease in the terms most familiar to the pump manufacturer, i.e., Saybolt Seconds Universal or Saybolt Seconds Furol; but if only a rough approximation could be given, it would be of great help.

For applications of this type, data taken from similar installations are most valuable. Such information should consist of type, size, capacity, and speed of already installed pumps, suction pressure, and temperature at the pump-inlet flange, total working suction head, and above all the pressure drop in a specified length of piping. From the latter, an excellent approximation of viscosity under actual operating conditions can be obtained.

Suction Conditions Suction lift occurs when the total suction head at the pump inlet is below atmospheric pressure. It is normally the result of a static lift and pipe friction. Although rotary pumps are capable of producing a high vacuum, it is not this vacuum that forces the fluid to flow. As previously explained, it is atmospheric pressure that forces the fluid into the pump. Since atmospheric pressure at sea level corresponds to 14.7 psia, or 30 inHg, this is the maximum amount of pressure available for moving the fluid, and suction lift cannot exceed these figures. Actually, it must be somewhat less since there are always pump-inlet losses which must be taken into account. It is considered the best practice to keep suction lifts just as low as possible.

The majority of rotary pumps operate with suction lifts of approximately 5 to

15 inHg. Lifts corresponding to 24 to 25 inHg are not uncommon, and there are numerous installations operating continuously and satisfactorily in which the absolute suction pressure is within ½ in of the barometer. In the latter cases, however, the pumps are usually taking the fluid from tanks under vacuum and no entrained or dissolved air or gases are present. Great care must be taken in selecting pumps for these applications, since the inlet losses can very easily exceed the net suction head available for moving the fluid into the pumping elements.

There are many known instances of successful installations in which pumps were properly selected for high-suction conditions. There are also, unfortunately, many other installations with equally high suction lifts which are not so satisfactory. This is because proper consideration was not given, at the time when the installations were made, to the actual suction conditions at the pump inlet. Frequently, suction conditions are given as "flooded" simply because the source feeding the pump is above the inlet flange. Absolutely no consideration is given to outlet losses from the tank or pipe friction, and these can be exceptionally high when dealing with extremely viscous fluids.

When it is desired to pump extremely viscous fluids such as grease, chilled shortening, and cellulose preparations, care should be taken to use the largest possible size of suction piping, eliminate all unnecessary fittings and valves, and place the pump just as closely as possible to the source of supply. In addition, it may be found necessary to supply the fluid to the pump under some pressure, which may be supplied by elevation, air pressure, or mechanical means.

Speed   It was previously stated that viscosity and speed are closely linked and that it is impossible to consider one without the other. Although rotative speed is the ultimate outcome, the basic speed which the manufacturer must consider is the velocity of the fluid going through the pump; this is a function of pump type and design. Certain types, such as gear and vane pumps, carry the fluid around the periphery of the pumping elements, and as a result the velocity of the fluid through the pump can become quite high unless relatively low rotative speeds are used. On the other hand, in screw-type pumps the flow is axial and fluid speeds are relatively lower, with the result that higher rotative speeds can be used. On the basis of handling light fluids, say, 100 to 500 SSU, gear- or vane-type pumps rarely exceed a rotative speed of 1200 r/min except in the case of a very small unit or special designs for a particular use such as for aircraft purposes. Screw pumps, however, for which timing gears are not required, commonly operate without difficulty at speeds up to 5000 r/min, and IMO pumps have been run in the field to 24,000 r/min.

Although rotative speeds are relative and depend upon the pump type, they usually should be reduced when handling fluids of high viscosity. This is due not only to the difficulty of filling the pumping elements but also to the mechanical losses which result from the shearing action of these parts on the fluid handled. The reduction of these losses is frequently of more importance than relatively high speeds, even though the latter might be possible because of positive inlet conditions.

Rotary pumps do not in themselves create pressure; they simply transfer a quantity of fluid from the inlet to the outlet side. The pressure developed on the outlet side is solely the result of resistance to flow in the discharge line. If, for example, a pump were to be set up and run without a discharge line, a gauge placed at the pump outlet flange would register zero no matter how fast or how long the pump was run.

Pipe Size   Resistance usually consists of differences of elevation, fixed resistances such as orifices, and pipe friction. Nothing much can be done about the first two, since these are the basic reasons for using a pump. Something, however, can be done about pipe friction. Millions of dollars are thrown away annually because of the use of piping that is too small for the job. To be sure, all pipe friction cannot be eliminated as long as fluids must be handled in this manner, but every effort should be made to use the largest pipe that is economically feasible. There are numerous tables from which friction losses in any combination of piping may be calculated, among the most recent of which are those published by the Hydraulic Institute, also available in this handbook in condensed form.

Before any new installation is made, the cost of larger-size piping which will result in lower pump pressures should be carefully balanced against the cost of a less expensive pump, a smaller motor, and a saving in horsepower over the expected life of the system. The larger piping may cost a little more in the beginning, but the ultimate saving in power will often offset the original cost many times. These facts are particularly true of the handling of extremely viscous fluids, and although most engineers dealing with fluids of this type are conscious of what can be done, it is surprising how many installations are encountered in which considerable savings could have been effected if a little more study had been made initially.

Abrasives   There is one other point that we have not as yet touched, and that is the handling of fluids containing abrasives. Because rotary pumps depend upon close clearances for proper pumping action, the handling of abrasive fluids will usually cause rapid wear. Much progress has been made in the use of harder and more abrasive-resistant materials for the pumping elements, so that a good job can be done in some instances. It cannot be said, however, that performance is always satisfactory when handling fluid laden excessively with abrasive materials. On the whole, rotary pumps should not be used for handling fluids of this character unless shortened pump life and an increased frequency of replacement are acceptable.

Design Details

It is virtually impossible to include a discussion of the design details for the many varieties of rotary pumps within the framework of this handbook; therefore, this subsection will be limited to a brief discussion of IMO pump design with some reference to other types when applicable.

Basic Construction   The IMO pump, as well as other types and makes of rotary screw pumps, is available in two basic configurations: single- and double-end construction. The double-end construction (see Fig. 8-52) is probably the best-known version, as it was by far the most widely used, for many years, because of the relative simplicity and compactness of its design. As pressure requirements were raised, however, the single-end version developed increased usage, until today it accounts for by far the largest portion of total IMO pump annual production (see Fig. 8-56).

The general double-end screw-pump construction usually is limited to low- and medium-pressure applications, with 400 lb/in^2 a good practical limit for planning purposes. However, with special design features incorporated, applications up to 800 lb/in^2 can be handled. Double-end pumps generally are employed when large flows are required or very viscous fluids are handled.

There is in use one other principal variation of IMO pump construction which must be mentioned briefly, that is, the four-rotor design having three idlers which is sometimes used for low-pressure lubricating service. The introduction of the third idler, in effect, makes the pump nonpositive, which gives it additional capability for handling heavily air-laden lubricating oil wihout cavitation or the related heavy vibration. This design, however, is restricted to very low pressure use because of the resulting increased slip characteristic.

The single-end screw-pump construction (see Fig. 8-53) is most often employed for handling low-viscosity fluids at high pressure or hydraulic-type fluids at very high pressure. It is most practical to provide the additional number of moving seals or closures between inlet and outlet necessary to handle high pressure in the single-end construction. This is accomplished in IMO pumps by literally stacking a number of medium-pressure single-end pumping elements in series within one pump casing. The single-end construction also offers the best design arrangement for high-production manufacture even though the design itself is more complex than the relatively simple double-end construction.

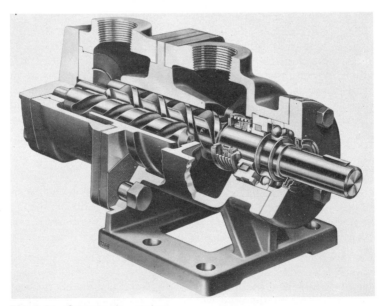

FIG. 8-56 Cutaway view of IMO pump.

The double-end type (see Fig. 8-52) is basically two opposed single-end pumps or pump elements of the same size with a common power rotor of double-helix design within one casing. As can be seen from the illustration, the fluid normally enters a common inlet, with a split flow going to the outboard ends of the two pumping elements, and is discharged from the middle or center of the pump elements. The two pump elements are, in effect, pumps connected in parallel. The design can also be provided with a reversed flow for low-pressure applications.

Axial Balance   Whichever design is employed, means must always be provided to absorb the mechanical and hydraulic axial thrust on the rotors of a screw pump. The double-end-design provides the simplest arrangement for accomplishing this, as both the power and the idler types of rotor (see again Fig. 8-52) are constructed with opposed-thread helices on the same shaft, which provides true axial balancing, both mechanically and hydraulically, since all thrust forces between the opposed pump elements are canceled out.

In single-end designs, special axial balancing arrangements must be employed for both the power and the idler rotors, and in this respect they are thus more complicated than the double-end construction. Mechanical thrust-bearing arrangements (see Fig. 8-56) are used for the idlers for 150-lb/in^2 differential pressures and below, while a hydraulic-balance arrangement (see Fig. 8-53) is used for pressures above 150 lb/in^2. Here hydraulic balance is accomplished by directing discharge pressure to a bearing area on the inlet end of the idler which is equal and opposite to the area exposed to discharge pressure on the outlet end of the same idler.

Hydraulic balance is provided for the power rotor through the balance piston (see Fig. 8-53) mounted on the power rotor between the outlet and seal chambers. This piston is exposed to discharge pressure on the outlet side and is equal and opposite in area to the exposed area of the power-rotor thread; thus the discharge-pressure hydraulic forces on the rotor threads are canceled out.

Seals   The IMO pump, like most other modern equipment, makes extensive use of mechanical-face seals for shaft sealing. Packing is now used only when absolutely necessary as dictated by the fluid handled. Seal technology has advanced rapidly, with many new materials such as Buna N, neoprene, Viton, and Teflon introduced for elastomers. Ni-Resist, carbon, carbide, and ceramics are now available in addition to the original standby pearlitic cast iron for use in the sealing faces. All this has made the use of packing virtually obsolete.

In all but some small low-pressure series, the IMO pump always has the seal located in a chamber connected to the suction side. To accomplish this in the single-end design in which the outlet is at the shaft end, the aforementioned power-rotor balance piston also serves as a breakdown bushing or flow restrictor between outlet and seal chambers to limit the pressure in the seal chamber. This seal chamber, in turn, is connected to the suction side of the pump through a small internal drilled conduit or through external tubing (see Fig. 8-53).

In most cases in which a mechanical seal is used in an IMO pump, an external grease-sealed ball bearing is employed on the power-rotor drive shaft to maintain precise shaft positioning. This assures long mechanical seal life. This bearing also serves to minimize flexible-coupling-misalignment conditions which can adversely affect the performance of high-speed equipment such as the IMO pump. The use of the ball bearing also provides a means for taking overhung loads, such as from belt drives, on certain models.

Inlet Pressures   Standard IMO pumps are normally designed to handle positive inlet pressures up to 40 psig. This limitation of pressure concerns the resulting thrust on the rotors, and design modifications can be made for much higher inlet pressures as required. The double-end design is ideal for adapting to high-inlet-pressure applications because the idlers are in thrust balance at all times and the power rotor can be thrust-balanced by making it double-ended so that both ends are exposed to the inlet pressure identically. The drawback is the need for two seals, but this is not very significant if the high inlet pressure really is important to obtain.

The above double-shaft arrangement also is used when two or more pumps are to be driven in tandem, which is quite advantageous in some applications. Shaft tapers are always used on larger IMO pumps for locating the coupling. The use of this taper helps to protect the mechanical seals and bearings from shock damage that can arise when installing a large coupling on a straight shaft.

Casings   Standard IMO pumps normally are provided with high-grade cast iron for the casing of low- and medium-pressure models. Standard high-pressure pumps employ ductile iron or cast steel for the casings with fabricated steel used for special orders when necessary. Casings also are made suitable for steam jacketing when absolutely necessary for high-viscosity applications; however, the use of heat tracing with either steam coils or electric tape covered with a good insulation blanket is the recommended preference.

IMO pumps can normally be mounted in virtually any position including vertical as well as all horizontal rotations. Double-end designs usually are arranged with opposed side inlet and outlet positions parallel to the foot mounting. Side inlet and top discharge can also be furnished if necessary.

Rotor Materials   The rotors and housings of the IMO pump can be made of various types and grades of hardened materials for use in handling corrosive-type fluids as well as those containing some abrasives. One of the popular material combinations in use in many of the medium- and high-pressure models is nitrided-steel power rotors and induction-hardened ductile-iron idlers with pearlitic gray-iron housings.

Most of the rotors of IMO pumps are finish-machined after hardening by thread grinding in order to obtain a high degree of accuracy. Very small and very large rotor sets at the extremes of the size range are finish-thread-milled. Thread forms are controlled very accurately to obtain the proper mating action of all rotor sets as well as to maintain running clearances between rotors to a minimum for limited internal leakage.

Installation and Operation

Rotary-pump performance can be improved by following the recommendations on installation and operation given below.

The pump should be placed on a smooth, solid foundation readily accessible for inspection and repair. It is essential that the power shaft and drive shafts be in perfect alignment. IMO practice normally requires a concentricity and parallelism of 0.003 FIR.

The suction pipe should be as short and straight as possible with all joints airtight. There should be no points at which air or entrapped gases may collect. If it is not possible to have the fluid flow to the pump, a foot or check valve should be installed at the end of the suction line or as far from the pump as possible. All piping should be independently supported to avoid strains on the pump casing.

A priming connection should be provided on the suction side and a relief valve set from 5 to 10 percent above the maximum working pressure on the discharge side.

Starting the unit may involve simply opening the pump suction and discharge valves and starting the motor, but it is always better to prime the unit on initial starting. On new installations, the system is full of air which must be removed. If it is not removed, the performance of the unit will be erratic, and in some cases air in the system can prevent the unit from pumping. Priming the pump should preferably consist of filling not only the pump with fluid but as much of the suction line as possible.

The discharge side of the pump should be vented on the initial starting. Venting is especially essential when the suction line is long or the pump is discharging against system pressure upon starting.

If the pump does not discharge after being started, the unit should be shut down immediately. The pump should then be primed and tried again. If it still does not pick up fluid promptly, there may be a leak in the suction pipe or the trouble may be traceable to excessive suction lift from an obstruction, throttled valve, or other causes. Attaching a gauge to the suction pipe at the pump will help find the trouble.

Once the pump is in service, it should continue to operate satisfactorily with practically no attention other than an occasional inspection of the mechanical seal or packing for excessive leakage and a periodic check to be certain alignment is maintained within reasonable limits for prolonged periods.

NOTE: Although mechanical seals are becoming more widely used, there are some applications in which packing will continue to be preferred, and it is therefore necessary to make some brief comment concerning the proper installation and care of packing. The packing gland should never be set up too tightly. Packing properly used will require some leakage to maintain correct lubrication. The recommended leakage rate is somewhat dependent upon the type of fluid being handled but should never be less than several drops per minute.

Excessive gland pressure on the packing causes scoring of shaft and rapid deterioration of the packing itself. The best practice is to keep the gland stud nuts about finger-tight.

Should the pump develop a noise after satisfactory operation, this usually indicates either excessive suction lift due to cold fluid, air in the fluid, misalignment of the coupling, or, in the case of an old pump, excessive wear.

Whenever the unit is shut down, if the operation of the system permits, both suction and discharge valves should be closed. This is particularly important if the shutdown is to be for an extended period because leakage in the foot valve, if the main supply is below the pump elevation, could drain the fluid from the unit and necessitate repriming as in the initial starting of the system.

Effect of Entrained or Dissolved Gas on Performance

A very important factor in rotary-pump applications is the amount of entrained and dissolved air or gas in the fluid handled. This is especially true if the suction pressure is below atmospheric. Such air or gas is generally neglected since rotary pumps are of the displacement type and hence are self-priming. If the entrained or dissolved air and gases are a large percentage of the volume handled and if their effect is neglected, there may be noise and vibration, loss of liquid capacity, and pressure pulsations.

The amount of entrained air or gas is extremely variable, depending upon the viscosity, the type of liquid, and the time and manner of agitation that it may have received.

There is little information available covering the solubility of air and other gases in liquids, especially all those handled by rotary pumps. About 1930 Dr. C. S. Cargoe of the National Bureau of Standards developed the following formula on the basis of available literature data to show the solubility of air at atmospheric pressure in oils, both crude and refined, and in other organic liquids:

$$\log_{10} A = \frac{792}{t + 460} - 4 \log_{10} \text{sg} - 0.4$$

where A = dissolved air, in^3/gal
   t = temperature, $°F$
   sg = specific gravity of the liquid

This equation is plotted as Fig. 8-57, as taken from a paper on rotary pumps by R. J. Sweeney in the February 1943 issue of the *Journal of the American Society of Naval Engineers*. The equation and curve should be considered as approximate only, since some liquids have a higher affinity for air and gases. For example, gasoline at atmospheric pressure will dissolve as much as 20 percent of air by volume.

This actual displacement is measured in terms of volume of fluid pumped and will be the same whether it is a liquid, a gas, or a mixture of both as long as the fluid can get to and fill the pump moving voids.

If the fluid contains 5 percent entrained gas by volume and no dissolved gas and the suction pressure is atmospheric, the mixture is then 95 percent liquid and 5 percent gas. This mixture fills up the moving voids on the inlet side, but 5 percent of the space is filled with gas and the remainder with liquid. Therefore, in terms of the amount of liquid handled, the output is reduced directly by the amount of gas present, or 5 percent. The liquid displacement as a function of the theoretical displacement when the suction pressure is atmospheric then becomes

$$D' = D(1 - E)$$

where D = theoretical displacement
   D' = liquid displacement
   E = percent entrained gas by volume at atmospheric pressure, divided by 100

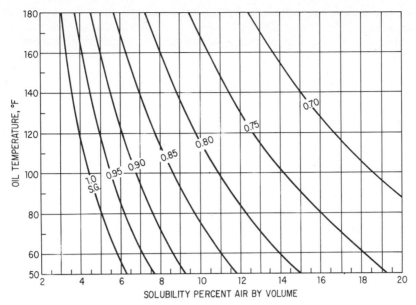

FIG. 8-57 Solubility of air in oil.

Assume that the fluid handled is a liquid mixture containing 5 percent entrained gas by volume at atmospheric pressure and no dissolved gas, but with the inlet pressure at the pump p_i in psia which is below atmospheric. The entrained gas will increase in volume as it reaches the pump in direct ratio to the absolute pressures. The new mixture will have a greater percentage of gas present, and the portion of theoretical displacement available to handle liquid becomes

$$D' = \frac{D(1 - E)}{(1 - E) + Ep/p_i}$$

where p = atmospheric pressure, psia
p_i = inlet pressure, psia

Note that p_i depends upon the vapor pressure of the liquid, the static lift, and the friction and entrance losses to the pump.

In the above equation, if the atmospheric pressure is 14.7 psia, the pump-inlet pressure 5 psia, and the vapor pressure very low, the liquid displacement is 86.6 percent of the theoretical.

If dissolved gases in liquids are considered, the effect on the liquid-displacement reduction is the same as that due to entrained gases, since in the latter case the dissolved gases come out of solution when the pressure is lowered. For example, assume a liquid free of entrained gas but containing gas in solution at atmospheric pressure and the pumping temperature. As long as the inlet pressure at the pump does not go below atmospheric pressure and the temperature does not rise, gas will not come out of solution. If pressure below atmospheric does exist at the pump inlet, gas will evolve and expand to the pressure existing. This will have the same effect as entrained gas taking up available displacement capacity and will reduce the liquid displacement accordingly. The liquid displacement then will be

$$D' = \frac{D}{1 + y(p - p_i)/p_i}$$

where the symbols have the meanings given above and y is the percentage of dissolved gas by volume at pressure p divided by 100. If the operating conditions are 9 percent of dissolved gas at 14.7 psia with a pump-inlet pressure of 5 psia, the liquid displacement will be 85.2 percent of the theoretical displacement.

If both entrained and dissolved gases are considered as existing in the material to be pumped, the liquid displacement becomes

$$D' = \frac{Dp_i(1 - E)}{(1 - E)[p_i + y(p - p_i)] + Ep}$$

where the symbols have the meanings given above. For operating conditions of 5 percent entrained gas, 9 percent dissolved gas at 14.7 psia, and a pump-inlet pressure of 5 psia, the liquid displacement is 75.2 percent of the theoretical. Figure 8-58 shows graphically the reduction in liquid displacement as a function of pump-inlet pressure, expressed in terms of suction lift, for different amounts of dissolved gas, neglecting slip.

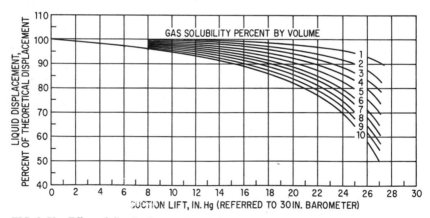

FIG. 8-58   Effect of dissolved gas on liquid displacement.

Figure 8-59 shows the reduction in liquid displacement as a function of pump-inlet pressure, expressed as suction lift, for different amounts of entrained air only, neglecting slip. From this figure it may be noted that a very small air leak can cause a large reduction in liquid displacement, especially if the suction lift is high.

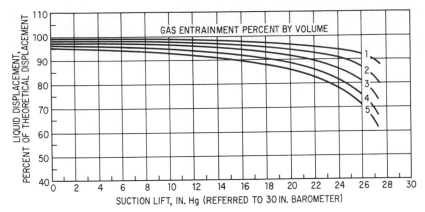

FIG. 8-59 Effect on entrained gas on liquid displacement.

From these few examples and curves it would appear that the problem of entrained and dissolved gases could be cared for by providing ample margins in pump capacity. Unfortunately, capacity reductions from the causes mentioned are attended by other and usually more serious difficulties.

The operation of a rotary pump is such that as rotation progresses, closures which fill and discharge in succession are formed. If the fluid pumped is compressible, such as a mixture of oil and air, the volume within each closure is reduced as it comes in contact with the discharge pressure. This produces pressure pulsations, the intensity and frequency of which depend upon the discharge pressure, the number of closures formed per revolution, and the speed of rotation. Under some conditions the pressure pulsations are of high magnitude and can cause damage to piping and fittings or even the pump, and they will almost certainly be accompanied by undesirable noise.

The amount of dissolved air or gas may be reduced by lowering the suction lift. This may often be controlled by pump location, suction-pipe diameter, and piping arrangement.

Many factors are associated with the amount of entrained air that can exist in a given installation. It is prevalent in systems in which the liquid is handled repeatedly and during each cycle is exposed to or mechanically agitated in air. Unfortunately in many cases the system is such that air entrainment cannot be entirely eliminated, as in the lubrication system of a reduction gear. Considerable work has been done by oil companies on foam dispersion, and while it has been recommended that special oils which are inhibited against oxidation and corrosion be used, all agree that the best cure is to remove or reduce the cause of foaming, namely, air entrainment.

Even though air entrainment cannot be entirely eliminated, in many cases it is possible, by adhering to the following rules, to reduce it and its ill effects on rotary-pump performance.

1. Keep liquid velocity low in the suction pipe to reduce turbulence and pressure loss. Use large and well-rounded suction bell to reduce entrance loss.

2. Keep suction lift low. If possible, locate the pump to provide positive head on the inlet.

3. Locate the suction piping within a reservoir to obtain maximum submergence.

4. Submerge all return lines particularly from bypass and relief valves, and locate them away from the suction.

5. Keep the circulation rate low, and avoid all unnecessary circulation of the fluid.

6. Do not exceed rated manifold pressures on machinery lubricating systems since the increased flow through sprays and bearings increases the circulation rate.

7. Heat the fluid when practical to reduce viscosity and as an expedient to drive off entrained air. Fluids of high viscosity will entrain and retrain more air than fluids of low viscosity.

8. Avoid all air leaks no matter how small.

9. Provide ample vents; exhauster fans to draw off air and vapors have been used with good results.

10. Centrifuging will break a foam and remove foreign matter suspended in the oil, which promotes foaming.

11. Use a variable-speed drive for the pump to permit an adjustment of pump capacity to suit the flow requirements of the machinery.

section 9

Compressors

DYNAMIC COMPRESSORS

Compression of Gases

Dynamic Compression   Dynamic compressors develop a pressure differential by the action of rotating blading that imparts velocity and pressure to the flowing medium. Both centrifugal and axial machines are of this general type.

Dynamic compressors operate at relatively high speeds and can be direct-connected to steam or gas turbines. Electric motor or reciprocating-engine drivers usually drive through a speed-increasing gear. Because dynamic compressors are smaller and lighter than positive-displacement compressors, for an equivalent gas volume, they require less floor space, their foundations are reduced, and they can be serviced by lighter cranes. They also operate without flow pulsations and therefore do not need receivers to even out the flow. They have the additional advantage of not adding oil to the gas stream and can pass moisture or dirt particles with less damage.

Figure 9-1 shows a cross section through a typical multistage centrifugal compressor. The gas enters the compressor through the initial suction nozzle and passes into the inlet volute. From the volute the flow enters the stationary inlet vanes, which guide the flow uniformly into the first-stage impeller. Once the gas enters the impeller inlet, it is picked up by the rotating curved vanes and thrown outward from the impeller into the diffuser, leaving a lower gas pressure at the impeller inlet and thus permitting more gas to be forced in by the external pressure in the suction line. Usually the diffuser is a single vaneless passage formed by the diaphragms; other types of diffusers may occasionally be used. The diffuser's purpose is to lower the high gas velocity leaving the impeller, thereby converting the kinetic energy imparted by the rotating impeller into pressure.

After leaving the diffuser in a multistage compressor, the gas passes through the crossover and enters the return vanes, which act to guide the flow uniformly into the next impeller inlet. At this point the gas has gone through one complete stage of compression. As the gas passes through subsequent stages, its pressure is further increased until it finally leaves through the discharge volute and nozzle at the discharge pressure of the unit.

Machines of the type shown in Fig. 9-1 are split horizontally at the centerline to facilitate maintenance. For higher pressures the casings may be built as complete cylinders, as shown in Fig. 9-2, and the internals removed axially. This type of compressor is generally referred to as either a vertically split or a barrel type. Many variations of multistage compressors are available in terms of multiple connections for external cooling and induction and extraction points.

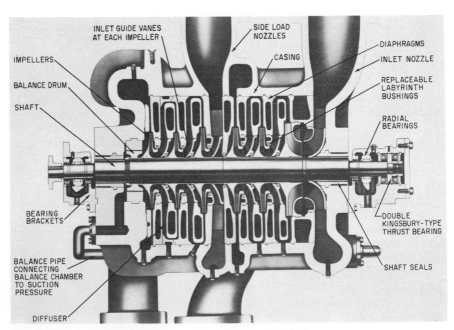

FIG. 9-1   Cross section of compressor with nozzles for intercooling.

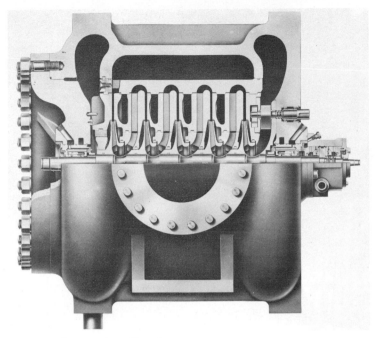

FIG. 9-2   Cross section of barrel compressor.

For pressure ratios within the range of one impeller, single-stage designs are available. A typical unit of this type is shown in Fig. 9-3. In this type of compressor, the flow enters the case from the inlet nozzle, but the passage from nozzle to impeller is different from that in a multistage compressor because of the nature of the single-stage case. In some designs, the inlet nozzle is located axially, directly in line with the impeller inlet, thereby eliminating the inlet losses associated with changing the gas direction in a side-suction case.

The flow in axial compressors is parallel to the machine axis, and the compression cycle involves passing the gas through alternating rows of rotating and stationary blading. In some axials, the function of the stationary blades is only to give proper direction to the gas approaching each rotating stage. This design, which is referred to as *100 percent reaction,* has an exceptionally high efficiency and a broad stability range. In other axial designs, the stationary blades may share equally in the energy conversion of the gas; this type of design is referred to as *50 percent reaction.* Various combinations between 50 and 100 percent reaction are possible. A typical axial compressor is shown in Fig. 9-4.

Thermodynamics of Compression   The relationship between the volume, absolute pressure, and absolute temperature of a perfect gas, based on Boyle's and Charles' laws, is

$$PV = WRT \quad \text{or on a mass basis} \quad Pv = RT$$

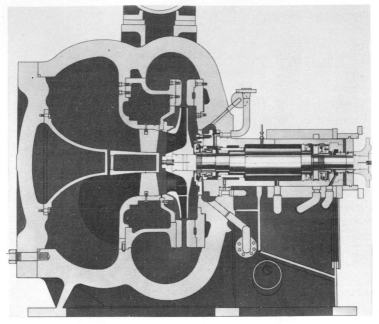

FIG. 9-3   Cross section of single-stage pipeline compressor.

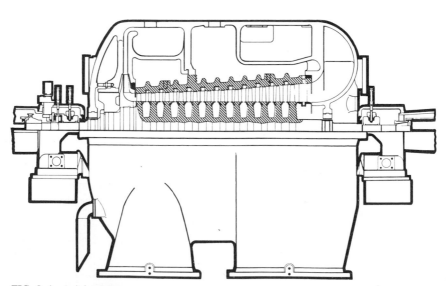

FIG. 9-4   Axial compressor.

where P = absolute pressure
V = volume of gas
W = weight of gas
R = gas constant
T = absolute temperature
v = specific volume

The *specific heat* of a gas is the amount of heat required to raise the temperature of one unit mass of gas one degree. If the volume of the gas is kept constant while the heat is added, all the heat is used in increasing the internal energy, i.e., in raising the temperature. This specific heat at constant volume is denoted C_v. If the pressure is kept constant and the volume is allowed to vary while the heat is added, more heat will be required. In addition to raising the temperature, the gas expands, and external work equal to $\int Pdv$ is done. This specific heat at constant pressure is called C_p.

The external work done when a unit mass of gas is heated at constant pressure $\int Pdv$ is equal to the gas constant R, or $C_p - C_v = R/J$.

$$J = 778 \ \text{ft} \cdot \text{lb/Btu}$$

The ratio of the specific heat at constant pressure to the specific heat at constant volume is known as k; hence $k = C_p/C_v$.

For all gases, the gas constant R is equal to the universal gas constant divided by the gas molecular weight, or $1544/M$. The specific gravity of air by definition is unity, and its molecular weight is 28.966. Hence the specific gravity of any gas of molecular weight M is $M/28.966$. R for any gas can then be defined as $1544/M$, or $53.34/sg$, where sg = specific gravity.

In the following discussion the subscripts i and f indicate respectively the initial and the final condition of the gas.

Isentropic process. If heat is neither added to nor removed from the gas during the compression, the process is reversible; the process is then called adiabatic. (In the compressor industry, the terms *isentropic* and *adiabatic* are generally used interchangeably.) The relationship between pressure and volume follows the law PV^k = a constant. The total work done on a unit mass of gas is

$$\text{Work} = C_p\,(T_f - T_i) = V_iP_i\left(\frac{k}{k-1}\right)\left[\left(\frac{P_f}{P_i}\right)^{(k-1/k)} - 1\right]$$

or

$$\text{Work} = RT_i\left(\frac{k}{k-1}\right)\left[\left(\frac{P_f}{P_i}\right)^{(k-1/k)} - 1\right]$$

Work in a compressor is commonly called *head*, and the equation is usually stated in terms of specific gravity, as follows:

$$\text{Head} = \frac{53.34T_i}{sg}\left(\frac{k}{k-1}\right)\left[\left(\frac{P_f}{P_i}\right)^{(k-1/k)} - 1\right]$$

where head = $\text{ft} \cdot \text{lbf/lbm}$
T_i = °R
P = psia

The ratio of isentropic work to actual work is called the *isentropic efficiency* η. The final temperature after compression may be found from the equation

$$T_f = T_i + \left(\frac{T_i}{\eta}\left[\left(\frac{P_f}{P_i}\right)^{(k-1/k)} - 1\right]\right)$$

The horsepower required for the compression is found by

$$hp = \frac{\text{head} \times \text{mass flow (lb/min)}}{33,000 \times \eta}$$

Isothermal process. By abstracting heat from the gas during compression, the temperature may be held constant; this process is called *isothermal*. The relationship between pressure and volume is $PV = $ a constant. The work done, or head, is

$$\text{Head} = P_i V_i \left(\log_e \frac{P_f}{P_i}\right) = RT_i \left(\log_e \frac{P_f}{P_i}\right) = \frac{53.34}{sg} T_i \left(\log_e \frac{P_f}{P_i}\right)$$

The isothermal head generally is not used in compression calculations, although it does take on some significance in multibody compressor trains with several points of intercooling.

Polytropic process. The actual compression path seldom follows either the isentropic or the isothermal process but is generally of the form $PV^n = $ a constant. In such cases the work, or head, is given by

$$\text{Head} = \frac{53.34}{sg} T_i \left(\frac{n}{n-1}\right)\left[\left(\frac{P_f}{P_i}\right)^{(n-1/n)} - 1\right]$$

where n is the exponent of polytropic compression. This factor may be found from the equation

$$\frac{n-1}{n} = \frac{k-1}{k}\frac{1}{\eta_p}$$

where η_p is the polytropic efficiency. Also, the relation between final pressure and temperature can be stated as

$$\frac{T_f}{T_i} = \left(\frac{P_f}{P_i}\right)^{(n-1/n)}$$

For the same actual performance, the value of the polytropic efficiency will be somewhat higher than that of the isentropic efficiency. A comparison of these two efficiencies is shown in Fig. 9-5.

To calculate horsepower by using the polytropic values, the equation used in the isentropic process is employed, except that polytropic values are substituted. Note that since the horsepower is the actual work required for compression and must be the same regardless of the process used, the following relationship exists:

$$\frac{\text{Adiabatic head}}{\text{Adiabatic efficiency}} = \frac{\text{polytropic head}}{\text{polytropic efficiency}} = \frac{\text{isothermal head}}{\text{isothermal efficiency}}$$

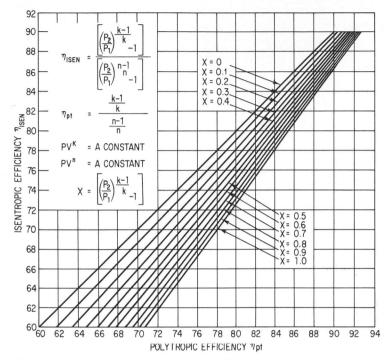

FIG. 9-5 Relationship between isentropic efficiency and polytropic efficiency based on a perfect gas.

<u>Compressibility</u>. The relationship of specific volume to pressure and temperature for a perfect gas is defined by the equation $Pv = RT$. However, many gases do not obey this perfect-gas law exactly. The deviation from the perfect-gas law is referred to as *compressibility* and is stated as a ratio of actual gas volume at a given pressure and temperature to the volume calculated by the theoretical law, in which case $Pv = ZRT$, where Z is the compressibility factor.

The compressibility factor Z can be derived from the rule of corresponding states by using reduced temperature and pressure. These reduced values are ratios of actual conditions to critical conditions:

$$\text{Reduced temperature} = T_r = \frac{T}{T_c}$$

$$\text{Reduced pressure} = P_r = \frac{P}{P_c}$$

Values of the critical constants T_c and P_c for individual gases are given in Table 9-1. As an example, T_r and P_r are calculated for a gas mixture with a volumetric composition of 14 percent ethane, 85 percent methane, and 1 percent nitrogen:

Gas	V volume %	T_c	VT_c	P_c	VP_c
C_2H_6	14	550.09	77.01	708.3	99.16
CH_4	85	343.5	292.00	673.1	572.19
N_2	1	227.2	002.27	492.0	4.92

For mixture $T_c = 371.28$, $P_c = 676.27$

By using the above values of T_c and P_c and assuming the gas at conditions of 100°F and 350 psia,

$$T_r = \frac{100 + 460}{371.28} = 1.507$$

$$P_r = \frac{350}{672.27} = 0.517$$

TABLE 9-1 Physical Constants of Gases

Compound	Formula	Mol. wt. M	c_p and c_p/c_v at 14.7 psia and 60°F		Critical constants		Mc_p at 60°F	Mc_p at 100°F	Mc_p at 200°F
			c_p	c_p/c_v	Pressure, psia P_c	Temp, °R T_c			
Acetylene.	C_2H_2	26.036	0.3966	1.238	905.0	557.4	10.33	10.69	11.53
Air.	N^+O_2	28.966	0.2470	1.395	547	238.7	6.96	6.96	6.99
Ammonia.	NH_3	17.032	0.5232	1.310	1,657	731.4	8.91	8.57	9.02
Benzene	C_6H_6	78.108	0.2404	1.118	714	1,013.0	18.78	20.47	24.46
1,2-Butadiene . . .	C_4H_6	54.088	(0.3458)	(1.12)	653	799.0	18.70		
1,3-Butadiene . . .	C_4H_6	54.088	(0.3412)	1.12	628	766.0	18.45		
N-Butane	C_4H_{10}	58.120	0.3970	1.094	550.7	765.6	23.07	24.51	26.16
Isobutane.	C_4H_{10}	58.120	0.3872	1.097	529.1	734.9	22.50	23.96	27.62
N-Butene.	C_4H_8	56.104	0.3703	1.105	583	755.6	20.77	22.09	25.18
Isobutene.	C_4H_8	56.104	0.3701	1.106	579.8	752.5	20.76		
Butylene	C_4H_8	56.104	0.3703	1.105	583	755.6	20.78	21.94	24.86
Carbon dioxide . .	CO_2	44.010	0.1991	1.300	1,073	548.0	8.76	9.00	9.35
Carbon monoxide .	CO	28.010	0.2484	1.403	510	242.0	6.96	6.96	6.98
Chlorine	Cl_2	70.914	0.1149	1.366	1,120	751	8.15		
Ethane	C_2H_6	30.068	0.4097	1.193	708.3	550.1	12.32	12.96	14.68
Ethyl alcohol. . . .	C_2H_5OH	46.069	0.3070	1.130	927.0	929.6	14.14		
Ethylene	C_2H_4	28.052	0.3622	1.243	742.1	509.8	10.16	10.68	12.08
N-Hexane.	C_6H_{14}	86.172	0.3984	(1.062)	439.7	914.5	34.33	36.23	41.08
Helium	He	4.003	1.2480	1.6598	188.0	510	5.00		
Hydrogen.	H_2	2.016	3.408	1.408	188.0	60.2	6.87	6.90	6.95
Hydrogen sulfide .	H_2S	34.076	0.254	1.323	1,306	672.7	8.66	8.18	8.36
Methane	CH_4	16.042	0.5271	1.311	673.1	343.5	8.46	8.65	9.30
Methyl alcohol. . .	CH_3OH	32.042	0.2700	1.203	1,157.0	924.0	8.65		
Nitrogen.	N_2	28.016	0.2482	1.402	492.0	227.2	6.95	6.96	6.963
N-Octane	C_8H_{18}	114.224	0.3998	(1.046)	362.1	1,025.2	45.67		
Oxygen	O_2	32.00	0.2188	1.401	730	278.2	7.00	7.03	7.120
N-Pentane	C_5H_{12}	72.146	0.3972	1.074	489.5	845.9	28.66	30.30	34.41
Isopentane	C_5H_{12}	72.146	0.3880	1.075	483.0	830.0	27.99	29.90	34.44
Propane.	C_3H_8	44.094	0.3885	1.136	617.4	666.2	17.13	18.21	20.90
Propylene	C_3H_6	42.078	0.3541	1.154	667	657.4	14.90	15.77	17.88
Sulfur dioxide . .	SO_2	64.060	0.1470	1.246	1,142	775.0	9.42		
Toluene.	C_7H_8	92.134	0.2599	1.091	611	1,069.5	23.95		
Water	H_2O	18.016	0.4446	1.335	3,206	1,165.4	8.01	8.03	8.12
Hydrogen chloride	HCl	36.465	0.1939	1.410	1,199.2	584.5	7.07		

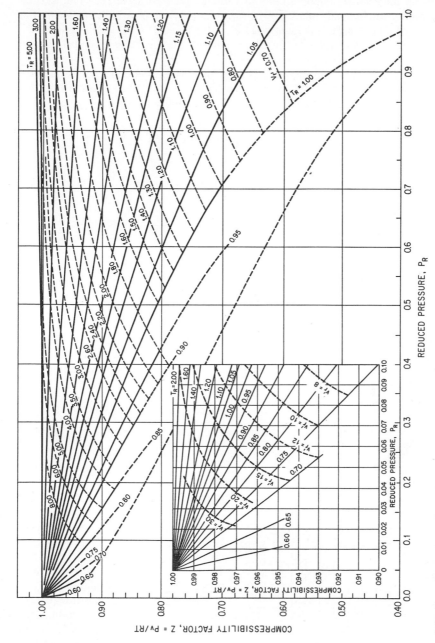

FIG. 9-6 Gas-compressibility factor in function of reduced pressure P_R.

Once the reduced temperature and reduced pressure have been found, compressibility can be read from Fig. 9-6, a generalized compressibility diagram. For the foregoing example, it is found that $Z = 0.95$.

Figure 9-7 gives values of the compressibility factor Z for a typical natural gas.

In the foregoing subsections on the isentropic, isothermal, and polytropic processes, the equations for work or head are true only for perfect gases. To correct for deviation from perfect-gas laws, the compressibility factor Z must be used in the equations, so that the equations become:

$$\text{Isentropic head} = \frac{53.34 Z T_i}{\text{sg}} \left(\frac{k}{k-1} \right) \left[\left(\frac{P_f}{P_i} \right)^{(k-1/k)} - 1 \right]$$

$$\text{Isothermal head} = \frac{53.34 Z T_i}{\text{sg}} \log_e \frac{P_f}{P_i}$$

$$\text{Polytropic head} = \frac{53.34 Z T_i}{\text{sg}} \left(\frac{n}{n-1} \right) \left[\left(\frac{P_f}{P_i} \right)^{(n-1/n)} - 1 \right]$$

In all these equations, the units in the U.S. customary system are

$$T_i = {}^\circ\text{R} = {}^\circ\text{F} + 460$$
$$P_i \text{ and } P_f = \text{psia}$$
$$\text{Head} = \text{ft} \cdot \text{lbf/lbm}$$

The common industry practice is to refer to head in terms of feet, since in the U.S. customary system pound-force and pound-mass are equal numerically and cancel each other. In the metric system, the correct units for head are joules/gram (J/g). The conversion is

$$\text{Head (ft} \cdot \text{lbf/lbm)} \times 0.0029891$$

$$= \text{head} \left(\frac{\text{joules}}{\text{gram}} \right)$$

Properties of gas mixtures. Before a compression cycle can be calculated, it is first necessary to know the specific heat ratio k, molecular weight M, and compressibility factor Z of the gas to be compressed. For a pure gas, this information can be taken directly from Table 9-1. For a gas mixture these properties must be calculated.

The chemical composition of the mixture is customarily given on a volumetric basis. The properties of the mixture are determined by the composite properties of the constituent gases. Each constituent exerts a partial pres-

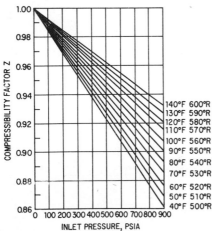

FIG. 9-7 Compressibility factor of natural gas.

sure which is determined by the amount of the constituent, the volume occupied, and the temperature, just as if the constituent alone were present. The total pressure is the sum of the partial pressures exerted separately by the constituents. The partial pressures are the same fractions of the total pressure as the volumes of the constituents, as given by a volumetric analysis, are of the whole volume.

The specific heats are the same for each constituent as if it alone were present. The values of the specific heats of the mixture are equal to the summation of the products of the proportional weight of each constituent and its specific heat; i.e.,

$$CG = c_1 g_1 + c_2 g_2 + \text{etc.}$$

where
C = specific heat of mixture
G = weight of mixture
c_1, c_2, etc. = specific heat of constituents
g_1, g_2, etc. = percentage weights of constituents

In a similar manner, the molecular weight of the mixture is equal to the summation of the products of the proportional volume of each constituent and its molecular weight; i.e.,

$$MV = m_1 v_1 + m_2 v_2 + \text{etc.}$$

where
M = molecular weight of mixture
V = total volume
m_1, m_2, etc. = molecular weights of constituents
v_1, v_2, etc. = volumes of constituents

The molecular weight of a gaseous substance is the weight in pounds that will occupy 379 ft^3 at 14.7 psia and 60°F.

The calculation of the properties of a mixture can best be done in tabular form. To illustrate, determine the properties of a natural gas having the following volumetric percentage composition: 14 percent ethane, 85 percent methane, and 1 percent nitrogen.

Gas	v volume fraction	m molecular weight	vm	$G = vm$ M fraction	C_p at 60°F	GC_p	C_v at 60°	GC_v
C_2H_6	0.14	30.07	4.21	0.233	0.4097	0.0955	0.3434	0.0800
CH_4	0.85	16.04	13.63	0.752	0.5271	0.3964	0.4021	0.3024
N_2	0.01	28.02	0.28	0.015	0.2482	0.0037	0.1773	0.0027
Total	1.00		$M = 18.12$	1.000	$C_p = 0.4956$		$C_v = 0.3851$	

$$\text{Molecular weight} = 18.12 \qquad \text{Specific gravity} = \frac{M}{28.966} = \frac{18.12}{28.966} = 0.6$$

$$\text{Mixture specific heat at constant pressure} = 0.4956$$

$$\text{Mixture specific heat at constant volume} = 0.3851$$

$$R = \frac{1544}{M} = \frac{1544}{18.12} = 85.2$$

$$k = \frac{C_p}{C_v} = \frac{0.4956}{0.3851} = 1.287$$

$$\frac{k - 1}{k} = \frac{1.287 - 1}{1.287} = 0.223$$

A simplified method for finding k makes use of the molal specific heat MC_p, as expressed in the following relationship:

$$k = \frac{MC_p}{MC_p - 1.99}$$

With this formula it is necessary to know only the value of MC_p for a gas or a gas mixture in order to calculate its k value, and C_v can be disregarded. The specific heat of a gas varies with temperature. Table 9-1 gives the value of MC_p for various gases at 60°F, 100°F, and 200°F. Since the calculated head varies only slightly with changes in k value, it is sufficient for most estimates to use MC_p at 60°F. On applications for which an accurate value of discharge temperature is required, however, such as a process in which polymerization may occur, the MC_p values should be estimated on the basis of the average temperature of the compression cycle. A straight-line interpolation of MC_p values can be made for temperatures not given in Table 9-1.

The k value of the same gas mixture used in the preceding example may be calculated in tabular form as illustrated below:

Gas	V volume %	MC_p at 60°F	$V \times MC_p$
C_2H_6	14	12.32	1.725
CH_4	85	8.46	7.191
N_2	1	6.95	0.069
	100		8.985

$$k = \frac{8.985}{8.985 - 1.99} = \frac{8.985}{6.995} = 1.284$$

Volumetric flow. The foregoing subsections have illustrated how to compute the required work of compression (head) by using the basic gas values of M (or sg), k, and Z. The computation of head is fundamental to the selection of the proper compressor. However, equally important to the selection of any compressor is the determination of inlet flow. This flow directly affects the size of the inlet nozzle, volute, and first-stage impeller.

To compute the inlet flow, we must return to the perfect-gas law, as corrected with the compressibility factor:

$$Pv = ZRT$$

solving for v,

$$v = \frac{ZRT}{P}$$

where v = specific volume
= volume per unit mass

The inverse of specific volume is density:

$$\rho = \frac{1}{v} = \frac{P}{ZRT}$$

In U.S. customary units, the specific volume is in cubic feet per pound-mass, or:

$$v = \frac{ft^3}{lbm}$$

where $v = \dfrac{ZRT}{144P}$
$R = 53.34/sg$
$T = {}^\circ R$
$P =$ psia
(144 is a conversion from in^2 to ft^2)

The flow rates specified by a user for a compressor application generally are not given in terms of volume. Rather, they are usually specified in terms used by the process engineer, such as *mass flow* or *molal flow*. To convert to inlet volumetric flow, the specific-volume equation above must be used with the specified inlet conditions.

To convert mass flow to volume flow, the mass flow must be multiplied by the specific volume:

$$m \times v = Q$$

$$\frac{Mass}{Unit\ time} \times \frac{volume}{mass} = \frac{volume}{unit\ time}$$

In U.S. customary units, this would resolve to

$$\frac{Pound\text{-}mass}{Minute} \times \frac{ft^3}{pound\text{-}mass} = \frac{ft^3}{minute} = ft^3/min\ (cfm)$$

When this flow is calculated at inlet conditions, it is usually referred to as inlet ft^3/min, or icfm. Also, the term *actual ft^3/min* is sometimes used to denote flow at the actual specified conditions.

To convert molal flow to volume flow, it must first be converted into mass flow:

$$Mass\ flow = molal\ flow \times M$$

Sometimes flow is specified in terms of some set of standard conditions and is then referred to as *standard flow*. In the U.S. customary system, this standard flow is usually referred to as scfm (standard cubic feet per minute).

The standard conditions generally used in this instance are 14.7 psia pressure and 60°F (520°R). To convert from scfm to acfm:

$$\text{acfm} = \text{scfm} \times \frac{P_s}{P_i} \times \frac{T_i}{T_s} \times Z_i$$

(Note that at standard conditions Z is considered to be unity.)

Sometimes other sets of "standard" conditions are used, such as the "normal" conditions of 760 mmHg and 0°C.

There are various other ways in which flow rates can be specified, and care should be taken that all units are kept correct. Once this inlet flow has been calculated, the compressor can be selected or "sized" on a preliminary basis.

Mollier method. For many pure gases, thermodynamic data are readily available in the form of Mollier diagrams, which graphically represent the relationships between pressure, temperature, specific volume, enthalpy, and entropy. A typical Mollier diagram is shown in Fig. 9-8.

When such data are readily available, an alternative method for calculating both head and flow can be used. First, the inlet conditions P_i and T_i are located on the diagram. From this point, the inlet specific volume can be read and then be used to calculate volume flow as shown in the preceding subsection.

Also, the inlet enthalpy and entropy can be read from the diagram:

$$h_i = \text{inlet enthalpy}$$
$$S_i = \text{inlet entropy}$$

For the next step, it is necessary to look again at the isentropic process. In this process the compression path follows a line of constant entropy. This is illustrated by the dotted line on the diagram, where the isentropic compression path is followed to the point at which it intersects the discharge-pressure line. At this point, the enthalpy is again read:

$$h_{fs} = \text{final enthalpy at constant entropy}$$

The difference between the two enthalpies at constant entropy is related to the work or head:

$$\Delta h_s = h_{fs} - h_i$$
$$\text{Adiabatic head} = \Delta h_s \times \text{conversion factor}$$

For example, in the U.S. customary system h is usually in Btu per pound, so that adiabatic head $= \Delta h_s$ Btu/lb $\times$ 778 ft·lb/Btu $=$ ft·lbf/lbm.

At this point, the flow and head have been computed, and the compressor can now be selected. To determine the discharge temperature, the actual enthalpy difference is calculated:

$$\Delta h \text{ actual} = \frac{\Delta h_s}{\eta_s}$$

where η_s is isentropic efficiency. Then, the actual final enthalpy is found:

$$h_f \text{ actual} = h_i + \Delta h \text{ actual}$$

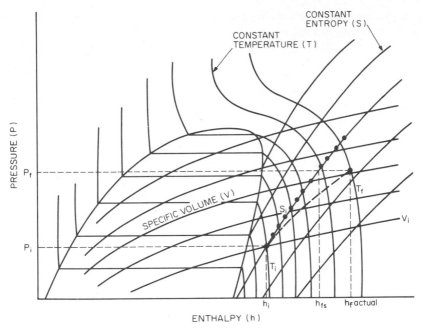

FIG. 9-8   Mollier method.

At the point where this actual final enthalpy intersects the final-pressure line, the actual final temperature can be read from the diagram.

The Mollier method illustrates the fact that the actual compression cycle does not follow any theoretical path (such as constant entropy). The actual path can be defined only in terms of its end points. One possible path is illustrated by the dashed line on Fig. 9-8. This illustration demonstrates the need to model the compression path on some theoretical process (i.e., isentropic, isothermal, or polytropic) and then correct that process by means of the efficiency.

Summary. This subsection on the thermodynamics of compression is intended to give sufficient information for a basic understanding of the subject. It must be noted that, in addition to the gas laws and data given here, there are many more sophisticated equations of state which define gas parameters more precisely for a certain set of circumstances.* These equations of state are particularly important when considering compressor applications which approach any extreme (for example, high pressure, high temperature, or low temperature). However, the data included here are considered sufficient for the majority of the applications of industrial compressors.

Performance Characteristics   The relationships between the inlet volume flow, head, speed, efficiency, and power of a dynamic compressor are often referred to

---

*Some of the other equations of state are Beattie-Bridgman, Benedict-Webb Rubin (BWR), Redlich-Kwong, and Starling–modified BWR.

as its *characteristic*. Typical characteristics for a centrifugal and an axial compressor are shown in Figs. 9-9 and 9-10 respectively.

At constant speed, as the system resistance increases, causing the head requirement to increase, the volume decreases until a peak head is reached. This point is referred to as *surge*. At this point a flow instability takes place, causing the flow to pulsate. The severity of operation in surge depends on many factors, including the pressure levels, the machine configuration, and the system configuration, but in general compressor operation should be limited to flows greater than the surge flow.

At the other end of the constant-speed curve there occurs a point of maximum flow and minimum head. This point is sometimes referred to as *carryout, stonewall,* or *choke flow*. Actually, these terms are not necessarily synonymous. The ultimate choke or stonewall flow occurs at a point at which the characteristic is nearly vertical (i.e., a small increase in flow produces a large decrease in head). This flow is a function mainly of impeller-inlet geometry; the impeller simply can-

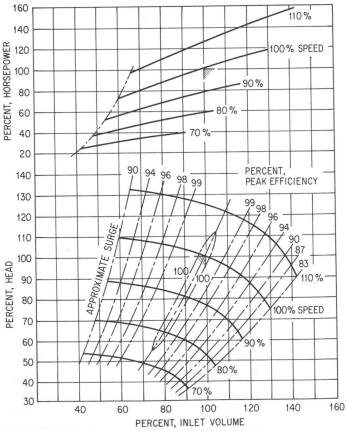

FIG. 9-9 Typical multistage-centrifugal-compressor characteristic.

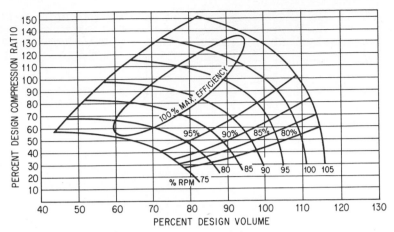

FIG. 9-10 Typical axial-compressor characteristic.

not "swallow" any more flow. However, at some flow less than the choke flow it generally becomes difficult to predict the shape of the characteristic curve, so most industrial-compressor-performance curves (especially those of multistage types) are not extended past this carryout area.

Variable speed. The effect of speed changes on centrifugal-compressor performance is similar to that on pump performance. The flow is directly proportional to the speed, the head is proportional to the speed squared, and the horsepower to the speed cubed, while the efficiency stays practically constant. These relationships are known as the fan laws:

$$Q \propto N$$
$$H \propto N^2$$
$$HP \propto N^3$$

where Q is flow, H is head, HP is power, and N is speed.

Reducing the impeller's outside diameter will generally have an effect similar to that of speed, except that variable impeller-blade geometry makes this relationship quite imprecise.

It should be realized that the fan laws will be less accurate as the speed change increases, particularly in multistage compressors. For example, it would be unreasonable to correct data taken at half speed up to full speed by means of fan laws.

Varying inlet conditions. As has been mentioned, the pressure delivered by any given dynamic unit depends on the density of the gas being compressed. This is demonstrated by Fig. 9-11, in which centrifugal characteristic curves for constant speed have been drawn for varying inlet conditions. Curve FG represents the characteristic for a centrifugal compressor designed to handle air at an inlet pressure of 14.4 psia, an inlet temperature of 60°F, a molecular weight of 28.95 (dry air), and a K value of 1.398. This unit develops 15 psig at an inlet capacity of 20,000 ft³/min. If the inlet temperature increases to 100°F, all other conditions remaining the same, the discharge pressure developed at 20,000 ft³/min is 13.7 psig. Likewise, if the inlet pressure drops to 12.4 psia but the inlet temperature and

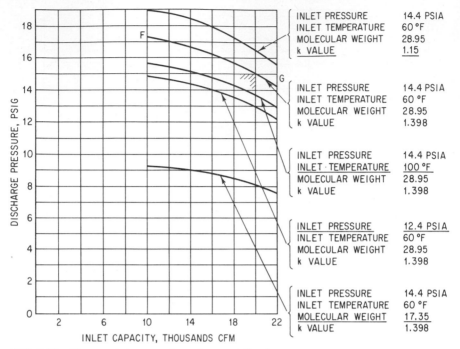

FIG. 9-11 Characteristic curves for a given centrifugal compressor.

other conditions remain as first specified, the discharge pressure at 20,000 ft³/ min is 12.9 psig. Further, if the molecular weight is the only variable, then with a 17.35-molecular-weight gas and 20,000 ft³/min, the discharge pressure developed will be 8.2 psig. A decrease in the k value to 1.15, all other conditions being unchanged, will result in a discharge pressure of 16.4 psig.

Performance Parameters   Various parameters have been established within the compressor industry to classify and quantify the different types of compressors available. The first of these parameters is called *specific speed,* which is a term used to classify compressor impellers on the basis of their performance and proportions regardless of their actual size or the speed at which they operate. Since specific speed is a function of impeller proportions, it is constant for any series of impellers having the same proportions or for one particular impeller operating at any speed.

Specific speed can be defined as the speed in revolutions per minute at which an impeller would rotate if reduced proportionately in size so as to deliver one cubic foot of gas per minute against a total head of one foot. It is found from the equation

$$\text{Specific speed} = N_s = \frac{N\sqrt{Q}}{H^{3/4}}$$

where N = operating speed, r/min
Q = design flow, icfm
H = design head, ft

It is not necessary to grasp the physical significance of specific speed to make use of it. Specific speed should be considered to be a type of characteristic of the impeller which specifies its general proportions and behavior.

Although specific speed is a useful parameter for comparison purposes, it is not always a useful tool for making compressor computations. The most commonly used parameters for compressor calculations are *flow coefficient* and *head coefficient*. Both these coefficients are nondimensional, and they can be used to define completely an impeller's performance.

The flow coefficient is defined as

$$\phi = \frac{3.056Q}{UD^2} = \frac{700Q}{ND^3}$$

where Q = inlet flow, icfm
U = tip speed, ft/s
D = impeller outside diameter, in
N = r/min

The flow coefficient can be used to put any compressor impeller's flow-handling capability in nondimensional terms.

Figure 9-12 presents a comparison of four compressor impellers, showing both specific speeds and flow coefficients. Figures 9-13, 9-14, and 9-15 show three different compressor rotors illustrating various impeller types.

Head coefficient is defined as

$$\Psi = \frac{Hg}{U^2}$$

where H = head, ft (per stage)
g = gravitational constant
U = tip speed, ft/s

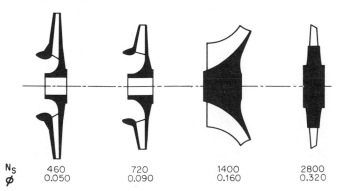

| N_S | 460 | 720 | 1400 | 2800 |
| ϕ | 0.050 | 0.090 | 0.160 | 0.320 |

FIG. 9-12 Impeller specific speeds and flow coefficients.

FIG. 9-13 Typical small-capacity low-specific-speed multistage rotor.

FIG. 9-14 Combined mixed-flow and medium-specific-speed centrifugal-compressor rotor.

FIG. 9-15 Typical axial-compressor rotor.

The head coefficient relates the head output of an impeller nondimensionally. By using a combination of flow coefficient and head coefficient, any impeller is defined in terms that can be easily manipulated.

Operational Control During the operating cycle of a typical compressor, process requirements may change, causing compressor conditions to vary. If the compressor has a variable-speed driver, these process changes can generally be accommodated by changing the speed of the compressor. In this situation, some type of control system usually produces the proper speed change on the basis of process requirements so that the compressor is kept running in a stable fashion. For example, if the flow requirement decreases while the pressure ratio (i.e., head) stays constant, the compressor could be slowed down to match the new lower flow.

However, if the compressor is driven by a constant-speed machine (such as a motor, for example), the compressor must be controlled by some other means. One such method employs adjustable inlet guide vanes. Figure 9-2 shows a compressor with such adjustable vanes. The purpose of any inlet guide vanes is to "guide" the flow smoothly into the impeller inlet. In the case of adjustable inlet guide vanes, the vanes are mechanically rotated so that the flow direction is changed prior to the impeller inlet. This change in direction in turn causes a change in the characteristic-curve shape of that stage. By setting the position of these guide vanes at various angles, several characteristic curves can be obtained, as shown on the typical curve in Fig. 9-16. These guide vanes can be moved by a number of methods, such as chain drives, gears, or other mechanical means.

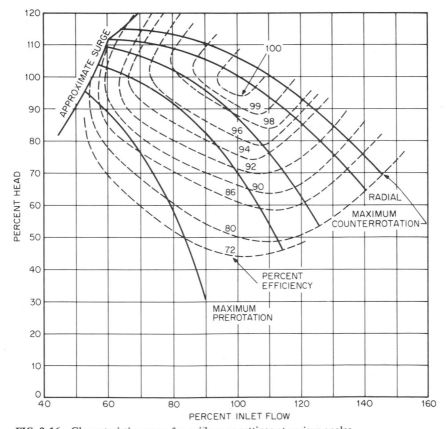

FIG. 9-16 Characteristic curves for guide-vane settings at various angles.

Adjustment of the vanes can be made an automatic part of the control system so that a change of guide-vane setting accomplishes essentially the same goal as a change in speed.

For some constant-speed compressors, adjustable inlet guide vanes either are not available or are not desirable, and other methods of control must be found. Two such methods are suction and discharge throttling. In suction throttling, a throttling device such as a butterfly valve is placed in the process inlet piping just prior to the compressor inlet nozzle. When this throttling valve is closed to some extent, it introduces a pressure drop which lowers the suction pressure. Usually the throttle-valve operation is connected to a control system which will cause it to react to changes in process conditions. For example, if the pressure required is held constant at the compressor discharge and the flow rate is reduced, the throttle valve will close slightly, causing the suction pressure to drop. This in turn causes the inlet volume and head requirements to increase, and the compressor then operates at a new point on its characteristic curve.

With discharge throttling, the valve is placed in the process discharge piping. In this case, if the flow rate is reduced into the compressor, the discharge pressure developed by the compressor increases. Then, after the flow leaves the compressor, the discharge throttle valve lowers the pressure to the required level.

Usually, several circumstances determine the type of driver and control that is used, among them the types of energy available, operator preference, and capital investment. One factor to be considered in this regard is comparison of the various energy requirements by using the different process controls mentioned above. For example, consider a typical compressor application in which the required flow periodically decreases by 20 percent while the head requirement stays constant. For the various alternatives discussed above, a comparison of the power needs at the lower flow is given:

Driver type	Control type	Relative power, percent
Turbine	Variable speed	100
Motor	Adjustable guide vanes	102
Motor	Suction throttling	104
Motor	Discharge throttling	106

Mechanical Design

Casings   The basic components of a dynamic compressor (impellers, diaphragms, diffusers, etc.) are housed in a casing which also serves to contain the pressures developed by the machines. (The casing must also maintain its alignment under extremes of conditions internal and external to the machine.) Figure 9-1 shows an axial section through a typical multistage compressor.

When possible, machines of the type shown in Fig. 9-1 are split horizontally at the centerline to facilitate maintenance. For certain gas compositions and higher pressures, the casing may be built as a complete cylinder, as shown in Fig. 9-2, and the internals removed axially. This type is commonly referred to as a vertically split or barrel design. Connections for bringing the gas into the compressor and removing it after it has passed through the impellers are part of the casing. Every casing of either type contains all inlet and discharge connections or nozzles. Other variations include reentry connections for external cooling and side-stream connections for adding or subtracting flow at intermediate points in the compression cycle.

Casings may be supplied as castings (Fig. 9-17), fabricated of sheet or plate (Fig. 9-18), or supplied as forgings (Fig. 9-19). Cast casings may be made of iron for low-pressure applications or of steel for higher pressures. Inlet and discharge nozzles and volutes are cast integrally with the casing.

Fabricated casings are supplied for a wide range of services and quite high pressures. The materials used are limited only by the service intended and suitability for welding. For higher pressures, a fabricated casing may be produced from a combination of sheet, plate, and forged parts. Volutes are formed as part of the fabrication, and nozzles may be castings or forgings welded to the casing.

Forged casings are supplied for special applications and extremely high pressures. For these applications the casing is usually a simple forged cylinder. The end closures are manufactured from forged disks. Inlet and discharge connections may be of special design, and volutes are incorporated in the inner assembly.

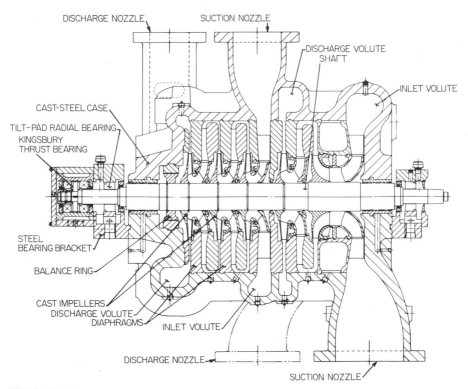

FIG. 9-17 Cast case.

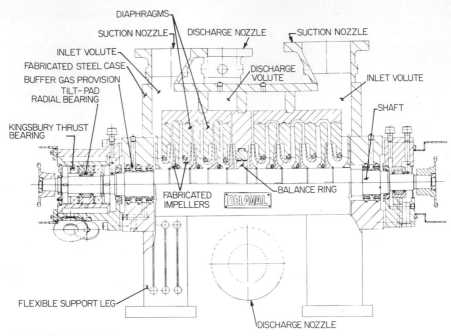

FIG. 9-18 Fabricated case.

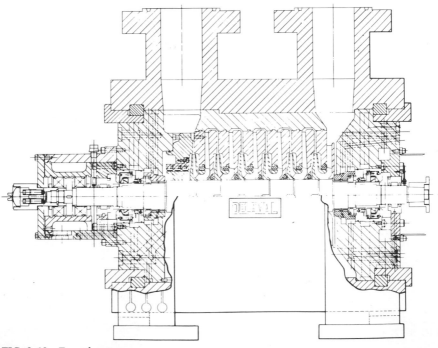

FIG. 9-19 Forged case.

Axial compressors are generally supplied for very high flows and for lower pressures. Options on case construction are essentially the same as for centrifugal machines except that vertically split, or barrel, designs are not available. (See Fig. 9-4.)

Internal Design   The stationary aerodynamic assembly for centrifugal compressors consists of several diaphragms which define the stationary gas-flow passages. In a horizontally split casing, these diaphragms may be mounted directly into the case halves, as in Fig. 9-1, or installed in an inner casing, which is then installed in the main casing. When the compressor uses a vertically split, or barrel, casing, the diaphragms are usually installed in an inner casing. This inner casing, in two halves, is bolted around the rotor. This assembly is then inserted axially in the main casing.

Several rotor and thus aerodynamic arrangements are possible. The basic configuration is the straight-through arrangement, in which the gas flow is directed into one end of the casing, through the impellers, and out the opposite end of the case (see Fig. 9-19). Variations of the straight-through arrangement include connections for adding or subtracting flow (see Fig. 9-20).

Another common configuration consists of connections for external cooling. With this arrangement the impellers may be arranged for straight-through flow, as in Fig. 9-1, or back to back, as in Fig. 9-18. The opposed impellers in the back-to-back design help balance the aerodynamic thrust developed in the rotor.

High flows can be accommodated with a double-flow rotor (see Fig. 9-21). In this arrangement the inlet flow is split, and half is directed to each of two identical rotor sections. The aerodynamic thrust is balanced because of the opposed identical rotor sections.

For pressure ratios within the range of one impeller, single-stage designs are available. This type of unit is typical of high-pressure gas-transmission-pipeline compressors (see Fig. 9-3). Similar types of units are supplied for high-pressure synthesis-gas recycle applications (see Fig. 9-22).

The flow in an axial compressor, as its name implies, is parallel to the machine axis, and the compression cycle involves passing the gas through alternating rows of rotating and stationary blading. The stationary blades are mounted in a carrier or inner casing that is mounted around the rotor in the main casing (see Fig. 9-4). Axial compressors are available in a straight-through arrangement, in a straight-through arrangement with a radial last stage, and in a back-to-back arrangement with a radial last stage after each section of compression (see Fig. 9-23).

A radial last stage replaces two or three axial stages. This feature results in a shorter bearing span for a given pressure ratio, which is a desirable effect for high-pressure ratios. To achieve good efficiency in very large compressors with high compression ratios, an intercooled back-to-back arrangement is provided. This configuration has a radial last stage at the end of each section (see Fig. 9-23).

For very high pressure ratios the intercooled type of axial compressor may be supplied with an additional single radial stage overhung from the compressor's low-pressure-end-bearing bracket. This arrangement provides a second intercooling point with a resulting improvement in overall isothermal efficiency. (See Fig. 9-24.)

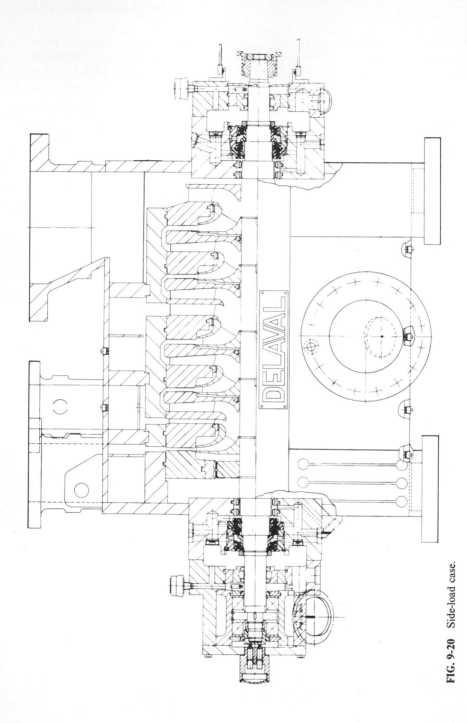

FIG. 9-20 Side-load case.

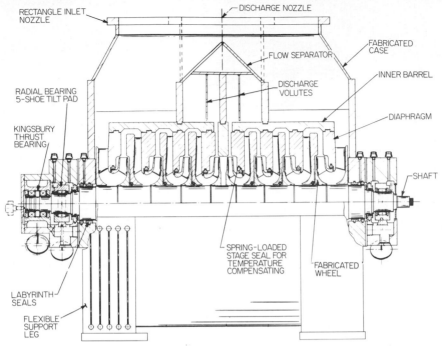

FIG. 9-21   Double-flow case.

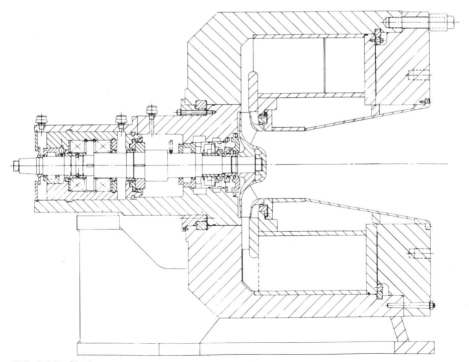

FIG. 9-22   Single-stage case.

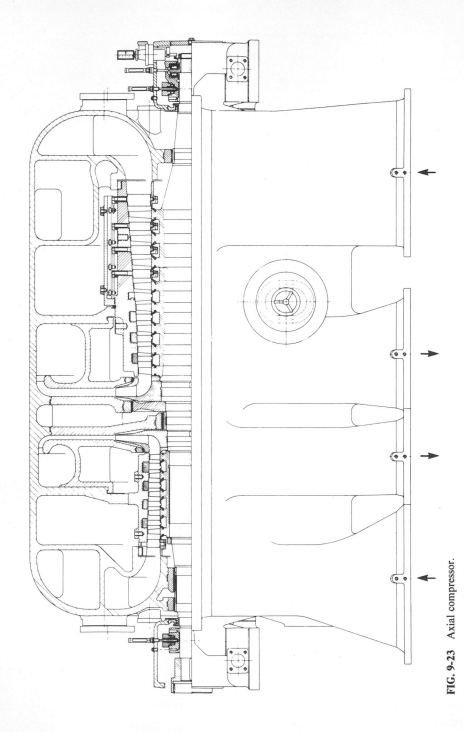

FIG. 9-23   Axial compressor.

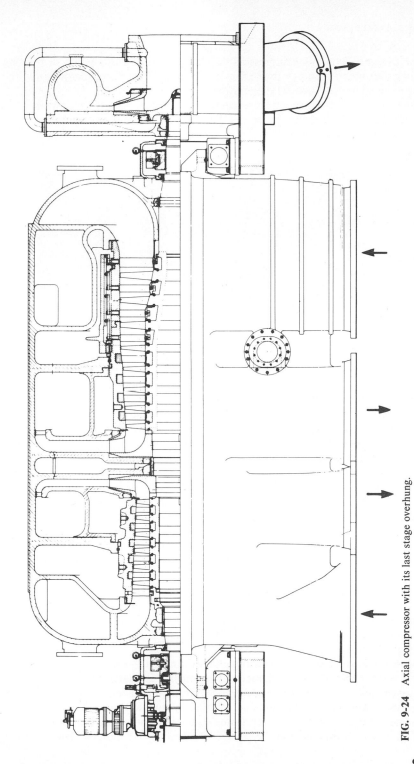

FIG. 9-24 Axial compressor with its last stage overhung.

Rotor Design Dynamic compressors develop pressure by the action of rotating blading that imparts energy to the flowing medium. Therefore, the design of the rotor is significant to successful operation of the machinery.

The rotor of a dynamic machine consists of a shaft, wheels (or impellers), a balance drum, and a thrust collar (see Fig. 9-25). The shaft is generally manufactured from a forged steel bar.

FIG. 9-25 Rotor.

Impellers for a centrifugal machine are manufactured in several ways (see Fig. 9-26*a*, *b*, and *c*). The method depends on many things including the design of the impellers, their size, and the material. Impellers may be cast in one piece, or they may be produced by machining the blades on a hub and welding them on a side plate. Small impellers may be plug-welded, while larger sizes are welded internally. Impellers of very high flow and large diameters may be manufactured in three pieces. In this method blades made from plate are welded to a machined hub and side plate. Impellers of very small diameter and low flow are being manufactured successfully by a chemical milling process.

The rotor for an axial compressor is often machined from a single forging. The rotor assembly is completed by mounting individual blades.

Centrifugal-compressor rotors are built up step by step. The impellers are mounted on the shaft individually, several methods of mounting being used. Hydraulically mounted impellers are common on single-stage overhung compressors such as those used in pipeline applications. Small rotors are assembled by using a polygon fit. For this arrangement a matching polygon of three or more sides is machined on the shaft and in the wheel bore. The bore fits on the shaft with a slight clearance. After the wheel has been positioned on the shaft, it is rotated with an appropriate tool and locked in place. Removal is a simple reversal of the procedure.

Most rotors are built with their impellers shrink-fitted to the shaft. The wheel bore is smaller than the shaft diameter. The impeller is heated to expand the bore and assembled over the shaft before it cools. Shrink-fitted impellers are generally keyed to prevent axial movement and rotation relative to the shaft.

Some rotors are assembled by using stub ends and a tie bar instead of a through shaft. In this method the parts are individually balanced. The tie bar is

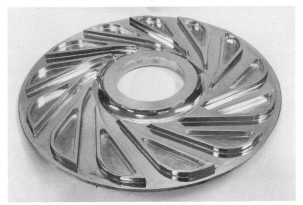

FIG. 9-26a Low-flow impeller showing how the plug-welding technique, used to attach covers to blades in this type of wheel, leaves smooth flow passages. *b* Display model of a Delaval wedge wheel without cover illustrates its unique blade profile. *c* Impeller prior to finish machining shows the thickness of the hub and cover material at this stage, which prevents distortion during welding and stress relieving.

attached to one stub end. The impellers are added individually, accurately keyed to each other and to the stub ends. The second stub end is added and the tie bar tightened to the design preload.

Rotor components are individually balanced and oversped before the rotor assembly is started. As the assembly of shrink-fitted rotors progresses, shaft run-out and assembly unbalance are monitored and corrected after the addition of each disk (impeller or balance drum). Limits of runout and balance are established, and if at any point in the assembly these limits are exceeded, the assembly is delayed until the cause has been determined and corrected.

Single-stage rotors using a hydraulically mounted impeller or rotors with polygon-fit impellers are not assembly-balanced because of the limits of the design. These rotors are assembled from components that are carefully balanced individually.

The balance of a dynamic-compressor rotor is critical to reliable operation of the machine. Unbalance in a rotating system may be thought of as the result of uncompensated centrifugal forces in the rotor. During operation high levels of unbalance may be evident from the amplitude and direction of motion measured at the bearings. Different types of rotors behave in different ways, and this behavior due to unbalance is referred to as *rotor response.* When the unbalance is large enough, the motion in the bearings may be greater than the designed-in mechanical clearance between shaft and bearing. When this happens, metal-to-metal contact occurs, resulting in severe damage to the compressor's stationary parts and rotating assembly. Therefore, the magnitude of the unbalance must be controlled for satisfactory operation.

The causes of unbalance and thus of response vary with different types of rotors. Rotor types may be described as either *rigid* or *flexible.* Rigid rotors are those whose condition of unbalance does not change significantly up to service speed; flexible rotors, those whose condition of unbalance changes with rotational speed.

The unbalance condition in a rigid rotor may be classified as *static unbalance, couple unbalance, quasi-static unbalance,* and *dynamic unbalance.* Static unbalance is caused when the center of gravity of the rotor is not on the axis of rotation. It may be corrected with one correction plane at the center of gravity.

Couple unbalance is caused by two equal and opposite nonsymmetrical masses located in two different radial planes. This type of unbalance may be corrected in two planes, the corrections being equal and opposite.

Quasi-static unbalance is similar to static unbalance except that the unbalance mass is not located on the plane containing the center of gravity. It corresponds to a condition of static unbalance and couple unbalance. It can be corrected in a single correction plane if the plane is chosen properly. This plane will also correct the couple unbalance.

Dynamic unbalance is the generalized condition of unbalance of a rotor and consists of a combination of a static and a couple unbalance, usually at differing angular positions. Static unbalance, quasi-static unbalance, and couple unbalance are all special cases of dynamic unbalance. Generally two balancing planes are used to correct dynamic unbalance.

Since almost all rotors require balancing, the cause of unbalance is significant. Unbalance is present if the mass distribution of a rotor with respect to its shaft axis is not symmetrical. This asymmetry can be caused by design, material distribution, or manufacturing and assembly tolerances. Design unbalance results when parts lack rotational symmetry. This condition can result from unmachined surfaces on cast or forged parts, loose fits, or open keyways. Material can vary in density and thickness.

The unbalance condition in a flexible rotor is quite different from that in a rigid rotor. This unbalance, which is due to speed, results from a deformation in the rotor that is asymmetric to the shaft axis. Deformations are classified as *plastic deformation* (permanent deformation at high speed that does not decrease as speed is decreased) and *elastic deformation* (one that decreases as speed is decreased). The latter type may be subdivided into *elasticity of the body* and *elasticity of the shaft*.

Rotors with plastic deformation often reach a state of equilibrium at high rotational speeds which then remains unaltered at lower speeds. By spinning the rotor at a speed which is known to be above the service speed, it is usually possible to achieve a stable condition of unbalance at all rotational speeds up to the service speed. After spinning such a rotor at high speed, it is possible to balance at any rotational speed below the service speed.

Rotors with body elasticity are those with masses whose center of gravity is neither on nor near the shaft axis and which undergo elastic deformation because of speed-dependent centrifugal force. The process has no tendency to reverse with increased speed, but because of certain design features it may reach a stable unbalance condition. Rotors of this type must be balanced at their service speed or at a speed above the limit at which stable unbalance conditions are achieved.

When elastic deformation due to centrifugal forces occurs in a rotor whose center of gravity lies on or near the shaft axis, shaft elasticity is involved. The state of unbalance changes at an increasing rate as speed is raised. Deformation reaches a maximum and then decreases. It has the appearance of a resonance. To balance a rotor with shaft elasticity it is necessary to apply sufficient correction masses to maintain rigid rotor balance and at the same time reduce deformation at critical speeds.

Seals   The many applications of rotary equipment handling air or gases require various methods for shaft sealing. Dynamic compressors and exhausters employ labyrinths, carbon rings, bushings, or contact seals either separately or in combination to effect the desired sealing.

Labyrinth seals. The labyrinth seal is the simplest, cheapest, and most reliable sealing device used. It consists of a series of annular orifices through which pressure breakdown is accomplished by the dissipation of velocity head.

Three types are illustrated in Fig. 9-27: an axial straight-pass labyrinth (Fig. 9-27a), an axial staggered labyrinth (Fig. 9-27b), and a running-knife labyrinth (Fig. 9-27c). The straight-pass labyrinth is used when pressures are relatively low, while the more efficient staggered labyrinth is preferred for higher pressures and/or equipment in which space limitations are important. The running-knife labyrinth is a straight-pass labyrinth which can be installed with small original clearance, and its performance approaches that of the staggered type.

Clearances selected for fixed labyrinth seals are governed by manufacturing tolerances, journal-bearing clearances, vibration of the rotor during operation, and relative thermal expansion of rotor and labyrinth through the operating-temperature range. Minimum practicable radial clearances vary from a few thousandths of an inch for small seals to $\frac{1}{32}$ in or more for very large seals. When a large range of temperatures is encountered, a segmental labyrinth mounted on

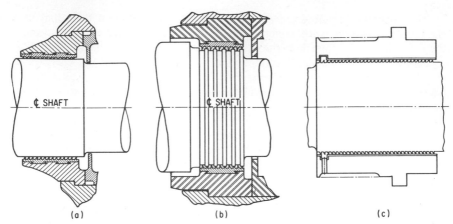

(a) (b) (c)

FIG. 9-27 Labyrinth types. (*a*) Axial straight-through labyrinth. (*b*) Axial staggered labyrinth. (*c*) Running-knife labyrinth.

springs is used to maintain more nearly constant clearance as the temperature changes.

Carbon rings. Shown in Fig. 9-28 is a typical carbon-ring seal in which each ring is characterized by a radial sealing surface, positive suspension within gland housing, and low operating clearance over the shaft. This design allows radial motion of the carbon ring when contacted by the shaft but maintains pressure between the ring and the radial sealing surface at all times.

The thermal coefficient of expansion of carbon is approximately one-fourth that of steel. Therefore, the hot clearance is in most cases less than the cold clearance. When determining the minimum practical clearance, consideration is given to shaft deflection, transverse journal motion, and thermal deformation of the shaft. Operating clearances of 0.003 to 0.004 in on the diameter are usually feasible.

Leakage through well-made and properly mounted carbon rings will range from 15 to 25 percent of the leakage from a well-designed knife-edge labyrinth utilizing the same shaft length. The ordinary laws of gas flow apply to carbon-ring seals. Carbon rings sealing clean gases having no highly oxidizing component can be expected to give approximately 20,000 h of satisfactory service. Dirt-laden atmosphere drastically reduces seal life.

Bushing seals. Basically a bushing seal is simply a close-clearance sleeve surrounding the shaft. Sealing is effected by the sealing face between the end of the bushing and its housing and by the restriction of the small clearance area between the sleeve and the shaft.

A typical bushing seal is illustrated in Fig. 9-29. Freedom of radial movement is inherent in the design, and motion in this plane occurs during start-up as the shaft journals rise in their bearings as well as in response to equalizing pressures from within the flow-path annulus.

Hydraulic pressure acting on exposed surfaces of the floating bushing exerts sufficient closing force to prevent leakage past the sealing face. Springs are pro-

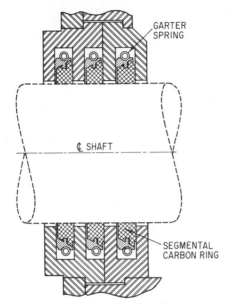

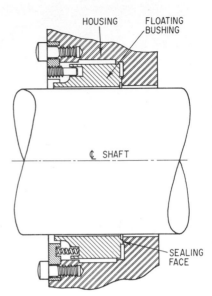

FIG. 9-28   Segmental carbon ring.

FIG. 9-29   Bushing seal.

vided to keep the bushing in place against the seal face until sufficient hydraulic pressure is applied.

The rate of liquid leakage through a bushing seal is a function of viscosity, temperature, differential pressure across the seal, seal geometry, and concentricity of shaft and seal. The rate of leakage will vary directly with the differential pressure and the wetted perimeter, with the cube of the clearance, and with the square of the eccentricity; and it will vary inversely with viscosity and length. Shear work done on the sealing fluid during its passage through the bushing raises its temperature.

Clearance between the bushing and the shaft and the length of the bushing must be selected to obtain minimum leakage without exceeding temperature limitations for the fluid. Flatness, parallelism, and surface finish of the mating seal faces must also be carefully controlled to obtain maximum seal effectiveness.

Contact seals. A contact seal (Fig. 9-30) is a mechanical seal consisting of a rotating collar and a nonrotating seal ring. Contact between the collar and the seal ring is maintained by calibrated springs and a small designed closing force resulting from unbalanced hydraulic pressure acting on the exposed surfaces of the seal ring.

The seal ring is suspended so that it may float in both the radial and the axial directions. This feature enables it to compensate for displacement of the collar due to rotor vibration, thermal expansion, or minute out-of-true location on the shaft.

An elastic O ring or a pair of accurately fitted Teflon piston rings comprises the secondary sealing element between the seal ring and the stationary mounting

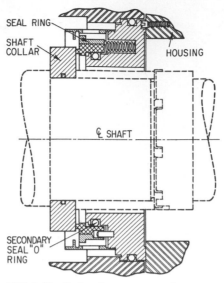

SEAL RING

SHAFT COLLAR

HOUSING

$\oint$ SHAFT

SECONDARY SEAL "O" RING

FIG. 9-30 Carbon face-contact seal.

piece. This element is carefully proportioned and mounted so as to offer minimum resistance to motion of the seal ring.

The effectiveness of this type of seal depends upon the maintenance of parallelism and the flatness of the mating faces on the collar and the seal ring. The collar is made of stable abrasion-resistant die steel. It is specially heat-treated and lapped to a surface flatness of 15×10^{-6} in. The seal ring is made of a refined grade of carbon, shrunk within a steel shroud. The ring is geometrically so contoured that dimensional stability is ensured and surface flatness preserved for long life.

Figure 9-31 shows a combination of an oil-contact seal and a bushing seal adapted for low- and medium-pressure sealing service. In this design the air-side seal-oil leakage is used to lubricate the journal bearing. Figure 9-32 shows a combination of an oil-contact seal and a bushing seal adapted for high-pressure service. A separate oil supply is provided to the seal and the journal bearing in this design. Two bushing seals can be combined as shown in Fig. 9-33; in all respects operation is similar to that of the combined face-and-bushing seal shown in Fig. 9-32.

Leakage comparison. The relative effectiveness of the several types of seals described for dry sealing may be seen from the following table:

Seal	Leakage index
Straight-pass labyrinth	100
Staggered labyrinth	56
Segmental carbon rings	20
Dry contact	2

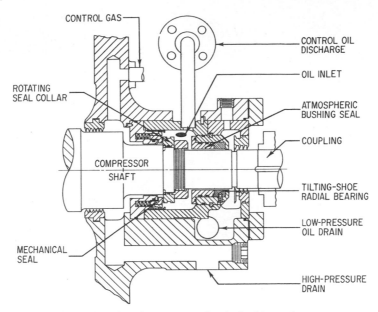

FIG. 9-31 Combination of a contact seal and a bushing seal.

A typical contact oil seal, with a mean face diameter of 6¾ in and a 30-lb/in² differential pressure, operating at 5000 r/min, will leak approximately 0.021 gal/h. Under similar operating conditions, a ⅝-in-long, 5½-in-diameter bushing seal with an 0.007-in diametral clearance and an oil temperature rise of 60°F will leak approximately 1¾ gal/h. The diametral clearance in a bushing seal may be reduced to a point at which leakage approaches that of a contact seal, but the oil-temperature rise becomes so great that the leakage oil is unfit for further use and must be discarded.

Sealing System

Single seals. The simplest sealing system is one in which a single dry seal is used at the point at which the compressor shaft emerges from the casing or the point at which the shaft passes between two internal chambers at different pressures. A certain amount of leakage is allowed, and the type of seal selected depends on the permissible leakage.

Evacuation systems. In an evacuation system two dry seals are employed at each sealing point, and the chamber between these two seals is connected to an ejector which creates a low pressure in the chamber. Leakage through both seals is withdrawn and discarded or returned to the system at a lower-pressure location.

Figure 9-34 illustrates a typical gas-compressor evacuation system in which the inner seals are labyrinths and the outer seals are carbon rings. The ejector is powered by gas from the discharge of the compressor, and the evacuated gas and air mixtures from the seal chambers are returned to the compressor suction. The

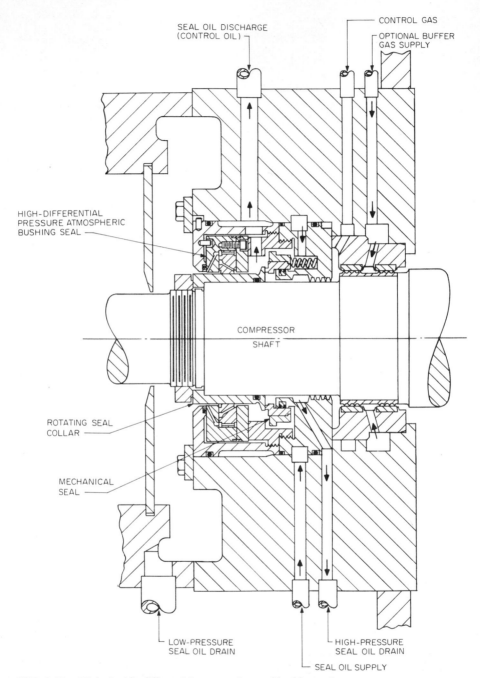

SEAL OIL DISCHARGE
(CONTROL OIL)

CONTROL GAS

OPTIONAL BUFFER
GAS SUPPLY

HIGH-DIFFERENTIAL
PRESSURE ATMOSPHERIC
BUSHING SEAL

COMPRESSOR
SHAFT

ROTATING SEAL
COLLAR

MECHANICAL
SEAL

LOW-PRESSURE
SEAL OIL DRAIN

HIGH-PRESSURE
SEAL OIL DRAIN

SEAL OIL SUPPLY

FIG. 9-32. High-air-side differential-pressure face-and-bushing seal.

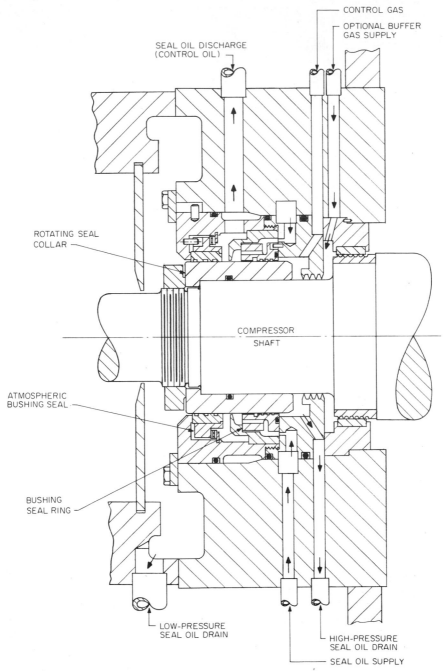

CONTROL GAS

OPTIONAL BUFFER
GAS SUPPLY

SEAL OIL DISCHARGE
(CONTROL OIL)

ROTATING SEAL
COLLAR

COMPRESSOR
SHAFT

ATMOSPHERIC
BUSHING SEAL

BUSHING
SEAL RING

LOW-PRESSURE
SEAL OIL DRAIN

HIGH-PRESSURE
SEAL OIL DRAIN

SEAL OIL SUPPLY

FIG. 9-33 Double-bushing seal.

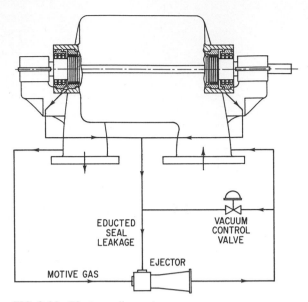

FIG. 9-34   Ejector sealing system.

gas required to power the ejector and the evacuated leakage, all recirculated through the compressor, represent a power loss which must be considered in selecting this type of system. The advantages of such a system are relatively low cost and the fact that it is a simple self-contained entity with few moving parts.

Injection systems. In an injection system two seals are again used at each sealing point, but in this case a sealing fluid is injected into the chamber between the seals at a pressure above that to be sealed. With dry seals the leakage from the seal chamber through the inner seal goes into the compressor, while the leakage from the seal chamber through the outer seal escapes to the atmosphere. With wet seals the leakage in both directions is collected as it leaves the seals and returned to the system.

Figure 9-35 illustrates a typical oil-injection seal for a high-pressure gas compressor. The inner seal, between the seal chamber and the compressor, is a wet-contact seal. The outer seal is a floating-bushing seal. Cooled and filtered oil is pumped into the seal chamber at a pressure slightly higher than the gas pressure in the compressor. A very small quantity of this oil leaks past the contact seal and is collected and returned to the reservoir through a high-pressure drain system. A larger quantity of oil leaks through the bushing seal to atmospheric pressure and is also drained back to the reservoir. Enough oil to maintain proper temperatures is circulated through the seal chamber at all times. This circulating oil is led out of the chamber, through a differential-pressure controller, and back to the reservoir.

In Fig. 9-35 the inward oil leakage through the contact seal is exposed to the compressed gas and may absorb appreciable amounts of this gas. In some appli-

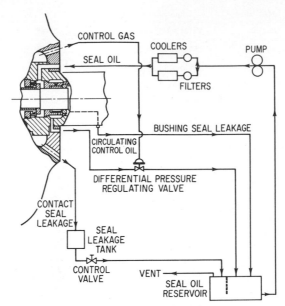

FIG. 9-35 Typical sealing system using an injected medium between compressor gas and atmospheric seal.

cations this oil becomes contaminated and must be discarded rather than returned to the system. In such cases the extremely low leakage rates obtainable with the contact seal minimize the cost of replacing the discarded seal leakage. The outward leakage through the bushing seal is not exposed to the gas and therefore is not subject to contamination.

Materials for Centrifugal Compressors

Consideration of materials for use in centrifugal compressors must begin with a careful study and appraisal of expected operating conditions (see Table 9-2). In addition to the nominal conditions of operation, it is necessary to take into account expected variations in these conditions. It is necessary to know whether variations are expected to be transient or of long duration.

The operating stresses of rotating parts are determined by the design, size, and speed of rotation and can influence material selection. The operating stress of the casing and other stationary machine elements must also be known, and for these parts stress is determined primarily by size, design, and internal pressure. The gas to be handled by the compressor must be evaluated for aggressiveness with respect to the material of construction. The expected temperature of operation also must be determined.

Several of the particularly specialized areas for compressor material selection are discussed below. A table of various component materials has been included (see Table 9-2).

TABLE 9-2 Materials of Construction for Compressor Components

Component	Material	Specification	Temperature limits (°F)	Remarks
Cast casings	Gray cast iron	ASTM A278, Class 40	−20–500	Air, nonflammable gases
	Ductile cast iron	ASTM A395	−20–500	Air, nonflammable gases
	Cast carbon steel	ASTM A216, Grade WCB	−20–750	
Fabricated castings	Steel plate	ASTM A515, Grade 55	50–650	
	Steel plate	ASTM A516, Grade 55	−20–650	Low temperatures
	Steel plate	ASTM A516, Grade 55	−50–650	Low temperatures (with impact test)
	Steel plate	ASTM A543 (HY-80)	−175–650	Low temperatures
	Steel plate	ASTM A553, Type II	−275–650	Low temperatures
	Steel plate	ASTM A553, Type I	−320–650	Low temperatures
Forged casings	Steel forgings	ASTM A266, Class 2	−20–650	High pressure
Diaphragms	Gray cast iron	ASTM A48, Class 40	−320–650	Low pressures only
	Cast carbon steel	ASTM A216, Grade WCB	−320–750	High pressure
	Fabricated steel	ASTM A515, Grade 55	−150–650	
	Cast carbon steel plus nickel	ASTM A352, Grade LC3	−320–650	
Impellers	Fabricated steel	ASTM A473 (12 percent chromium)	−20–650	
	Fabricated steel	ASTM A543 (HY-100)	−75–650	
	Fabricated steel	ASTM A543 (HY-80)	−75–650	Low yield; H_2S service
	Fabricated steel	ASTM A553, Type II	−275–650	
	Fabricated steel	ASTM A553, Type I	−320–650	
	Fabricated steel	Monel K-500	−175–650	Special corrosive conditions
	Fabricated steel	17-4PH/15-5PH	−100–650	Corrosive conditions; moisture
	Cast steel	17-4PH/15-5PH	−100–650	Corrosive conditions; moisture
Shafts	Forging	AISI 4145	−20–750	
	Forging	AISI 4345	−175–800	Low temperature
	Forging	17-4PH	−100–650	Oxygen
	Forging	9 percent nickel steel	−320–650	Low temperature
Labyrinth seals	Aluminum alloy	Alloy CJ90	−320–600	Limited corrosion resistance
	Bronze	ASTM B584, Alloy 976	−150–650	Oxygen service
	Fluorosint	Ceramic-filled Teflon	−350–500	Excellent corrosion resistance
	Tin-base babbitt	ASTM B23, Class 2	−175–400	
	Felt metal	. . .	Above 350	High-temperature service
Shaft sleeves	AISI-410 stainless steel	ASTM A276, Type 410	−100–750	
O rings	Buna N	. . .	250 maximum	Limited corrosion resistance
	Silicone	. . .	350 maximum	Excellent corrosion resistance
	Viton A	. . .	300 maximum	Excellent corrosion resistance

Hydrogen Sulfide Service Two corrosion factors which must be considered in sour-gas (hydrogen sulfide) service are sulfide stress-corrosion cracking and hydrogen embrittlement. Studies have shown that low-alloy steels having a maximum yield strength of 90,000 lb/in^2 and a maximum hardness of Rockwell C-22 are not susceptible to sulfide stress-corrosion cracking. The yield strength of a material also determines the degree of hydrogen embrittlement that can occur.

Low-alloy steel is usually employed for impellers. The optimum material which resists sulfide stress-corrosion cracking and results from heat treatment by quenching and tempering is tempered martensite. Yield strength and hardness are controlled to 90,000 lb/in^2 and Rockwell C-22 maximums respectively. The martensitic transformation occurs by rapid cooling or quenching. The martensitic phase is formed from a retained-high-temperature phase at lower temperatures; essentially the martensite is considered to be a supersaturated solution.

The hardness of iron-carbon martensite is measured with its carbon content. Prior to tempering, the martensite is extremely hard and brittle. By tempering, ductility is increased and hardness and strength are decreased, making the alloy tougher for more practical application. Steels are usually tempered after quenching.

AISI-410 stainless steel containing approximately 12 percent chromium is usually selected for the first-stage impeller to provide more erosion-corrosion resistance than is available from a low-alloy steel. The additional erosion-corrosion resistance is required because compressor suction could contain sour gas saturated with moisture. The remaining impeller stages would not be affected by moisture due to higher temperatures after the first stage.

Welding may result in local composition changes and produce local stresses. Stress relief after welding is required to relieve internal stresses and to bring hardness to the Rockwell C-22 maximum.

A study of the factors influencing sulfide cracking has confirmed that the phenomenon can best be explained in terms of hydrogen embrittlement. The primary variables which are known to influence hydrogen-induced delayed failure, i.e., the strength level of the steel, the stress level on the steel, and the concentration of hydrogen in the steel, also influence susceptibility to sulfide cracking. Delayed failure of stressed specimens exposed to a hydrogen sulfide content occurs when hydrogen, as a result of corrosion, embrittles steel to such an extent that the fracture stress reaches the stress under load. In general, resistance to cracking diminishes as strength increases. With the exception of carbon, changes in alloy content do not appear to influence sulfide cracking behavior. Cracking susceptibility appears to increase as the carbon content increases. Normalized or normalized and tempered steels are less resistant than quenched and tempered steels. Small differences in microstructure can have a profound effect on cracking resistance.

For the above reasons, studies of the influence of composition must take strict account of microstructural and strength changes accompanying changes in composition as well as heat treatment. Consequently, empirical relationships designed to evaluate changes in composition must be treated with caution. Austenite-containing steels are considerably more resistant than steel with body-centered structures. Therefore, an increase in elements such as nickel and manganese which lower the m.s. temperature* and stabilize the face-centered structure can be very beneficial.

Chlorine Service Perhaps the most persistent problem encountered by many plant operators is caused by the corrosive nature of chlorine. Decreasing plant output over a period of time can result from a buildup of material in process components including the stationary internal-flow passages of the compressor. The width of diffuser and return channels in the compressor is reduced, adversely affecting its performance characteristics.

The buildup on critical compressor surfaces has been prevented with the development of materials and methods of application suitable for coating the basic iron or steel stationary internal compressor parts. Polyvinylidene chloride resin has been applied with the dimensional accuracy required for compressor internals.

The internal seals of centrifugal compressors are made by positioning stationary elements with a small clearance adjacent to the rotor. During normal operation, these seals do not touch, but in periods of upset or malfunction they can rub and generate considerable heat. When a compressor is handling dry chlorine gas, this heat is potentially dangerous. The rotating element of the seal is made of alloy steel, and the stationary element is stabilized Teflon material. The ignition temperature of alloy steel in dry chlorine gas can be reached when the seals rub. This can result in an internal fire in the compressor and considerable damage.

Inlaying the alloy-steel seal areas on the rotor with a metal that has a higher ignition temperature such as Inconel 82 will reduce the possibility of internal fires when rubs occur. Such an inlay material has better corrosion resistance to dry chloride gas than does alloy steel, and the life of the seals is extended when the inlay is used. This inlay is covered under United States patent Serial No. 4,060,250, dated November 29, 1977, and held by Transamerica Delaval Inc.

Carbon Monoxide Service Carbon monoxide does not ordinarily affect steels at low pressures, but it can cause severe attacks at high pressures if hydrogen is present. At ordinary pressures, this gas is inert to most constructional materials even at high temperatures. However, at high pressures corrosion is so severe that special materials must be used; carbon steel severely corrodes even at low pressures and temperatures of 500°F. Thirteen percent chromium steel, which severely resists attack at high pressures and 500°F, should be utilized for carbon monoxide applications.

Low-Temperature Service When compressors are required to operate at subzero temperatures, consideration must be given to the problem of brittle failure. As temperatures fall, all material becomes stronger and less ductile, and some material becomes increasingly susceptible to brittle failures. Materials with this susceptibility include most of those commonly employed in compressor construction.

For a brittle failure to be initiated, there must be a stress exceeding the yield strength. While the nominal stress is always well below the yield strength, it is not possible to design and manufacture any engineering structure without stress concentrations, and it is from these sites that brittle failures propagate. Therefore,

---

*The temperature at which martensite starts to form.

special materials with increased impact strength at low temperatures must be used.

Systems and Instrumentation

Oil Systems The primary function of an oil system is to provide the proper quantities of cooled and filtered oil at the required regulated pressure levels to the driven and driving equipment. This oil can be used for lubrication, shaft-sealing, and/or control-oil purposes. The oil system is designed to furnish the oil required at all operating conditions of the equipment. A basic oil system is described below and shown in Fig. 9-36.

A fabricated steel reservoir tank serves to store a volume of oil sufficient for 5 to 8 min at normal flow. The tank is fitted with both a dipstick and a sight-glass level gauge. Removable heating elements, either steam or electric, are usually provided. These heaters are sized to heat the oil from the minimum site ambient temperature to the equipment's minimum required oil temperature (usually 70°F) within 12 h. The tank is furnished with a temperature indicator and with a level switch which activates an alarm when the oil level is below the minimum operating level. The purge and vent connections on the tank provide a means to exhaust any gases that are released from the oil.

Oil is drawn from the bottom of the reservoir through suction strainers by motor-driven or steam-turbine-driven pumps. The main and auxiliary pumps are identical and supply a constant flow of oil. Each pump discharge line has a relief valve which protects the equipment from any overpressure caused by a system malfunction. The relief valves are sized to pass the full pump capacity. A pressure gauge and a check valve are furnished in each pump discharge line. A block valve is placed downstream of the check valve for maintenance purposes.

The flow of oil then passes through a transfer valve which can "transfer" the flow from one filter-cooler set to the other set without interrupting the flow. Out-of-service units can be opened for cleaning or maintenance while the other units are in service. Two identical coolers, each capable of handling the system's maximum flow and heat load, are furnished. Water flow to the coolers is regulated to maintain the desired oil outlet temperature of 120°F (49°C). Two identical filters are also supplied. One filter is placed downstream of each cooler to remove particulate material as small as 5 μm. Filters, with clean cartridges, are sized to handle the maximum system flow and pressure with a pressure drop no greater than 5 lb/in². A differential-pressure indicator and a differential-pressure switch are placed across the filter-cooler combination to warn of the need to change the filter cartridges.

To avoid oil-pressure surging and interruption of the oil supply, the out-of-service filter-cooler combination should be filled with oil before it is put on the line. The vent valves on the filters and coolers are used to vent air from the units, while the oil cross-connect line is opened to fill them. This cross-connect line is left open to keep the out-of-service units pressurized. Thermometers are fitted upstream and downstream of the coolers to check the unit's performance.

A back-pressure regulator valve is supplied to establish and control the header pressure after the filter-cooler units. This valve can be either self-operated or

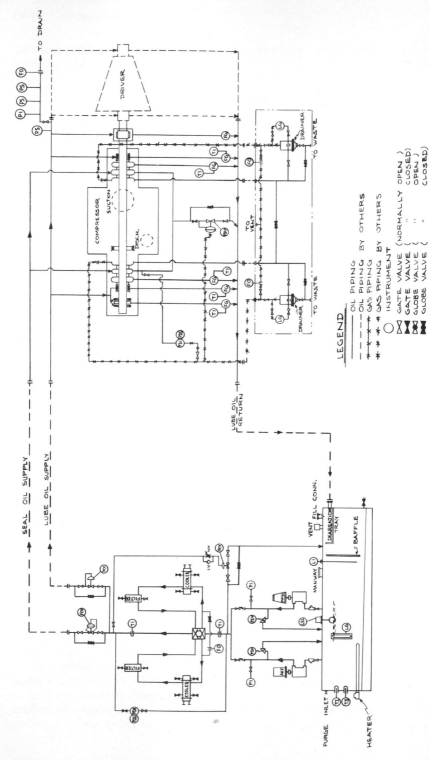

FIG. 9-36 Typical oil system.

pneumatically operated, whichever method is dictated by duty. Oil is taken from a point before the filter-cooler units and bypassed back to the reservoir. The valve is sized to control a wide range of flows, with either one pump or both pumps in operation. In the latter case, the valve would pass a maximum flow of the two pump capacities less the flow required by the system.

Several different pressure levels are often required to carry out the various functions of an oil system. A pressure-reducing valve is used to reduce the header pressure (established by the back-pressure regulator) to the required pressure level for lubricating oil, control oil, etc. One valve is used for each required pressure level.

When oil seals are used on a compressor, the flow to the seals is set by a flow-control valve. This valve is designed to maintain a constant volume of oil regardless of how the oil pressure may vary.

All control valves—flow, pressure, or differential-pressure—have bypass provisions. In case of malfunction, the control valve can be isolated and the flow or pressure adjusted manually through the bypass globe valve.

The unit lubricating-oil line receives oil from the console and delivers it at 20 psig to the compressor thrust and journal bearings, to the coupling, and to the driving-equipment bearings. The line is fitted with a pressure indicator and several pressure switches at the farthest extreme of the header to ensure an adequate oil supply at this outermost point. One switch is set to trigger an alarm, and a second switch starts the auxiliary oil pump, when the lubricating-oil pressure falls to 12 psig. The third switch is set to trip the unit at 8 psig decreasing.

The lubricating-oil drain system collects the oil used by the compressor, coupling, and driving equipment and returns it to the reservoir. Each atmospheric drain line from the bearings and seals is fitted with a sight flow indicator and a thermometer.

Seal oil is supplied to each compressor through a separate header. The seal-oil pressure is controlled from the downstream side of the seal. A differential back-pressure regulator valve is used for mechanical seals, while a head tank is used for bushing seals.

When mechanical seals are used, the oil-side pressure is set 45 lb/in^2 above the gas side. A differential back-pressure regulator senses gas suction pressure and regulates the seal oil to maintain the proper differential pressure. The seal-oil system is fitted with a differential-pressure indicator and several switches. These are connected between the seal-oil return line and the control-gas connection. The switches activate an alarm and start the auxiliary pump when the seal-oil–gas differential falls to 35 lb/in^2 ΔP.

Bushing seals have a head tank that is mounted above the compressor to maintain a 5-lb/in^2 differential above suction pressure. A level-control valve is supplied downstream of the seal-oil return lines to maintain the proper level in the head tank: approximately 14 ft above the centerline of the compressor.

The overhead seal-oil head tank is sized to have the proper capacity and rundown time for emergency operation, coastdown, and block-in. A pneumatic level transmitter sends a signal to a level-indicating controller which operates the level-control valve. Level switches and gauges are furnished for monitoring, alarming, and trip functions.

Oil drainers are supplied for both mechanical contact and bushing seals. They collect the contaminated seal oil and provide an automatic means of discharging this oil for reuse or disposal. An oil drainer is essentially a tank with a float-operated drain valve arranged so that the seal-oil leakage can be drained without releasing any gas. Each drainer can be isolated for servicing by closing three valves. The nonoperational drainer is bypassed, and both seals are allowed to drain into the remaining drainer by opening the valves in the crossover lines. A level glass is fitted on the drainer to indicate the operating level.

The majority of the components in the oil system are usually mounted on a preassembled console such as that shown in Fig. 9-37. Consoles of this type are piped and tested prior to shipment; they can generally accommodate a wide selection of valves, flanges, fittings, and other components to meet various specifications. Piping can be entirely carbon steel, entirely stainless steel, or any combination of the two, depending mostly on user preference.

FIG. 9-37 Typical oil console.

For compressors with gas-turbine drivers, compressor oil is usually supplied by the turbine oil system. In such cases, a small seal console which contains seal booster pumps, filters, and the necessary seal-oil controls and instrumentation is provided. On some units, the main seal-oil pump is driven from the compressor shaft with an auxiliary pump on the seal console. An emergency seal tank can be mounted directly on the compressor, under pressure with oil circulating through it at all times.

In most units the seal oil is combined with the lubricating oil in one system, but separate lubricating-oil and seal-oil systems can be provided if necessary because of potential contamination of the lubricating oil.

Compressor Monitoring Systems Industrial compressors can be instrumented for maximum reliability by incorporating state-of-the-art concepts and equipment. Mechanical protective systems are designed into most units, and fail-safe principles usually are employed whenever applicable.

The mechanical integrity of a compressor can be continuously monitored by using noncontact eddy-current types of proximity probes to measure shaft vibration and axial position and movement. This instrumentation permits continuous monitoring of the actual dynamic conditions of the machine in addition to providing a capability of signature comparison and analysis over a period of operation. The actual readout and monitoring instruments can be furnished for mounting either in a local panel or in a remote-control room. The manufacturer can coordinate design with the customer to assure the compatibility of the remote monitoring and readout instruments with the machine-mounted sensors and transducers.

Temperatures of both thrust and journal bearings can be measured by using sensors embedded in the bearing metal. The drain-oil temperature can be measured by placing sensors in the oil drain lines. Resistance-type temperature elements or thermocouples may be used to provide compatibility with the user's temperature-measuring systems. Temperature readout and alarming instruments can be furnished for local-panel mounting along with annunciators and vibration monitors as well as other local instruments. A typical local instrument panel is shown in Fig. 9-38.

Operational and Antisurge Systems   In the majority of compressor applications, a control system of some sort is used to maintain stable compressor operation under a variety of conditions. The method of control can be variable speed, adjust-

FIG. 9-38 Typical instrument panel.

able guide vanes, suction or discharge throttling, or bypass (recycle). The method chosen usually depends on the type of process and the type of driver. In any case, this control system is an integral part of overall process control and, as such, is usually the responsibility of the user or the contractor.

To protect the compressor from surging, an antisurge system is usually supplied. A basic antisurge system consists of a bypass line from compressor discharge to inlet. This line has a valve which is opened when the unit nears surge, allowing some of the flow to be diverted from the process stream and back into the compressor and thus maintaining a minimum flow in the unit at some point at which stable operation is ensured. This bypass flow is usually cooled in order not to increase the compressor discharge temperature gradually past its limit. Operation of the bypass valve can be a function of parameters such as flow, pressure, and speed. In many instances the antisurge system is integrated into the overall process control system. In other cases a separate antisurge system can be furnished either by the user or by the manufacturer.

Industry Standards

In the early years of dynamic compressors, every manufacturer had its own standard design philosophy, and there was not necessarily any agreement between manufacturers and users on a common set of specifications. Consequently, a user would have to evaluate each manufacturer's designs strictly on their merits. The result was a lack of consistency within the industry.

As time progressed, various groups with common interests joined to establish standards organizations for most industrial products including compressors. Today the compressor industry is guided by a great number of such standards, which allow both user and manufacturer to maintain consistency within the industry.

The following is a partial list of standards organizations that apply to dynamic compressors. It should be understood that in most cases users consider these standards to be minimum requirements. Frequently users will augment these standards with their own specifications, while in some instances manufacturers may have specifications which differ from the standards.

ANSI	American National Standards Institute	Codes for piping; safety standards
API	American Petroleum Institute	Specifications for all major equipment in refineries
ASME	American Society of Mechanical Engineers	Boiler and pressure-vessel codes; performance-test codes
ASTM	American Society for Testing and Materials	Materials specifications
CAGI	Compressed Air and Gas Institute	Vibration standards
ISO	International Standards Organization	
NEMA	National Electrical Manufacturers Association	
NFPA	National Fire Protection Association	National Electrical Code
OSHA	Occupational Safety and Health Administration	
TEMA	Tubular Exchanger Manufacturers Association	Heat-exchanger codes and standards

Application of Dynamic Compressors

Compressors are available in a wide variety of sizes and designs and can be used for a multitude of applications. The methods by which they are selected for a particular application vary from manufacturer to manufacturer. A common practice is to establish a "family" or "line" of compressors which utilizes a basic design philosophy in a variety of compressor sizes. Essentially this results in a group of machines which are of different sizes yet are proportionally similar. This family concept can be carried further by establishing a family of impellers that are of different diameters but again are similar proportionally. The use of flow and head coefficients makes this concept easily applied. An advantage of this concept is that once the performance characteristics of a particular impeller design have been established for one impeller size, they can essentially be applied to the same design in any size.

Often a manufacturer will end up with several standard equipment lines to suit various applications. Some of these applications are highly specialized, while others may be handled by several types of compressors. A few typical examples are given to illustrate the wide variety of compressor applications that exist today (Figs. 9-39 through 9-44).

FIG. 9-39 Multistage compressor driven by a 14,600-hp gas turbine in mainline service for a southeastern pipeline.

FIG. 9-40 Natural-gas-transmission single-stage beam-type compressor to accommodate a double-ended drive. A 12,000-hp gas turbine presently drives through the inboard side, while the outboard shaft end awaits future connection to a power-recovery turbine.

FIG. 9-41 A fabricated, rather than cast, casing made it possible to design this chlorine compressor. Shown just prior to shipping, this model has all six nozzles in the upper half of the casing.

FIG. 9-42 Hydrogen-recycle compressor driven by an electric motor through a speed-increasing gear.

FIG. 9-43 Gas-gathering compressor assembled as an integrated package for rapid installation.

FIG. 9-44 Two compressors in tandem for separate services in a gas-treating plant. The double side-loaded unit compresses propane for refrigeration, while the high-pressure barrel recompresses the natural gas. A 21,000-hp gas turbine is geared up to the compressors, which in turn are connected through a speed-reducing gear to an electric generator.

RECIPROCATING COMPRESSORS

General

Reciprocating gas compressors are required for a great many widely varied industrial services. Their basic function is to raise the pressure level of the gas being compressed. Raising the pressure level is desirable for any of the following reasons:

1. *Storage.* Natural-gas storage in natural underground reservoirs, bottled gases for industrial uses, shop air compression and storage.

2. *Transmission.* Natural-gas-transmission pipelines.

3. *Process.* Chemical reactions that take place at elevated pressures (for example, ammonia synthesis at 5000 lb/in^2).

4. *Energy conversion.* Mechanical to thermal energy conversion (refrigeration systems).

Reciprocating compressors have inherent advantages over other compressors in their ability to adapt to a wide range of load, speed, pressure conditions, and pressure ratios ($P_{discharge}/P_{suction}$). The load may be varied from 0 to 100 percent; the speed may have a wide range, depending on the driver.

Pressures may vary from a few inches of mercury absolute suction pressure in the case of a vacuum pump to 30,000 lb/in^2 or more discharge pressure for process-gas compressors. Pressure ratios may vary from slightly over 1, in the case of natural-gas-transmission pipeline service, to 8 or more in the case of shop air compressors. Several stages of compression are often used when the overall pressure ratio is high.

There are many available prime movers suitable for driving reciprocating compressors. These include electric motors, turbines, natural-gas engines, diesel engines, and dual-fuel engines. Many electric motors and reciprocating engines have rotative speeds similar to those of reciprocating compressors and can be direct-connected, eliminating speed reduction (or multiplication) between the prime mover and the compressor.

Description of an Ideal Compressor Cycle   Positive-displacement compressors are machines in which successive volumes of gas are confined within a closed space and elevated to a higher pressure. The reciprocating compressor is a special type of positive-displacement compressor that elevates the pressure of the trapped gas by decreasing the volume that the trapped gas occupies. A piston moving in a cylinder is used to reduce the volume of the trapped gas (see Fig. 9-45).

Compression (a–b). The cylinder has filled with gas at the suction pressure with the piston at position *a*. The piston moves from *a* toward *b*, compressing the gas isentropically (with no heat transfer and no turbulent or frictional losses) until the pressure within the cylinder reaches the discharge-line pressure.

Discharge (b–c). At this point the discharge valve opens and permits gas to flow from the cylinder into the discharge line until the piston has reached the end of its stroke at point *c*.

Expansion (c–d). Since it is impossible to build a compressor with zero clearance volume, gas remains in the cylinder's clearance volume at the end of the discharge stroke. The gas remaining expands isentropically to suction pressure as the piston moves from *c* to *d*.

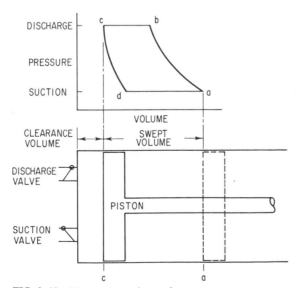

FIG. 9-45   Ideal compression cycle.

Suction (*d*–*a*). When the pressure within the cylinder reaches the suction pressure, the suction valve opens and permits gas at suction pressure to enter as the piston moves from *d* to *a*.

Since points *b* and *d* are determined by the pressures during the cycle, the cycle is described as having a suction stroke (piston moves from *c* to *a*) and a discharge stroke (piston moves from *a* to *c*).

Compressor Arrangements and Their Application

Small air compressors are usually single-acting, while most of the higher-horse-power units are double-acting. Double-acting compressors have pistons that compress gas on both ends so that one end is on its suction stroke while the other end is on its discharge stroke (see Figs. 9-46 and 9-47). The force resulting from the pressure and area differential across the piston is referred to as the *piston-rod load*. Reciprocating compressors are rated in terms of their rod-load capability rather than by horsepower. Rod-load ratings range up to 175,000 lb. Higher-horse-power compressors are built with a basic frame, with a wide range of cylinders that are interchangeable on the frame. The cylinders

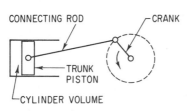

FIG. 9-46 Single-acting design.

range from small-diameter high-pressure cylinders to large-diameter low-pressure cylinders. A line of cylinders is usually designed so that each cylinder matches the rod-load capability of the compressor frame at the maximum working pressure of the cylinder and the expected pressure ratio of the applications for which the cylinder is intended.

Double-acting compressor cylinders of the type described above are mounted on several types of frames. The two basic classifications of compressor frames are the balanced-opposed type and the integral type (Figs. 9-48 and 9-49).

The balanced-opposed frame is characterized by an adjacent pair of crank throws 180° out of phase and separated by a crank web only. With this configuration the inertia forces are balanced if the reciprocating weights of opposing throws are balanced. The balanced-opposed design is a separable frame so that the basic compressor may be driven by any number of prime movers, including

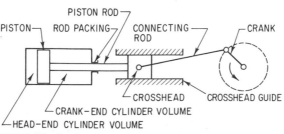

FIG. 9-47 Double-acting design.

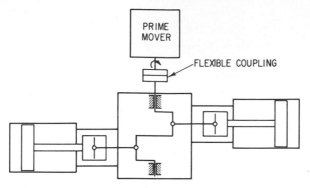

FIG. 9-48 Balanced-opposed compressor.

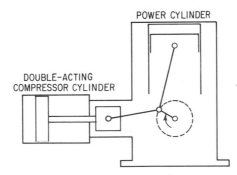

FIG. 9-49 Integral-type gas-engine compressor.

diesel, natural-gas, and dual-fuel engines, gas and steam turbines, and electric motors.

The integral compressor has compressor cylinders and power cylinders mounted on the same frame and driven by the same crankshaft. Some integrals have in-line power cylinders mounted vertically and compressor cylinders out one or both sides in the horizontal plane. Others are V engines with compressor cylinders out one or both sides.

Figure 9-50 shows a balanced-opposed four-throw reciprocating compressor with many of the parts identified.

Type of Service Reciprocating compressors are presently used for a large variety of services. The following is a description of some of the more common applications.

Pipeline service. This service is characterized by large flow rates, very low pressure ratios (1.1 to 1.5), and moderately high pressures (1000 psig).

Process-gas compression. A wide range of cylinders can be applied to process-gas work. The process itself dictates the requirement of compression. Gasoline plants may use low pressure (400 psig), and ammonia synthesis plants may require 5000 psig.

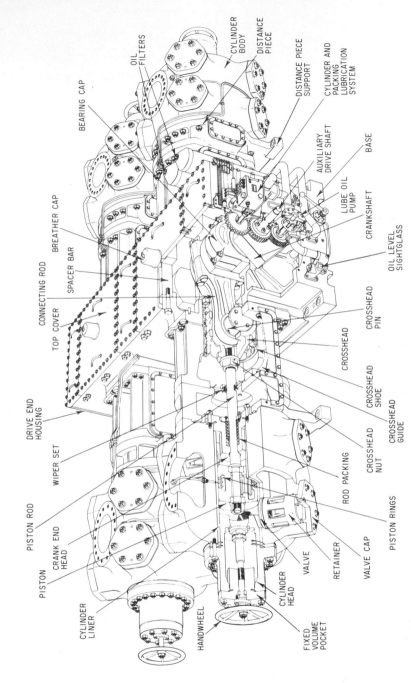

CYLINDER BODY

DISTANCE PIECE

DISTANCE PIECE SUPPORT

CYLINDER AND PACKING LUBRICATION SYSTEM

OIL FILTERS

BEARING CAP

AUXILIARY DRIVE SHAFT

BASE

LUBE OIL PUMP

CRANKSHAFT

OIL LEVEL SIGHTGLASS

BREATHER CAP

SPACER BAR

CONNECTING ROD

TOP COVER

CROSSHEAD PIN

CROSSHEAD

CROSSHEAD SHOE

CROSSHEAD GUIDE

DRIVE END HOUSING

CROSSHEAD NUT

WIPER SET

ROD PACKING

PISTON ROD

PISTON RINGS

PISTON

CRANK END HEAD

VALVE CAP

RETAINER

VALVE

CYLINDER HEAD

CYLINDER LINER

HANDWHEEL

FIXED VOLUME POCKET

FIG. 9-50 Engine-compressor.

9-58

Gas gathering. This consists of collecting the output of many gas wells and compressing the gas so that it will flow to a plant for further refinement or to a pipeline.

Gas injection. This technique is used for natural-gas storage in large natural underground storage volumes. Gas is injected at pressures up to 4000 psig during periods of small demand and withdrawn during periods of high demand (cold winter days).

Industrial air compressors. These supply air for industrial purposes such as spray painting, drilling, grinding, and riveting. Air is taken at atmospheric pressure, filtered, and compressed to 100 to 150 psig for most industrial uses.

The selection of compressor cylinders is based on the requirements of the application, including operating pressures, type of gas, range of pressure, etc. Cylinders for pipeline service are designed with large clearance volumes and gas passages, large valve areas, and no water cooling. The large clearance volumes are desirable from the standpoint of keeping horsepower absorption reasonably constant over a wide range of operating pressures. Large gas passage and large valve areas are necessary to keep the frictional losses low. Water cooling is not required in view of the low temperature rise that results from the very low pressure ratios involved in pipeline gas compression. Most of the other compression services require low clearance volumes and water cooling because of the higher temperatures which accompany higher pressure ratios.

Valve velocities are generally higher in general-purpose cylinders because of the low-clearance-volume requirements for higher-pressure-ratio applications.

Compressor cylinders and packings are normally lubricated by a high-pressure force-feed lubrication system supplying lubricant to individual points in each cylinder bore and packing. When contamination of the compressed gas is detrimental, the lubrication can be eliminated from the cylinder bore or the cylinder bore and packing. Nonlubricated compressors must be designed with materials and surface finishes that will operate without lubrication.

Performance Characteristics

Performance calculations are made to determine the throughput of a reciprocating compressor and the horsepower absorbed by the process. The capacity throughput is determined by the displacement rate of the compressor, the specific volume at suction conditions, and the suction volumetric efficiency. The piston-displacement rate is determined by the piston area and the piston speed.

The specific volume is a function of the pressure, temperature, and gas composition. The suction volumetric efficiency VE can be determined by thermodynamic calculation as VE percent $= 100 - R - $ (percent clearance)$(R^{1/k} - 1)$, where R is the ratio of absolute discharge to absolute suction pressure P_d/P_s and the percent clearance is (clearance volume/swept volume) $\times$ 100 percent. Figure 9-51 shows volumetric efficiency versus pressure ratio for varying clearance.

Horsepower absorbed by the compressor is the product of capacity throughput times horsepower per unit capacity, usually hp/lb, or horsepower per million cubic feet per day (hp/million ft^3/day).

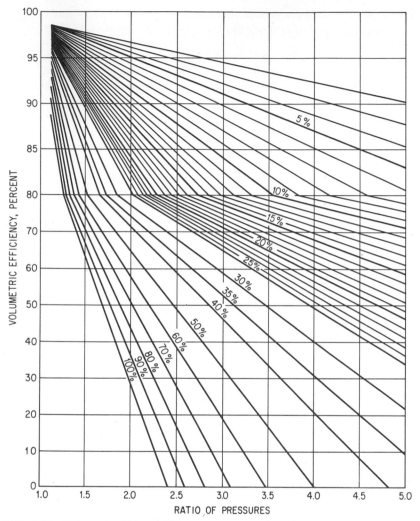

FIG. 9-51   Volumetric efficiency *VE* versus pressure ratio *R* for varying clearance volume ($k = 1.30$).

Figures 9-52 and 9-53 show horsepower versus pressure ratio for varying clearances with constant discharge pressure and constant suction pressure, respectively. These two curves are used not quantitatively but to indicate trends in horsepower due to changes in suction or discharge pressures. The adiabatic efficiency used in the curves was obtained from Fig. 9-54.

Compressor Valves and Losses   In Fig. 9-45 is shown an *ideal* compression cycle. The cycle includes (1) an isentropic-compression process from *a* to *b*, (2) a con-

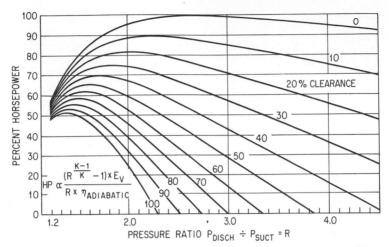

FIG. 9-52 Horsepower characteristic curves: constant discharge pressure ($k = 1.26$).

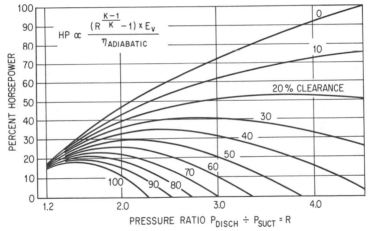

FIG. 9-53 Horsepower characteristic curves: constant suction pressure ($k = 1.26$).

stant-pressure discharge process from b to c, (3) an isentropic expansion of the gas trapped in the cylinder-clearance volume from c to d, and (4) a constant-pressure suction process. Figures 9-55 and 9-56 show a typical *actual* compression process complete with valve losses.

The suction and discharge valves are actually spring-loaded check valves that permit flow in one direction only. The springs require a small differential pressure to open. For this reason, the pressure within the cylinder at the end of the suction

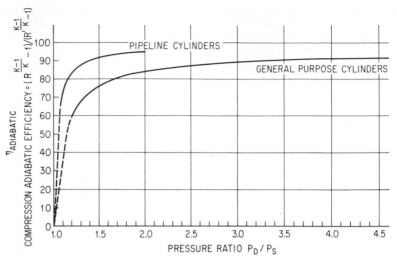

FIG. 9-54   Adiabatic efficiency versus pressure ratio.

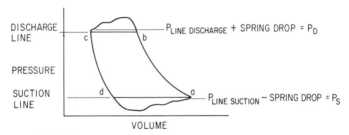

FIG. 9-55   Compressor PV diagram.

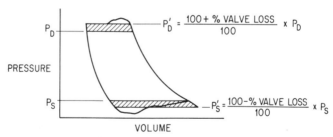

FIG. 9-56   Compressor PV diagram showing valve-loss pressures P_d' and P_s'.

stroke is lower than the line suction pressure by the amount of the differential pressure required to hold the valve off its seat. Likewise, the pressure at the end of the discharge stroke is higher than the line discharge pressure (see Fig. 9-55). The suction and discharge processes are not constant-pressure processes because of the pressure losses associated with forcing the gas through the discharge and suction valves at high velocities (see Fig. 9-56). Since the area of the PV diagram is directly related to the work required, the adiabatic efficiency is determined by dividing the ideal work (area $abcd$) by the area of the actual PV diagram.

For ease of calculation, valve losses are determined either analytically or experimentally and are used to determine fictitious pressures P_d' and P_s' and a fictitious pressure ratio R'.

$$P_d' = P_d \frac{100 + \text{percent discharge-valve loss}}{100}$$

$$P_s' = P_s \frac{100 - \text{percent suction-valve loss}}{100}$$

$$R' = \frac{P_d'}{P_s'}$$

The fictitious pressure ratio R' is then used to determine the horsepower required for compression by applying it in place of R in the isentropic-compression horsepower formula

$$\frac{\text{ichp}}{\text{million ft}^3/\text{day}} = 43.67 \frac{k}{k-1} (R'^{(k-1)/k} - 1)$$

where ichp/million ft^3/day = isentropic horsepower per million cubic feet (14.4 psia and suction temperature) per day

k = ratio of specific heats c_p/c_v

$$R' = \frac{P_d'}{P_s'} = \frac{P_d}{P_s} \frac{(100 + \text{percent discharge-valve loss})/100}{(100 - \text{percent suction-valve loss})/100}$$

$$= R \frac{100 + \text{percent discharge-valve loss}}{100 - \text{percent suction-valve loss}}$$

Capacity and Horsepower Control   For a given set of temperature and pressure conditions, the horsepower input of a reciprocating compressor is proportional to the capacity throughput. For this reason, it is desirable to be able to control the capacity. The capacity equation is

$$Q \text{ (capacity in million ft}^3/\text{day at 14.4 psia and } 60°\text{F)} = \frac{PD \times VE \times P_s}{10^4 \times Z_s}$$

where PD is the piston-displacement rate, ft^3/min, and Z_s is the deviation of the gas from the ideal-gas laws. This shows that for fixed conditions the capacity can be altered by changing the piston-displacement rate or the volumetric efficiency. Piston displacement can be varied by changing the compressor revolutions per minute. Volumetric efficiency can be varied by changing the clearance volume within the compressor cylinder. Clearance volume can be added in fixed steps by opening fixed-volume pockets such as those shown in Fig. 9-50. The handwheel in the figure is turned to open the fixed-volume-pocket valve, which adds a fixed volume to the existing head end-clearance volume. Variable-volume pockets are also used to control capacity. They provide infinitely variable control over a wide range. With variable-volume pockets, the head end clearance can be varied from about 15 to over 100 percent. Suction-valve lifters can also be used to hold the suction valves open on one end of a cylinder, so that the gas is moved back and forth through the suction valve rather than compressed and pushed out the discharge valve. In this way, the capacity of that end of the compressor cylinder is reduced to zero. These three methods of unloading add greatly to the versatility of the reciprocating compressor.

Power Transmission

HELICAL GEARS

General

Gears are associated with nearly every human activity in the modern world. They come in all kinds of sizes, shapes, and materials. They go by such names as spur, helical, bevel, hypoid, worm, skew, internal, external, epicyclic, and so on.

The following material is presented to assist an engineer who is not a gear specialist in determining the basic size and requirements of a gearset for one specific type of gearing: high-speed, high-power parallel-axis gears. The industry definition of high speed is 3600 r/min and/or 5000-ft/min pitch-line velocity. In this instance, *high power* means from 1000 to 2000 hp at the low end and upward of 50,000 hp at the high end. The kinds of applications that generally require high-speed gearing are those involving steam and gas turbines, centrifugal pumps and compressors, and marine propulsion equipment.

High-Speed Gears

Gears for high-speed service are usually of the helical type. They can be either single- or double-helical and can be used in either single or double stages of increase or reduction, depending on the required ratio. The ratio of a single stage is usually limited to about 8 to 1. There is a very small difference in frictional loss at the teeth, depending on whether the pinion or the gear is driven, but for all practical purposes no distinction need be made between speed increasers and speed reducers.

Most high-speed gearing operates at pitch-line velocities of 25,000 ft/min or less. At higher speeds, up to about 33,000 ft/min, special consideration must be given to many aspects of the gearset and housing. Speeds of over 33,000 ft/min should be considered developmental.

As gears go faster, the need for gear accuracy becomes greater. The following can be used for guidelines for high-speed gearing. Tooth-spacing errors should not exceed about 0.00015 in; tooth-profile errors, about 0.0003 in; and helix or lead error, as reflected by tooth contact over the entire face, about 0.0005 in. The usual range of helix angles on single-helical gears is between 12 and 18°. For double-helical gears, the helix is generally between 30 and 40°. Pressure angles are usually found between 20 and 25° (in the plane of rotation).

In addition to the requirement for extreme accuracy, a characteristic of high-speed helical gears that sets them apart from other helicals is the design objective of infinite life, which in turn results in fairly conservative stress levels.

Overload and Distress

If a gear set is overloaded from transmitting more than the design power, or by being undersized, or as a result of misalignment, the teeth are likely to experience distress. The three most probable forms of distress are *pitting, tooth breakage,* and *scoring.*

Pitting is a surface-fatigue phenomenon. It occurs when the hertzian, or surface, compressive stresses exceed the surface-endurance strength.

Tooth breakage is exactly what the name implies: sections of gear teeth literally break out. It occurs when the bending stresses on the flank or in the root of the teeth exceed the bending-fatigue strength of the material.

Scoring, sometimes called *scuffing,* is actually instantaneous welding of particles of the pinion and gear teeth to each other. It occurs when the oil film separating the teeth becomes so hot that it flashes or so thin that it ruptures, thereby permitting metal-to-metal sliding contact. The heat generated as the pinion and gear teeth slide on each other is sufficient to cause localized welding. These tiny welds are immediately torn loose and proceed to scratch the mating surfaces—hence the name *scoring.*

Neither pitting nor scoring causes immediate shutdown. If allowed to progress, however, they can produce a deterioration of the involute profiles in addition to producing stress risers. If permitted to continue too long, pitting or scoring can lead to tooth breakage.

Basic Sizing

The basic sizing of a gearset, or what can be called the *preliminary design,* is based on resistance to pitting. Since the surface endurance strength is a function of the material hardness, preliminary sizing of a gearset is relatively simple. It should be understood, however, that the final design requires the efforts of a competent gear engineer to investigate and attend to such matters as:

1. The selection of materials and processing
2. The determination of the number of teeth on the pinion and gear, which is a function of the pitch, which in turn determines the tooth bending strength
3. An investigation of the scoring resistance of the gearset, which is a function of the gear-tooth geometry, the surface finish of the teeth, and the properties of the lubricant
4. Rotor proportions and bearing design, with particular interest in related vibration characteristics
5. Gear-case features, including such things as running clearances, proper drainage, venting, mounting, doweling, and, in particular, maintenance of internal alignment
6. The many system considerations such as lateral and torsional vibration, external alignment with associated forces and moments on shaft ends, torque pulsations, etc.

The American Gear Manufacturers Association (AGMA) fundamental equation for surface durability (pitting resistance) of helical gear teeth is

$$s_c \left[\frac{C_L C_H}{C_T C_R} \right] \leqq \left[\frac{1/\pi}{\dfrac{1 - \mu_P^2}{E_p} + \dfrac{1 - \mu_G^2}{E_G}} \right]^{1/2} \times \left[\frac{W_t C_o}{C_v} \frac{C_s}{dF} \frac{C_m C_f}{I} \right]^{1/2} \quad (10\text{-}1)$$

where s_c = contact-strength number
C_L = life factor
C_H = hardness-ratio factor
C_T = temperature factor
C_R = factor of safety
μ_p, μ_G = Poisson ratio for pinion and gear
E_p, E_G = modulus of elasticity for pinion and gear
W_t = transmitted tangential load at pitch diameter
C_o = overload factor

C_v = dynamic factor
C_s = size factor
d = pinion pitch diameter
F = face width
C_m = load-distribution factor
C_f = surface-condition factor
I = geometry factor

Substituting appropriate values for high-speed gears and rearranging Eq. (10-1) results in

$$K \leqq [s_c \times 10^{-4}]^2 \times \frac{1.56}{C_o C_m} \tag{10-2}$$

where K is an index of hertzian stress. It is defined mathematically as:

$$K = \frac{W_t}{Fd} \times \frac{R + 1}{R} \tag{10-3}$$

where R = ratio (D/d)
D = gear pitch diameter
W_t = tangential tooth load
= $\dfrac{126{,}000 \times \text{hp}}{N_p \times d}$

and

hp = transmitted, or design, horsepower
N_p = pinion r/min

The term C_o, the overload factor, is accounted for by application service factors (SF), shown in Table 10-1.

The term C_m, the load-distribution factor, accounts for maldistribution of load across the face width of the gearset due to lateral bending and torsional twisting of the pinion, thermal distortion of the pinion and/or the gear, and centrifugal deflection of the gear. If the length-diameter ratio (L/d) of the pinion is kept within reasonable limits, usually less than 2.2 for double-helical and 1.5 for single-helical gears, and proper attention is paid to cooling and gear-band deflection, the magnitude of the C_m factor will probably lie between 1.2 and 1.4. If the higher value is used in the interest of conservatism, Eq. (10-2) can be further simplified to

$$K \leqq \frac{(s_c \times 10^{-4})^2 \times 1.11}{\text{SF}} \tag{10-4}$$

By using Eqs. (10-3) and (10-4) and Tables 10-1 and 10-2, the basic size of a high-speed gearset can be determined in several minutes on a hand calculator.

Table 10-2 gives values of s_c for commonly used gear materials. The rating of a gearset should be based on the softer of the two members, which is normally the gear. The Brinell hardness of the pinion in through-hardened gearsets is usually about 1.2 times that of the gear.

TABLE 10-1 Service-Factor Values

Application	Service factor		
	Prime mover		
	Motor	Turbine	Internal-combustion engine (multicylinder)
Blowers			
Centrifugal	1.4	1.6	1.7
Lobe	1.7	1.7	2.0
Compressors			
Centrifugal: process gas except air conditioning	1.3	1.5	1.6
Centrifugal: air-conditioning service	1.2	1.4	1.5
Centrifugal: air or pipeline service	1.4	1.6	1.7
Rotary: axial flow—all types	1.4	1.6	1.7
Rotary: liquid piston (Nash)	1.7	1.7	2.0
Rotary: lobe-radial flow	1.7	1.7	2.0
Reciprocating: three or more cylinders	1.7	1.7	2.0
Reciprocating: two cylinders	2.0	2.0	2.3
Dynamometer: test stand	1.1	1.1	1.3
Fans			
Centrifugal	1.4	1.6	1.7
Forced-draft	1.4	1.6	1.7
Induced-draft	1.7	2.0	2.2
Industrial and mine (large with frequent-start cycles)	1.7	2.0	2.2
Generators and exciters			
Base-load or continuous	1.1	1.1	1.3
Peak-duty cycle	1.3	1.3	1.7
Pumps			
Centrifugal (all service except as listed below)	1.3	1.5	1.7
Centrifugal: boiler feed	1.7	2.0	. . .
Centrifugal: descaling (with surge tank)	2.0	2.0	. . .
Centrifugal: hot oil	1.5	1.7	. . .
Centrifugal: pipeline	1.5	1.7	2.0
Centrifugal: waterworks	1.5	1.7	2.0
Dredge	2.0	2.4	2.5
Rotary: axial flow—all types	1.5	1.5	1.8
Rotary: gear	1.5	1.5	1.8
Rotary: liquid piston	1.7	1.7	2.0
Rotary: lobe	1.7	1.7	2.0
Rotary: sliding vane	1.5	1.5	1.8
Reciprocating: three cylinders or more	1.7	1.7	2.0
Reciprocating: two cylinders	2.0	2.0	2.3
Marine service			
Ship's service turbine-generator sets	. . .	1.1	. . .
Turbine propulsion	. . .	1.25	. . .
Diesel propulsion	. . .	. . . .	1.35

TABLE 10-2 s_c Values

	Gear hardness	s_c
Through-hardened	229 BHN	112,000
	248 BHN	117,500
	302 BHN	135,000
	340 BHN	152,000
Nitrided	55 R_c	207,000
	58 R_c	218,700
	60 R_c	226,800
	63 R_c	239,400
Case-carburized	55 R_c	230,000
	58 R_c	243,000
	60 R_c	252,000
	63 R_c	266,000

NOTE: BHN = Brinell hardness number; R_c = Rockwell number.

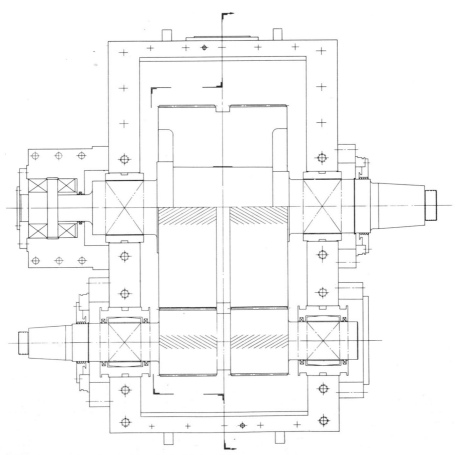

FIG. 10-1 Plan cross section, typical industrial gear.

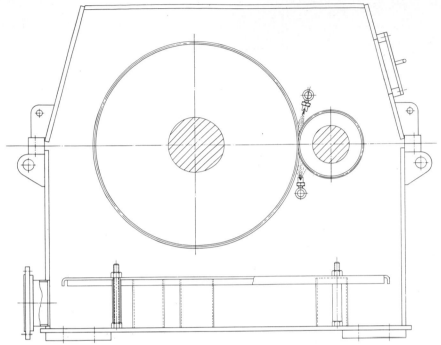

FIG. 10-2   End cross section, typical industrial gear.

Arrangements

Figures 10-1 and 10-2 show sections through a typical industrial high-speed-gear unit.

Figures 10-3 and 10-4 show sections through a typical turbine-driven marine propulsion reduction gear. It will be noted that the high-speed pinions each mesh with two first-reduction gears, thereby splitting the power from each turbine. These twin-power-path gears, or so-called locked-train gears, are popular in the horsepower range of 30,000 shp and up.

Figures 10-5 and 10-6 show sections through a typical diesel-driven marine propulsion reduction gear. In this arrangement, each pinion is fitted with a pneumatically operated clutch which permits either engine to be operated singly or one engine ahead and one astern for fast maneuvering.

Horsepower Losses

Prediction of gear-unit losses is an inexact science at best. The total power loss of a gear unit is made up of (1) the frictional loss in the oil film separating the teeth as they slide over one another, (2) bearing losses, and (3) windage and pumping losses.

Empirical equations have been developed for most types of gears to calculate these losses. Often rule-of-thumb estimates are as good as the calculations. Tooth-

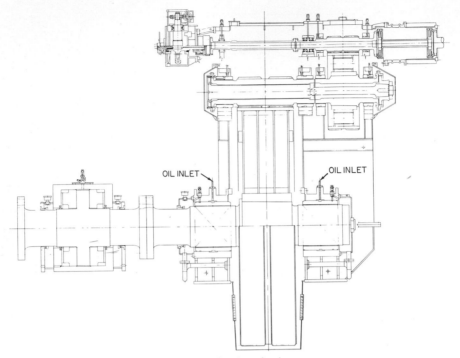

OIL INLET

OIL INLET

FIG. 10-3   Plan cross section, typical locked-train reduction gear.

mesh losses usually amount to between 0.5 and 1 percent of the transmitted horse-power at each mesh. Bearing losses may vary a bit more, depending primarily on the bearing type, operating clearance, and sliding velocity. They usually fall in a range of 0.75 to 1.5 percent of transmitted power.

Windage losses depend primarily on the clearance between rotating parts and the housing, the smoothness of the surfaces and the peripheral velocities.

Pumping loss, the displacement of the air-oil mixture from the tooth space as engagement takes place, is influenced by tooth size, helix angle, rotative speed, and location of the oil sprays. Losses of this type are the biggest variable and can fall anywhere from 0.5 to about 2 percent of transmitted power.

The most important consideration is that a realistic view be taken of gear losses when selecting a pump, cooler, and filters for the lubrication system. These should be large enough to do the job.

Lubrication

The oils normally used in high-speed-gear applications are rust- and oxidation-inhibited turbine oils in the viscosity range of 150 to 300 SSU at 100°F. As a general rule, the higher the pitch-line speed of the gear, the lower the viscosity oil required. In marine units, in which the propeller shaft turns at a relatively low

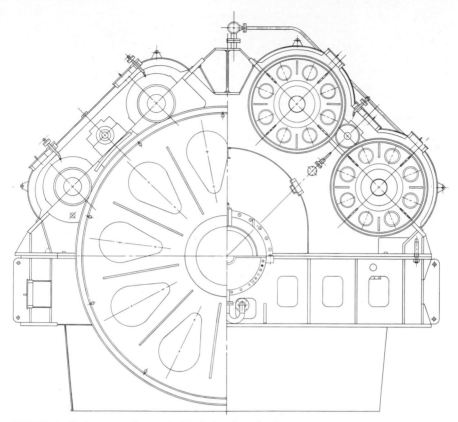

FIG. 10-4 End cross section, typical locked-train reduction gear.

speed, pitch-line speeds are frequently found below 5000 ft/min. In these cases, it is generally desirable to use a more viscous oil. The viscosity of the oils frequently found in turbine-driven propulsion plants is in the range of 400 to 700 SSU at 100°F. In diesel propulsion gearing, in which the engine and the gear are on separate systems, the viscosity of the gear oil is frequently in the range of 600 to 1500 SSU at 100°F.

Regardless of the application, the scoring or scuffing resistance of the gear teeth should be investigated. In many cases, it will be desirable to use an oil with appropriate extreme-pressure additives that greatly increase the antiweld or antiscoring characteristics of the lubricant.

Installation and Maintenance

If a gear unit is correctly sized, properly installed, and properly maintained, it can be expected to last indefinitely. Proper installation includes (1) proper initial alignment, both internal and external; and (2) a rigid foundation that will not

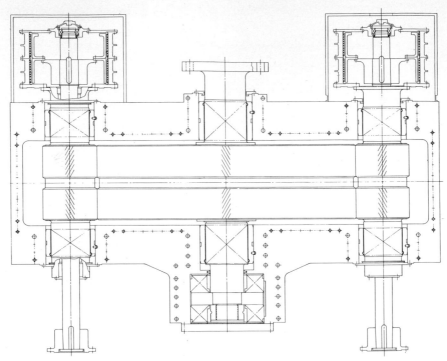

FIG. 10-5 Plan cross section, typical diesel propulsion reduction gear.

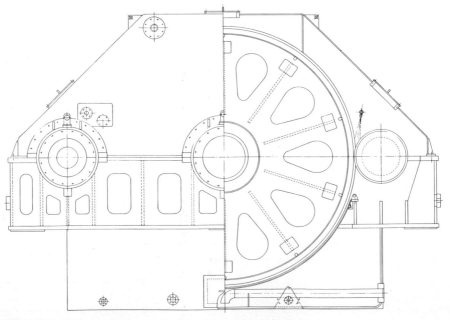

FIG. 10-6 End cross section, typical diesel propulsion reduction gear.

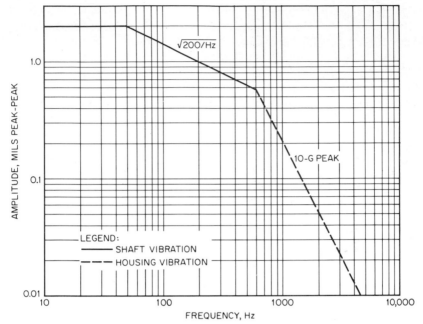

FIG. 10-7   Acceptable vibration levels.

settle, crack, or elastically or thermally deform under operating conditions in amounts greater than the gear-alignment tolerance.

For those interested in additional information on systems considerations (overloads, system vibration, alignment, foundations, piping, and lubrication), AGMA Information Sheet 427.01, *Systems Considerations for Critical Service Gear Drives,* is recommended.

Proper maintenance consists primarily of providing a continuous supply of the correct lubricant at the right temperature, pressure, and condition. Obviously, alignment and balance must be maintained. Vibration monitoring is a good preventive-maintenance tool. Figure 10-7 can be used as a guide for acceptable lateral-vibration limits. Additional information regarding vibration instruments, interpretation, tests, etc., may be found in AGMA Standard 426.01, *Specification for Measurement of Lateral Vibration on High Speed Gear Units.*

WORM GEARS

General

The use of high-speed drivers for efficient operation makes speed reduction necessary for many applications.

Worm-gear reducers are very compact, requiring less space than belts, chains, or trains of open gearing. The right-angle drive often permits compact placement of the driving and driven machines. Since three or more teeth are always in contact, there is an even flow of torque which reduces vibration, prolongs the life of the driven machinery, and provides quiet power transmission. There are few moving parts (hence few bearings), and these are enclosed in a dustproof housing which contributes to long life and avoids danger of injury to workers.

Worm gearing consists of an element known as the *worm,* which is threaded like a screw, mating with a gear whose axis is at a 90° angle to that of the worm. The gear is throated and partially envelops the worm. The worm may have one or more independent threads, or "starts."

The ratio of speeds is determined by dividing the number of teeth in the gear by the number of threads in the worm. Since a single-threaded worm acts like a gear with one tooth and a double-threaded worm as a gear with two teeth, very large ratios can be designed into one set of gearing. Ratios between 3 to 1 and 100 to 1 are common for power transmission purposes, and even higher ratios are employed for index devices.

Mechanical Elements

Dimensions of the worm and worm gear are defined as follows (see Fig. 10-8):

Outer diameter of worm is the diameter of a cylinder touching the tops of the threads.

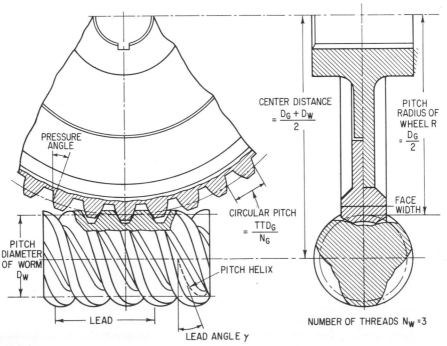

FIG. 10-8

Pitch diameter of worm is the diameter of a circle which is tangent to the pitch circle of the mating gear in its midplane.

Outer diameter of gear is the diameter over the tips of the teeth at their highest points.

Throat diameter of gear is the diameter over the tips of the teeth at the middle plane which is perpendicular to the axis of the gear shaft and passes through the axis of the worm.

Pitch diameter of gear is the diameter of the pitch circle at the midplane of the gear which would roll upon the pitch line of the worm if the latter were used as a rack.

Circular pitch is the distance from a point on one gear tooth to the same point of the succeeding tooth measured circumferentially on the midplane pitch circle. It is equal to the axial pitch of the worm, that is, the distance from any point on a thread of the worm to the corresponding point on the next thread, measured parallel to the axis.

Lead of worm is the distance parallel to the axis of the worm from a point on a given thread to the corresponding point on the same thread after it has made one turn around the worm. If the worm has only one thread, this distance is equal to the circular pitch, but if the worm has multiple threads, it is equal to the circular pitch multiplied by the number of threads. It is the distance that a point on the pitch circle of the gear is advanced by one revolution of the worm.

One revolution of the worm advances the gear by as many teeth as there are threads on the worm. Therefore, the *ratio of transmission* is equal to the number of teeth on the gear, divided by the number of threads on the worm, without regard to the pitch.

Lead angle of the worm threads is the angle between a line tangent to the thread helix at the pitch line and a plane perpendicular to the axis of the worm. The pitch lines of the worm threads lie on the surface of a cylinder concentric with the worm and of the pitch diameter. If this cylinder is thought of as unrolled or developed on a plane, the pitch line of the thread will appear as the hypotenuse of a right-angled triangle, the base of which will be the circumference of the pitch circle of the worm and the altitude of which will be the lead of the worm. In

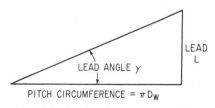

FIG. 10-9   Lead angle.

Fig. 10-9 the lead angle is γ, and the tangent of this angle is equal to the lead L divided by π times the pitch-line diameter D_w of the worm, $\tan \gamma = L/\pi D_w$.

Pressure angle is defined as the angle between a line tangent to the tooth surface at the pitch line and a radial line to that point.

Classification

A large number of arrangements are available, permitting flexibility in application to a wide variety of driven machinery. Some of the typical arrangements manufactured are shown in Figs. 10-10 to 10-16.

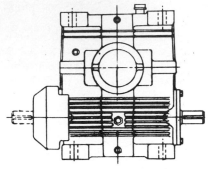
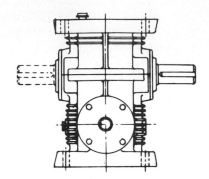

FIG. 10-10   Single worm reduction.

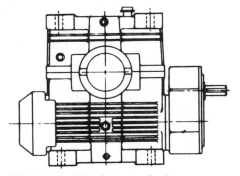
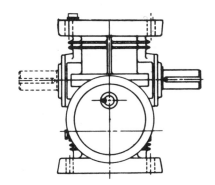

FIG. 10-11   Helical worm reduction.

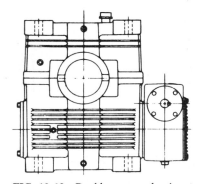
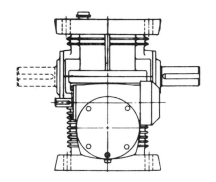

FIG. 10-12   Double worm reduction.

Motorized units may be furnished for:

Horizontal-shaft units
   Single worm reduction
   Helical worm reduction
   Double worm reduction

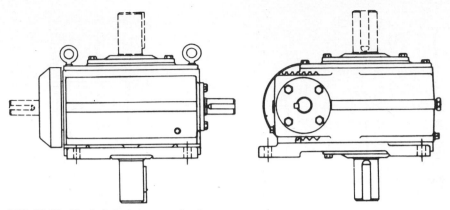

FIG. 10-13 Vertical single worm reduction.

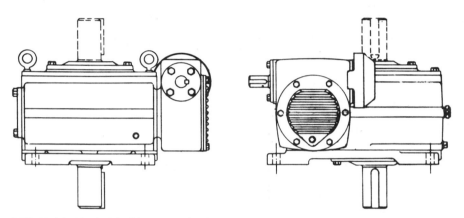

FIG. 10-14 Vertical double worm reduction.

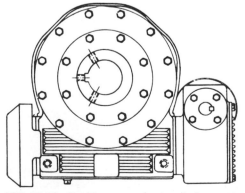
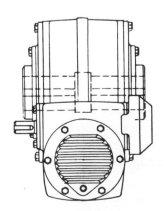

FIG. 10-15 Double-worm-reduction shaft-mount unit.

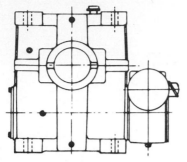

FIG. 10-16   Motorized worm reduction.

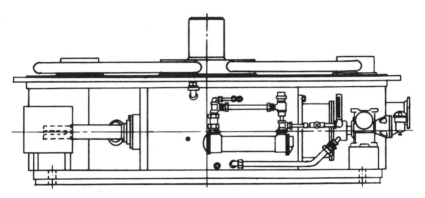

FIG. 10-17   Large vertical-shaft single worm reduction.

Vertical-output-shaft units
  Single worm reduction
  Helical worm reduction
  Double worm reduction
Shaft-mount units
  Single worm reduction
  Helical worm reduction
  Double worm reduction

Special Reducers   Special reducers in various combinations are also available. An example is presented in Fig. 10-17, which shows a large vertical-output-shaft unit with a single worm reduction, having 38-in gear centers, which is used in pulverized-coal service.

Efficiency of Worm Gearset

To determine the approximate efficiency of a worm gearset in which the worm threads are of hardened and ground steel and the gear teeth of nickel bronze or phosphor bronze, lubricated with a steam-cylinder oil, Figs. 10-18 and 10-19 may

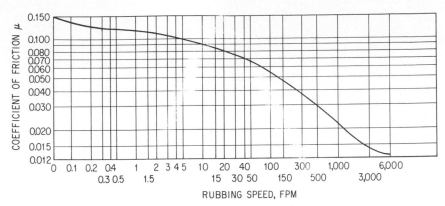

FIG. 10-18   Coefficient-of-friction curve.

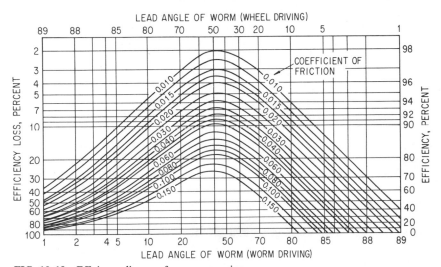

FIG. 10-19   Efficiency diagram for worm gearing.

be used. To use the coefficient-of-friction curve, calculate the rubbing speed of the worm from the following formula:

$$\text{Rubbing speed, ft/min} = \frac{\text{pitch diameter of worm} \times 0.262 \times \text{r/min}}{\cos \text{lead angle}}$$

(See Fig. 10-9 for a definition of lead angle.) With this rubbing speed noted at the bottom of the diagram, read vertically upward until you intersect the coefficient-of-friction curve. Read the value of the coefficient of friction from the left-hand side of the diagram.

When the worm is the driver, enter the efficiency diagram with the lead angle of the worm at the bottom of the diagram. Read upward to the intersection of the curve with the correct coefficient of friction. The efficiency of the gearset may be read from the right-hand side of the diagram or the efficiency loss on the left-hand side of the diagram.

When the gear is the driver, enter the efficiency diagram with the lead angle of the worm at the top of the diagram, reading down to the curve with the correct coefficient of friction. Find the efficiency as before.

These efficiencies, while approximate, are very close to the operating efficiency of the gearset alone. When the gearset is enclosed in a housing with bearings, seals, and oil reservoir, some allowance must be made for bearing loss, seal drag on the shaft, and churning of oil.

Self-Locking Gearset

A self-locking gearset is one which cannot be started in motion by applying power at the gear. Theoretically, this can be obtained when the lead angle of the worm is less than the friction angle. For normal static conditions the friction angle would be approximately 8°30′, and therefore it might be deduced that gearsets having a worm lead angle less than this value would be self-locking.

However, it is impossible to determine the point of positive self-locking for several reasons. The value of the static coefficient of friction varies considerably because of the effect of a number of variables. Furthermore, if a source of vibration is located near a self-locked gearset, a very slight motion might occur at the gear contact. Since the coefficient of friction decreases rapidly with an increase in rubbing velocity from the static condition, the friction angle may become smaller than the lead angle. Once this occurs, motion will continue and the gearing will accelerate under the action of the power applied to the gear.

Figure 10-20 indicates the rapid increase in efficiency with increase in rubbing speed from the static condition for both the worm driving and the gear driving. For this particular example at a rubbing velocity of 500 ft/min, there are only a few points of efficiency difference between the two curves.

The best way to obtain locking is to use a brake, released electrically when the motor is started. With worm gears of high ratios, the braking effect need be only

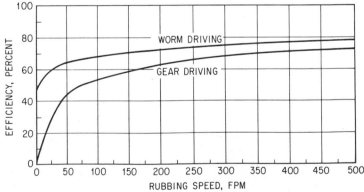

FIG. 10-20   Comparison of efficiencies at tooth contact (ratio 50 on 20-in-center distance).

a fraction of full-load motor torque. A solenoid brake is usually best suited for this operation since the braking effect may be adjusted by weights which can be proportioned to stop the load gradually and avoid damage. Dashpots can be employed to ensure gradual setting of the brake.

Tooth Form

The tooth form used by Delaval is the involute helicoid. Figures 10-21 and 10-22 show the straight generating line tangent to the base circle and the convex axial section of thread.

Worm-gear performance is judged in terms of load capacity, smooth, silent running, and high efficiency. The attainment of these goals requires accurate methods of producing and inspecting the worm and gear.

Since the involute helicoid worm is based on generation of a straight line tangent to the base circle, the accuracy of this line is very simple to check (Fig. 10-23). This thread form lends itself to accurate manufacture, inspection, and interchangeability, as all worms can be checked to calculated measurable dimensions.

All wheels are checked with a master worm to ensure interchangeability and correctness of form (Fig. 10-24).

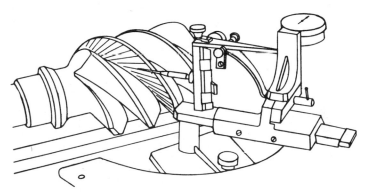

FIG. 10-21   Generation of tooth form.

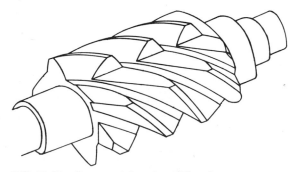

FIG. 10-22   Convex axial section of thread.

FIG. 10-23   Inspection of tooth form.

FIG. 10-24   Checking with master worm.

Tooth Contact

The involute helicoid thread form is a calculated form, and the theoretical contact is maintained more accurately and is more easily determined than that of any other worm thread, particularly a concave thread flank.

Figure 10-25 shows theoretical "lines" of contact that exist between two worm threads and two gear teeth at a given angular position of the worm. As the worm rotates in the direction shown, these contact lines move progressively across the flanks of the worm and gear teeth and are inclined at an angle to the direction of

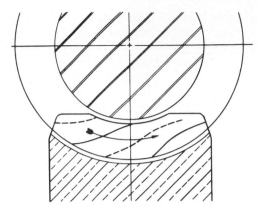

FIG. 10-25   Tooth contact.

sliding. This inclined effect is known to give a highly efficient form of surface lubrication and a low coefficient of friction as compared with a gear form in which the lines of contact are in the approximate direction of sliding. The contacting surfaces are always freshly lubricated and are not subject to the undesirable effects of double contact.

Depending on the relative radii of curvature between the two contacting surfaces and the load applied, these lines of contact actually have some width, thereby providing area contact. In spite of claims to full area contact, line contact occurs on all other thread forms including the double-enveloping thread form. Only the involute helicoid thread form provides the necessary control of the geometry of thread form in design and manufacture to obtain optimum contact conditions.

All gears, bearings, and housings deflect and distort to some extent when operating under load as compared with conditions under no load. A correction in the tooth-contact pattern is provided to assure proper contact under loaded conditions. This correction is accomplished by producing gears with leaving-side contact as shown in Fig. 10-26. This is the ideal contact pattern to aim for when assembling a worm gearset under a no-load condition. This contact pattern allows a lubricant-entry gap in tooth contact. When the gear deflects under load, the contact tends to move to a more central position on the bronze gear face, still allowing a lubricant-entry gap.

A contact pattern such as that shown in Fig. 10-27 is the worst possible contact pattern under a no-load condition. This contact does not allow a lubricant-entry

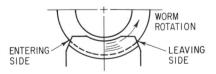

FIG. 10-26   Tooth contact: good.

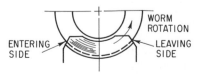

FIG. 10-27   Tooth contact: poor.

gap, and deflection under load will aggravate this condition. A gearset mounted in this manner may cause a temperature rise in oil 20 percent higher than that of the same gearset mounted as shown in Fig. 10-26. The remedy is to move the gear axially to the left (adjusting by shims or other adjustments provided) until a contact similar to that of Fig. 10-26 is obtained.

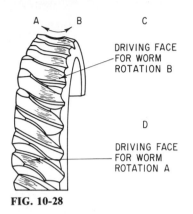

FIG. 10-28

When assembling worm gears which will run in both directions of rotation, it is necessary to consider both driving faces of the gear and to aim for contact as shown in Fig. 10-28. When the worm is rotating in direction A in Figure 10-28, contact should be at D on the leaving side. When the worm is rotating in direction B in Fig. 10-28, contact should be at C on the leaving side. For gears that will run in one direction only, it is necessary to obtain a contact pattern that is correct for the driving-side flank of gear teeth only.

Assembly Adjustment   The gear should be mounted approximately on the centerline of the worm. The worm threads should be coated with prussian-blue dye. A section of the gear teeth should be coated with an orange-colored lead paste. The worm and gear should be rotated in both directions of rotation by hand. The blue markings from the worm threads will show the contact against the orange coating on the gear teeth. If the contact pattern is not as desired, the gear should be adjusted axially until a correct pattern is obtained.

Design Considerations for Worm and Gearset

It is assumed that at the start of this design sequence the center distance for this gearset is known.

Minimum Recommended Number of Gear Teeth for
General Design

Center distance, in.	Minimum number of teeth
2	20
3	25
5	27
10	29
14	35
20	40
24	45

The maximum number of teeth selected will be governed by high ratios of reduction and consideration of strength and load-carrying capacity.

Number of threads in worm. The minimum number of teeth in the gear and the reduction ratio will determine the number of threads for the worm. Generally 1 to 10 threads are used.

Gear ratio

$$\text{Gear ratio} = \frac{\text{number of teeth in gear}}{\text{number of threads of worm}}$$

Pitch. Axial pitch of worm = circular pitch of gear. Keep the fraction simple so that accurate factoring can be used to determine change gears.

Worm pitch diameter. The pitch diameter of the worm is assumed to be at the mean working depth of the worm thread. The following factors should be considered when selecting worm pitch diameter:

1. Smaller pitch diameters provide higher efficiency and reduce the magnitude of the tooth loading.

2. The root diameter that results from pitch-diameter selection must be sufficiently large to prevent undue deflection and stress under load.

3. For low ratios the minimum pitch diameter is governed by the desirability of avoiding too high a lead angle. Lead angles up to 50° are practical.

Gear pitch diameter

$$\text{Gear pitch diameter} = 2 \times \text{center} - \text{pitch diameter of worm}$$

Recommended pressure angle. For general usage, pressure angles of from 20 to 25° are common. Smaller values of pressure angle decrease the separating force, extend the line of action, making the amount of backlash less sensitive to change in center distance, and are used in index gearing. Larger values of pressure angle provide stronger gear teeth and assist in preventing undercutting of teeth with large lead angle. They are used in extremely heavily loaded applications.

Gear-face width (Fig. 10-29). Maximum effective face width is the length of a line tangent to the mean worm diameter, to a point at which the outside diameter of the worm intersects the gear face. Any face width larger than this effective face width is of very little value and is wasteful of material.

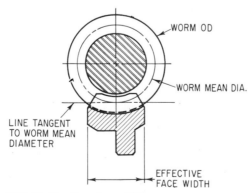

FIG. 10-29   Gear-face width.

Gear-throat diameter = gear pitch diameter + 2 × gear addenda. Gear outside diameter = gear throat diameter + 1 addendum of worm rounded off to the nearest fraction of an inch.

<u>Gear blank under rim diameter (Fig. 10-30)</u>

$$h_t = \text{tooth depth of gear}$$

Underrim dimension for bronze gear block = gear-root diameter

$$- \ 2 \ \text{to} \ 2\tfrac{1}{2} \times \text{gear-tooth depth}$$

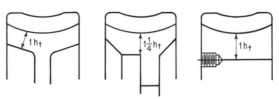

FIG. 10-30 Gear-blank shapes.

<u>Worm face</u>

Minimum worm face
$$= 2\sqrt{(\text{gear-throat diameter}/2)^2 - (\text{gear pitch diameter}/2 - \text{gear addendum})^2}$$

Allowable shaft stresses. All shafting in accord with AGMA Practice 260.01, March 1953.

Allowable bolt stresses. All bolts in accord with AGMA Practice 255.02, November 1964.

Bearing loading. All bearings selected in accord with AGMA Practice 265.01, March 1953. (See Fig. 10-31.)

Ball and roller bearings are selected on the basis of supporting loads equal to the maximum basic rating of the gear reducer and allow a minimum bearing life of 5000 h or an average life of 25,000 h.

TABLE 10-3 Bearing Loads

Resulting from	Bearing No. 1	Bearing No. 2	Bearing No. 3	Bearing No. 4
P	$Pa/L_W = P_1$	$Pb/L_W = P_2$	$Pr_G/L_G = U_3$	$Pr_G/L_G = U_4$
S	$Sa/L_W = S_1$	$Sb/L_W = S_2$	$Sd/L_G = S_3$	$Sc/L_G = S_4$
T	$Tr_W/L_W = U_1$	$Tr_W/L_W = U_2$	$Td/L_G = T_3$	$Tc/L_G = T_4$
Radial load	$\sqrt{P_1^2 + (S_1 - U_1)^2} = R_1$	$\sqrt{P_2^2 + (S_2 + U_2)^2} = R_2$	$\sqrt{T_3^2 + (U_3 - S_3)^2} = R_3$	$\sqrt{T_4^2 + (S_4 + U_4)^2} = R_4$
Thrust load	. . . .	T	. . . .	P

Performance

Mechanical Ratings of Cylindrical Worm Gears   The practice for this rating follows AGMA Practice 440.03, September 1959.

The ratings which are cataloged according to this practice are *wear* ratings which the gearset will satisfactorily permit, at the load shown, provided the driven

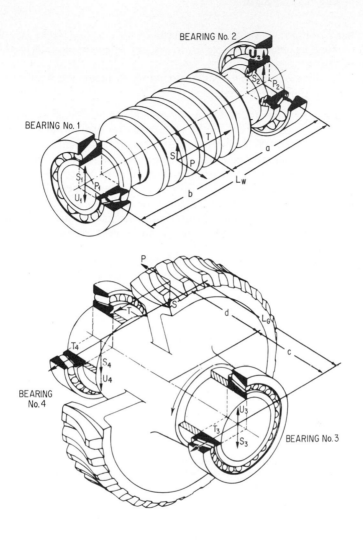

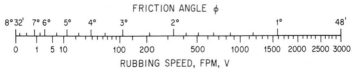

FIG. 10-31 Principal forces and bearing loads in a worm and gearset. D_w = pitch diameter of worm, in; r_w = pitch radius of worm, in; r_G = pitch radius of gear, in; γ = lead angle of worm, °; P = tangential force on worm, lb; Q = torque input to worm, in·lb; S = separating force, lb; T = axial thrust of worm, lb; NPA = normal-pressure angle: ϕ = friction angle for worm driving; rpm (r/min) = worm speed; V = rubbing speed, fpm (f/min); $P = Q/r_w$; $S = P \tan \text{NPA}/\sin(\gamma + \phi)$; $T = P/\tan(\gamma + \phi)$; $V = 0.262$ D_w rpm/cos γ. *(Courtesy of the Timken Roller Bearing Company.)*

machine has a uniform load requirement free of shock loading, 10 h/day. This is the basic rating by which worm-gear drives are selected, subject to thermal limitations.

Service factors are applied to this basic rating to factor the wear rating for shock loading or intermittent service.

Thermal Ratings of Cylindrical Worm Gears Thermal ratings above 100- to 200-r/min worm speed represent the input horsepower and output torque which will provide a stabilized 100°F oil-temperature rise over ambient air temperature when the machine is operated continuously. For example, if the ambient air temperature is 70°F, a reducer carrying rated thermal horsepower will operate with an average oil temperature of 170°F. Since normal worm-gear lubricants will deteriorate rapidly, require frequent replacement, and may not support the gear-mesh loads when the machine is operating continuously at 210 to 220°F, the practical maximum ambient air temperature for worm-gear reducers carrying full thermal rating horsepower is 100°F.

For operation at higher ambient-air temperatures, a larger unit with a higher thermal rating must be selected for continuous operation, or a cooling system must be employed. For example, if a unit is to operate in an ambient-air temperature of 150°F, the increase in oil temperature must be limited to 50°F to keep the oil temperature from rising above 200°F. This means that the heat generated in the reducer must be one-half of the heat generated when the machine is operating at the catalog thermal rating; or since bearing and oil losses remain constant for a given speed, applied horsepower must be less than one-half of the catalog thermal rating.

For operation at ambients of less than a maximum of 100°F or when artificial or natural air drafts are present, catalog thermal ratings can be exceeded. For a proper evaluation, all data on ambient conditions should be determined.

Allowable Starting Load Worm-gear reducers have a momentary overload strength rating + 300 percent of mechanical-wear rating. Peak starting load of the driven machine should not exceed 300 percent of the mechanical-wear rating.

Lubrication

General Because of the nature of worm-gear sliding and rolling action, lubricants used for other types of gearing are not satisfactory. All units are shipped without oil, but reducer instructions and lubrication nameplates refer to the use of AGMA lubricants. Generally speaking, suppliers of *industrial lubricants, not* service stations, should be contacted and should be able to supply suitable lubricants from stock to meet these AGMA specifications. The units should be filled with the proper lubricant before operating.

These lubricants are basically a steam-cylinder oil. A list of trade names of the various manufacturers of oils which meet the AGMA 7 Compounded and AGMA 8 Compounded specifications is maintained by Delaval. These lubricants are basically petroleum-base oils but with 4 to 5 percent acidless tallow additives which provide additional film strength. They are heavy oils, much heavier than normal motor oils. The viscosity of AGMA 7 Compounded is approximately 135

TABLE 10-4 Basic Lubricant Recommendations (AGMA)

Worm speed, rpm	Size 60 units and smaller, ambient temp, °F		Size 70 units and larger, ambient temp, °F	
	15–60	50–125	30–60	50–125
Up to 400	7 Comp.	8 Comp.	7 Comp.	8 Comp.
Above 400	. . . . . .	. . . . . .	7 Comp.	7 Comp.

SSU at 210°F, and that of AGMA 8 Compounded is approximately 150 SSU at 210°F. This heavy viscosity plus the plating action of the additives on the worm and gear contact surfaces is required to ensure the long trouble-free life that the gearing is designed to provide.

Lubricants Not Recommended   The following lubricants should never be used for worm gearing:

1. Ordinary motor oils, no matter what their viscosity.

2. Automotive rear-end oils.

3. Extreme-pressure lubricants containing compounds of sulfur or phosphorus. It may be claimed that these lubricants are noncorrosive to steel, but they are extremely corrosive to bronze and will not provide the necessary plating action required.

4. Greases of any kind. These do not flow sufficiently to provide the necessary cooling.

Cold-Weather Lubricants   If ambient temperatures below 15°F are expected, a winter, or cold-weather, lubricant must be selected, since the AGMA 7 Compounded or 8 Compounded will solidify and the motion of the gears will channel the solidified oil until no lubricant is present at the gear mesh. For this condition, a minimum ambient temperature to be expected must be estimated and a reputable supplier consulted to recommend an oil with a channel point well below the expected minimum ambient temperature. This will require a lighter-viscosity oil, but the oil should still contain additives. The best selection is usually the mild extreme-pressure oils containing lead naphthanate with the following viscosities:

For min ambient temp, °F	Use a mild extreme-pressure oil containing lead naphthanate and having a viscosity of
0	120 SSU at 210°F
−10	100 SSU at 210°F
−20	75 SSU at 210°F
−30	53 SSU at 210°F

The lubricant should be changed to the heavier-viscosity oils when the ambient temperature again goes above 15°F.

Frequency of Oil Changes   The frequency of oil changes varies with the type of service. After the initial 50 to 100 h of running, a change should normally be made to remove the particles of bronze burnished off the gear during the run-in period. Thereafter, a general rule is that the oil should be changed every 6 months of normal service and every 3 months of severe service. However, if the unit is in a dusty or moist atmosphere, dirt or water accumulation in the oil reservoir may require more frequent changes. Many oil suppliers will test a lubricant after a period of use free of charge and determine its useful life for a specific application.

Procedure for Long Shutdown Periods   If the unit is to be idle for any length of time, particularly outdoors, something must be done to prevent rusting of the bearings, gears, and other internal parts. The easiest solution is usually to fill the unit completely with clean oil. Of course, before the unit is started again, the oil should be drained and refilled to its proper level.

Installation and Operation

Installation   Normal good practice must be followed when handling the unit, choosing a foundation, checking alignment, and mounting couplings, pulleys, gears, sprockets, etc. Couplings should be pressed or shrunk on the reducer shafts. Do not drive couplings on shafts, as this may damage the bearings and also cause the shafts to spring. This, in turn, may result in failure of the bearings, vibration, and oil leakage. Sprockets, pulleys, and pinions should be mounted as close to the case as possible in order to avoid undue bearing-load and shaft deflection.

Operation   The unit is shipped from the factory *without oil* but is slushed internally with a rust-preventive compound, which need not be removed since it is oil-soluble. *Make certain that the reducer is filled to the correct level before start of operation in accordance with lubrication specifications.* The unit must be filled to, but not above, the oil-level gauge. The oil level will, of course, change with the mounting arrangement. It should be checked periodically and only at a time when the unit is not operating. A dipstick is provided in the oil-level gauge.

All units have been subjected to test before shipment, but it takes additional hours of running under full gear load to attain highest efficiency. The gear may, if necessary, be put to work immediately on full load, but if circumstances permit, it is better for the ultimate life of the gear to run it under gradually increasing load. Immediate application of full load concentrates high unit pressures on tooth surfaces. When new driven equipment requires operation to achieve freedom and minimum friction loss, use precaution in the early stages of operation to prevent the reducer from taking an overload. When overload tests are specified on a machine before it is shipped, it is better to make preliminary runs under part load before building up to full load and overload. A reasonable running-in procedure is half load for a few hours, building up to full load, in two stages if possible.

Temperature rise on the initial run will be higher than that eventually attained after the gear is fully run in.

Some slight wear and/or pitting of the bronze gear teeth may be observed after a short period of initial operation. This condition is normal, as some initial wear is necessary for the hardened-steel worm to seat itself properly with the bronze gear.

Steam Condensers

STEAM CONDENSERS

General

Steam condensers are used:

1. In the exhaust system of steam-turbine power plants, to augment the cycle temperature range by condensing the steam discharging from the turbine and thus creating a low absolute pressure at the turbine exhaust.

2. In process plants, for better heat balance, recovery of condensate, low-temperature evaporation, etc. Condensers similar to steam condensers are used in chemical, refrigeration, and other fields to liquefy vapors other than steam.

Steam condensers are classified as:

11-1

1. *Direct-contact condensers* in which the steam and cooling water are intermingled (often designated as *barometric* condensers or low-level direct-contact condensers).

2. *Surface condensers* equipped with a shell-and-tube structure in which steam is segregated from the cooling water by metal walls through which heat exchange occurs. The latter type of condenser permits recovery of the condensate for boiler-feed purposes.

Steam Surface Condenser   In the following paragraphs only steam *surface condensers* are discussed in greater detail.

Basic elements. In simple terms a steam surface condenser is a gastight chamber fitted with heat-conductive tubes through which cooling water is circulated and with means provided for continuously removing the condensed steam and the noncondensable gases. The noncondensables enter the chamber via leakage through the turbine-shaft seals, through subatmospheric drains and returns to the condenser, and from any system subatmospheric-pressure vessels or lines (see Fig. 11-1).

Purpose. The purpose of the surface condenser as used in steam power generation is to reduce the back pressure on the turbine to the lowest practical absolute pressure. This is accomplished by circulating cooling water through the tubes, thus condensing the steam on the tubes. The volume of the steam is reduced to that of the liquid. At 2-inHg absolute pressure, the ratio of vapor specific volume to liquid specific volume is about 1 to 21,000. It is this large reduction in volume that causes the reduction in pressure. The condensate thus formed is usually

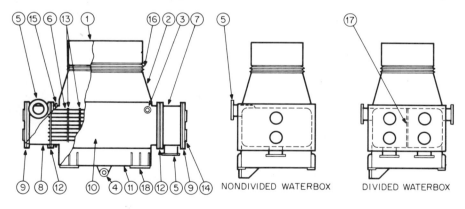

NONDIVIDED WATERBOX   DIVIDED WATERBOX

1. Exhaust connection	7. Inlet waterbox	13. Tube plates
2. Exhaust neck	8. Outlet waterbox	14. Handholes or manholes
3. Venting outlet	9. Waterbox covers	15. Shell expansion joint
4. Condensate outlet	10. Condenser shell	16. Exhaust neck expansion joint
5. Circulating-water inlet or outlet	11. Hot well	17. Waterbox-dividing partition
6. Tubes	12. Tube sheets	18. Support feet

FIG. 11-1   Single-pass surface-condenser nomenclature. More than one of the various parts may be included in a single contract. Exact locations of connections will vary from one condenser to another.

removed from the condenser hot well by a special type of centrifugal pump designed specifically for this service. The noncondensable gases present are removed by a steam-jet air ejector or a mechanical air pump. In fossil-fuel and pressurized-water-reactor cycles, the noncondensables are mostly air. However, in boiling-water-reactor nuclear cycles, large quantities of free hydrogen and oxygen generated by radiolytic decomposition are also present and must be removed.

In a well-designed power plant the noncondensable gases constitute a very low percentage in the steam and have little effect on its condensing temperature (partial-pressure effect is nil up to 0.1 percent). In certain special cycles, specifically geothermal and ocean-thermal-energy-conversion cycles, the noncondensable gas-to-condensable-vapor ratio is significantly higher than in previously defined cycles, and its effect on both the condensing overall heat-transfer rate and the condensing steam temperature and on the design, sizing, and selection of venting equipment for removal must be considered. In these cycles the simplified approaches defined in this subsection do not apply except in a very general way. Accumulation of noncondensable gases and/or inadequate venting arrangements in a condenser may be the single most influential factor in poor condenser performance.

Heat-Transfer Principles

Before attempting to work out a preliminary condenser calculation, it is necessary to have a clear understanding of basic heat-transfer principles and definitions. To condense steam, L, the latent heat of condensation, must be removed. This is the amount of heat involved in the change from vapor state to liquid state at constant temperature and varies slightly according to the absolute pressure or vacuum. For convenience, in the absence of specifically defined values based on heat balances, the value of L may be taken as 950 Btu/lb for turbine exhaust steam and 1000 Btu/lb for engine exhaust steam. Since the heat removed from the steam must be absorbed by the cooling water (with radiation losses neglected), each pound of engine steam condensed will heat 1000 lb of cooling water 1°F or 100 lb of water through a range of 10°F.

To transfer this heat to the cooling water requires a positive temperature gradient in the direction of heat flow from the condensing steam to the circulating water. If the water were not colder than the steam, the heat would not flow and the steam could not be condensed. The rate of heat flow depends not only upon temperature gradients but also upon the amount of heat-transfer surface and the resistance to heat flow consisting of condensing film resistance, external fouling resistance, tube-wall resistance, internal fouling resistance, and internal cooling-water film resistance. The specific determination of these values is beyond the scope of this general explanation but may be found in heat-transfer texts. For use in condenser-design formulas the effect of these resistances to heat flow is combined in a factor defined as U_o, or overall heat-flow coefficient, the overall U_o being the reciprocal of the sum of the individual resistances. The comparative effect of these resistances may be seen in Fig. 11-2. The way in which the temperature gradient ΔT and the heat-flow coefficient affect the amount of surface

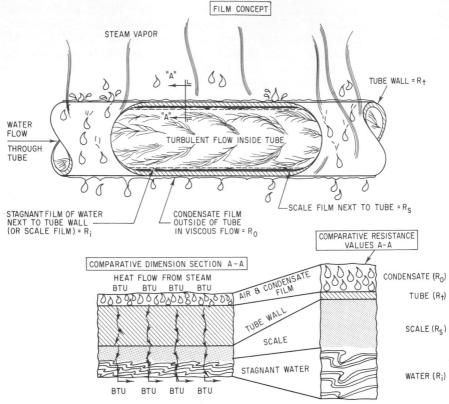

FIG. 11-2 Heat absorbed in cooling water.

or the condensing capability of a given surface is shown by the following simple expression:

$$Q = S\Delta TU = WL \qquad (11\text{-}1)$$

where Q = rate of heat flow, Btu/h

S = outer-surface area of heat-flow circuit, ft^2

ΔT = effective mean heat potential or mean temperature difference, °F, between steam and cooling water

U = overall heat-flow coefficient, Btu/(ft^2·°F·h)

W = lb steam condensed/h

L = latent-heat content of steam, Btu/lb

Since the water becomes heated as the steam condenses, it is necessary to consider ΔT as the average or mean temperature difference throughout the condenser.

For the purpose of rough approximation, mean ΔT may be taken as the arithmetic average of the temperature differences between steam and inlet cooling water and steam and outlet cooling water, where arithmetic average equals

$$\frac{(t_s - t_1) + (t_s - t_2)}{2}$$

where t_s is the steam temperature, °F.

This relationship is approximately correct for high-temperature differences; for the low heat potentials in high-vacuum surface condensers, it has been found necessary to use the natural logarithmic average or log mean temperature difference (LMTD):

$$\text{LMTD} = t_m = \frac{\theta_1 - \theta_2}{\log_e \dfrac{\theta_1}{\theta_2}}$$

t_m = LMTD
θ_1 = greater terminal difference
θ_2 = lesser terminal difference

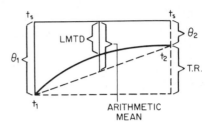

Cooling-Water Requirements

Since all the heat of the steam goes into the cooling water, the cooling-water requirement is as follows:

$$Q = \text{gal/min} \times 500 \times (t_2 - t_1) \tag{11-2}$$

$$\text{gal/min} = \frac{Q}{500 \times (t_2 - t_1)}$$

where Q = Btu/h (rate of heat flow)
gal/min = gallons per minute of cooling water
   500 = 8.33 × 60, which converts gal/min to lb/h
   t_1 = inlet water temperature, °F
   t_2 = outlet water temperature, °F

Influencing Factors   From Eq. (11-1) it follows that

$$S = \frac{WL}{U \times \text{LMTD}} = \frac{Q}{U \times \text{LMTD}} \tag{11-3}$$

Equation (11-3) expresses mathematically the fact that the size of the condenser (tube surface) is directly proportional to the steam flow times its latent-heat content and inversely proportional to the log mean temperature difference (or heat potential) as well as inversely proportional to the U value or overall heat-conductivity constant. It is obvious that for a given heat load and vacuum the size of the condenser is then determined by the amount of water, its inlet temperature, and the U value.

As a rule, the water inlet temperature is beyond the control of the condenser designer, and to a lesser degree water flow and temperature rise are also likely to be established, at least within limits. Therefore, most of the ingenuity of condenser designers is focused on obtaining the highest possible U values in their condensers.

There are a great many factors affecting U, the most important of which are as follows:

1. Cooling-water velocity through tubes
2. Viscosity of cooling water (temperature)
3. Thermal conductivity of water
4. Viscosity of condensate
5. Thermal conductivity of condensate
6. Extent of loading (both quantity and distribution) of condensing surface with condensate
7. Quantity of noncondensable gases present
8. Extent of blanketing of surface by noncondensable gases
9. Characteristics of vapor condenser-cooler and air-removal equipment
10. Arrangement of condensing surface
11. Pressure loss through main tube bank
12. Diameter of tubes
13. Thickness of tube walls
14. Thermal conductivity of tube material
15. Condition of tube surface (wettable, polished)
16. Cleanliness of tube surfaces

The flow of heat through the tube walls of a surface condenser may be likened to the flow of electricity through a number of parallel resistances, each of which is made up of several conductive sections of different metals in series. The flow of heat through any individual tube is illustrated in Fig. 11-2.

We combine these resistances to heat flow (reciprocals of heat conductance) to determine the overall resistance R, which is generally expressed in terms of its reciprocal U, the overall heat-transfer coefficient in Btu/(ft$^2 \cdot$ °F$\cdot$h).

It has been found that the overall heat-transfer coefficient U_0 is generally determined by the use of the Heat Exchange Institute (HEI)* curves found in the standards group (see Fig. 11-3). Heat-transfer approaches contained in thermodynamic or heat-transfer texts may also be used.

The noncondensable-gas content of the steam has also been found to be a very important factor in commercial condenser operation and is responsible for much of the poor performance of improperly designed condensers, especially at partial load. Because of the noncondensable-gas content of the steam, provisions must be made in the condenser to ensure deaeration of the condensate. This is necessary to minimize damage to the power plant boiler-cycle auxiliaries and the piping

*Heat Exchange Institute, 1230 Keith Building, Cleveland, Ohio 44115.

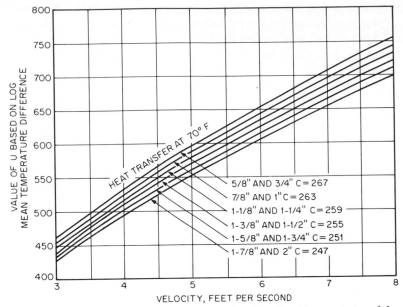

FIG. 11-3   Heat transfer through condenser tubes. *(Reprinted by permission of the Heat Exchange Institute.)*

system. It is essential that the venting sections of a condenser be properly designed and located and that noncondensable-gas-removal equipment be properly sized.

Tube surface, or S, is determined from Q, or Btu heat input, the U value, and the LMTD, or °F logarithmic mean temperature difference between condensing steam and cooling water, by using Eq. (11-3):

$$S = \frac{Q}{U \times \text{LMTD}}$$

Tube surface is only a preliminary design element leading to good condenser performance. Proper arrangement of the tube surface is vital to performance. It is only by providing for proper steam distribution and penetration that large multitube power plant condensers can be made to approximate the performance of the single-tube (ideal) condenser.

Engineering Design

Introduction   Having reviewed some basic theories of heat transfer as applied to steam condensation, we may now consider the engineering design and economics of surface condensers.

1. The first and primary function of the condenser is to maintain the lowest economically justified absolute pressure at the turbine-exhaust connection.

2. Its next most important function is to produce oxygen-level condensate at a substantially saturated exhaust-steam temperature. A commonly defined oxygen level in large utility power plant cycles is 7 parts per billion.

3. Secondary functions, such as heating and deaerating boiler-feedwater makeup, supplying warm water for process uses, and acting as a low-pressure drain sump, must also be met as required by cycle operation and within allowable space limitations.

Importance of Good Design   The factors which determine the excellence of a surface-condenser design are numerous, and their relative importance varies according to the size of the installation and the underlying economic considerations.

In large base-load steam electric-generating stations, the entire heat balance is very carefully and fully developed to promote maximum overall efficiency. Because of the size of the equipment and the tremendous output, it is economically justified to add refinements to reduce the Btu input per kilowatthour output.

Such features are economizers, air preheaters, waterwalls, superheaters, and stage heating of the feedwater. The pressure and temperature rating of steam generators is also increased to the maximum associated with the state of the art of high-temperature metallurgy. The relative steam flow through the turbogenerator is greatly reduced by high-energy-content steam at the throttle, reheating between stages, dividing the flow, cooling the generator windings, and other features. The condenser is designed for the highest practical vacuum even at partial loads, and other cycle auxiliaries add their proportional improvement to the overall economy. The net result is that such highly perfected large-scale power generators regularly produce a kilowatthour of electric energy with less than half of the fuel consumption of the relatively small condensing plant. But with all this efficiency the surface condenser still must dissipate to the cooling water more than twice as many heat units as are converted to useful electric energy (see heat-balance diagram in Fig. 11-4), so that under these conditions the importance of good condenser selection and design cannot be negated. (The low thermal efficiency of the Rankine cycle is well known but often is not fully appreciated.)

The heat-balance diagram in Fig. 11-4 gives the actual distribution of heat units in percentage of heat released for a medium-pressure marine-type power plant. The simplified diagram summarizes the distribution figures to show the three principal outlets for heat: (1) stack, (2) useful work, and (3) cooling water.

A modern large-scale central station might show perhaps a 30 percent overall thermal efficiency instead of the 20 percent illustrated. In any event, the surface condenser still must be depended upon to remove the bulk of the heat units from the system.

Yardstick for Good Condenser Design   From a practical viewpoint, the following combination of results usually characterizes the product of a good normal-balanced design for an important installation:

1. Relatively high water velocity through the tubes equaling a high heat-transfer coefficient

2. Economical consideration of cooling-water-tube surface metallurgy

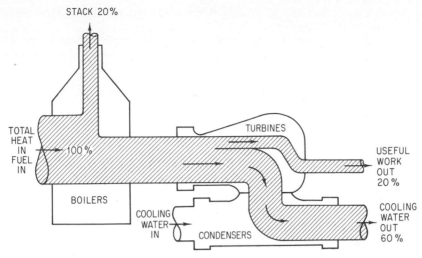

FIG. 11-4   Heat-balance diagram.

3. Balance between pumping costs, water quantity, and tube surface

4. Maintenance of design vacuum at full load with normal tube cleanliness

5. Low-pressure drop through tube bank at full load

6. High vacuum at light load and with cold water

7. Simple and economical air-removal system

8. Adequate air-removal capacity at design and partial loads

9. Condensate delivered at temperature equal to steam temperature

10. Deaeration of condensate

11. Leak-free construction

12. Properly sized vent section for noncondensable-gas cooling

13. Low maintenance

14. Initial cost well balanced economically with turbine characteristics and station-operating costs

15. Reliable—easy to operate and to keep clean

16. Meeting space, weight, and cost limitations

Not all the foregoing factors have equal or relative importance in all cases. It is wise, therefore, to know the value of the influencing factors before attempting to design a condenser. Certain factors are incompatible with others, and engineering compromises are often necessary, but these should be made with considered judgment based on available data. The success of the final design and ultimately of the condenser itself often depends upon how completely and accurately this information is transmitted to the designer.

Certain intangible considerations do not permit a definite quantitative measure, but they may be assigned a weighting to evaluate the final design. On other occasions, missing factors may be arbitrarily assigned reasonable typical values and a series of selections offered.

Although it is not possible to assign exact values to all the various factors except in the light of individual applications, certain values may be suggested as typical of good practice:

1. Water velocity from 6 to 8 ft/s, U from 650 to 800 Btu/(ft$^2 \cdot$ °F·h), with clean tubes.

2. Water flow widely variable (from 20 to 100 times steam flow), with an average of 60 to 80 times steam flow.

3. Waterpower-surface ratio dependent upon relative values.

4. Tubes (nonferrous tubes) normally assumed to be 85 percent clean; i.e., 0.85 × HEI overall clean-tube value.

5. Pressure drop of over ⅒ inHg across tube bank at full load normally considered excessive.

6. Single system of twin two-stage ejectors with isolating valves and inter-after condensers.

7. Single two-stage element should maintain ½ inHg absolute with full air load.

8. 0°F condensate temperature depression during normal operating conditions.

9. An oxygen content in condensate of not over 7 parts per billion.

10. 22°F is the average initial terminal difference between inlet water and condensing steam.

11. 1 inHg absolute is a popular condensing pressure in temperate climates where natural water sources are used. Somewhat higher pressures are encountered in the south. When cooling towers and/or ponds are used, economic evaluations of the best design pressure for the system are necessary.

12. Divided water boxes are used on most of the larger installations to permit half operation for tube-end joint or tube-damage repair while the unit is operating.

13. ⅞-in- and 1-in-outside-diameter tubes are used in the larger condensers.

14. Two half-capacity circulating pumps and two full-capacity condensate pumps are used with almost all large condensers.

Standard of the Industry Overall Heat-Transfer Coefficient U

The most important standard adopted by the industry set forth definitely the relationship between water velocity and the overall heat-transfer coefficient U. This was the outcome of coordinating theory with observed results from the best commercial installations. It took the form of a family of curves with water velocity through various standard diameters of condenser tubes as the independent variable for determining directly the overall heat-transfer coefficient, or U value, for clean new tubes. A *temperature curve* was added to correct variations observed in U with changes in the temperature of inlet cooling water.

Although obviously an approximation, the HEI graph for determining U value (Fig. 11-3) has proved satisfactory in practice and is quite generally accepted by manufacturers and users of surface condensers.

In view of its wide acceptance and generally conservative results for clean new tubes, manufacturers currently use the HEI standard graph for U-value determination for performance guarantees in proposals (see Figs. 11-3 and 11-5). It is carefully checked, however, by far more accurate data for actual design purposes, and suitable allowances are made for fouling.

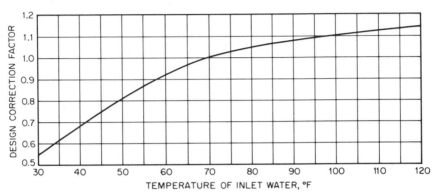

FIG. 11-5   Inlet-water-temperature correction factor C_w. *(Reprinted by permission of the Heat Exchange Institute.)*

The values shown in Fig. 11-3 are based on use of the formula

$$U = C \sqrt{V}$$

where C = factor for tube size (see Fig. 11-3).

For condenser-design evaluation, the formula is modified to include tube material, tube thickness, water temperature, and fouling factor. The applicable formula now takes the form of

$$U = C \sqrt{V} \, C_t C_w C_f$$

where C_t = tube material and gauge factor (see Table 11-1)
   C_w = inlet-water-temperature correction factor (see Fig. 11-5)
   C_f = fouling or scaling factor (cleanliness factor)

Typical Condenser Construction

Typical condensers are shown in Figs. 11-6 and 11-7. Figures 11-8, 11-9, and 11-10 are typical condenser constructions.

Delaval Reverse-Flow On-Line Self-Cleaning Condensers   Delaval reverse-flow condensers have been developed, patented, and manufactured by Transamerica Delaval. Reverse-flow condensers list among their advantages the following:

1. Reverse flow flushes out grass, sticks, leaves, and other foreign matter which clog the inlet ends of condenser tubes.

TABLE 11-1   Material and Gauge Factor C_t

Tube materials	Tube-wall gauge (BWG)						
	24	22	20	18	16	14	12
Admiralty metal	1.06	1.04	1.02	1.00	0.96	0.92	0.87
Arsenical copper	1.06	1.04	1.02	1.00	0.96	0.92	0.87
Copper iron 194	1.06	1.04	1.02	1.00	0.96	0.92	0.87
Aluminum brass	1.03	1.02	1.00	0.97	0.94	0.90	0.84
Aluminum bronze	1.03	1.02	1.00	0.97	0.94	0.90	0.84
90-10 copper-nickel	0.99	0.97	0.94	0.90	0.85	0.80	0.74
70-30 copper-nickel	0.93	0.90	0.87	0.82	0.77	0.71	0.64
Cold-rolled low-carbon steel	1.00	0.98	0.95	0.91	0.86	0.80	0.74
Stainless steels, Type 304/316	0.83	0.79	0.75	0.69	0.63	0.56	0.49
Titanium	0.85	0.81	0.77	0.71	. . . .	. . . .	. . . .

Reprinted by permission of the Heat Exchange Institute.

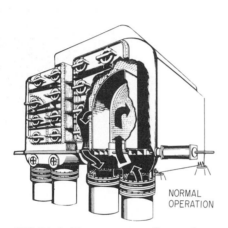
NORMAL
OPERATION

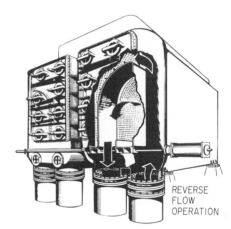
REVERSE
FLOW
OPERATION

FIG. 11-6   Two-pass reverse-flow condenser.

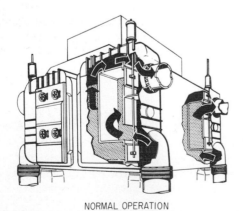
NORMAL OPERATION

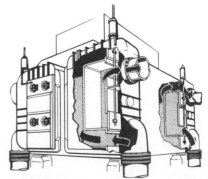
REVERSE FLOW OPERATION

FIG. 11-7   Single-pass reverse-flow condenser.

11-12

FIG. 11-8 Shop view of two-pass reverse-flow condenser.

FIG. 11-9 Shop view of condenser construction.

11-13

FIG. 11-10   Shop view of small condenser.

2. Reverse flow aids in the removal of sand, silt, etc., from the tubes' inner walls before this residue can harden into a damaging coating.
3. Reverse flow causes a sudden change in water temperature. Although this variation amounts to only the degrees of temperature rise, it is enough to destroy many types of marine organisms which might otherwise grow into troublesome blocks in the system.
4. Reverse flow minimizes fouling and maintains high heat-transfer rates.
5. Reverse flow permits reconsideration of sites with debris-laden cooling water.
6. Reverse flow permits utilization of certain on-line cleaning systems. One such system is the reversing-brush on-line cleaning system, shown below in Fig. 11-21.

CONDENSER AUXILIARIES

Atmospheric Relief Valves

Most industrial-application surface condensers are equipped with an atmospheric relief valve to protect the condenser against overpressure in case of failure of the cooling-water supply. They are usually set to open at 10 psig. Larger units are usually protected by blowout diaphragms on the turbine casing. In current-day

TABLE 11-2   Atmospheric-Relief-Valve Sizes

| Maximum steam flow lb/h | Sizes of atmospheric relief valves, in | |
	For protection*	For maximum noncondensing operation*
Up to 7,500	6	8
7,501– 11,800	8	10
11,801– 17,000	8	12
17,001– 20,000	8	14
20,001– 23,100	10	14
23,101– 30,200	10	16
30,201– 38,200	12	18
38,201– 45,000	12	20
45,001– 47,200	14	20
47,201– 62,000	14	24
62,001– 68,000	16	24
68,001– 82,000	16	30
82,001–106,000	18	30
106,001–120,000	18	
120,001–170,000	20	
170,001–250,000	24	
250,001–380,000	30	
380,001–550,000	36	

Reprinted by permission of the Heat Exchange Institute.
*The sizes listed under "For protection" are normally used under ordinary condensing operation. However, if it is desired to operate the turbine temporarily noncondensing at its maximum noncondensing capacity, the sizes listed under "For maximum noncondensing operation" should be used.

practice for prime-mover applications, condensers range in size from relatively small shipboard or industrial-application turbine-generator units of 300 ft^2 to large nuclear-application utility-power-generation-system units greater than 1,000,000 ft^2 in surface.

The size of atmospheric relief valves is somewhat dependent upon local operating conditions. It is always understood that the valves must be of a size to pass all the steam which can be admitted to a turbine or engine through any openings except from lines which are already protected by relief valves set to open at a pressure not exceeding 10 psig. As an example, an extraction or bleeder turbine would normally require an atmospheric relief valve of a size to take care of the full-throttle flow to the turbine. It should not be based on the steam flow to the condenser under normal operation only.

In Table 11-2, the column "For protection" will give the size of atmospheric relief valves used for normal operation of condensing turbines or engines. For temporary full-load noncondensing operation, the sizes listed under "For maximum noncondensing operation" should be used.

Air-Pump Equipment

Steam Ejectors   Effective operation of a surface condenser requires that air leakage and noncondensable gas be kept to a minimum in the condenser. Air leakage into a system under vacuum is subject to considerable variation because leakage occurs principally through the turbine gland seals, valve stem glands, and pipe points and connections. Modern construction with welded piping tends to keep leakage to a minimum.

Required capacity. The total capacity required for venting a condenser depends upon the type of system. It may consist only of air leakage into the system, or it may consist of air leakage plus one or more of the following:

1. Noncondensable gases or vapors liberated during the process, which must be removed by the ejector in addition to the air leakage (nuclear-geothernal service)

2. Water or other condensable vapors associated with the air or noncondensable gases when the ejector is handling a saturated mixture, as in the case of an ejector serving a surface or barometric jet condenser

3. Air and gases liberated from the injection water when the ejector is serving a barometric or low-level jet condenser

Calculation of total capacity of ejectors serving suface condensers. The air withdrawn from a surface condenser is saturated with water vapor, and the amount of water vapor depends upon the temperature and absolute pressure at the air outlet of the condenser.

In most condensers, the pressure drop through the condenser can be considered negligible and the absolute pressure at the air outlet assumed to be the same as that at the entrance of the condenser. To provide for sufficient total capacity, the temperature at the air outlet of a well-designed surface condenser is generally assumed to be about 7.5°F below the temperature of saturated steam at the absolute pressure at design suction pressure of the venting equipment.

Calculation of the amount of water vapor associated with the air is best illustrated by the following example.

Example: An ejector is required to remove 20 lb of dry-air leakage per hour plus the associated water vapor of saturation from a surface condenser at an absolute pressure of 1 inHg, the temperature of the mixture at the air outlet of the condenser being 71.5°F. What is the total amount of air and vapor mixture to be removed?

Total pressure of mixture is 1.000 inHg.

It is found from Fig. 11-11 that 2.18 lb of water vapor is required to saturate 1 lb of dry air under the specified conditions, and the total air and vapor mixture to be handled by the ejector is accordingly

$$20 + (2.18 \times 20) = 63.6 \text{ lb/h}$$

When the noncondensable gas is other than air, reference may be made to HEI *Condenser Standards,* or Dalton's law should be used to calculate the saturation component of this design noncondensable load.

Recommended ejector capacity. Table 11-3 shows accepted values for sizing venting equipment for condensers serving turbines. Leakage values for various industrial-process applications can be determined only by careful consideration of all details and of previous experience with similar installations. However, actual leakage values alone should not be used to size venting equipment. For proper operation condensers must be adequately vented, and venting equipment

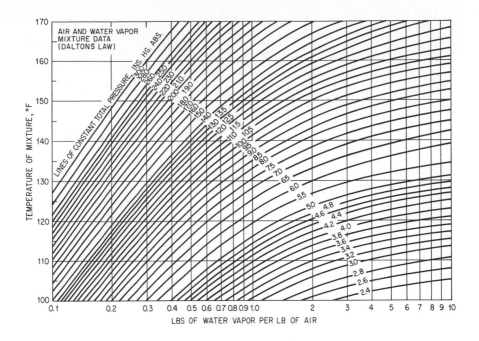

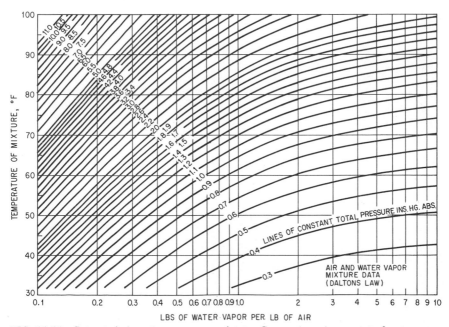

FIG. 11-11 Saturated air and water-vapor mixture. Curves show the amount of water vapor required to saturate 1 lb of dry air at various temperatures and absolute pressures.

TABLE 11-3 Venting-Equipment Capacities Recommended for Surface Condensers (One Shell)*

Effective steam flow, each main exhaust opening, lb/h		Total number of exhaust openings								
		1	2	3	4	5	6	7	8	9
Up to 25,000	scfm†	3.0	4.0	5.0	5.0	7.5	7.5	7.5	10.0	10.0
	Dry air, lb/h	13.5	18.0	22.5	22.5	33.8	33.8	33.8	45.0	45.0
	Water vapor, lb/h	29.7	39.6	49.5	49.5	74.4	74.4	74.4	99.0	99.0
	Total mixture, lb/h	43.2	57.6	72.0	72.0	108.2	108.2	108.2	144.0	144.0
25,001–50,000	scfm†	4.0	5.0	7.5	7.5	10.0	10.0	10.0	12.5	12.5
	Dry air lb/h	18.0	22.5	33.8	33.8	45.0	45.0	45.0	56.2	56.2
	Water vapor, lb/h	39.6	49.5	74.4	74.4	99.0	99.0	99.0	123.6	123.6
	Total mixture, lb/h	57.6	72.0	108.2	108.2	144.0	144.0	144.0	179.8	179.8
50,001–100,000	scfm†	5.0	7.5	10.0	10.0	12.5	12.5	15.0	15.0	15.0
	Dry air, lb/h	22.5	33.8	45.0	45.0	56.2	56.2	67.5	67.5	67.5
	Water vapor, lb/h	49.5	74.4	99.0	99.0	123.6	123.6	148.5	148.5	148.5
	Total mixture, lb/h	72.0	108.2	144.0	144.0	179.8	179.8	216.0	216.0	216.0
100,001–250,000	scfm†	7.5	12.5	12.5	15.0	17.5	20.0	20.0	25.0	25.0
	Dry air, lb/h	33.8	56.2	56.2	67.5	78.7	90.0	90.0	112.5	112.5
	Water vapor, lb/h	74.4	123.6	123.6	148.5	175.1	198.0	198.0	247.5	247.5
	Total mixture, lb/h	108.2	179.8	179.8	216.0	251.8	288.0	288.0	360.0	360.0

250,001–500,000	scfm†	10.0	15.0	17.5	20.0	25.0	25.0	30.0	30.0	35.0
	Dry air, lb/h	45.0	67.5	78.7	90.0	112.5	112.5	135.0	135.0	157.5
	Water vapor, lb/h	99.0	148.5	173.1	198.0	247.5	247.5	297.0	297.0	346.5
	Total mixture, lb/h	144.0	216.0	251.8	288.0	360.0	360.0	432.0	432.0	504.0
500,001–1,000,000	scfm†	12.5	20.0	20.0	25.0	30.0	30.0	35.0	40.0	40.0
	Dry air, lb/h	56.2	90.0	90.0	112.5	135.0	135.0	157.5	180.0	180.0
	Water vapor, lb/h	123.6	198.0	198.0	247.5	297.0	297.0	346.5	396.0	396.0
	Total mixture, lb/h	179.8	288.0	288.0	360.0	432.0	432.0	504.0	576.0	576.0
1,000,001–2,000,000	scfm†	15.0	25.0	25.0	30.0	35.0	40.0	40.0	45.0	50.0
	Dry air, lb/h	67.5	112.5	112.5	135.0	157.5	180.0	180.0	202.5	225.0
	Water vapor, lb/h	148.5	247.5	247.5	297.0	346.5	396.0	396.0	445.5	495.0
	Total mixture, lb/h	216.0	360.0	360.0	432.0	504.0	576.0	576.0	648.0	720.0
2,000,001–3,000,000	scfm†	17.5	25.0	30.0	35.0	40.0	45.0	50.0	55.0	60.0
	Dry air, lb/h	78.7	112.5	135.0	157.5	180.0	202.5	225.0	247.5	270.0
	Water vapor, lb/h	173.1	247.5	297.0	346.5	396.0	445.5	495.0	544.5	594.0
	Total mixture, lb/h	251.8	360.0	432.0	504.0	576.0	648.0	720.0	792.0	864.0
3,000,001–4,000,000	scfm†	20.0	30.0	35.0	40.0	45.0	50.0	55.0	60.0	65.0
	Dry air, lb/h	90.0	135.0	157.5	180.0	202.5	225.0	247.5	270.0	292.5
	Water vapor, lb/h	198.0	297.0	346.5	396.0	445.5	495.0	544.5	594.0	643.5
	Total mixture, lb/h	288.0	432.0	504.0	576.0	648.0	720.0	792.0	864.0	936.0

Reprinted by permission of the Heat Exchange Institute.
NOTE: These tables are based on air leakage only and the air-vapor mixture at 1 inHg absolute and 71.5°F.
*HEI Condenser Standards has additional venting-equipment-capacity recommendations.
†scfm (cubic feet per minute of gas flow at specified standard conditions of temperature and pressure) = 14.7 psia at 70°F.

must in many instances be larger than that sized by actual leakage. The noncondensable-removal equipment must, in addition to removing noncondensables, provide an adequate flow of noncondensable gas and vapor from all sections of the condenser to avoid stagnation zones. In most instances this flow is greater than that which would be obtained from actual leakage rates. Guidelines for sizing venting equipment taken from HEI *Condenser Standards* are contained in part in Table 11-3.

Determination of air leakage in existing vacuum systems. The rate of air leakage into a vacuum system can readily be determined by pulling a vacuum on the system and then noting the time required for the vacuum to drop a certain amount after closing the valve in the suction line to the vacuum pump. Any convenient vacuum can be selected as long as the final vacuum at the end of the test is not less than 15 inHg. The rate of leakage can be determined from the following formula:

$$W = \frac{0.15 \times VD}{t}$$

where W = rate of air leakage, lb/h
V = net volume of system under vacuum, ft^3
D = drop in vacuum during observation, inHg (vacuum at start minus vacuum at end)
t = time, min, for vacuum to drop amount D

Example: A 500-ft^3 vacuum system dropped from 25 to 23 in vacuum in 10 min. From the foregoing formula, the leakage rate is found to be 15 lb of air per hour.

Single-stage ejectors. A basic ejector assembly, illustrated in Figure 11-12, consists of a steam chest, a steam nozzle, a suction chamber, and a diffuser. High-pressure steam, in passing though the divergent nozzle, emerges at a high velocity, thereby continuously entraining the gases or vapors surrounding the jet of steam in the suction chamber. Kinetic energy transferred to the gases or vapors thus produces the required suction effect, the mixture then being discharged into a convergent-divergent diffuser. The divergent portion of the diffuser converts the input velocity flow into pressure, thereby elevating the pressure at the diffuser discharge by a predetermined amount above the suction-inlet pressure.

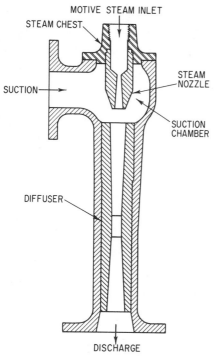

MOTIVE STEAM INLET
STEAM CHEST
STEAM NOZZLE
SUCTION
SUCTION CHAMBER
DIFFUSER
DISCHARGE

FIG. 11-12 Section through a single-stage ejector.

The degree of compression that can be accomplished satisfactorily in a single stage of compression depends upon such factors as steam pressure available for operating the ejector, absolute pressure to be produced, and efficiency of the ejector design.

An ejector is a fixed-capacity machine by reason of its construction; that is, an increase or decrease in the quantity of gas or vapor being handled under constant suction and discharge conditions cannot be accomplished in the basic assembly. The capacity at any given suction pressure is essentially constant and cannot be altered by changing motive-steam flow or pressure.

Single-stage ejectors are used for a wide variety of services for which moderate vacuums are required. They are normally designed for maintaining absolute pressures down to 3.5 inHg. However, when conditions of required capacity and available operating steam pressure are favorable, they can be designed for a somewhat lower absolute pressure. They are furnished in any required size from 1½ inch up to extremely large sizes for unusual process applications.

Single-stage ejectors are normally designed to discharge to atmospheric pressure at any available operating steam pressure above 50 lb psig. These ejectors are used in many applications, such as maintaining a vacuum in condensers, stills, evaporators, etc., and are also employed extensively as priming or exhausting ejectors for the rapid initial evacuation of equipment.

Two-stage ejectors. Commercial ejectors are arranged in a variety of forms to meet limitations with respect to required degree of compression and flexibility. When the required degree of compression is beyond the capabilities of a basic single-stage assembly, two stages are arranged to operate in series, each stage effecting part of the total compression. When flexibility in capacity is required, two or more ejectors, either single-stage or multistage as required by the degree of compression, can be arranged to operate in parallel so that each set effects full compression of a portion of the total capacity.

It is evident that ejectors can readily be arranged in any desired combination to suit specific requirements.

Intercondenser. In a two-stage ejector, a condenser may be placed at the discharge of the stage that exhausts to intermediate vacuum, thereby condensing a large portion of the motive steam and vapors and leaving only air and noncondensable gases and their saturation component to be handled by the subsequent stage. This arrangement has the effect of greatly reducing the amount of motive steam required for total compression from suction to final-stage discharge and accordingly permits obtaining large capacities at high vacuum without excessive steam consumption. When condensers are used in this manner, they are called *intercondensers.*

Since intercondensers operate under a vacuum, it is necessary to provide a means of draining the condensed ejector steam from surface-type intercondensers. In power plant applications the condensed steam is usually drained back to the equipment being evacuated.

Aftercondenser. The use of an aftercondenser for condensing the steam discharged by a single-stage ejector or the final stage of a multistage assembly has no direct bearing on the performance of the ejector itself as to capacity and steam consumption. It is an optional feature determined by the requirements of the installation. For example, surface aftercondensers are used exclusively when ejectors are serving power plant condensers, for the reason that complete heat recovery of the ejector steam is essential and recovery of motive steam instead of loss to the atmosphere is desired.

The type of ejectors most often used in power plant applications will be discussed here.

Two-stage ejectors with surface inter-after condensers. These units are used principally in connection with surface condensers serving prime movers in power plants and are capable of maintaining the vacuum produced in modern well-designed surface condensers (Fig. 11-13).

Modern power plant turbines are designed to utilize effectively the highest vacuum that can be produced economically in the surface condenser commensurate with the temperature of the cooling water, and it is not uncommon to find condensers operating at absolute pressures less than 1 inHg under cold-water conditions.

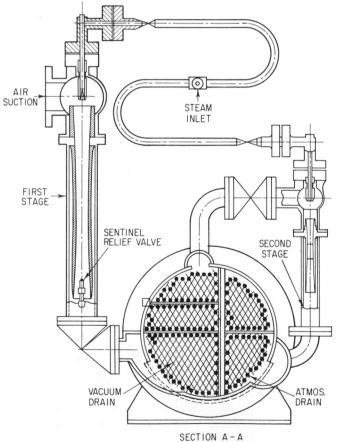

SECTION A - A

FIG. 11-13   Cross section through a typical two-stage ejector.

Main-condenser condensate is used for cooling the inter-after condensers of these units, thus recovering practically all the heat in the ejector steam and thereby improving the overall efficiency of the plant. The small amount of condensate from the inter-after condenser (condensed ejector steam) is returned to the condenser hot well and is thus returned to the boilers with deaerated feedwater. When the amount of condensate from the main condenser is inadequate for properly cooling the inter-after condenser, as might be true during light loads or starting, a portion of the condensate is recirculated back to the main condenser, thus limiting the temperature rise to the required amount.

Figure 11-14 shows in diagrammatic form the installation and piping for a surface condenser equipped with two-stage ejectors and an inter-after condenser for a typical power plant application.

Vacuum Pumps   In place of steam ejectors, mechanically driven vacuum pumps can be employed for air removal in surface condensers. Calculation of required capacity is the same as described under "Steam Ejectors."

Figure 11-15 shows a typical mechanical vacuum pump used for this service.

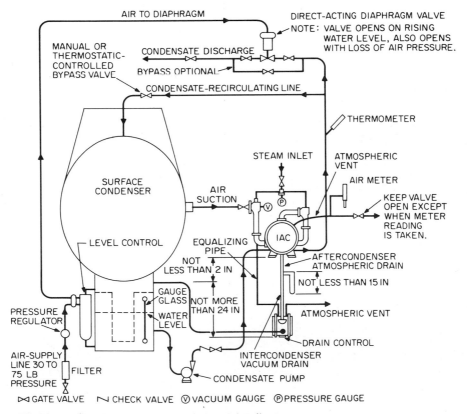

FIG. 11-14   Typical condensate and airflow piping diagram.

FIG. 11-15   Nash AT-3004E main-condenser exhauster.

Condensate Pumps

Condensate pumps are used to remove the water from the condenser hot well and to pump it to the deaerating heater in the feed system. In many instances they are sized to handle approximately 125 percent of the condensed-steam flow; i.e., a condenser handling 100,000 lb/h of steam would require a condensate pump rated at 250 gal/min. Condensate pumps usually operate with a small submergence and pump liquid at a temperature equal to saturation temperature at the condenser pressure. For this reason they are of special design.

Single-stage horizontal pumps are used for smaller power plants with heads up to 125 ft, while multistage pumps, suitable for heads of 200 ft or more, are used for larger installations (see Fig. 11-16). When the available submergence is extremely low, vertical pit-type condensate pumps are frequently used in modern installations (see Fig. 11-17).

Condensate pumps are usually motor-driven and the pump discharge throttled to control the condenser hot-well level. A diagram of this control setup is shown in Fig. 11-18.

Condenser Circulating Pumps

Condenser circulating pumps are used to circulate cooling water through the condenser. Their required capacity has been established at the beginning of this section. Depending on the temperature of the available water, a rough rule for modern condensers calls for 0.1 to 0.25 gal/min of circulating water per pound of steam condensed per hour.

The head required for these pumps is low, consisting mainly of the friction loss through the condenser tubes and the water boxes. In many installations the static

FIG. 11-16 Two-stage single-suction condensate pump with a horizontally split case.

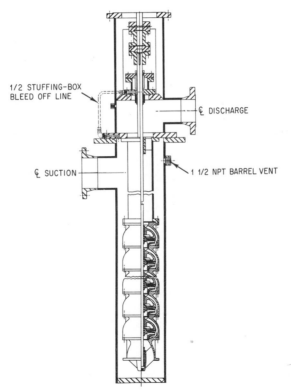

FIG. 11-17 Vertical five-stage condensate pump. *(Courtesy of Byron Jackson Pump Division.)*

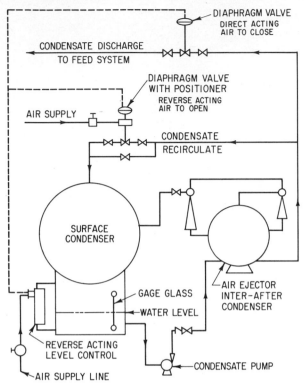

FIG. 11-18 Typical condenser level-control arrangement.

head must be overcome only during starting because of siphon action of the piping, and in this case the pump must overcome only the friction head of the system when operating at rated capacity.

The head required is in the neighborhood of 25 to 50 ft.

Vertical submerged units of the propeller or mixed-flow type are used in modern large installations (see Fig. 11-19).

On-Line Cleaning Systems

To maintain the design overall heat-transfer state U_0, many utility power plants use continuous on-line tube-cleaning systems. These systems permit continuous operation of the plants without the need to resort to costly shutdowns in order to clean the tubes manually. Frequent cleaning reduces the susceptibility of the tubes to corrosion. Two currently available systems are shown in Figs. 11-20 and 11-21).

The sponge-ball system operates on a closed cycle, arranged as shown in Fig. 11-20. Elastic sponge-rubber balls, oversized in comparison with the tube diameter, are injected into the cooling-water flow and forced through the condenser

FIG. 11-19 Vertical mixed-flow circulating-water pump. *(Courtesy of Byron Jackson Pump Division.)*

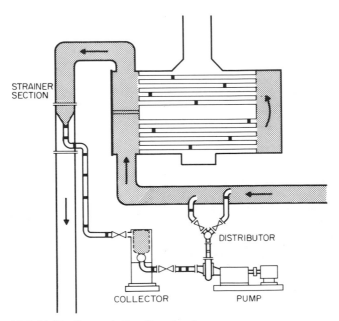

STRAINER SECTION

DISTRIBUTOR

COLLECTOR

PUMP

FIG. 11-20 Sponge-ball on-line cleaning system.

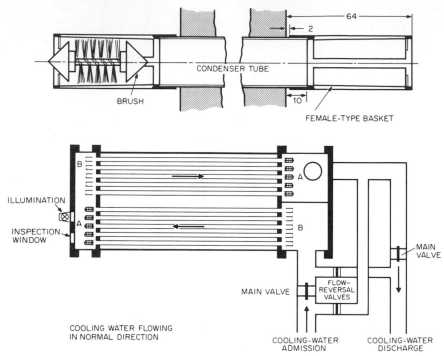

FIG. 11-21   Reversing-brush on-line cleaning system.

tubes by the natural pressure differential between condenser inlet and condenser outlet.

A special screening device installed in the condenser discharge line routes the sponge-rubber balls together with a small quantity of cooling water through a recirculating unit to the condenser inlet pipe, thus completing the cycle. With this constant recirculation of balls, fouling is removed from the tube surface.

The reversing-brush system is shown in Fig. 11-21. It uses circulating-water flow reversal to drive cleaning brushes through the tubes. Cooling-water pipes run to each condenser half and are provided with cross headers with suitably controlled valving for flow reversal. Each condenser tube is provided with a basket at each end. The cleaning brush is held in one of the two baskets by the normal water flow (see position *A* in Fig. 11-21). If the direction of the cooling-water flow is reversed, all brushes are swept by the pressure drop through the tubes into the baskets at the opposite end (see position *B* in Fig. 11-21). This operation is repeated when the cooling-water flow is again reversed and the brushes are carried back to their original position. This completes the cleaning cycle. It may be repeated as often as necessary to keep the tubes clean.

section 12

Filtration Systems

Introduction

Filtration is the process of separating suspended solids from liquid by means of a porous medium.

Solid materials can be grouped as follows:

1. A soluble phase in which the material is soluble in the liquid as a true solution

2. A soluble or insoluble colloidal phase having a diameter of 0.001 to 1.0 μm

3. A supracolloidal phase of particles 1.0 to 100 μm in diameter

4. A coarse-particle phase which settles or floats

Filtration is concerned with Groups 2 and 3, for Group 1 requires chemical separation and Group 4 is usually treated by settling or flotation.

The mechanism of separation within all filter media is not wholly understood but is generally accepted to be any combination of the following: straining, sedimentation, agglomeration, adsorption, molecular cohesion, streaming potential, zeta potential, and ionization.

When the filter medium loses porosity because of retention of solids to the point at which further filtration is ineffective or uneconomical, it must be cleaned or replaced.

12-1

Classification of Filters

Filters may be classified by combinations of:

1. Method of liquid movement through the filter, e.g.,
   a. Pressure
   b. Gravity
   c. Vacuum
2. Method of medium handling, e.g.,
   a. Permanent
   b. Periodic replacement
   c. Terminal (end-of-cycle) replacement
3. Type of medium, e.g.,
   a. Screen
   b. Cartridge
   c. Cloth or paper
   d. Coarse granular
   e. Filter aid

Simple Screen Filters

Screen filters have fixed pore sizes generally equal in size and not of great depth. Therefore, all solids above the fixed pore size are prevented from passing through. Screen filters may be made of wire mesh, perforated metal screens, porous membranes, or the edges of a series of disks or spiral-wound wire. Screens are generally cleaned and reused.

Cartridge Filters

Cartridges are removable filter elements, preassembled or molded into the proper shape. They depend on depth filtration, removing solids within the labyrinth path of flow. They are made of fibrous material, metal wool, porous stone or sintered metal, filter aids, or combinations of these items. Cartridges are generally thrown away but may be cleaned or occasionally washed in place.

Cloth Filters

Filter cloths are made of paper, textiles, plastics, rubber, and various other materials or combinations thereof. Although filter cloths are used in cartridge filters, they are more generally used in filter presses or flat-bed indexing filters. Filter presses are a series of chambers formed by recessed filter plates or alternate frames and plates, forced together by a closing device. Filter cloths are placed over the filter plates and removed to be discarded or cleaned at the end of each cycle

Flat-bed indexing filters use a roll of cloth which is indexed or pulled across the filtering area at the end of the cycle or on a continuous basis.

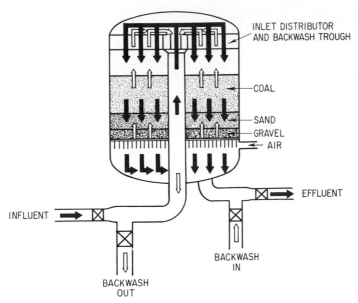

FIG. 12-1 Flow schematic: sand filter.

Coarse Granular Filters

Granular filters (see Fig. 12-1) all depend to a varying degree on depth filtration. The granules should be of uniform size and density to get the most effective filtration. Some filters use more than one filter medium, the size being successively reduced so as to give series filtration. In this case, different densities must be used to allow restratification after backwashing. Flow is normally downward but may be upward. Some filters have a moving bed, which is externally cleaned and returned continuously.

Good filtration depends on even distribution of flow across the filter area. This can be a function of the underdrain system which supports and retains the filter medium while allowing the liquid to flow through. The underdrain system and the overflow or backwash troughs must evenly distribute the backwash for complete cleaning.

Slow sand filters operate at very low flow rates of about 0.5 gal/ft^2·min and are not backwashed but scraped on the surface. Rapid sand filters produce high clarity up to about 8 gal/ft^2·min by using multiple media. High-rate filters operate up to 20 gal/ft^2·min or higher and normally produce a lesser degree of clarity.

Precoat Filters

A very fine, porous material such as diatomaceous earth, cellulose fibers, or perlite is used as a filter aid or medium in precoat filters (see Figs. 12-2 and 12-3). The

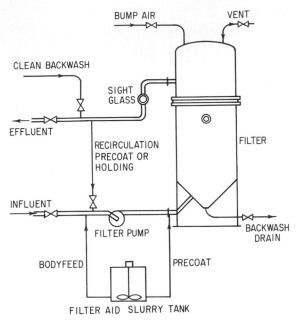

FIG. 12-2   Flow schematic: precoat filter.

filter aid is distributed on and held by a septum or filter element. A very thin coating (about ¹⁄₁₆ in) is used initially and forms a *precoat*. The septa may be tubular or flat but generally are vertical and present a large amount of surface area within a small volume. The filter aids have the property of bridging across openings much larger than their own particle size. The cake is distributed and held on by the velocity of the liquid through it. Therefore, flow must never be stopped, and when filtration ceases temporarily, a recirculating or holding flow is established.

The filter aid itself forms the filter medium, and use of varying sizes or grades of particles will change the effluent clarity. Most removal is accomplished on the surface, but some is due to the depth filtration of the many labyrinth passages. Most filter aids are also added continuously to the incoming stream as *body feed*. This allows the cake to grow, entrapping the removed contaminants and maintaining porosity to increase the cycle length.

Figure 12-4 shows the installation of a precoat filter for municipal-water treatment, and Fig. 12-5 shows the control panel for such a system.

Filtration with Otherwise Active Media

Sometimes filtration is accomplished as an adjunct to another type of removal such as ion exchange in high-rate demineralizers or powdered-resin filters.

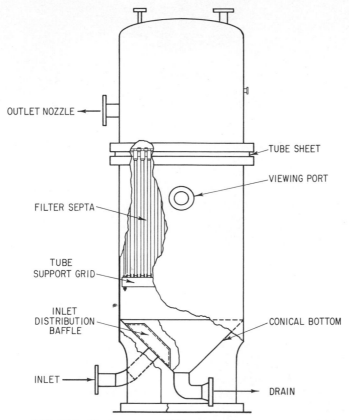

FIG. 12-3 Pressure precoat filter.

Because of the expense of the other process, filtration is usually treated as secondary and should not be the controlling design factor.

Backwashing

Backwashing is the term applied to cleaning or renewing the filter medium by using liquid and/or air. For permanent media, backwashing dislodges the entrapped contaminants and transports them to waste. With replaceable media, the medium and the contaminants are both removed.

Liquid is used both as a washing and as a transporting medium. Generally a substantial velocity is required to perform these functions, although care must be exercised to prevent the loss of any permanent medium.

Air is used either for agitation and scouring or to create a pressure release and impart high initial velocity to break the contaminants free.

Backwashing is one of the most important aspects of filtration. The ability to

FIG. 12-4 3500-gal/min automatic precoat filter for municipal-water treatment.

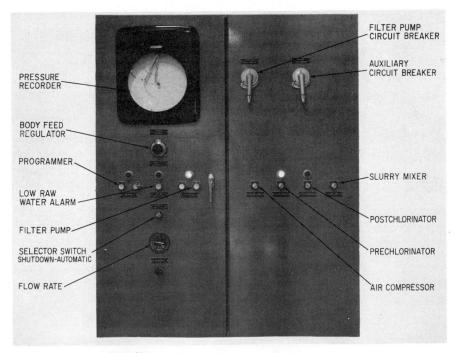

FIG. 12-5 Typical water-filter control panel.

renew full filtration capability after each cycle contributes to sustained effluent quality, reduced maintenance, and economical operation.

Pretreatment

Pretreatment is used with filtration either to condition the particulate matter by agglomeration so as to simplify or improve filtration or to precipitate soluble compound into filterable particulate material.

One of the most common pretreatments is with coagulants, such as alum or ferric sulfate, which form a charged floc to attract small particles electrostatically.

Filter Selection

The selection of a filter to suit a specific process requires evaluation of many factors that are generally interrelated:

1. Required clarity of effluent
2. Maximum size of particle in effluent
3. Variation of inlet flow and/or solids concentration
4. Requirements for medium replacement or backwashing
5. Disposal of removed solids and backwash liquid
6. Use of pretreatment
7. Operating economics

Since a greater degree of filtration requires higher operating costs, it is most important that care be exercised in determining this requirement. If a cutoff particle size is absolutely necessary, a screen or fixed-pore-size filter will be needed since depth filtration can give only a nominal rating based on average percentage removal for various sizes.

Filters work best under constant conditions. Variations affect effluent, equipment sizing, and economics.

Also of major importance is tying the filter application with the backwashing of equipment. The external requirements for backwashing and the disposal of the materials all affect the selection.

The effect of any pretreatment on the operation of a filter must be considered. Finally, when filters are capable of performing the requirements, a full evaluation of economics must be undertaken.

Economics

The costs of filtration are interrelated. They cover:

1. Power to overcome friction and filter pressure drop and to perform backwash
2. Backwash liquid
3. Medium replacement

4. Pretreatment costs, if used
5. Amortization of investment
6. Maintenance
7. Labor

Power is utilized by pumps, compressors, agitators, controls, etc. An increase in maximum pressure drop will increase power consumption but is offset by less backwash cost and medium replacement because of longer runs and by a reduction of first cost.

Backwash liquid is an expense if it costs to produce or handle it. Sometimes it is expressed in net output. Backwash air is reflected in power and investment cost.

Medium replacement is reduced by increasing the length of run with higher terminal pressure drop or lower filtration rates. However, these measures involve other offsetting costs.

With filter aid, precoat usage is a function of surface area and therefore of filtration rate as well as length of run. An increase in body-feed usage will increase length of run and can result in reduced power usage or increased filtration rate with an associated reduction in first cost.

When pretreatment is used for precipitation or treatment as an adjunct to filtration, it is an independent cost. When presettling or conditioning of particles is involved, costs must be weighed against the resultant reduction in other filtration costs.

For true evaluation, first cost should be amortized as a cost of the final product. This should reflect interest rate on money, required time of depreciation, and equipment usage rate.

Maintenance includes not only equipment upkeep but also replacement of permanent filter media.

Most filtration systems can be automated, which may sizably reduce labor costs.

Test Instrumentation

TEST CODES

Nationally recognized test codes, standards, and test specifications all serve to guide manufacturer and customer in the evaluation of a piece of machinery either at the manufacturer's test facility or under actual service conditions at the customer's site. These codes provide a common ground for communication by clearly establishing the goals of a test program, the procedure under which the test will be conducted, and the presentation of final results.

The purpose of a test covered by a code may be simply to verify the mechanical integrity of a machine, or it may include exhaustive testing to verify that performance guarantees have, in fact, been met. Test codes seldom specify absolute levels of performance but often do limit the tolerances, fluctuations, and deviations of both measurements and test conditions.

Most test codes were prepared as a cooperative effort by representatives of the manufacturer, the user of the machine, and the educator. The codes are generally thorough, well-written texts, which are periodically revised and updated by the issuing authorities. As is the case with any document, these test codes require judicial interpretation at times. This is generally accomplished formally or informally through consultations between the parties concerned with the test. Several of the codes even provide detailed lists of items to be discussed prior to the actual test.

By specifying that a machine is to be tested in accordance with a nationally recognized test code, purchasing-contract language is simplified and the chance of ambiguity is reduced.

Test codes, therefore, are a convenient, authoritative means of conveying a test program to those responsible for the test by those specifying the test.

Copies of the most common codes pertaining to Delaval products can be obtained from the following issuing authorities:

1. American Bureau of Shipping. Rules covering shipboard propulsion machinery, turbogenerators, and auxiliary marine machines. 65 Broadway, New York, New York 10006.

2. American Gear Manfacturers Association, 1901 North Fort Myer Drive, Arlington, Virginia 22209.

3. American National Standards Institute. Codes pertaining to measurement of airborne noise. 1430 Broadway, New York, New York 10018.

4. *American Petroleum Institute Guide for Inspection of Refinery Equipment.* Guide presenting requirements for pumps, compressors, turbines, gears, and heat exchangers. American Petroleum Institute, 2101 L Street, N.W., Washington, D.C. 20037.

5. American Society of Mechanical Engineers. Power test codes dealing with pumps, compressors, turbines, and heat exchangers. 345 East Forty-seventh Street, New York, New York 10017.

6. Heat Exchange Institute. Codes covering heat exchangers and steam-jet ejectors. 1230 Keith Building, 1621 Euclid Avenue, Cleveland, Ohio 44115.

7. *Hydraulic Institute Standards.* Codes dealing with centrifugal and positive-displacement pumps. Hydraulic Institute, 712 Lakewood Center North, 14600 Detroit Avenue, Cleveland, Ohio 44107.

8. Institute of Electrical and Electronics Engineers. Codes pertaining to generators and generator drives. 345 East Forty-seventh Street, New York, New York 10017.

9. National Electrical Manufacturers Association. Codes dealing with generators and their drives. 2101 L Street, N.W., Washington D.C. 20037.

TEMPERATURE MEASUREMENT

Introduction

Measurement of temperature is generally considered to be one of the simplest and most accurate measurements performed in engineering. The desired accuracy in the measurement can be obtained, however, only by observing suitable precautions in the selection, installation, and use of temperature-measuring instruments and in the proper interpretation of the results obtained with them.

Four phenomena form the basis for most measuring instruments:

Change in physical dimensions or characteristics of liquids, metals, or gases
Changes in electrical resistance
Thermoelectric effect
Radiant energy

The following types of instruments are available for use under appropriate conditions:

Liquid-in-glass thermometer
Resistance thermometer
Thermocouple thermometer
Filled-system thermometer
Bimetallic thermometer
Radiation thermometer
Optical pyrometer

Liquid-in-Glass Thermometer

A liquid-in-glass thermometer consists of a thin-walled glass bulb attached to a glass capillary stem closed at the opposite end, with the bulb and a portion of the stem filled with an expansive liquid, the remaining part of the stem being filled with the vapor of the liquid or a mixture of this vapor and an inert gas. Etched on the stem is a scale in temperature degrees so arranged that, when calibrated, the reading corresponding to the end of the liquid column indicates the temperature of the bulb. The three types of liquid-in-glass thermometers (Fig. 13-1) are partial-immersion, total-immersion, and complete-immersion thermometers.

1. A partial-immersion thermometer is designed to indicate temperature correctly when used with the bulb and a specified part of the liquid column in the stem exposed to the temperature being measured; the remainder of the liquid column and the gas above the liquid are exposed to a temperature which may or may not be different.

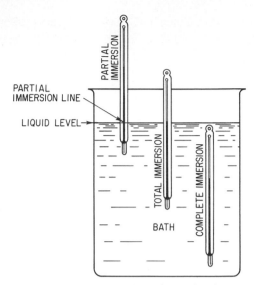

FIG. 13-1 Partial-, total-, and complete-immersion thermometers.

2. A total-immersion thermometer is designed to indicate the temperature correctly when used with the bulb and the entire liquid column in the stem exposed to the temperature being measured and the gas above the liquid exposed to a temperature which may or may not be different.

3. A complete-immersion thermometer is designed to indicate the temperature correctly when used with the bulb, the entire liquid column in the stem, and the gas above the liquid exposed to the temperature being measured.

Tables 13-1 and 13-2 show National Bureau of Standards (NBS) certification tolerances for laboratory thermometers. The term *tolerance in degrees* means acceptable limits of error of uncertified thermometers. *Accuracy in degrees* is the limit of error to be expected when all necessary precautions are exercised in the use of thermometers. The limits to which NBS certification values are rounded off are shown in the column "Corrections stated to."

The operation of a liquid-in-glass thermometer depends on having the coefficient of expansion of the liquid greater than that of the bulb glass. As a consequence, an increase in temperature of the bulb causes the liquid to be expelled from the bulb, resulting in a rise in position of the end of the liquid column. The capillary stem attached to the bulb serves to magnify this change in volume on a scale.

The most frequently encountered source of error when using liquid-in-glass thermometers is the misuse or complete neglect of the emergent-stem correction. This correction derives from the use of the thermometer with a portion of the stem exposed to a different temperature from that of calibration. A common example is the use of partial immersion of a thermometer calibrated for total immersion. For detailed information on this correction, see the American Society of Mechanical Engineers' *Power Test Codes: Temperature Measurement.*

TABLE 13-1 Tolerances for Fahrenheit Mercurial Total-Immersion Laboratory Thermometers

Temperature range in degrees	Graduation interval in degrees	Tolerance in degrees	Accuracy in degrees	Corrections stated to
Thermometers for low temperatures				
−35 to 32	1 or 0.5	1	0.1 −0.2	0.1
−35 to 32	0.2	0.5	0.05	0.02
Thermometers not graduated above 300°				
32 up to 300	2	1	0.2 −0.5	0.2
32 up to 300	1 or 0.5	1	0.1 −0.2	0.1
32 up to 212	0.2 or 0.1	0.5	0.02−0.05	0.02
Thermometers not graduated above 600°				
32 up to 212	2 or 1	1	0.2 −0.5	0.2
Above 212 up to 600		2	0.5	0.2
Thermometers graduated above 600°				
32 up to 600	5	4	0.5 −1.0	0.5
Above 600 up to 950		7	1 −2	0.5
32 up to 600	2 or 1	3	0.2 −1.0	0.2
Above 600 up to 950		6	0.5 −1.0	0.2

TABLE 13-2 Tolerances for Fahrenheit Mercurial Partial-Immersion Laboratory Thermometers

Temperature range in degrees	Graduation interval in degrees	Tolerance in degrees	Accuracy in degrees	Corrections stated to
Thermometers for low temperatures				
−35 to 32	1	1	0.3−0.5	0.1
Thermometers not graduated above 300°				
32 up to 300	2 or 1	2	0.2−1.0	0.2
Thermometers not graduated above 600°				
32 up to 212	2 or 1	2	0.2−0.5	0.2
Above 212 up to 600	2 or 1	3	1 −2	0.5
Thermometers graduated above 600°				
32 up to 600	5 or 2	5.0	1 −2	1
Above 600 up to 950		10	2 −3	1

Resistance Thermometer

A resistance thermometer is a temperature-measuring instrument in which electrical resistance is used as a means of temperature measurement. The instrument consists of a resistor, a resistance-measuring instrument, and electrical conductors connecting the two. The resistor may be metallic (usually in wire form) or a thermistor (a thermally sensitive variable resistor made of ceramiclike semiconducting material).

The basis for resistance thermometry is the fact that most metals and some semiconductors change in resistivity with temperature in a known, reproducible manner. Several materials are commonly employed for resistance thermometers, the choice depending on the compromises that may be accepted. Although the actual resistance-temperature relation must be determined experimentally, for most metals the following empirical equation holds very closely:

$$R_t = R_0(1 + AT + BT^2) \tag{13-1}$$

where R_T = resistance at temperature T
  R_0 = resistance at $0°C$
  T = temperature, K
A and B = constants depending on material

The temperature-resistance function for a thermistor is given by the following relationships:

$$R = R_0 e^k \tag{13-2}$$

$$k = \beta\left(\frac{1}{T} - \frac{1}{T_0}\right) \tag{13-3}$$

where R = resistance at any temperature T, K
  R_0 = resistance at reference temperature T_0, K
  e = base of napierian logarithms
  β = a constant (which usually has a value between 3400 and 3900, depending on the thermistor formulation or grade)

Types of Resistance Thermometers   Platinum thermometer. This thermometer is known for its high accuracy, stability, resistance to corrosion, and other characteristics. It has a simple relation between resistivity and temperature, shown in Eq. (13-1).

Precision platinum thermometer. This thermometer is used to define the International Practical Temperature Scale from -297.3 to $1168.3°F$. The purity and physical properties of the platinum of which the thermometer is made are prescribed to meet close specifications. Different procedures are used for making precision thermometers to cover different temperature ranges.

Industrial platinum resistance thermometer. The requirements for reproducibility and limit of error for thermometers of this type are lower than those for standard thermometers; so are the manufacturing precautions lowered for these thermometers.

Nickel resistance thermometer. This thermometer has been adapted satisfactorily in industrial applications for a temperature range from -100 to $300°F$. The nickel resistance thermometer is less stable than platinum thermometers, but its low cost favors its usage.

Copper resistance thermometer. Copper is an excellent material for resistance thermometers. Its availability in a pure state makes it easy to match with established standards. The resistivity curve of copper is a straight-line function of temperature between -60 and $400°F$, and that makes copper resistance thermometers suitable for the measurement of temperature differences with high accuracy. Copper resistance thermometers are reliable and accurate means of temperature measurement at moderate temperature levels.

Thermistors (nometallic resistance thermometers). Thermistors are characterized by a negative coefficient of resistivity, and their temperature-resistivity curve is exponential. Modern thermistors are very stable; they have high temperature sensitivity and very fast response. Because thermistors are high-resistance circuits, the effect of the lead wires is minimized, and regular copper wires can be used throughout the circuit. Noninterchangeability owing to the difficulty of reproducing resistance properties and the nonlinearity of the resistivity curve limits the use of thermistors.

Information on important characteristics of different classes of resistance thermometers is included in Table 13-3.

Accessories Some forms of Wheatstone-bridge circuits are used for the measurement of temperature with base-metal or industrial platinum resistance ther-

TABLE 13-3 Typical Characteristics of Resistance Thermometers

	Noble metal		Nonmetallic
	Precision	Industrial	Thermistor
Sensitivity	$0.1 \ \Omega/°C$	$0.22 \ \Omega/°F$	Varies with units
Precision	$\pm 0.001°C$	$\pm 0.3°F$	$\pm 0.02°F$ up to $200°F$
Accuracy	$\pm 0.01°C$	$\pm 3.0°F$ standard	$\pm 0.5°F$ standard
		$\pm 1.5°F$ special	$\pm 0.2°F$ special
Response: bare		15 s	Fast
Response: with well		30 s	
Resistance	$25.5 \ \Omega$ at $0°C$	$25 \ \Omega$ at $32°F$	Varies with units
Linearity	$70.1°C/50°C$ span	$70.1°C/50°C$ span	Exponential
Range	-452.2 to	-297.3 to	-100 to $500°F$
	$1168.3°F$	$1950°F$	$(-75$ to $260°C)$
	$(-269$ to	$(-182.96$ to	
	$630.74°C)$	$1064°C)$	

	Base metal		
	10 Ω	Copper, 100 Ω	Nickel, 100 Ω
Sensitivity	$0.22 \ \Omega/°F$	$0.22 \ \Omega/°F$	$0.186 \ \Omega/°F$ ($0.213 \ \Omega/°F$)
Precision	$\pm 0.1°F$	$\pm 0.1°F$	$\pm 0.1°F$
Accuracy	$\pm 0.5°F$ standard	$\pm 0.5°F$ standard	$\pm 0.5°F$ standard
	$\pm 0.2°F$ special	$\pm 0.2°F$ special	$\pm 0.2°F$ special
Response: bare	20 s	40 s	40 s
Response: with well	60 s	90 s	90 s
Resistance	$10 \ \Omega$ at $77°F$	$100 \ \Omega$ at $77°F$	$100 \ \Omega$ at $77°F$
Linearity	Excellent	Excellent	Excellent
Range	-100 to $300°F$	-325 to $300°F$	-100 to $300°F$
	$(-75$ to $150°C)$	$(-200$ to $150°C)$	$(-75$ to $140°C)$

mometers, while the Mueller bridge is used with precision platinum resistance thermometers.

Thermocouple Thermometer

A thermocouple thermometer is a temperature-measuring instrument in which the electromotive force developed in a circuit composed of two dissimilar metals is used as a means of temperature measurement. It consists of a device for measuring electromotive force, a sensing element (thermocouple), and electrical conductors operatively connecting the two.

The thermocouple thermometer operates on the principle that an electric current will flow in a closed circuit of two dissimilar metals when the junctions of the metals are at two different temperatures. Thermocouple materials are available for use within the approximate limits of -300 to $3200\,°F$. Platinum is the generally accepted standard material to which the thermoelectric characteristics of other materials are referred. The emf-temperature relations of conventional thermoelements versus platinum are shown in Fig. 13-2. Reference tables of temperature versus electromotive force as well as polynomial equations expressing the temperature-voltage relationship for different types of thermocouples are available in technical literature.

The iron-Constantan thermocouple is used most widely in industrial applications. The copper-Constantan thermocouple is used widely in industrial and laboratory thermometry.

The platinum–10 percent rhodium versus platinum (Type S) thermocouple serves as an instrument for defining the International Practical Temperature

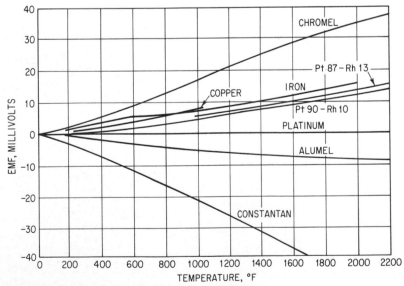

FIG. 13-2   Emf's of various materials versus platinum.

Scale from 630.74 to 1064.43 °C. It is being used in industrial laboratories as a standard for base-metal thermocouples and other temperature-sensing devices.

Table 13-4 lists the seven commonly used thermocouples and some of their characteristics.

The electrical conductors connecting the thermocouple and the measuring instrument may use the actual thermocouple wires, extension wires, or connecting wires (see Fig. 13-3). When it is not possible to run the thermocouple wires to the reference junction or to the measuring instrument, extension wires can be used. To assure a high degree of accuracy, extension wires should have the same thermoelectric properties as the thermocouple wires with which they are used. Significant uncertainties are introduced when extension wires are not matched properly. Calibration of the instrument with extension wires helps to minimize these uncertainties. Connecting wires are a pair of conductors which connect the reference junction to the switch or potentiometer. They are usually made of copper. They do not cause uncertainty in measurements when the reference junction is kept at constant temperature, for example, the ice point.

Indicating potentiometers are recommended by the *ASME Power Test Codes* for performance-test work, although recording potentiometers are used for industrial-process temperature measurement.

Thermocouples may be joined in series. The series connection, in which the output is the arithmetic sum of the emf's of the individual thermocouples, may be used to obtain greater measurement sensitivity and accuracy. A series-connected thermocouple assembly is generally referred to as a *thermopile* and is used primarily in measuring small temperature differences. A schematic diagram of a series-connected thermocouple is shown in Fig. 13-4.

TABLE 13-4   Limits of Error of Thermocouples

Thermocouple type	Temperature range, °C	Limits of error	
		Standard	Special
T (copper-Constantan)	−184 to −59	. . .	±1 percent
	−101 to −59	± 2 percent	±1 percent
	−59 to +93	±0.8°C	±0.4°C
	+93 to +371	±¾ percent	±⅜ percent
J (iron-Constantan)	0 to 277	±2.2°C	±1.1°C
	277 to 760	±¾ percent	±⅜ percent
E (Chromel-Constantan)	0 to 316	±1.7°C	. . .
	316 to 871	±½ percent	. . .
K (Chromel-Alumel)	0 to 277	±2.2°C	±1.1°C
	277 to 1260	±¾ percent	± ⅜ percent
R and S (platinum vs. 13 percent rhodium-platinum)	0 to 538	±1.4°C	. . .
(platinum vs. 10 percent rhodium-platinum)	538 to 1482	±¼ percent	. . .
B (platinum vs. 30 percent rhodium-platinum vs. 6 percent rhodium)	871 to 1705	±½ percent	. . .

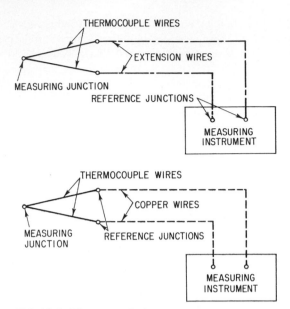

FIG. 13-3   Thermocouple thermometer systems.

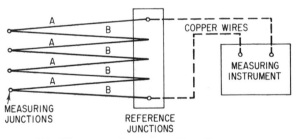

FIG. 13-4   Thermocouples connected in series.

Thermocouples may also be joined in parallel. In the parallel-connected thermocouple circuit, a mean value of the individual thermocouples is indicated, and it will be the true arithmetic mean if all thermocouple circuits are of equal resistance. A schematic diagram of a parallel-connected thermocouple circuit is shown in Fig. 13-5.

The installation of extensive thermocouple equipment requires the services of qualified instrument technicians, and special attention should be given to extension wires, reference junctions, switches, and terminal assemblies.

Opposed thermocouple circuits are sometimes used to obtain a direct reading of a temperature difference between two sets of thermocouples reading two levels of temperature. The number of thermocouples in each set is the same. This method is considered to provide the highest degree of accuracy in the measurement of the critical temperature difference.

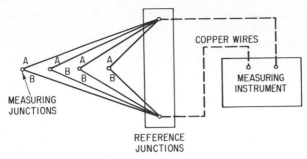

FIG. 13-5 Thermocouples connected in parallel.

Filled-System Thermometer

A filled-system thermometer (Fig. 13-6) is an all-metal assembly consisting of a bulb, a capillary tube, and a Bourdon tube and containing a temperature-responsive fill. Associated with the Bourdon is a mechanical device which is designed to provide an indication or record of temperature.

The sensing element (bulb) contains a fluid which changes in physical characteristics with temperature. This change is communicated to the Bourdon through a capillary tube. The Bourdon provides an essentially linear motion in response to an internally impressed pressure or volume change.

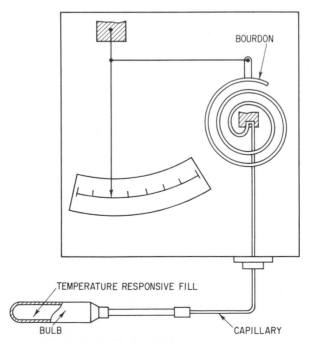

FIG. 13-6 Filled-system thermometer.

Filled-system thermometers may be separated into two types: those in which the Bourdon responds to volume changes and those which respond to pressure changes. The systems that respond to volume changes are completely filled with mercury or other liquid, and the system that responds to pressure changes is either filled with a gas or partially filled with a volatile liquid.

Bimetallic Thermometer

A bimetallic thermometer (Fig. 13-7) consists of an indicating or recording device, a sensing element called a bimetallic-thermometer bulb, and a means for operatively connecting the two. Operation depends upon the difference in thermal expansion of two metals. The most common type of bimetallic thermometer used in industrial applications is one in which a strip of composite material is wound in the form of a helix or helices. The composite material consists of dissimilar metals which have been fused together to form a laminate. The difference in thermal expansion of the two metals produces a change in curvature of the strip with changes in temperature. The helical construction is used to translate this change in curvature to rotary motion of a shaft connected to the indicating or recording device.

A bimetallic thermometer is a relatively simple and convenient instrument. It comes in industrial and laboratory versions.

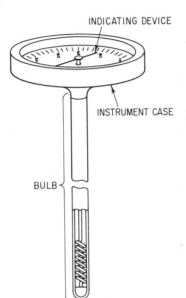

FIG. 13-7   Bimetallic thermometer.

Pyrometry

There are two distinct pyrometric instruments, the radiation thermometer and the optical pyrometer, which are described in greater detail in the following subsections. Both pyrometers utilize radiation energy in their operation. Some of the basic laws of radiation transfer of energy will be described briefly.

All bodies above absolute-zero temperature radiate energy. This energy is transmitted as electromagnetic waves. Waves striking the surface of a substance are partially absorbed, partially reflected, and partially transmitted. These portions are measured in terms of absorptivity α, reflectivity ρ, and transmissivity τ, where

$$\alpha + \rho + \tau = 1 \tag{13-4}$$

For an ideal reflector, a condition approached by a highly polished surface, $\rho \rightarrow 1$. Many gases represent substances of high transmissivity, for which $\tau \rightarrow 1$, and a blackbody approaches the ideal absorber, for which $\alpha \rightarrow 1$.

A good absorber is also a good radiator, and it may be concluded that the ideal radiator is one for which the value of α is equal to unity. In referring to radiation as distinguished from absorption, the term *emissivity* ϵ is used rather than *absorptivity* α. The Stefan-Boltzmann law for the net rate of exchange of energy between two ideal radiators A and B is

$$q = \sigma(T_A^4 - T_B^4) \qquad (13\text{-}5)$$

where q = radiant-heat transfer, Btu/h·ft^2
$\quad \sigma$ = Stefan-Boltzmann constant
T_A, T_B = absolute temperature of two radiators

If we assume that one of the radiators is a receiver, the Stefan-Boltzmann law makes it possible to measure the temperature of a source by measuring the intensity of the radiation that it emits. This is accomplished in a radiation thermometer.

Wien's law, which is an approximation of Planck's law, states that

$$N_{b\lambda} = C_1\lambda^{-5}e^{-C_2/\lambda T} \qquad (13\text{-}6)$$

where $N_{b\lambda}$ = spectral radiance of a blackbody at wavelength λ and temperature T
C_1, C_2 = constants
$\quad \lambda$ = wavelength of radiant energy
$\quad T$ = absolute temperature

The intensity of radiation $N_{b\lambda}$ can be determined by an optical pyrometer at a specific wavelength as a function of temperature, and then it becomes a measure of the temperature of a source.

Radiation Thermometer

A radiation thermometer consists of an optical system, used to intercept and concentrate a definite portion of the radiation emitted from the body whose temperature is being measured, a temperature-sensitive element, usually a thermocouple or thermopile, and an emf-measuring instrument. A balance is quickly established between the energy absorbed by the receiver and that dissipated by conduction through leads, convection, and emission to surroundings. The receiver equilibrium temperature then becomes the measure of source temperature, with the scale established by calibration. An increase in the temperature of the source is accompanied by an increase in the temperature of the receiver that is proportional to the difference of the fourth powers of the final and initial temperatures of the source.

The radiation thermometer is generally designated as a total-radiation thermometer which utilizes, as an index of the temperature of a body, all the energy (all wavelengths) per unit area per unit time radiated by the body. Radiation thermometers are classified according to the method of collecting the radiation and focusing it on the receiver: single mirror, double mirror, and lens.

The radiation thermometer can be classified not as a primary laboratory instrument but rather as an industrial instrument. Its practical useful range extends from ambient temperature to 7500°F, although different thermometers must be used to cover this range.

Optical Pyrometer

Optical pyrometers use a method of matching as the basis of operation. Generally, a reference temperature is provided in the form of an electrically heated lamp filament, and a measure of temperature is obtained by optically comparing the visual radiation from the filament with that from the unknown source. In principle, the radiation from one of the sources, as viewed by the observer, is adjusted to match that from the other source. Two methods are employed: (1) the current through the filament may be controlled electrically, through a resistance adjustment; or (2) the radiation accepted by the pyrometer from the unknown source may be adjusted optically by means of some absorbing device such as an optical wedge, a polarizing filter, or an iris diaphragm. The two methods are referred to, respectively, as the method using the variable-intensity comparison lamp and the method using the constant-intensity comparison lamp. In both cases the adjustment required is used as the means for temperature readout. Figure 13-8 illustrates schematically an arrangement of a variable-intensity pyrometer.

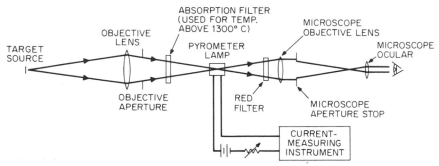

FIG. 13-8   Schematic diagram of an optical pyrometer.

A typical optical pyrometer consists of a power supply and an optical system. The optical system incorporates a telescope, a calibrated lamp, a filter for viewing nearly monochromatic radiation, and an absorption glass filter (see Fig. 13.8). The filament of the lamp and the test body are viewed simultaneously. The filament current is adjusted until the filament image disappears in the image of the test body.

Visual optical pyrometers should not be used for the measurement of temperatures below 1400°F. Automatic optical pyrometers can be used for the measurement of lower temperatures, and they are of great value in the measurement of very high temperatures.

Calibration

To compare or to measure temperature, a temperature scale is necessary. Two ideal temperature scales were proposed: the thermodynamic scale of Kelvin and the ideal-gas scale. The International Committee on Weights and Measures came up with a more practical temperature scale, the International Practical Temperature Scale of 1968 (IPTS-68), which is based on 11 fixed, reproducible temperature points.

There are two widely used temperature scales in engineering practice. The first, the Celsius scale, derives directly from IPTS-68; it has 100 units (degrees) between the ice point and the steam point of water. The second, the Fahrenheit scale, has 180 units (degrees) between these two fixed temperature points. In the first case the freezing point is marked 0, while in the second case this point is marked 32. The relationship between the two scales is as follows:

$$F = \% \, C + 32, \text{ degrees Fahrenehit}$$
$$C = \% \, (F - 32), \text{ degrees Celsius}$$

Calibration at fixed points is a complex process. Standard platinum resistance thermometers and standard platinum-rhodium-platinum thermocouples are calibrated at fixed points for use as primary standards. It is recommended that calibration be done by the NBS or other qualified laboratory. The narrow-band optical pyrometer is another primary standard; its range over the freezing point of gold is obtained through extrapolation. Ordinary calibration of temperature-measuring instruments is effected by comparison of their readings with those of primary or secondary standards at temperatures other than fixed points. Comparators are used to produce those temperatures.

Secondary standards are liquid-in-glass thermometers and base-metal thermocouples. They are calibrated by comparing them with primary-standard platinum resistance thermometers or standard platinum-rhodium versus platinum thermocouples at temperatures generated in comparators. These secondary standards are used in turn for the calibration of other devices, such as liquid-in-glass thermometers, bimetallic thermometers, filled-system thermometers, and base-metal thermocouples, in which the highest degree of accuracy is not required. Optical pyrometers as secondary standards are compared with primary-standard optical pyrometers, and they are then used for calibration of regular test pyrometers.

There is ample literature by the American Society of Mechanical Engineers (ASME), the American Society for Testing and Materials, the National Bureau of Standards, and others that deals with calibration methods, specifications for construction and usage of measuring instruments and temperature comparators, and processing of calibration data. It is advisable in each case to have the major components of the system (primary and secondary standards), potentiometers, and Mueller bridges calibrated periodically by the NBS or other qualified laboratory.

Other Considerations

The preceding presentation on temperature measurement shows clearly how complex the subject is and what precautions must be taken to obtain a meaningful temperature measurement. The proper use of the right temperature-measuring instruments is very important. Calibration for instrumental errors is mandatory for temperature-sensing devices and other temperature-measurement-system components; periodic checking of the calibration is also very important.

If for reasons of protection of the sensitive temperature-measuring element against corrosive atmosphere or excessive mechanical stress, the use of thermometer wells is prescribed, such wells should be designed and installed with the utmost care to avoid damage and the introduction of additional errors. The *ASME Power Test Codes* should be followed in this respect. The most important precautions in using a thermometer well are to keep the sensing element in intimate contact with the well and to have the exposed parts of the well as small as possible and insulated from their surroundings.

The nature of heat transfer between the medium, the temperature of which is being measured, and the sensing element and the sources of temperature errors due to conduction, radiation, and aerodynamic heating are described below. The temperature-sensing element indicates its own temperature, which may not be the exact temperature of the fluid in which it is inserted. The indicated temperature is established as a result of heat-flow equilibrium of convective-heat transfer between the sensing element and the fluid on one side and heat flow through conduction and radiation between the element and its surroundings on the other side. This applies closely to fluids at rest or to fluids moving with low velocities. The conditions are more complex for fluids moving at higher velocities (corresponding to a Mach number greater than 0.3), in which the aerodynamic heating effect plays a greater part in heat balance.

Conduction Error   Conduction error, or immersion error, is caused by temperature gradients between the sensing element and the measuring junction. This error can be minimized by high heat convection between fluid and sensor and low heat conduction between sensor and measuring junction. In the thermocouple this would mean a small diameter, low conductivity, and long immersion length of the wires.

Radiation Error   When the sensing element (other than radiation thermometer) is placed so that it can "see" surfaces at a much lower temperature (a sink) or at a much higher temperature (a source), a radiant-heat interchange will result between the two, causing the sensing element to read an erroneous temperature.

Radiation error may be largely eliminated through the proper use of thermal shielding. This consists in placing barriers to thermal radiation around the probe, which prevent the probe from seeing the radiant source or sink, as the case may be. For low-temperature work, such shields may simply be made of sheet metal appropriately formed to provide the necessary protection. At higher temperatures, metal or ceramic sleeves or tubes may be employed. In applications in which gas temperatures are desired, care must be exercised so as not to cause stagnation of flow around the probe.

Measurement of Temperature in a Rapidly Moving Gas   When a probe is placed in a stream of gas, the flow will be partially stopped by the presence of the probe. The lost kinetic energy will be converted to heat, which will have some bearing on the indicated temperature. Two "ideal" states may be defined for such a condition. A true state would be that observed by instruments moving with the stream, and a stagnation state would be that obtained if the gas were brought to rest and its kinetic energy completely converted to heat, resulting in a temperature rise. A fixed probe inserted into the moving stream will indicate conditions lying between the two states.

An expression relating stagnation and true temperatures for a moving gas, with adiabatic conditions assumed, may be written as follows:

$$t_s - t_t = \frac{V^2}{2g_c J c_p} \tag{13-7}$$

This relation may also be written as

$$\frac{t_s}{t_t} = 1 + \frac{(k-1)M^2}{2} \tag{13-8}$$

where t_s = stagnation or total temperature, °F
t_t = true or static temperature, °F
V = velocity of flow, ft/s
g_c = gravitational constant, 32.2 ft/s²
J = mechanical equivalent of heat, ft·lb/Btu
c_p = mean specific heat at constant pressure, Btu/lb·°F
k = ratio of specific heats
M = Mach number

A measure of effectiveness of the probe in bringing about kinetic-energy conversion may be expressed by the relation

$$r = \frac{t_i - t_t}{t_s - t_t} \tag{13-9}$$

where t_i = temperature indicated by the probe, °F
r = recovery factor

If $r = 1$, the probe would measure the stagnation temperature, and if $r = 0$, it would measure the true temperature.

By combining Eqs. (13-7) and (13-9), the following relationships are obtained:

$$t_t = t_i - \frac{rV^2}{2g_c J c_p} \tag{13-10}$$

or

$$t_s = t_i + \frac{(1-r)V^2}{2g_c J c_p} \tag{13-11}$$

The value of the recovery factor r depends on the type and design of the temperature-measuring probe; it can be anywhere between 0 and 1.0. Often it is specified by the manufacturer for the specific designs of the temperature probes, or it should be determined experimentally. The difference between stagnation and static temperature increases rapidly as the flow Mach number increases. It is important therefore to know the value of the recovery factor in order to get as accurate as possible evaluation of the temperatures of the moving gas.

BIBLIOGRAPHY

ASME Power Test Codes, part 3, PTC 19.3-1974, *Temperature Measurement.*
Benedict, R. P.: *Fundamentals of Temperature, Pressure, and Flow Measurements,* 2d ed.,
   John Wiley & Sons, Inc., New York, 1977.

PRESSURE MEASUREMENT

General Principles and Definitions

1. *Pressure* is defined as the force per unit area exerted by a fluid on a containing wall.
2. *Pressure relationships* (Fig 13-9).

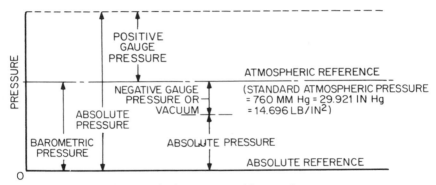

FIG. 13-9   Relations between absolute, gauge, and barometric pressure.

Differential pressure is the difference between any two pressures.

Absolute pressure is the force per unit area exerted by a fluid on a containing wall.

Gauge pressure is the difference between absolute pressure and ambient-atmospheric pressure.

Vacuum pressure is negative gauge pressure.

3. *Flow-stream pressures.*

Static pressure is pressure measured perpendicularly to the direction of flow. This is the pressure that one would sense when moving downstream with the fluid.

Total pressure is pressure in the direction of flow, where pressure as a function of direction is at a maximum. Total pressure would be sensed if the stream were brought to rest isentropically.

Velocity pressure is the difference between static and total pressure measured at a specific region in the direction of flow. It is called *velocity head* when measured in height of fluid. Velocity pressure is equal to $\frac{1}{2}\rho V^2$, where ρ is the fluid density and V is the fluid velocity.

Pressure Connections

1. *Sources of errors.*

Flow errors—Leakage errors can be eliminated by proper sealing of connections. Errors due to friction, inertia, and lag errors in the gauge piping, encountered in dynamic flow, can be minimized by using short large-diameter connecting tubes.

Turbulence errors—The static-pressure tap on the wall parallel to the flow should not be too large in order to prevent a disturbance in the flow that would cause an inaccurate static reading; the tap, however, should be large enough to give a proper response. The area surrounding the pressure tap should be smooth to ensure that a burr or other obstruction will not affect the reading. The edge of the hole should be sharp and square. When the pressure is fluctuating, a damping device can be used to improve readability, although the accurate method would be to use a suitable recording instrument and determine the average pressure over a period of time.

2. *Static taps.* Static taps (Fig. 13-10*a*) should be at least 5 diameters downstream from symmetrical pipe fittings and 10 diameters downstream from unsymmetrical fittings, according to the *ASME Power Test Code.* When possible, a Weldolet or pipe coupling should be welded to the outside of the pipe and the hole then drilled through to the main pipe. Since the error increases with velocity pressures, care must be taken in high-velocity areas to ensure sharp, square holes that are as small as possible (down to $\frac{1}{16}$ in) to keep disturbance and error to a minimum. In low-velocity areas, larger holes should be used to improve dynamic-pressure response and prevent clogging. When flow is nonuniform, several taps should be used along the periphery of the pipe.

3. *Static tubes.* Static tubes (Fig. 13-10*b*) are used for measurement of static pressure in a free stream as on a moving plane. Static taps in the wall are preferable, since static tubes disturb the flow, making calibration necessary for accurate measurement. Unless one expects a static-pressure distribution, wall taps should be used.

4. *Impact tubes or pitot tubes.* An impact tube (Fig. 13-10*c*) faces directly into the flow, giving a total-pressure reading. Velocity pressure is determined by taking a static reading, preferably along a wall, and taking the difference; impact tubes can be used to get a velocity profile by traversing. Maximum-velocity direction can be determined by rotating the tube.

5. *Piping arrangement.* Connecting piping (Fig. 13-10*f*) should be arranged to avoid liquid pockets in gas-filled lines and air pockets in liquid-filled lines. This is accomplished by having gas-filled lines sloping up to the measuring instrument and liquid-filled lines sloping downward to the instrument. Both types should have vents close to the instrument to bleed lines. More vents might be needed if lines must have dips or twists in them. For vacuum pressures an air bleed allowing very small flow should be provided near the instrument to keep lines purged of condensate, etc., between readings. When using a manometer, this can be accomplished by a valve, or a very small hole can be drilled near the top of the manometer, which would be closed or covered when taking a reading. For mechanical or electrical transducers as measuring instruments, the same procedures hold true,

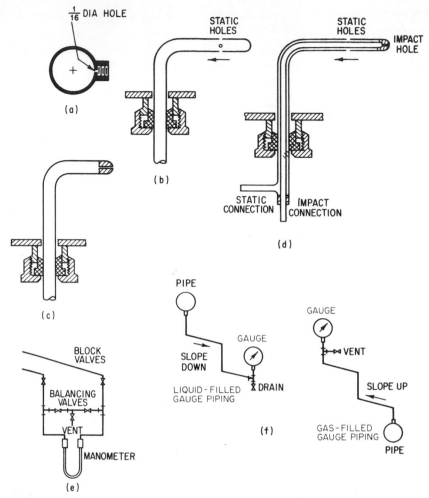

FIG. 13-10   (*a*) Static-pressure connection. (*b*) Static tube. (*c*) Impact tube. (*d*) Combination pitot-static tube. (*e*) Cross connection. (*f*) Typical pressure-gauge piping arrangements.

but in hot-steam lines it might be necessary to loop the line and fill with water up close to the instrument to protect the instrument from the high temperature. For differential measurement the arrangement shown in Fig. 13-10*e* should be used to prevent and detect leakage.

6. *Calibration and error analysis.* Pressure measurements are referred to primary standards of pressure which can be calibrated in terms of mass, length, and time. All pressure-measuring devices have an associated error that must be considered in making a pressure measurement. In a field environment of noise, vibration, moisture, temperature fluctuation, pressure fluctuation, pressure-tap geom-

etry, connecting tubing, etc., other errors or uncertainties must be considered in evaluating the pressure measurement.

Liquid-Level Gauge

1. A manometer measures pressure by balancing it against a column of liquid with a known density and height. Selection of the liquid depends on test conditions; however, the liquid always must be denser than the flowing fluid and immiscible with it. Other factors to consider are the specific gravity, the useful temperature range, the flash point, the viscosity, and the vapor pressure. The basic manometer liquids used are water (specific gravity, 1), mercury (specific gravity, 13.57), red oil (Meriam; specific gravity, 0.827), tetrabromoethylene (specific gravity, 2.95), and carbon tetrachloride (specific gravity, 1.595). Special fluids are also available with specific gravities of 1.20 and 1.75.[1] The manometer fluid used must be kept pure to ensure that the specific gravity remains constant.

2. The basic types of manometers are U-tube and cistern (Fig. 13-11).

In the U-tube manometer the pressure on one leg balances the pressure in the other leg. By performing a fluid balance and knowing the density of all fluids and their height, one can calculate the pounds per square inch difference between the two. Often the second leg is open to atmospheric pressure so that the pressure difference represents gauge pressure and must be added to barometric pressure to find the total pressure. When the second leg is connected to a pressure other than atmospheric, it is called a *differential pressure* and represents the direct difference between the two pressures.

In the well- or cistern-type manometer one leg has a cross section much larger than that of the other leg. The zero adjustment in the cistern is usually made manually with an adjusting screw. Then the pressure is found by the following formula

$$P = P_a + Z_1 g \rho_2 - Z_3 g \rho_1$$

where g = acceleration due to gravity
ρ = density of liquid

Special types of manometers sometimes used for more accurate measurement include the inclined manometer, the micromanometer, and U tubes installed with hook gauges, as well as many special types of manometers for vacuum measurement, which will be mentioned later.

3. An inclined manometer (Fig. 13-12) is a manometer inclined at an angle with the vertical. Although the vertical displacement is still the same, the movement of liquid along the tube is greater in proportion to the secant of the angle. The common form of inclined manometer is made with a cistern, as shown in Fig. 13-12.

The scale can be graduated to take account of the liquid density, inclination, and cistern-level shift so that readings will be in convenient pressure units such as equivalent vertical inches or centimeters of water. A spirit level and leveling screws are usually provided so that the designed angle can be reproduced in installation.

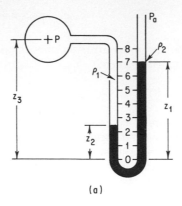

(a)

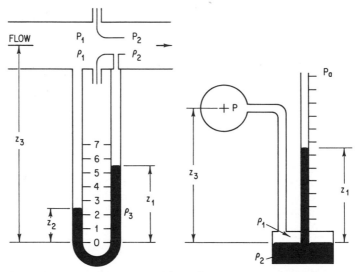

FIG. 13-11   Manometer types. (a) U-tube manometer, open to the atmosphere. (b) Differential U-tube manometer. (c) Cistern manometer.

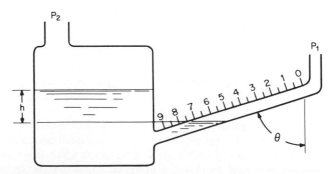

FIG. 13-12   Inclined manometer.

This form of manometer is useful for gas pressures, as for draft gauges. The graduation intervals are commonly 0.01 in of water (0.25 mm of water) with spans up to about 10 in (25 cm).

4. Barometers are a special case of manometers to measure atmospheric pressure. A primary barometer would be a U tube with one end open to the atmosphere and the other end connected to a continuously operating vacuum pump.

In many cases a Fortin-type barometer (Fig. 13-13) is suitable. In this case the mercury in the well is exposed to the atmosphere with the other end evacuated and sealed. All barometer readings should be corrected for temperature, local gravity, and capillary effect. Atmospheric pressure can also be measured by an aneroid barometer, which is a special type of elastic gauge. It is sometimes used in place of a monometer-type barometer because of the ease of transportation.

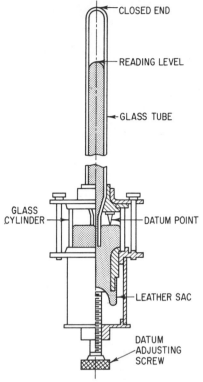

FIG. 13-13   Fortin barometer.

Deadweight Tester and Gauges

1. *Principle, design, and operation testers.* Deadweight testers are the most common instrument for calibrating elastic gauges with pressures in the range from 15 to about 10,000 lb/in^2 or higher. Deadweight testers (Fig. 13-14) have a piston riding in a cylinder with a close clearance. The total weight on the piston including that of the platform and the piston itself and any additional weights, divided by the cross-section area of the piston (which is usually an even fraction of an inch such as ⅛ in^2), determine the pressure on the gauge being tested. The piston must be in a vertical position and spinning freely when the measurement is being taken. The inertia created by spinning minimizes the viscous drag on the piston by spreading oil around the diameter. Maximum error is usually 0.1 percent of the pressure measured.

To operate, put the desired weight on the piston, close the pressure-release valve, and pressurize the tester's fluid with the displacer pump or screw-type ram until the weights are lifted and the piston is floating. Then slowly spin the piston, and take the gauge reading and compare it with the equivalent pressure created by the piston and weights. The gauge reading must then be corrected accordingly.

Special testers include high-pressure, low-pressure, and lever types. For very high pressure (above 10,000 lb/in^2) it is necessary to use a tester that makes adjustments to minimize the leakage and to correct for deformation of the piston

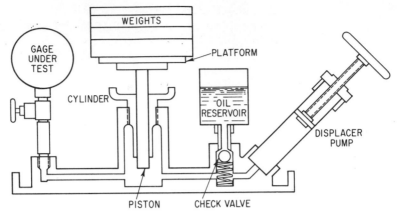

FIG. 13-14   Deadweight tester.

and cylinder. Low-pressure testers (0.3 up to 50 lb/in^2 are covered) use air as the working fluid for a more accurate measurement. Lever-type testers use a force-amplifying linkage to apply weight to the piston with an inertial wheel on a motor to keep the piston spinning freely.

2. *Deadweight gauges*. Deadweight gauges are mainly used to measure a relatively stable pressure so that it can be maintained. These gauges give very precise measurements but are not practical for a test with a wide range of pressures since many weight changes would be necessary.

3. *Corrections*. Corrections include those necessary for local gravity, weight measurement, effective area, head, and buoyancy adjustments. The head correction is usually the only one necessary when accuracy of ¼ percent is satisfactory.

Elastic Gauges

1. In elastic gauges, an elastic member is caused to stretch or move by a given pressure. The movement is amplified through a linkage and usually is employed to rotate a pointer indicating the pressure reading in relation to atmospheric pressure.

2. Bourdon gauges (Fig. 13-15) contain a hollow tube curved in an arc that tends to straighten as internal pressure is applied, moving the linkage and pointer to indicate the pressure reading. Differential as well as compound, vacuum, and straight-pressure Bourdon gauges are available. Differential-pressure gauges have either the Bourdon tube enclosed in a seal-pressurized case or two Bourdon gauges, one subtracting from the other. Ranges go from 0 to 15 psig up to 0 to 100,000 psig as well as the vacuum range.

3. Bellows gauges (Fig. 13-16) have a bellows or elastic chamber expanding to actuate the guage. They are usually used in low-pressure applications with a maximum reading of about 50 psig.

4. Diaphragm gauges (Fig. 13-17) use a flexible diaphragm as the inducer. This type is suitable for ranges from 0 to 1 inHg up to 200 lb/in^2. Variations of this gauge are valuable in special cases in which the process fluid must be kept

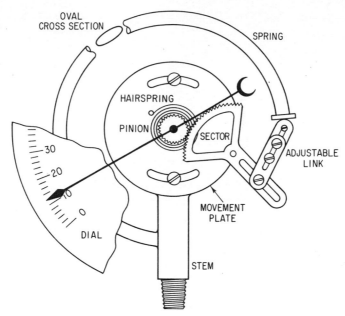

FIG. 13-15   Bourdon gauge.

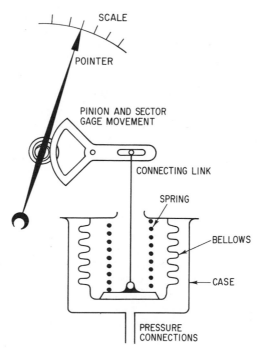

FIG. 13-16   Bellows gauge.

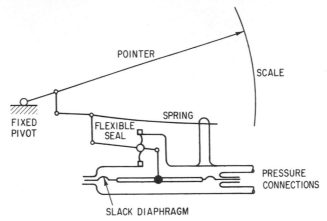

FIG. 13-17 Slack-diaphragm gauge.

separate from the gauge, as when the fluid is very hot (up to 1500°F with special modifications) or corrosive, or when the fluid would tend to clog other gauges.

5. All elastic gauges must be calibrated continually to ensure accuracy. Accuracy to 0.5 percent or better of full scale can be obtained.

6. Gauges must be bled for assurance that neither air nor water bubbles are present in the lines. To obtain a gauge reading, first make sure that the linkage is free. This is done with a light tap to the gauge. When damping the gauge needle by closing down on the inlet line, the needle is left fluctuating slightly to indicate that the line is still open.

Special Measuring Devices

For low-pressure measurement the McLeod gauge (Fig. 13-18) is a primary measuring device. The calibration depends only upon dimensional measurements. Other direct-reading pressure gauges, for low pressure, are the mercury micromanometer, the Hickman butyl phthalate manometer, and the consolidated diaphragm comparator. Gauges measuring properties directly convertible to pressure are the thermal-conductivity gauges (thermocouple gauge, Pirani gauge), ionization gauges (Philips-Penning gauge, alphatron gauge), and the molecular-vacuum gauge.

Electric Transducers Devices that convert a pressure into a mechanical analog of that pressure, such as a manometer which exhibits a difference in the height of a liquid column, were discussed in the preceding subsections. Practical reasons make it difficult to transmit these mechanical signals over large distances, but modern control systems require this capability. Transmitting information over great distances is easily accomplished by electronic instrumentation. Transmission of signals representing measured pressure can be accomplished by varying an electric current through wires to the remote location.

The device used to obtain an electronic signal that is related to a pressure is an electric transducer. An electric transducer consists of the following:

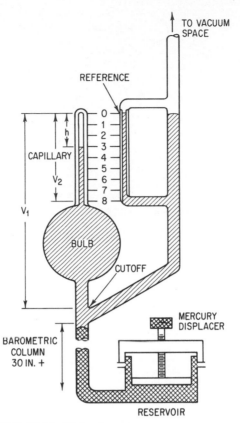

FIG. 13-18   McLeod gauge.

1. *Sensing element.* A device which receives a pressure signal and converts it to a force or displacement.

2. *Transmitter.* A device which contains a sensing element, detects the force or displacement in the sensing element, and sends an electric signal (related to the force or displacement) to a receiver.

Implicit in the use of an electric transducer is a receiver that detects the electric signal and indicates the pressure.

Many of these transducers measure the resistance change of a wire or strain gauge deformed by pressure. These instruments can be calibrated to measure pressure directly. There are two basic ways of mounting these gauges. With bonded strain gauges

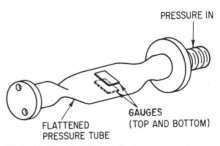

FIG. 13-19   Bonded strain gauge.

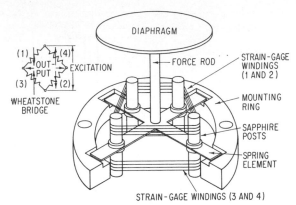

FIG. 13-20 Unbonded strain gauge.

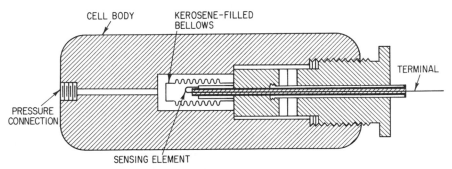

FIG. 13-21 Section through a bulk-modulus pressure gauge.

(Fig. 13-19) they are usually mounted on a diaphragm or tube that will deform as pressure is applied, changing the resistance of the gauge. With unbonded strain gauges (Fig. 13-20) a thin wire is usually wrapped around a sensing element that deforms and stretches the wire, changing its resistance, which can also be converted to pressure. Another special-type resistance gauge for high-pressure reading is the bulk-modulus pressure gauge (Fig. 13-21), which uses direct pressures on a loosely wound coil of fine wire to get a resistance change. The sensing mechanism is separated from the process fluid by a bellows. Other specialized types use differences in inductance or capacitance and correlate them to pressure.

REFERENCES

1. *Meriam Standard Indicating Fluid Bulletin,* IM-11.

BIBLIOGRAPHY

ASME Power Test Codes: Instrument and Apparatus, part 2, *Pressure Measurement,* American Society of Mechanical Engineers, New York, 1964.

Beckwith, T. G., and N. Lewis Buck: *Mechanical Measurements,* Addison-Wesley Publishing Company, Inc., Reading, Mass., 1961.

Benedict, R. P.: *Fundamentals of Temperature, Pressure, and Flow Measurements,* 2d ed., John Wiley & Sons, Inc., New York, 1977.

Consolidated Electrodynamics, *Transducer,* General Catalog Bulletin 1322, December 1965.

FLOW MEASUREMENT

General

The three most extensively used types of flow-metering devices are the thin-plate square-edged orifice, the flow nozzle, and the venturi tube. They are differential-head instruments and require secondary elements for measurement of the differential pressure produced by the primary element. The Supplement to *ASME Power Test Codes: Instruments and Apparatus,*[1] describes construction of the above primary flow-measuring elements and their installation as well as installation of the secondary elements. The method of flow measurement, the equations for flow computation, and the limitations and accuracy of measurements are discussed. Diagrams and tables showing the necessary flow coefficients as a function of Reynolds number and diameter ratio β are included in the standards. Diagrams of the expansion factor for compressible fluids are given.

Some characteristic features of various types of primary elements are listed in the following:

Orifice. Simple, inexpensive, well-established coefficient of discharge, high head loss, low capacity for given pipe size, danger of suspended-matter accumulation; requires careful installation of pressure connections.

Flow nozzle. High capacity, more expensive, loss comparable with that of the orifice; requires careful installation of pressure connections.

Venturi tube. High capacity, low head loss, most expensive, greater weight and size; has integral pressure connection.

Nomenclature

a = throat area of primary element, in^2
C = coefficient of discharge
d = throat diameter of primary element, in
D = pipe diameter, in
E = $1/\sqrt{1 - \beta^4}$, velocity-of-approach factor
F_a = thermal-expansion factor
h = manometer differential pressure, in
h_w = manometer differential pressure, in H_2O at 68°F
k = ratio of specific heats
K = CE = combined-flow coefficient for orifices, velocity-of-approach factor included
n = numerical factor dependent upon units used
q = capacity of flow, gal/min
Q_i = capacity of flow, ft^3/min, at conditions i

R_d =   Reynolds number based on d

r = $\dfrac{P_2}{P_1}$ = pressure ratio across flow nozzle, where P_1 and P_2 are absolute pressures

M =   rate of flow, lb/s

M_h =   rate of flow, lb/h

M_m =   rate of flow, lb/min

Y =   net-expansion factor for square-edged orifices

Y_a =   adiabatic-expansion factor for flow nozzles and venturi tubes

β = $\dfrac{d}{D}$ = diameter ratio

ρ =   specific weight of flowing fluid at inlet side of primary element, lb/ft^3

ρ_i =   specific weight of flowing fluid at conditions i

Primary-Element Construction and Installation

The primary element may be installed within a continuous section of pipe flowing full or at the inlet or exit of a pipe or a plenum chamber. Orifice and venturi tube are installed within the pipe in a closed-loop test. The flow nozzle may be installed within, at inlet, or at outlet of the pipe.

It is normal practice to use a venturi tube installed within a continuous section of pipe in pump-acceptance tests and a flow nozzle at the exit of the discharge pipe in compressor-acceptance tests. More closed-loop testing has recently been required in compressor testing. Industry normally uses the nozzle configuration shown in Fig. 13-26 with a closed loop. The construction of the primary elements and examples of their installation are given in the following paragraphs.

Orifice   (Fig. 13-22)   The recommended diameter ratio $\beta = d/D$ is from 0.20 to 0.75. The thickness of the orifice plate shall be not less than shown in Table 13-5.

Three types of pressure connections may be used: vena contracta taps, 1-D and ½-D taps, and flange taps. Appropriate discharge coefficients have to be used in each case.

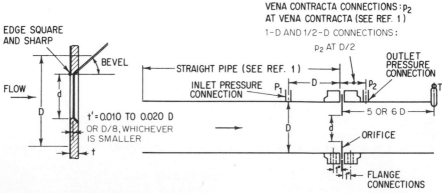

FIG. 13-22   Orifice construction and installation.

TABLE 13-5   Minimum Recommended Thicknesses of Orifice Plates, in

Differential pressure, in H_2O	Internal diameter of pipe, in				
	3 and less	6	10	20	30
$\beta < 0.5$					
< 1000	⅛	⅛	3/16	⅜	½*
< 200	⅛	⅛	⅛	¼	⅜
< 100	⅛	⅛	⅛	¼	⅜
$\beta > 0.5$					
< 1000	⅛	⅛	3/16	⅜	½
< 200	⅛	⅛	⅛	3/16	⅜
< 100	⅛	⅛	⅛	3/16	¼

*For ½-in plate in 30-in line, maximum differential = 500 in.

Flow Nozzle   The flow nozzle should be of either the low- or the high-ratio long-radius type as shown in Fig. 13-23. The diameter ratio β should be from 0.15 to 0.80, although *ASME Power Test Codes: Compressors and Exhausters* (page 22)[2] recommends the β range between 0.40 and 0.60.

Different test arrangements for the measurement of subcritical flow with the flow nozzle are shown in Figs. 13-24 to 13-26.

Venturi Tube   The proportions of the standard-form (Herschel-type) venturi tube and its installation are shown in Fig. 13-27. The diameter ratio $\beta = d/D$ should be between 0.4 and 0.75 for the best results. Special forms (venturi-nozzle tube for high-pressure feedwater application, venturi-insert nozzle) may be used. They need individual calibration, however, while graphs may be used for the standard form.

Calculation of Flow Rates

The basic relationships for flow computation were obtained from PTC 19.5, 4-1959, Supplement to *ASME Power Test Codes,* "Flow Measurement of Quantity of Materials."[3] The nomenclature in these relationships was, however, adapted to the more recent terminology of Interim Supplement 19.5, *Fluid Meters.*

Incompressible Fluids.   The flow of any liquid through an orifice, flow nozzle, or venturi tube is determined by the following equation:

$$M = \frac{CaE_a n\rho}{\sqrt{1 - \beta^4}} \sqrt{2gh} \qquad (13\text{-}12)$$

With the units in American practice, Eq. (13-12) is written

$$M_h = 359CEd^2F_a \sqrt{h_w\rho} \qquad (13\text{-}13)$$

Compressible Fluids   To compensate for the change in specific weight as the fluid passes through the primary element, the equation must be modified by the expansion factor Y for orifices or Y_a for nozzles.

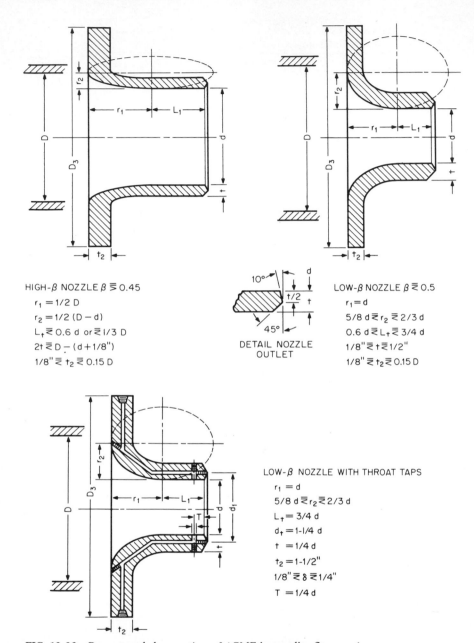

HIGH-β NOZZLE $\beta \lessgtr 0.45$

$r_1 = 1/2\,D$

$r_2 = 1/2\,(D - d)$

$L_t \gtrless 0.6\,d$ or $\gtrless 1/3\,D$

$2t \gtrless D - (d + 1/8'')$

$1/8'' \lessgtr t_2 \lessgtr 0.15\,D$

DETAIL NOZZLE
OUTLET

LOW-β NOZZLE $\beta \gtrless 0.5$

$r_1 = d$

$5/8\,d \lessgtr r_2 \lessgtr 2/3\,d$

$0.6\,d \lessgtr L_t \lessgtr 3/4\,d$

$1/8'' \lessgtr t \lessgtr 1/2''$

$1/8'' \lessgtr t_2 \lessgtr 0.15\,D$

LOW-β NOZZLE WITH THROAT TAPS

$r_1 = d$

$5/8\,d \lessgtr r_2 \lessgtr 2/3\,d$

$L_t = 3/4\,d$

$d_t = 1\text{-}1/4\,d$

$t = 1/4\,d$

$t_2 = 1\text{-}1/2''$

$1/8'' \lessgtr \delta \lessgtr 1/4''$

$T = 1/4\,d$

FIG. 13-23 Recommended proportions of ASME long-radius flow nozzles.

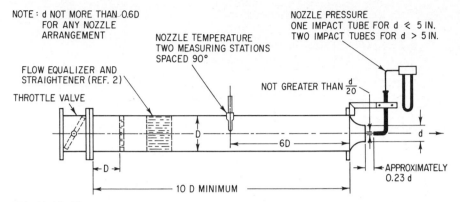

FIG. 13-24 Test arrangement *A*.

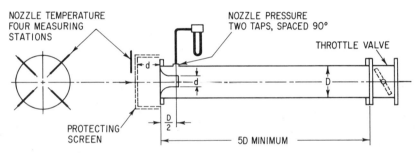

FIG. 13-25 Test arrangement *B*.

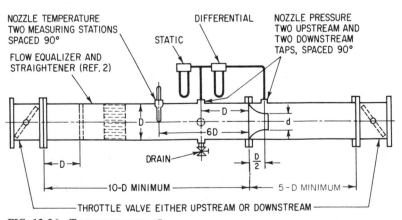

FIG. 13-26 Test arrangement *C*.

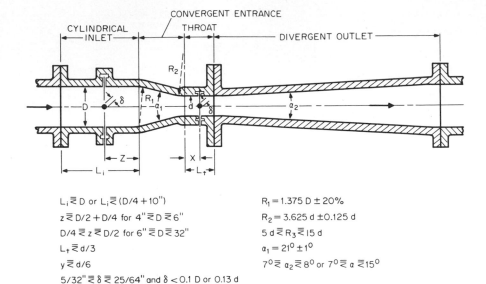

$L_i \gtreqless D$ or $L_i \gtreqless (D/4 + 10")$

$z \gtreqless D/2 + D/4$ for $4" \gtreqless D \gtreqless 6"$

$D/4 \gtreqless z \gtreqless D/2$ for $6" \gtreqless D \gtreqless 32"$

$L_t \gtreqless d/3$

$y \gtreqless d/6$

$5/32" \gtreqless \delta \gtreqless 25/64"$ and $\delta < 0.1\, D$ or $0.13\, d$

$R_1 = 1.375\, D \pm 20\%$

$R_2 = 3.625\, d \pm 0.125\, d$

$5\, d \gtreqless R_3 \gtreqless 15\, d$

$a_1 = 21° \pm 1°$

$7° \gtreqless a_2 \gtreqless 8°$ or $7° \gtreqless a \gtreqless 15°$

FIG. 13-27 Dimensional proportions of classical (Herschel) venturi tubes with a rough-cast convergent-inlet cone.

Orifices

$$M_h = 359CEd^2F_aY \sqrt{h_w\rho} \tag{13-14}$$

Flow nozzles and venturi tubes

$$M_h = 359CEd^2F_aY_a \sqrt{h_w\rho} \tag{13-15}$$

Quite often the flow coefficient $K = CE$ is specified for use in Eqs. (13-13) and (13-14) for orifices.

Delaval Applications   With particular reference to the methods used at Delaval, the above equations assume the forms shown below.

Venturi tube. For measurement of the flow rate of water at 68°F with a *venturi tube* with mercury used as the manometer liquid and a value of 0.984 used for the flow coefficient, Eq. (13-13) becomes

$$q = 19.79 \frac{d^2}{\sqrt{1 - \beta^4}} \sqrt{h_0} \tag{13-16}$$

where h_0 = monometer pressure, inHg at 68°F

If water at a temperature other than 68°F is metered, a temperature correction should be made.

The coefficient of 0.984 is applicable for β values from 0.3- to 0.75-in pipes of 4 up to 32 in, provided the Reynolds number is greater than 200,000 (page 232).[1] The *Hydraulic Institute Standards*[4] require a minimum velocity of 20 ft/s in the throat for the rated flow of the pump.

Flow nozzle. For measurement of airflow with *flow-nozzle* arrangement A (see Fig. 13-24),

$$M_m = 5.983 CEd^2 Y_a \sqrt{h_w \rho} \qquad (13\text{-}17)$$

and

$$Q_i = \frac{M_m}{\rho_i} \qquad (13\text{-}18)$$

where $\quad Y_a = r^{1/k} \left[\dfrac{k}{k-1} \dfrac{1 - r^{(k-1)/k}}{1-r} \right]^{1/2} = $ adiabatic-expansion factor   (13-19)

The formula applies to subcritical flow with pressure ratio P_2/P_1 preferably greater than 0.7.

Unless the nozzle in this arrangement has been individually calibrated, the value of the flow coefficient should be taken as 0.99, provided that the throat Reynolds number exceeds 200,000. For nozzles within the pipe and using pipe taps, the coefficient of discharge for pipe Reynolds numbers above 100,000 varies from 0.980 to 0.993, depending on β. Exact coefficients can also be obtained from Tables II–III-5 in Supplement 19.5 on *Fluid Meters*.[1]

For a critical flow measurement, an arrangement similar to that of Fig. 13-24 with static-pressure taps 1 D upstream from the pipe exit but without an impact tube is used. Rules of the *ASME Power Test Codes: Compressors and Exhausters*[2] should be used for flow computation.

Table 13-6 gives approximate flow rates for different nozzle sizes. It is useful for air tests with flow-nozzle arrangements A and B, when using a long-radius low-β-series nozzle (see Fig. 13-23).

Orifice. Equation (13-13) is used for measuring the flow rate of water with orifice and Eq. (13-14) for measuring the rate of flow of air. Interim Supplement 19.5 on *Fluid Meters* gives the tables of the values of discharge coefficient C for

TABLE 13-6   Long-Radius Low-Ratio Nozzle

d	Approximate flow rates, ft³/min	
	10 in H₂O	40 in H₂O
2.000	250	500
2.500	400	800
3.000	575	1,150
4.000	1,000	2,000
5.000	1,600	3,200
6.000	2,250	4,500
8.000	4,000	8,000
10.000	6,250	12,500
12.000	9,000	18,000
18.000	20,000	40,000
24.000	36,000	72,000

SOURCE: *ASME Power Test Codes*, PTC 10-1965, reaffirmed 1973, *Compressors and Exhausters*.

TABLE 13-7   Values of Discharge Coefficient C for 8-In Pipe (Velocity-of-Approach Factor Not Included)

β \ R_d	Reynolds Number					
	14,000	25,000	50,000	100,000	500,000	1,000,000
	Flange taps					
0.250	0.6190	0.6091	0.6028	0.5997	0.5971	0.5968
0.500	0.6443	0.6259	0.6142	0.6084	0.6037	0.6032
0.750	. . . . . .	0.6749	0.6351	0.6153	0.5993	0.5973
	Taps at 1 D and ½ D					
0.250	0.6036	0.6012	0.5990	0.5975	0.5955	0.5950
0.500	0.6150	0.6112	0.6079	0.6055	0.6024	0.6017
0.750	. . . . . .	0.6337	0.6254	0.6196	0.6118	0.6099
	Vena contracta taps					
0.250	0.6071	0.6035	0.6003	0.5981	0.5952	0.5945
0.500	0.6185	0.6136	0.6094	0.6064	0.6024	0.6014
0.750	. . . . . .	0.6318	0.6240	0.6185	0.6112	0.6095

various pipe diameters as a function of diameter ratio β and pipe Reynolds number R_d based on orifice diameter d. Excerpts from the tables in the Supplement are given in Table 13-7 for the three types of pressure-tap locations. The values below the stepped line are extrapolations and are subject to larger tolerance as given in Table II-V-1 in the Supplement.[1]

REFERENCES

1. *Fluid Meters*, part II, *Application*, Interim Supplement 19.5, *ASME Power Test Codes: Instruments and Apparatus*, 6th ed., 1971, pp. 179–256.

2. *ASME Power Test Codes*, PTC 10-1965, reaffirmed 1973, *Compressors and Exhausters*.

3. Supplement to *ASME Power Test Codes: Instruments and Apparatus*, PTC 19.5, 4-1959, chap. 4, "Flow Measurement of Quantity of Materials."

4. *Hydraulic Institute Standards*, 13th ed., 1975, pp. 64–68.

Switches

TEMPERATURE SWITCHES

Temperature switches are used to monitor or control the temperature of various fluids. Their broadest use is in the range from $-65°$ to $+600°$F.

A temperature switch consists primarily of a temperature sensor, an electric switch, and a means of causing the switch to be operated by the sensor's reaction to temperature changes. In the switches commonly used in industrial applications the electric switch is mechanically actuated. The sensors are of four types: bimetallic elements, gas-filled bulbs, liquid-vapor-pressure bulbs, and liquid-filled bulbs.

Proper application of temperature switches requires knowledge of the proposed service. This knowledge should include the type of fluid to be sensed, the desired set-point temperature, the extremes of temperature excursions, the need for manual reset, and the maximum pressure level of the fluid. The extremes of ambient temperature and the vibration level at the sensor location should be known. Desired service factors such as accuracy, reliability, life, and response time should also be determined.

The characteristics of the various sensors should be known (see Table 14-1).

Bimetallic-Type Sensor

The bimetallic-type sensor operates on the differential expansion of two dissimilar metals that are bonded together. A common type of unit incorporates the bimetallic element wound in a helical configuration. One end is fixed so that, when heated, the element expands, causing a rotary motion of the free end. This movement actuates the switch element. One material is usually Invar because of its low coefficient of thermal expansion; the other is usually brass for low temperatures or nickel for higher temperatures. A variation of this type is a strut-and-tube thermostat composed of an outer shell of high-expanding metal and a strut assembly of low-expanding metal.

The application range is generally from $-100°$ to $+1000°F$. The bimetallic sensor has the advantage of fast response time and is relatively low-cost and easily adjusted in the field. It can only be local-mounted, has an accuracy of 2 to 5 percent, can be damaged by overheating, and has low repeatability.

Units based on this principle are used in wall thermostats and other space-heating applications. This control generally has wider *actuation values* than liquid-filled systems. Actuation value is the difference between the temperature at which a switch actuates and the temperature at which it releases upon temperature reversal.

Vapor-Pressure-Type Sensors

Vapor-pressure sensors contain a volatile liquid with a vapor pressure that increases with increasing temperature. The increase is independent of the volume of the liquid as long as liquid remains in the sensor to be vaporized.

This type is designed in both local-mount and bulb-and-capillary models. Its general application range is from $-60°$ to $+100°F$. An advantage is its insensitivity to ambient-temperature changes, and the sensor also has a fairly fast response time. But it has limited ranges, a low safety factor in high-proof temperatures, and larger bulb sizes to perform its function.

Pressure rises very rapidly in the higher portion of the temperature range. Thus, the rate of this sensor is not linear, a factor that severely limits the full range span and proof temperatures. Several types of fluids must be used to cover the full range of temperatures.

It is advisable to use a vapor-pressure sensor only when the ambient temperature is higher than the process-fluid temperature. Then the fluid in the capillary and bellows remains in the vapor state.

TABLE 14-1 Most Common Types of Commercially Available Temperature Switches

	Bimetallic	Liquid-filled	Vapor-pressure	Gas-filled	Thermocouples	Electroresistive
Range, °F	−100 to 1000	−150 to 1200	−100 to 550	−100 to 1500	−400 to 4500	−300 to 1800
Advantages	Inexpensive, easy to adjust in field	Moderate price, low on-off differential, fast response, small bulb size	Insensitive to ambient-temperature changes	Insensitive to ambient-temperature changes	Wide temperature, rapid response	High sensitivity
Limitations	Damaged by overheating, low repeatability	Ambient compensation required	Relatively expensive, large bulb required, slow response	Relatively expensive, large bulb required, slow response	Expensive accessories required	Expensive accessories required

14-3

Gas-Actuated-Type Sensors

The gas sensor is basically the same as the vapor-pressure type except it does not use a volatile liquid. The system is filled with inert gas, usually nitrogen. The gas pressure increases with temperature. Such systems are used in the range of $-100°$ to $+1000°F$.

A great advantage of the gas sensor is the long capillary lines which can be used. Although there is an ambient-temperature effect on these lines, it can be minimized by increasing the bulb volume. Gas-filled sensors are not widely used in industry because of their inherent slow response time, large bulb size, and relatively high cost.

Liquid-Filled Sensors: Bulb-and-Capillary Type

This type of sensor is shown in Fig. 14-1. It comprises a bellows and a capillary and bulb assembled to produce a perfectly hermetically sealed system. The assembly is evacuated in a vacuum chamber. The filling fluid, while under vacuum, is allowed to flow into the bellows and the capillary and bulb. The bulb is heated to a specific calibration temperature for the particular sensor. The fill hole is then sealed.

The basic system requirement is linear movement of the bellows with a change of bulb temperature. To achieve this movement the fluid must have a relatively constant coefficient of thermal expansion over the range of sensing.

The relationship of bulb size, bellows size, and coefficient of thermal expansion is important and must be closely controlled to obtain the desired characteristics. The bulb size changes slightly with temperature change. With proper selection of the fill liquid, the coefficient of thermal expansion changes with temperature so that these two variables are compensating and the linear characteristics are maintained.

The effect of ambient-temperature changes on the fill-fluid volume in the capillary and bellows is a set-point decrease on an ambient-temperature increase and the reverse on a temperature decrease. Compensating bimetallic washers are used to offset this effect (see Fig. 14-1). The total fluid change for the temperature range is absorbed by the bellows. A plunger transmits the bellows movement to the snap-action switch. The plunger spring provides overtravel to prevent damage to the switch.

Liquid-Filled Sensors: Local-Mount Type

In the local-mount, outside-fill type of sensor shown in Fig. 14-2 the filling surrounds the bellows. Decreases in temperature cause the fluid volume to be reduced. This causes the bellows to move toward the fill-tube end of the housing. Temperature increase causes the fluid to expand, and the bellows moves in the opposite direction. Thus, bellows action is reversed in respect to temperature change from the action of the bulb-and-capillary type.

The local-mount, inside-fill type, shown in Fig. 14-3, is a lower-cost, less responsive sensor. Construction includes an integral capsule (a bellows in a fit-

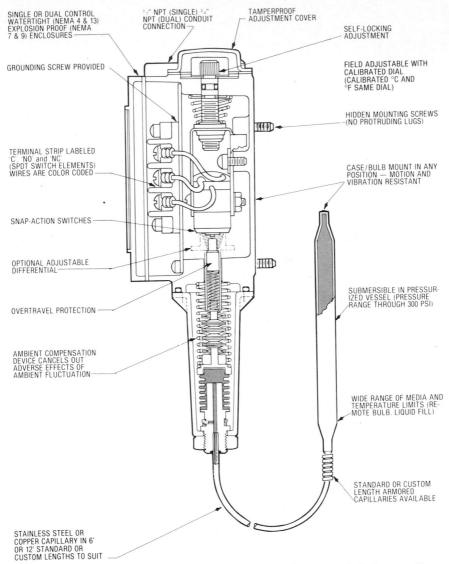

SINGLE OR DUAL CONTROL
WATERTIGHT (NEMA 4 & 13)
EXPLOSION PROOF (NEMA
7 & 9) ENCLOSURES

½" NPT (SINGLE) ¾"
NPT (DUAL) CONDUIT
CONNECTION

TAMPERPROOF
ADJUSTMENT COVER

SELF-LOCKING
ADJUSTMENT

GROUNDING SCREW PROVIDED

FIELD ADJUSTABLE WITH
CALIBRATED DIAL
(CALIBRATED °C AND
°F SAME DIAL)

HIDDEN MOUNTING SCREWS
(NO PROTRUDING LUGS)

TERMINAL STRIP LABELED
'C', 'NO' and 'NC'
(SPDT SWITCH ELEMENTS)
WIRES ARE COLOR CODED

CASE/BULB MOUNT IN ANY
POSITION — MOTION AND
VIBRATION RESISTANT

SNAP-ACTION SWITCHES

OPTIONAL ADJUSTABLE
DIFFERENTIAL

SUBMERSIBLE IN PRESSUR-
IZED VESSEL (PRESSURE
RANGE THROUGH 300 PSI)

OVERTRAVEL PROTECTION

AMBIENT COMPENSATION
DEVICE CANCELS OUT
ADVERSE EFFECTS OF
AMBIENT FLUCTUATION

WIDE RANGE OF MEDIA AND
TEMPERATURE LIMITS (RE-
MOTE BULB, LIQUID FILL)

STANDARD OR CUSTOM
LENGTH ARMORED
CAPILLARIES AVAILABLE

STAINLESS STEEL OR
COPPER CAPILLARY IN 6'
OR 12' STANDARD OR
CUSTOM LENGTHS TO SUIT

FIG. 14-1 Basic components of the bulb-and-capillary type of sensor are the bellows, capillary, and bulb shown.

ting) in which the inside of the bellows contains the fill fluid, rather than the fluid surrounding the bellows as in the outside-fill type.

The air space between the shell and the bellows presents high resistance to heat transfer, resulting in a longer response time.

Local-mount switches are not affected by ambient-temperature changes

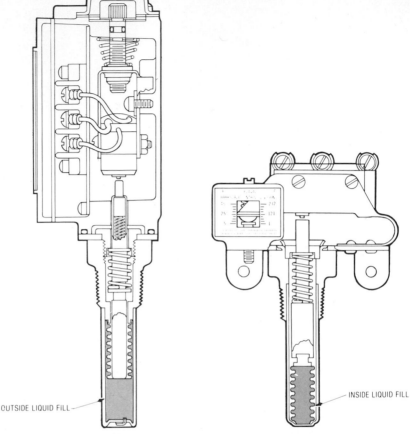

OUTSIDE LIQUID FILL

INSIDE LIQUID FILL

FIG. 14-2   Local-mount outside-fill type.   **FIG. 14-3**   Local-mount inside-fill type.

because all the fill fluid is exposed to the process-fluid temperature. For this reason, no compensating washers are required (see Fig. 14-4).

Response Time

Response time is the time required for the sensor to react to a change in the process-fluid temperature. It is a function of the heat transfer between the process fluid and the sensing fluid.

A standard method of expressing response time is the step-change curve (Fig. 14-5). Thirteen seconds is an excellent response time for filled systems. In general, rapid changes in temperature do not occur in process fluids.

Proof Temperatures

The normal proof temperatures required generally are from 25 to 50°F above the maximum set point. Therefore, high proof temperatures are usually established

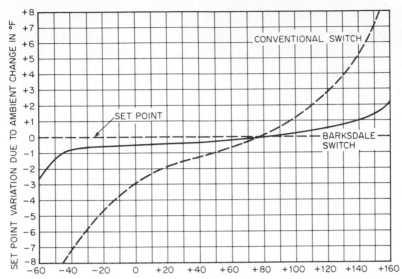

FIG. 14-4 Ambient-temperature compensation. Ambient-temperature range, −65 to +165°F for all models except -H600 series (0 to 165°F). Curves reflect averages of a series of tests. Barksdale switches are designed for ±½ percent accurate response through the mid-60 percent of the adjustable range (±2 percent at minimum or maximum adjustable set point).

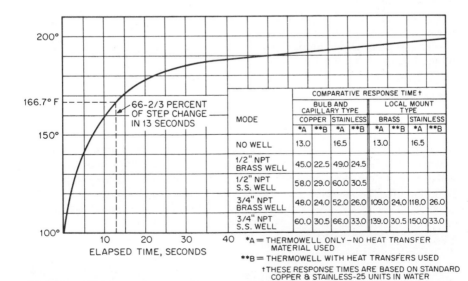

MODE	COMPARATIVE RESPONSE TIME †							
	BULB AND CAPILLARY TYPE				LOCAL MOUNT TYPE			
	COPPER		STAINLESS		BRASS		STAINLESS	
	*A	**B	*A	**B	*A	**B	*A	**B
NO WELL	13.0		16.5		13.0		16.5	
1/2" NPT BRASS WELL	45.0	22.5	49.0	24.5				
1/2" NPT S.S. WELL	58.0	29.0	60.0	30.5				
3/4" NPT BRASS WELL	48.0	24.0	52.0	26.0	109.0	24.0	118.0	26.0
3/4" NPT S.S. WELL	60.0	30.5	66.0	33.0	139.0	30.5	150.0	33.0

*A = THERMOWELL ONLY – NO HEAT TRANSFER MATERIAL USED
**B = THERMOWELL WITH HEAT TRANSFERS USED
†THESE RESPONSE TIMES ARE BASED ON STANDARD COPPER & STAINLESS-25 UNITS IN WATER

FIG. 14-5 Response time: 100°F step-change curve (copper bulb in water).

TABLE 14-2   Thermowell Pressure-Temperature and Velocity Limitations

Maximum fluid velocity, ft/s

Material	Insertion length—U							
	2½	4½	7½	10½	13½	16½	19½	22½
Brass	321 (150)	129 (83.5)	46.8	23.6	14.5	9.6	6.9	5.1
Carbon steel	410 (270)	249 (150)	90.3	45.6	27.8	18.5	13.2	9.8
AISI-304 and -316	483 (350)	272 (208)	97.3	49.7	30.4	20.3	14.5	10.7
Monel	396 (300)	214 (167)	77.5	39.2	23.8	16.0	10.3	7.7

NOTE: The values in parentheses represent safe values for water flow. Unbracketed values are for steam, air, gas, and similar low-density fluids.

Pressure-temperature rating, lb/in^2

Material	Temperature, °F				
	70°	200°	400°	600°	800°
Brass	5000	4200	1000	.... *	.... *
Carbon steel	5200	5000	4800	4600	3500
AISI-304	7000	6200	5600	5400	5200
AISI-316	7000	7000	6400	6200	6100
Monel	6500	6000	5400	5300	5200

*Stainless steel recommended.

at 50°F above the maximum set point. It is also possible, owing to unforeseen conditions, that the actual temperature reached is higher than the proof temperature. A safety factor must therefore be available in the proof-temperature ratings.

The low proof temperature is determined by the lowest temperature that the filling fluid can tolerate with safety.

Thermowells

Both bulb-and-capillary and local-mount types of sensors may use thermowells in their installation. These are employed to allow removal of the temperature switch without losing process fluid, to provide extra corrosion protection, to protect the temperature sensor from system pressures, and to protect the sensor from high velocity, wear, and abrasion. Table 14-2 shows velocity and pressure-temperature limitations for thermowells.

PRESSURE SWITCHES

A pressure switch is an electric switch designed to be actuated by a device that senses a change in pressure. All pressure switches sense the difference between two pressures, the reference pressure and the variable pressure. In most cases the reference pressure is ambient air, and the switch is then known as a *gauge-type instrument*. Some applications require a reference pressure other than ambient air. An instrument of that type is known as a *differential-pressure-actuated switch*.

In most industrial applications variations in barometric pressure are not significant. But some applications, such as aircraft with its extreme changes in altitude, require a constant reference pressure; so a reference pressure is sealed into the instrument. In an *absolute-pressure switch,* the sealed space is evacuated as closely as possible to a perfect vacuum.

When selecting a pressure switch, an analysis of the application should include expected service life, frequency and magnitude of pressure cycles, maximum pressure, the fluid to be controlled, and the desired electrical function. Consideration should be given to three basic types of pressure sensors: metal-diaphragm sensors, Bourdon-tube sensors, and piston-type sensors.

Metal-Diaphragm Sensors

A metal-diaphragm sensor has a diaphragm assembled into a pressure-containing capsule. The diaphragm acts as a spring that moves linearly with differential pressure applied. The design of the convolutions in the diaphragm and the thickness of the diaphragm determine the spring rate and the differential-pressure limits. Use of a base plate in the capsule designed with the same contour as the diaphragm supports and protects the diaphragm at pressures higher than its effective range. The maximum permissible pressure is called *proof pressure.*

Two materials commonly used for the diaphragm are beryllium copper and 17-7PH stainless steel. The material must have low fatigue characteristics, have a low hysteresis effect, and resist deterioration by the process fluid. (See Figs. 14-6 and 14-7.)

Bourdon-Tube Sensors

The Bourdon-tube type of sensor (see Fig. 14-8) is a hollow metal tube with an elliptical cross section bent in a C-shaped arc. One end is fixed and open to pressure, and the other end is closed and free to move. When pressure is applied internally, the elliptical cross section deflects toward a more circular form. This causes the metal in the outer radius to be in tension and the material in the inner radius to be in compression. The resultant forces produce a tendency for the tube to straighten, and the free end of the tube moves in an arc. The tube acts as a spring to oppose the pressure. The amount of movement of the free end is proportional to the magnitude of the pressure.

Bourdon tubes are very accurate sensors and are capable of reproducing identical free-end movements with repeated pressure cycles. Because of their long unsupported length, however, they are sensitive to external forces on the free end. The Bourdon tube must eventually operate a switch element. It is, therefore, very important that the external forces exerted on the free end of the tube by the switch element and any related operating mechanism be constant.

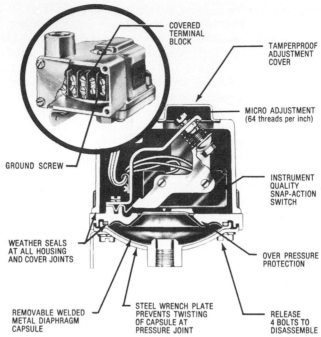

COVERED
TERMINAL
BLOCK

TAMPERPROOF
ADJUSTMENT
COVER

MICRO ADJUSTMENT
(64 threads per inch)

GROUND SCREW

INSTRUMENT
QUALITY
SNAP-ACTION
SWITCH

WEATHER SEALS
AT ALL HOUSING
AND COVER JOINTS

OVER PRESSURE
PROTECTION

REMOVABLE WELDED
METAL DIAPHRAGM
CAPSULE

STEEL WRENCH PLATE
PREVENTS TWISTING
OF CAPSULE AT
PRESSURE JOINT

RELEASE
4 BOLTS TO
DISASSEMBLE

FIG. 14-6 Diaphragm-model pressure-and-vacuum switch: a weld-sealed metal diaphragm direct-acting on a snap-action switch.

Precision snap-acting switches feature constant internal spring forces, which contribute to accuracy when applied to a Bourdon tube. The direct-acting design eliminates the chance of developing unpredictable and changing frictional factors which affect accuracy.

The amount of movement of the free end is a function of the following:

1. The ratio of the major to the minor cross-sectional axis. The higher this ratio, the greater the movement.

2. The tube length from the fixed to the free end. A longer tube will move to a greater extent.

3. The radius of curvature. The smaller the radius, the greater the movement.

4. The total angle of curvature. The greater the angle, the greater the movement.

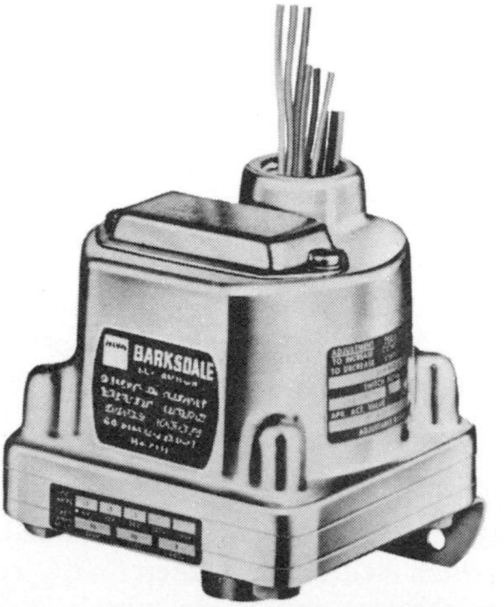

FIG. 14-7 Barksdale housed diaphragm pressure switch.

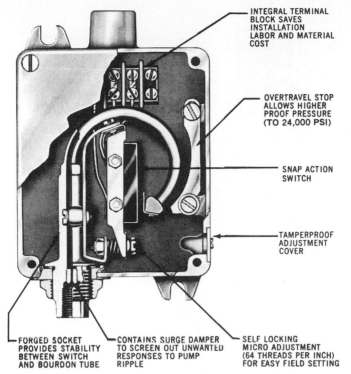

INTEGRAL TERMINAL
BLOCK SAVES
INSTALLATION
LABOR AND MATERIAL
COST

OVERTRAVEL STOP
ALLOWS HIGHER
PROOF PRESSURE
(TO 24,000 PSI)

SNAP ACTION
SWITCH

TAMPERPROOF
ADJUSTMENT
COVER

FORGED SOCKET
PROVIDES STABILITY
BETWEEN SWITCH
AND BOURDON TUBE

CONTAINS SURGE DAMPER
TO SCREEN OUT UNWANTED
RESPONSES TO PUMP
RIPPLE

SELF LOCKING
MICRO ADJUSTMENT
(64 THREADS PER INCH)
FOR EASY FIELD SETTING

FIG. 14-8 Bourdon-tube-model pressure switch: a weld-sealed Bourdon tube direct-acting on a snap-action switch.

5. The wall thickness of the tube. The thinner the wall, the greater the movement.

6. The modulus of elasticity of the tube material. The lower the modulus, the greater the movement.

Tubes are generally formed from steel or phosphor bronze seamless tubing.

Some Bourdon-tube pressure switches incorporate a tube stop to prevent movement of the free end of the tube beyond the amount required to achieve the maximum set point of the pressure switch. This increases the proof pressure.

Piston-Type Pressure Switches

A piston-type sensor (see Figs. 14-9 and 14-10) combines a rod and a piston to produce a relatively low-cost pressure switch with long life at high pressures and cycling rates. Piston switches are generally used in rigorous-service applications because of high cycling rates or system pressure surges.

The sensor is composed of a piston with an O ring and Teflon backup ring, rod, spring, and body. The amount of pressure required to produce movement is a function of the piston area, spring rate, and the extent to which the adjusting screw has compressed the spring.

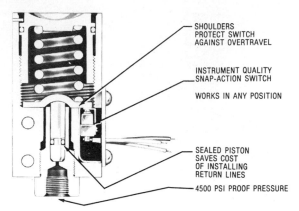

SHOULDERS
PROTECT SWITCH
AGAINST OVERTRAVEL

INSTRUMENT QUALITY
SNAP-ACTION SWITCH

WORKS IN ANY POSITION

SEALED PISTON
SAVES COST
OF INSTALLING
RETURN LINES

4500 PSI PROOF PRESSURE

FIG. 14-9 Sealed-piston pressure switch: an O-ring sealed piston acting on a snap-action switch.

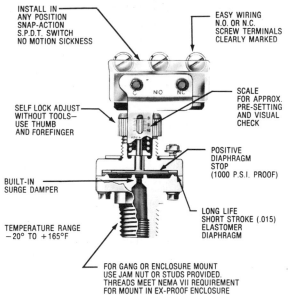

INSTALL IN
ANY POSITION
SNAP-ACTION
S.P.D.T. SWITCH
NO MOTION SICKNESS

EASY WIRING
N.O. OR N.C.
SCREW TERMINALS
CLEARLY MARKED

SELF LOCK ADJUST
WITHOUT TOOLS—
USE THUMB
AND FOREFINGER

SCALE
FOR APPROX.
PRE-SETTING
AND VISUAL
CHECK

POSITIVE
DIAPHRAGM
STOP
(1000 P.S.I. PROOF)

BUILT-IN
SURGE DAMPER

LONG LIFE
SHORT STROKE (.015)
ELASTOMER
DIAPHRAGM

TEMPERATURE RANGE
−20° TO +165°F

FOR GANG OR ENCLOSURE MOUNT
USE JAM NUT OR STUDS PROVIDED.
THREADS MEET NEMA VII REQUIREMENT
FOR MOUNT IN EX-PROOF ENCLOSURE

FIG. 14-10 Diaphragm sealed pressure or vacuum switch: a diaphragm sealed piston direct-acting on a snap-action switch.

Piston-type switches covering adjustable ranges from 15 to 12,000 lb/in^2 and proof pressures from 3000 to 20,000 lb/in^2 are commercially available.

Other Sensor Types

Other types of sensors with more limited applications include the brass bellows type used in conjunction with a loading spring inside the bellows. Pressure is introduced on the outside of the bellows, which is enclosed within a pressure-containing chamber. A plunger transmits movement of the bellows to the switch element.

But there are major weaknesses in bellows-type sensors, including relatively short life, limited proof pressure in relation to the maximum set point, sensitivity to vibration and shock, less resistance to surge pressure, and a requirement of the plunger to operate the switch element. They are generally offered for pressures up to 1500 lb/in^2 and are most often used when pressures do not exceed 500 lb/in^2.

Helical and spiral Bourdon tubes are two other types in limited use. The helical tube is similar in deflection behavior to a C-shaped tube. Since it is coiled into a multiturn helix with its total angle between 1800 and 3600°, its tip travel is proportionately greater than that of a tube with less than 360° curvature.

The spiral tube also amplifies tip travel because of its multiturn configuration (usually four to eight turns). The performance of both tubes is very similar, and choice of either tube is dictated by the limitation on the overall sensing-element diameter and length. A partly helical, partly spiral tube design has also been developed. Such tubes would be more sensitive to vibration than a C-shaped tube. Neither design is widely used in pressure switches.

Another type is the negative-spring-rate sensor called a *Belleville washer* (see Fig. 14-11). When pressure is applied to the sensor, the pressure plate acts against the cup, which exerts force on the washer. The washer snaps upward, and the snap-over deflection operates the snap-action switch. A major advantage is that the washer makes the pressure switch less sensitive to vibration.

Another type, the unsealed-piston design, offers the advantage of not requiring an O ring. But one disadvantage is the requirement for a drain line. Also, dirt in

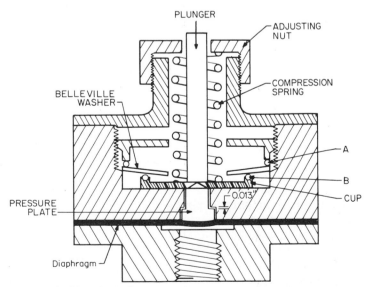

FIG. 14-11   Negative-spring-rate sensor.

the system can enter the grooves and destroy their ability to provide lubrication and pressure balance around the piston. When this occurs, rapid wear and high friction result.

Accuracy and Life

The ability of a pressure switch to operate repetitively at its set point is its accuracy, or repeatability. It can be expressed as a percentage of either the adjustable range or the set point. The factors most strongly influencing life are:

1. Number of cycles to which the sensor and switch element respond
2. Range of pressure fluctuations through which the sensor flexes
3. The pressure surges which exist in the system
4. The electrical load on the switch element
5. Corrosion resistance to the fluid in the sensor and the environment

Anticipated service life is generally the first consideration in selecting a pressure switch regardless of the pressure or sensitivity desired. If the service life (number of cycles during which the switch is expected to operate without replacement) is 1 million or less, a Bourdon tube or a diaphragm switch is indicated. If more than 1 million cycles is desired, a piston switch should be used. An exception is made when the pressure change in a system is quite slight (20 percent or less of the adjustable range). Under such conditions, a Bourdon tube or a diaphragm switch can be used for up to 2.5 million cycles before metal fatigue.

Application and Cost

There are two basic types of applications:

1. Applications which require a high degree of accuracy and in which pressure surges and cycling rates normally are relatively low. These are instrument-type applications.
2. Applications which require a moderate or low degree of accuracy and for which cycling rates and pressure surges can be relatively high. These are control-type applications.

The cost-accuracy relationship generally merits attention. Within limits, the higher the accuracy requirements, the higher the cost. In addition, the cost is generally proportional to the pressure required. Type 1 applications are generally served best by metal-diaphragm and Bourdon-tube sensors and Type 2 applications by lower-cost sealed-piston sensors.

LIQUID-LEVEL AND LIQUID-FLOW INSTRUMENTATION

Manufacturing and process machinery today utilizes many types of liquid media to ensure correct operation and function. Liquid fuels are used as a source of energy, viscous oils lubricate moving assemblies, and water is used to remove heat from many types of electronic and process equipment. These liquids and others can be considered integral to each machine, as any subassembly or electronic circuit would be. The control and monitoring of these liquids is, therefore, a very important factor in basic design.

Many techniques which employ basic engineering principles to accomplish this control and monitoring have been developed. Liquid-level and liquid-flow instrumentation plays a major role in satisfying the requirements of indication, control, and/or protection of the medium.

Level Sensing

Some of the different types of level medium-sensing techniques used today are listed below. This is by no means a complete list, and many slight modifications of the principles discussed also are used.

Magnetic-Float Principle   The liquid level is monitored by means of a magnetically equipped float that traverses an in-tank stem or probe. As the level in the tank changes, the float moves correspondingly, rising and falling along the stem. Inside the stem, a reed switch is positioned so that as the float moves by it, it actuates, signaling that the level is at that particular point.

Capacitance Method   The liquid level is measured by monitoring a change in capacitive reactance between two points. A conductor probe inserted in the tank and the tank wall itself serve as two conductor plates of a capacitor. The medium between the probe and the tank wall act as the dielectric. The capacitance between the probe and the tank wall will vary as a result of liquid-level change. This varying capacitance can be related to a varying level output.

Conductance Method   The liquid level is monitored by using the conduction of current through the liquid. Two conductor probes are placed within the tank. As the liquid level changes and the probes become immersed in the liquid, a voltage potential across the two probes causes current to flow from one probe through the liquid to the other probe and then out to the receiver or other sensing equipment. As current starts to conduct, it is known that the liquid level is at that point within the tank.

Ultrasonic Method   The liquid level is measured through the use of ultrasonic-frequency-wave generation. This system consists of a transmitter-receiver inside the tank and a remote indicator outside. The transmitter-receiver is mounted at the tank top and generates a high-frequency wave which is directed at the medium level in the tank. Thus, the liquid level is monitored by tracking the time required for the ultrasonic-frequency wave to make contact with the liquid level and return to the receiver sensing element in the head assembly. That signal is then transferred electrically to an indicator module.

Electromechanical (Displacement) Method   The liquid level is measured by using a weighted displacer float, which is connected to a cable length and placed inside a liquid. When the float is immersed in the liquid, it will displace a certain volume. This displaced volume exerts a force change on the connecting-cable arrangement above the displacer. That force change is detected through a sensor such as a strain gauge and converted to an indication of liquid-level change.

Differential-Pressure Method   The liquid level is measured by the indication of a pressure change between two points within the tank. A pressure-sensitive sensor such as a diaphragm, bellows, or strain gauge is connected to a high and a low port within the tank configuration. As liquid level changes, head pressure also changes in the tank. This pressure change can be proportional to a level change and indicated as such. For pressurized tanks, a reference pressure is established and connected to the sensor. This enables a change in pressure to be compared with the reference to indicate the level.

Flow-Measurement Sensing

Although not encompassing the entire range of flow-sensing techniques presently available, the techniques listed in the following paragraphs provide a guide which states the basic principles. Modifications as well as combinations of principles are used in flow measurement.

Pressure-Head Flow Measurement   This method of monitoring flow utilizes a change in pressure drop to a corresponding change in flow. A restriction positioned inside a flow line creates a pressure drop as the flow passes across it. The volumetric flow rate past the restriction is proportional to the square root of the pressure drop. Some current devices that use the pressure-differential principle are the venturi, the flow nozzle, and the orifice plate.

   In some cases, measure of a predetermined minimum or maximum flow rate is desired. One system uses the pressure drop across a restriction to cause a shuttle to move a magnet against a spring to actuate a reed switch. The reed switch closes an electric circuit to actuate a control or an alarm. Different flow-rate indications can be obtained by varying either the shuttle size or the spring rate. (See Fig. 14-12.)

Positive-Displacement Flow Measurement   This method uses the principle of volumetric displacement to determine the flow rate. It consists of a measuring

FIG. 14-12   Flow switch.

chamber and a displacer rotating within that chamber. As liquid flow passes through the chamber, it causes the displacer to rotate. Each rotation displaces a given volume of liquid, which is monitored. By simply counting the revolutions, the amount of liquid displaced in a given time frame, i.e., the flow rate, can be established. Examples of this principle are the rotating vane, the sealed drum, and lobed-impeller meters.

Variable-Area Flow Measurement   This method employs the principle of a variable-area orifice with a constant pressure differential. The construction is a vertical transparent tube with a tapered bore and the largest diameter at the top. The flow is always vertically upward through the tube. Inside the tapered bore is a float which moves vertically to an area corresponding to the flow rate. Since the weight of the float is constant, the only variable is the flow area. The outside of the tube is calibrated to show flow rate as a function of the position of the float.

Ultrasonic Flow Measurement   There are many modifications and versions of this method. The basic principle is that a frequency wave is established and directed at the medium. Therefore, a change in medium flow will cause a change in the established frequency. This change can be related to a change in flow rate.

The Doppler shift is used in one method. Frequency waves are transmitted in the same direction as the flow of the medium. As the fluid rate changes, the wave frequency will actually shift in output. This shift is proportional to the change in flow. By measuring this new frequency shift, the flow rate is monitored.

Vortex Flow Measurement   This method utilizes the principle of vortex shedding to monitor flow rates. The frequency of the vortices formed is proportional to flow rate. An element, usually of triangular shape, is placed perpendicularly to the flow. Fluid vortices are formed around the element and eventually are shed downstream. These vortices move downstream with a certain oscillatory frequency. This frequency will change as the flow is decreased or increased. Heated thermistors are placed in the path of the vortices; their cooling rate increases as a vortex passes. The frequency of the cooling cycles is an index of the flow rate.

Turbine Flow Measurement   This measurement principle states that a turbine wheel in a flow stream revolves at a speed proportional to the fluid-flow rate. A small turbine wheel is allowed to rotate on an axis inside the unit body. As the flow rate changes, the speed of rotation of the turbine will also change. To detect this change, a small magnet is located on a point on the wheel, thus generating a pulse for each revolution. This pulse is sensed and recorded by a pickup coil circuit on the outside of the body. The number of pulses detected can be translated into the amount of flow passing through the turbine wheel.

Reed Switches

Reed switches are used in many applications as well as in level- and flow-control devices. The reason for their extensive use is their reliability and long life. These qualities are due to the hermetically sealed reed blades and the overall simplicity of the design.

Construction   A reed switch is constructed of very simple material components. Two reed blades made of a soft, thoroughly magnetic material are situated in a gas-purged glass envelope which is hermetically sealed (see Fig. 14-13). The tips of the blades are plated with either gold or silver to ensure long life as well as low contact resistance. Rhodium and tungsten are also used for plating.

Contact arrangements of the reed switch are available in single-pole–single-throw (SPST) and single-pole–double-throw (SPDT) types. Reed switches are normally identified by contact configuration (see Figs. 14-14 and 14-15) as well as by resistive-power ratings.

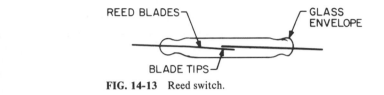

FIG. 14-13   Reed switch.

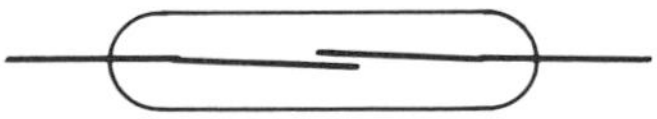

FIG.   14-14   Single-pole–single-throw (SPST) reed switch.

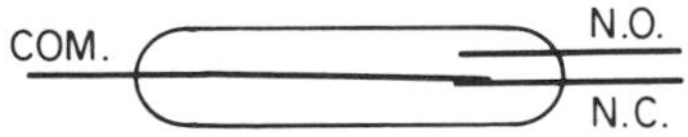

FIG.   14-15   Single-pole–double-throw (SPDT) reed switch.

Example: "SPST, 10 W" refers to a single-pole–single-throw reed switch with a 10-W resistive-power-rating factor.

Operation   When the reed switch is in proximity to a magnetic field, the blade tips will become magnetically opposed (north-south poles). At that point, the reed blades will snap together and remain in contact until the magnetic field has been removed or sufficiently decreased. The physical gap between the two reeds acts as magnetic resistance to the magnetic flux generated from the magnet. This gap resistance decreases as the reed blades come closer together.

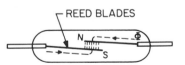

FIG. 14-16   Magnetic activation of reed switch.

The magnetic force produced by permanent magnets or electromagnets is inversely proportional to the square of this distance gap. Therefore, the reed-switch-blade closure will accelerate as the tips approach each other. The larger the magnetic field, the faster the blades snap together. (See Fig. 14-16.)

Index

Index

About the Editor

Harry J. Welch was educated at the University of Wisconsin (B.S. in electrical engineering) and the Illinois Institute of Technology. He is a registered professional engineer in Texas and New Jersey and a member of the American Society of Mechanical Engineers, which he has served as its chairman of the Trenton Section and secretary of Region III. After serving as a consulting engineer with the Allis Chalmers Manufacturing Co., he became in 1964 chief research engineer of the Engineering Research Department and Engineering Laboratory of De Laval Turbine Inc. Subsequently he became assistant to the manager of engineering of the Turbine Division, chief engineer of the Fluid Mechanics Department, manager of engineering of the Deltex Division and of the Special Products Department, and technical assistant to the manager of engineering of the Turbine/Compressor Division of Transamerica Delaval Inc.